# The Rise and Fall of Mass Production
# Volume I

# The International Library of Critical Writings in Business History

*Series Editor:* Geoffrey Jones
Professor of Business History,
University of Reading

1. The Growth of Multinationals
   *Mira Wilkins*
2. Government and Business
   *Steven W. Tolliday*
3. Mergers and Acquisitions
   *Gregory P. Marchildon*
4. Antitrust and Regulation
   *Giles H. Burgess Jr*
5. The Rise of Big Business
   *Barry E. Supple*
6. Marketing (Volumes I and II)
   *Stanley C. Hollander and Kathleen M. Rassuli*
7. Coalitions and Collaboration in International Business
   *Geoffrey Jones*
8. Business Elites
   *Youssef Cassis*
9. Technology Transfer and Business Enterprise
   *David J. Jeremy*
10. Big Business in Mining and Petroleum
    *Christopher Schmitz*
11. Organizational Capability and Competitive Advantage
    *William Lazonick and William Mass*
12. Entrepreneurship and the Growth of Firms (Volumes I and II)
    *Harold C. Livesay*
13. Family Business
    *Mary B. Rose*
14. Industrial Research and Innovation in Business
    *David E.H. Edgerton*
15. Human Resources and the Firm in International Perspective (Volumes I and II)
    *Robert Fitzgerald and Christopher Rowley*
16. The Rise and Fall of Mass Production (Volumes I and II)
    *Steven W. Tolliday*

Future titles will include:

Women in Business
*Mary A. Yeager*

# The Rise and Fall of Mass Production
# Volume I

*Edited by*

Steven Tolliday

*Professor of Economic and Social History*
*University of Leeds, UK*

THE INTERNATIONAL LIBRARY OF CRITICAL WRITINGS IN BUSINESS HISTORY

**An Elgar Reference Collection**
Cheltenham, UK • Northampton, MA, USA

Published by
Edward Elgar Publishing Limited
8 Lansdown Place
Cheltenham
Glos GL50 2HU
UK

Edward Elgar Publishing, Inc.
6 Market Street
Northampton
MA 01060
USA



A catalogue record for this book is available from the British Library.

**Library of Congress Cataloging in Publication Data**

The rise and fall of mass production / edited by Steven Tolliday.
(The international library of critical writings in business history : 16) (An Elgar reference collection)
Includes bibliographical references and index.
1. Production management. 2. Assembly-line methods.
3. Motor vehicle industry--Automation. I. Tolliday, Steven.
II. Series. III. Series : An Elgar reference collection.
TS155. R49 1998
670.42--dc21

98-11610
CIP

ISBN 1 85898 042 9 (2 volume set)

Printed and bound in Great Britain by
MPG Books Ltd, Bodmin, Cornwall

# Contents

# Acknowledgements

The editor and publishers wish to thank the authors and the following publishers who have kindly given permission for the use of copyright material.

*American Machinist*, a Penton publication, for article: Fred H. Colvin (1913), 'Building an Automobile Every 40 Seconds', *American Machinist*, **38**, 8 May, 757–62.

*Archiv für Sozialgeschichte* for article: Steven Tolliday and Jonathan Zeitlin (1988), 'Between Fordism and Flexibility: The Automobile Industry and Its Workers – Past, Present and Future', *Archiv für Sozialgeschichte*, **XXVIII**, 153–71.

British Sociological Association and Cambridge University Press for article: Karel Williams, Colin Haslam and John Williams (1992), 'Ford Versus "Fordism": The Beginning of Mass Production?', *Work, Employment and Society*, **6** (4), December, 517–55.

*Business and Economic History* for article: Donald Hoke (1989), 'Product Design and Cost Considerations: Clock, Watch, and Typewriter Manufacturing in the 19th Century', *Business and Economic History*, Second Series, **18**, 119–28.

Economic History Association and Cambridge University Press for article: Daniel Nelson (1987), 'Mass Production and the U.S. Tire Industry', *Journal of Economic History*, **XLVII** (2), June, 329–39.

Encyclopaedia Britannica, Inc. for excerpt: Henry Ford (1926), 'Mass Production', in *Encyclopaedia Britannica*, 13th Edition, Volume 15, 38–41.

Thomas P. Hughes for his own excerpt: (1985), 'Gropius, Machine Design and Mass Production', *Wissenschaftskolleg Jahrbuch 1983–4*, 171–80.

Institute of Industrial Engineers for articles: John H. Van Deventer (1922), 'Ford Principles and Practice at River Rouge', *Industrial Management*, **LXIV** (2), August, 64–5; John H. Van Deventer (1922), 'What Ford is Doing and How He Does It: An Editorial Announcement of First Importance', *Industrial Management*, **LXIV** (3), September, 129–30; John H. Van Deventer (1922), 'I – Links in a Complete Industrial Chain', *Industrial Management*, **LXIV** (3), September, 131–7; John H. Van Deventer (1923), 'IX – Machine Tool Arrangement and Parts Transportation', *Industrial Management*, **LXV** (5), May, 257–67; John H. Van Deventer (1923), 'XI – Machining Tractor Parts and Assembling Fordson Tractors', *Industrial Management*, **LXVI** (1), July, 19–27; John H. Deventer (1923), 'XII – The Manufacture of Ford Car Bodies', *Industrial Management*, **LXVI** (2), August, 85–95;

John H. Van Deventer (1923), 'XIII – Manufacturing and Assembling Body Parts', *Industrial Management*, **LXVI** (3), September, 151–9.

*Journal of Contemporary History* for article: Charles S. Maier (1970), 'Between Taylorism and Technocracy: European Ideologies and the Vision of Industrial Productivity in the 1920s', *Journal of Contemporary History*, **5** (2), 27–61.

M.I.T. Press for excerpts: David A. Hounshell (1985), 'Ford Eagle Boats and Mass Production during World War I', in Merritt Roe Smith (ed.), *Military Enterprise and Technological Change. Perspectives on the American Experience*, Chapter 4, 175–202; Merritt Roe Smith (1990), 'Eli Whitney and the American System of Manufacturing', in Carroll W. Pursell, Jr (ed.), *Technology in America: A History of Individuals and Ideas*, Chapter 5, Second Edition, 45–61.

Oxford University Press for article: Michael Storper (1989), 'The Transition to Flexible Specialisation in the US Film Industry: External Economies, the Division of Labour and the Crossing of Industrial Divides', *Cambridge Journal of Economics*, **13** (2), 273–305.

University of Chicago Press for articles: Stephen Meyer (1988), 'Technology and the Workplace: Skilled and Production Workers at Allis-Chalmers, 1900–1941', *Technology and Culture*, **29** (4), October, 839–64; Jonathan Zeitlin (1995), 'Flexibility and Mass Production at War: Aircraft Manufacture in Britain, the United States, and Germany, 1939–1945', *Technology and Culture*, **35** (1), January, 46–79; Regina Lee Blaszczyk (1995), '"Reign of the Robots": The Homer Laughlin China Company and Flexible Mass Production', *Technology and Culture*, **36** (4), October, 863–911.

University of Illinois Press for excerpt: Wayne Lewchuk (1989), 'Fordism and the Moving Assembly Line: The British and American Experience, 1895–1930', in Nelson Lichtenstein and Stephen Meyer (eds), *On the Line: Essays in the History of Auto Work*, Chapter 2, 17–41.

Every effort has been made to trace all the copyright holders but if any have been inadvertently overlooked the publishers will be pleased to make the necessary arrangement at the first opportunity.

In addition the publishers wish to thank the Library of the London School of Economics and Political Science and B & N Microfilm, London for their assistance in obtaining these articles.

# Introduction: The Rise and Fall of Mass Production?

*Steven Tolliday*

In many accounts, mass production is at the core of economic transformation in the 20th century. By harnessing the logic of the relentless repetitive machine, engineers unleashed a technological torrent, achieving unprecedented standards of productivity and an undreamed of range of mass consumption. There have been costs, in the form of de-skilling (or dehumanization) of workers and homogenization of products; it has been both dream and nightmare. But it is a central fact of economic life – and social scientists from the traditions of both Smith and Marx have built it into their fundamental conceptual models. Yet doubts remain. Mass production is a concept perhaps too easily grasped to have been carefully theorized. It is universal, yet strangely hard to find. This collection brings together some of the most important work on the history of mass production and highlights crucial debates in an attempt to understand what it has been, what it is, and what social and economic effects it has had.

The ramifications of mass production (from the economic calculation of production engineers to the psychology and theories of mass culture) are almost endless. But this collection focuses only on six critical themes of great importance. (1) Where did mass production come from? What were the mechanical, economic and social roots of the transformation of production that burst on the world in the early 20th century? (2) Why did 'Fordism' emerge when and where it did? And what were the fundamental elements of the new system? (3) Why, and how far did mass production spread into the wider economy, and how did it change in the process? (4) What was the impact of mass production on work and workers? (5) Has the success of Japan in the late 20th century been founded on a deeper and more ruthless exploitation of the principles of mass production, or has it gone beyond it to a new form of productive organization? (6) More generally, is the late 20th century witnessing a new 'industrial divide', the end of mass production as a dominant or viable paradigm? The introduction briefly locates the selected articles within these debates and provides a guide to the further reading available to pursue the issues.

## 1. Antecedents

The articles in Part I look at the controversies surrounding the prehistory of mass production. Standard accounts have traced a genealogy running directly from Eli Whitney, the originator of interchangeable parts and the 'father of mass production', through the 'American System of Manufactures', to Henry Ford. Recent work, however, has suggested that this picture is problematic. Investigative work by Woodbury (1960) and Battison (1966) demolished Whitney's claims to have pioneered the technology of interchangeable parts, and showed that, for example, the famous demonstration of interchangeable musket parts before Congress in 1801 was a fake. Subsequent work by Smith (1977) has re-constructed in detail the history of the genuine creation of systems of interchangeable parts. In Chapter 1, Smith synthesizes his re-evaluation of the rise of the American System. He shows that, while Whitney was not a charlatan, his contribution to the practice of mechanization and interchangeability was modest. However, his zealous advocacy of the engineering ideal of uniformity was a potent force among the military and engineering community and drove the rise of the American System. Genuine interchangeability was achieved only after long and costly endeavour at the Harpers Ferry and Springfield armouries, supported and subsidized by government contracts. From the 1840s, Smith argues, the US armouries became a 'clearing-house for the acquisition and dissemination of technical knowhow' to wider American manufacturing industry where they were a critical force in industrial innovation. In other words, he claims to have identified 'a mechanical genealogy that directly links Whitney and other early arms makers with the mass production industries of the 20th century' (compare also, Uselding, 1974; Mayr and Post, 1982; Rosenberg, 1963; Howard, 1978).

David Hounshell, however, in his monumental study of the evolution of American manufacturing techniques (Hounshell, 1984), has nuanced these earlier views. He shows that key sectors in Smith's genealogy of diffusion of 'armory practice' do not really fit the model. In sewing machines, Singer made only limited use of armoury practice and pursued a different manufacturing paradigm, relying on skilled work and quality in manufacture, known as the 'European method', until the 1880s and even then continued to use limited amounts of precision machinery and extensive hand fitting. Singer's success came more from marketing strategies than production innovations. The bicycle industry did indeed use armoury practice extensively, and in some senses can be seen as 'a bridge between the American system and mass production', but, even here the crucial aspect of bicycle manufacture was not simply the adoption of armoury practice, but its fusion with other manufacturing traditions in welding, punching, stamping and sheet metal work, which concentrated on volume and cost reduction rather than the high-cost precision of armoury practice. Thus Hounshell stresses a more diverse genealogy for mass production and also emphasizes that in all these cases the most advanced manufacturing techniques still emphasized diversity and premium products. In this sense, Ford broke more radically with the past than older accounts allow and the Ford system 'marks an entirely new epoch in the manufacture of consumer durables in America'.

Donald Hoke (Chapter 2 and see also Hoke, 1990), pushes the revisionist view still further. He argues that the entire historiography of the American System has been skewed towards the role of the federal armouries and the public sector. In contrast, he argues, the American System was primarily a private sector phenomenon, with its achievements of systematic production and interchangeability growing from rather different roots, notably innovations in design for manufacture, 'adjustable' interchangeability and the use of new materials and methods (compare also Thomson, 1989). Case studies of the clock, watch and typewriter sectors show how, rather than relying on the armoury practice path of increasingly precise machining, manufacturers used design innovations to build adjustability into their products to make them function with more generous tolerances. Although they rationalized assembly processes, they continued to rely on large numbers of skilled workers, and their innovations in design, method or materials were as important as refinements in production method. Rather than 'diffusion' from a government-favoured sector of advanced process engineering, the American System represented a coming together of a wide range of American engineering, designing and marketing traditions (compare also Rosenbloom, 1993). A similar emphasis on the broader structural roots of mass production is to be found in the work of economic geographers Page and Walker (1991) who stress the roots of the new system in the under-researched agro-industrial complex of the Midwest region (including the Chicago meat packers in their 'dis-assembly' lines or the Minnesota flour millers with their sophisticated flow processes), rather than in a single technological stream.

The literature on the antecedents of mass production has also paid a great deal of attention to the international diffusion of the 'American System'. Burn (1931) and Rosenberg (1969), have detailed the early reception and interpretation of American innovations in Europe, and some finely detailed studies have recorded the British response to these methods in small-arms and watches, stressing managerial conservatism (Fries, 1975), dynamic adoption (Lumley, 1989), or the rationale of rejection (Davies, 1993). In engineering, a number of excellent studies have detailed the diversity of experience within the industry and the significant levels of learning, selective imitation, and adaption that followed the emergence of the 'American challenge' (Saul, 1960; Floud, 1974; Zeitlin, 1997).

## 2. The 'Moment' of Fordism

As the literature on the antecedents of mass production clearly demonstrates, many firms implemented individual elements of the system before Henry Ford created his pioneering factory. But in a remarkable three year period after the opening of Ford's Highland Park factory in January 1910, Ford pulled together all these elements, and more, in a great deluge of process innovation that produced what the world at once acknowledged to be 'Fordism' or 'mass production'. The articles in Part II focus on this transformation.

The origins of this productive transformation have been studied by many scholars (in addition to the contributions in this volume see particularly, Hounshell, 1984; Lewchuk, 1987; Russell, 1978; Dassbach, 1991; Nevins, 1954; Williams, Haslam and Williams, 1993; Biggs, 1995). But despite the availability of oral and archival sources, our understanding of this crucial period, and of the subsequent development of the system at the River Rouge factory, rest heavily on classic contemporary studies by American engineers of the early Ford factories. This volume reprints crucial but inaccessible articles by these engineers on which all the more recent scholarship has drawn heavily.

Fred Colvin's article (Chapter 3) summarizes some of his key findings from a series of 15 technically detailed studies of Ford's Highland Park factory, subsequently published in the *American Machinist*. The study came at a critical moment and provides vital evidence on the emergence of mass production. Colvin studied the factory in the spring of 1913. The articles were written before the first assembly line was introduced, yet by the time they were published, Ford had already launched into extensive experimentation with moving assembly lines (the start of the first assembly line at Highland Park, in the magneto shop, is generally dated to 1 April 1913). Thus Colvin's work provides a snapshot of how far older notions of volume manufacturing could be pushed, and the setting for Ford's breakthrough. His work stresses the power, economy, continuity and speed of the Ford system, which already produced almost 200 000 cars per year by stationary methods (no European producer attained such volumes until the 1950s). The keys to this, Colvin shows, were the rigour of interchangeability, strict standardization, sequential organization of production flow, dedicated and innovative specialized tooling, and the radical simplification of tasks, all of which were in place *before* the assembly line. Perhaps above all, however, Colvin demonstrates the advances in machine-tool practice in the United States at the turn of the century, an area that still remains significantly under-researched. The new speed, accuracy and precision of the machine tool producers like Ingersoll, Cincinnati and Foot-Burte provided an essential foundation for the edifice of mass production.

The work of Arnold and Faurote (Chapter 4) has been the basis for all more recent accounts of the start-up of assembly line production (see especially Nevins, 1954; Hounshell, 1984). The rush of change in these months left little clear archival documentation, and even Henry Ford's own (ghost-written) account of the transition (Ford, 1922; and Chapter 6) rested heavily on Arnold and Faurote's study. The articles reprinted here document in particular the deluge of internal experimentation and learning in the factory as the new techniques came in and measure the initial 'productivity shock' of the moving line. Importantly, though less conspicuously, they also explain the logistics of the system (including its handling of work flow and inventory control) and the processes of work organization and task simplification.

While these accounts have been extensively used by historians to provide the fundamentals of our understanding of the origins of the assembly line, an equally valuable series of plant studies by John H. Van Deventer, documenting the consolidation of the Ford system into a colossal new purpose-built plant at the River Rouge, have been comparatively neglected (Chapter 5). In this series of articles, Van Deventer demonstrates how the functioning of the River Rouge plant rested even more on a remarkable degree of vertical integration, encompassing raw materials, railroads, cement and plate glass factories, and a huge power plant and foundry, than on the processes of machining and assembly more conventionally associated with Ford. In the articles included here, Van Deventer provides an overview of the practice of total vertical integration, the central role of plant transportation innovations, based around the 'high line' track system, the contribution of the world's largest foundry and largest manufacturing power plant, and the integration of machining and assembly into a complete continuous flow system. In the course of this account, Van Deventer highlights several points that historians have perhaps unduly neglected. First, after the initial heavy investment in specialized tooling described by Colvin and Arnold and Faurote, Ford moved towards the use of general-purpose tools customized by elaborate jigs and fixtures. Secondly, contrary to later images of mass production as a system depending on large inventories, Van Deventer documents how the River Rouge was a remarkable embodiment of low inventory, low waste practice, focused on the continuous movement of parts directly between processes. Recent writers like Williams, Haslam and Williams, and Shimokawa (Chapters 8 and 22) have traced in this a crucial anticipation of Japanese Just-In-Time methods. Thirdly, Van Deventer shows how significant areas of Ford work still relied on extensive hand work and skill, while body construction and much woodworking remained difficult to mechanize.

Fourthly, Van Deventer shows that initially the factory was not primarily an assembly factory. Although tractors and engines were assembled at the Rouge, the principal function of the plant was as a massive centralized manufacturing site for kits of components and parts that were then shipped in knocked-down form for final assembly at decentralized branch plants across America. The essence of mass production in this plant was not the rhythms of final assembly (indeed, finished cars did not come out of the plant) but the rhythms of the box-cars shuttling packs of manufactures to their final destinations.

Henry Ford's famous article in the 1926 edition of the *Encyclopaedia Britannica* (actually written by an assistant W.J. Cameron) (Chapter 6; see also Lyddon, 1996) was a decisive moment in the definition and widespread adoption of the term 'mass production'. Until then, the new methods had generally been popularly described as 'the Ford system' or 'Fordism'. The article emphasizes the sharpness of the break from what had gone before, differentiating mass production from earlier forms of volume production and from Taylorism. It also stresses the social role of mass production, as both creator and fulfilment of mass consumption, and as a path to a better life through improved earnings and newly available conveniences. In terms of production, it focuses quite narrowly on standardization, mechanization and the assembly line. The theme of the transformative social effects of Fordism or mass production were widely debated in the 1920s (see Filene, 1924; Ford, 1922) and were subsequently taken up by political and social theorists (Gramsci, 1971; see also Chapter 10).

Academic studies of the origin of Fordism have taken a variety of interpretative approaches. The accounts by Hounshell (1984), Nevins (1954) and the engineering contributions to this volume described above, primarily focus on the technological response to problems of large-scale production, and issues of manufacturing innovation. But another strong current, represented by the contribution of Lewchuk in this volume (Chapter 7) has argued that primacy should be given not to matters of technology, but to issues of management control (compare also Gartman, 1986; Dassbach, 1991; Meyer, 1981; Russell, 1978). Lewchuk stresses the problems that Ford faced in maintaining control over labour effort as output grew and labour turnover and the threat of labour unionization increased. Ford resolved these problems by the assembly line, which subjected all workers to the detailed control of line pacing of de-skilled and standardized jobs, in conjunction with payment and paternalist welfare systems to make the new conditions of work more acceptable. In contrast, Williams, Haslam and Williams (Chapter 8; see also Williams, Haslam and Williams, 1993) take issue with these interpretations and, in the process, challenge many features of the 'unexamined stereotype' of mass production. In the first place, they reconsider the sources of the dramatic productivity advances achieved at Highland Park between 1910 and 1916 and question the primacy of the assembly line in these developments. They show that most of the cost reductions of these years stemmed from three sources. First, Ford expanded the span of its operations and became a major manufacturer rather than simply an assembler, bringing previously outsourced work in-house and dramatically reducing the costs of these parts. Secondly, it did this by massive investment in manufacturig and machining capital goods. And thirdly, it massively reduced indirect labour in the factory by a range of changes in materials handling, the redesign of layouts for sequential production and the simplification of tasks. All of these changes brought massive cost reductions even before the assembly lines. The assembly lines were important, but their main function was to break a bottleneck, raising assembly capacity to match the hugely enhanced (and previously imbalanced) manufacturing and machining capacity which had been developed since 1910. Its direct cost savings, though significant, were not unique and its pre-eminence in accounts of the Ford system is unjustified.

Williams, Haslam and Williams build on this reinterpretation to question the validity of the common stereotype of mass production (see also Chapter 31). In particular, they argue that several features of the stereotype such as dedicated machine tools, Taylorized de-skilled workers, a rigid and inflexible work process, and a standard and unchanging product, *cannot* be found at Highland Park. Instead Ford widely applied standard and reusable machines, continuously reorganized work, and pursued continuous improvement and flexibility not only within the plant but also in a continually improved product. The authors note that the system became increasingly rigid over time, especially with the construction of the River Rouge plant, which depended heavily on insatiable output for its single product, and which turned away from productive flexibility in the pursuit of once-for-all cost savings. They conclude that the 'classic case' does not fit the stereotype, and thus lodge an important question mark over the wider analytic use of the stereotype.

## 3. Diffusion and Variations

This wider critique of the notion of 'mass production' is discussed more fully in later articles, but this line of approach also has important implications for our understanding of patterns of diffusion or variation of 'Fordist' or 'mass production' practices. These are considered in the articles in Part III.

The recent influential work of the International Motor Vehicle Project (IMVP), for example, argues a 'univeralistic' case about diffusion: what had emerged in Detroit was a superior system that could and should have changed the world. Where it failed to do so, it was largely a result of social or political obstacles that cost the world dear (Womack, Jones and Roos, 1990; see also Chapter 26). The historical parts of this work draw heavily on Lewchuk (Chapter 7 and Lewchuk, 1987), who argues that the techniques of Fordism were widely applicable not only in the USA but in other advanced countries like Britain as well. Market demand was not a major problem because the techniques of the assembly line could be applied at relatively low volumes. British firms, however, failed to apply these systems for nearly fifty years, and this failure was a root cause of the ultimate decline of the British motor industry. They failed to do so primarily because of social constraints, notably cultural attitudes and the resistance of labour. These views have been strongly criticized by Tolliday and others (Tolliday, 1987; Lyddon, 1996) who have reasserted the constraints imposed by markets outside the USA, and stressed that Ford-style mass production was not primarily based on the assembly line but on the large-scale manufacturing production that underlay it. Instead they have focused on the creativity and dynamism of the British tradition, involving the selective adaptation of mass production methods and their hybridization with other native engineering traditions (Tolliday, 1998).

In fact, though the classic example of mass production was widely admired and studied, few producers simply attempted to imitate it, and those who did often found it a route to disaster. Instead, diffusion proceeded in more adaptive and variegated forms, with different industries and sectors developing certain central notions associated with mass production in often contrasting ways. Tolliday and Zeitlin (Chapter 9) provide a long period overview of the diffusion and development of mass production in the international automobile industry. In particular, they show that the 'classic' Fordist model was speedily modified even in the United States, and was soon superseded by the 'Sloanist' model of General Motors, combining large-scale production with controlled variety and change driven by marketing requirements (Sloan, 1965; Snell, 1970; Kuhn, 1986). Abroad, notably in Europe, mass production was always more problematic, though it enjoyed at least a partial resurgence in Continental Europe after World War II. Rather than a history of imitation and diffusion, the story is one of continuing dilemmas and trade-offs between productivity and flexibility (Abernathy, 1978).

However, responses to Fordism and mass production were never solely determined by economic forces. Charles Maier (Chapter 10) discusses the social and political responses to Taylorism and Fordism (or 'Americanism') in Europe in the interwar period. From the start, the responses these systems evoked were as much the result of symbolic and cultural appeals as of any strictly technical features. Both left and right (for example, Mussolini, Hitler and Lenin) found interesting possibilities in engineering models of social management. Above all, 'Americanism' offered an escape from having to accept class confrontation and social division – through a 'mental revolution' the interests of employers and workers could be brought together in the common pursuit of social efficiency. Maier shows, however, that during the 1920s an originally multi-sided debate quickly became ideologically stunted as these ideas were taken up by conservative business communities and made to serve national and conservative ideologies. Ultimately, 'The supreme confidence in engineering as social redemption perished with the other dreams of the 20th century' (see also, Nolan 1994).

From the 1920s, mass production was alive in the world as a powerful general concept. It quickly shed its technical or precise meaning and became an abstract ideal of standardization, mechanization and repetitive production, allied to implications of order, rationality and universality, and notions of what Giedion calls: 'form purified by serial production' (Giedion, 1948, pp. 491–2). These ideas could be deployed in diverse cultural and artistic forms, just as they were in economics and politics. They had enormous technical, social and aesthetic consequences. Most famously, perhaps, the Modern Movement's enthusiasm for the machine as both an aesthetic value and a means to resolve social contradictions through the mass supply of cheap products has received considerable attention (Greenhalgh, 1990; Batchelor, 1994). These notions powerfully influenced Soviet as well as capitalist society. Both Lenin and Stalin were fascinated by the methods of mass production, and despite the ideological tensions involved, both Ford and Fordism enjoyed cult status in the Soviet Union in the mid-1920s (Chapter 10). According to one contemporary Soviet commentator: 'Cheap mass production is a Soviet goal, more precious from a political standpoint than world revolution' (White, 1986). The idea and image of Ford's methods as much as sheer commercial calculations pushed forward and structured the diffusion of Ford methods, partly remade in a new Soviet image (White, 1986; Sutton, 1968).

Thomas Hughes (Chapter 11; see also Hughes, 1989; Batchelor, 1994) stresses the central inspiration of Ford and mass production methods in European art and architecture in the 1920s. Focusing on Gropius and Le Corbusier, he stresses that, more than simply admiring the new products, these figures were also fascinated by its techniques. Gropius's imaginative deployment of mass production was not confined to his design influences on Behrens and AEG, or the dissemination of 'Fordist' ideas into the German rationalization movement. He also developed a project to design and develop mass-produced houses which could be selected and assembled 'like shoes from a catalog' out of standardized interchangeable parts. He founded a company (DEWOG) for this purpose and had some success with a 'housebuilding factory' for regional projects before the Depression.

Thus mass production resonated deeply into society, politics and culture, influencing product conception as well as execution. But the task of balancing innovation and changing demand with large-scale high throughput factory methods continued to challenge producers across a wide range of 'mass production' industries. The next three articles, by Blaszczyk, Meyer, and Nelson (Chapters 12, 13 and 14 respectively) examine these dilemmas in practice.

Regina Lee Blaszczyk's study of the Homer Laughlin China Co., the United States' and the world's largest pottery manufacturer in the first half of the 20th century, provides a fascinating case study of hybrid or crossover production methods. Homer Laughlin was the industry's low-cost high-volume producer selling to working class and lower middle class markets through mass retailers in the five-and-dime stores (notably Woolworths) and mail order catalogues. In the interwar period the company pursued a strategy which involved some remarkable experiments in the direct emulation of Fordist methods in the auto industry. Yet they never committed themselves unilaterally to a single 'mass production' strategy. Instead they saw 'mass production' as a grab bag of techniques to be drawn on as and when needed. Although observers talked of the 'reign of the robots' in their factories, complete mechanization was never pursued. Their perceived need to cater for varied and changing customer tastes made them aware of the need to manage complex trade-offs between the demands of manufacturing processes and the requirements of design, and not to lose sight of their 19th century roots as the 'Staffordshire of America'.

The industry was subject to the powerful demands of the mass retailers like Kresge and Woolworth. Homer Laughlin's leading competitor, Mount Clemens, was actually taken over by Kresge and converted into a fully mass production, non-union, de-skilled maker of cheap non-differentiated products for the bottom of the market. Homer Laughlin studied and learned from these experiences, but went a different way. Its leaders believed that rather than the mass market becoming increasingly homogeneous, it was, in fact, producing working class consumers with 'Tiffany tastes'. The company also recognized in the 1930s that cheap labour products from Japan would soon seize the low end of the market and that the future lay in combining design and variety with as much low cost mass production as possible, but maintaining a balance between the two. Blaszczyk's article demonstrates in detail how, often hesitantly and erratically, Homer Laughlin selectively developed mass production techniques in key operations. Walter Emerson, the architect of many of these changes, was a remarkable and innovatory apostle of Fordist practices in the industry – but he knew their limits. This is a case of the creative coexistence of mass production and traditional methods, and fits with the wider work of Philip Scranton (1991, 1994) and others who have warned that excessively rigid taxonomies separating mass and batch production may distract from an understanding of multiple hybrid or crossover methods of production (compare also Chapter 17).

Stephen Meyer's study of the Allis-Chalmers Company of Milwaukee, looks at a company that moved from craft production of heavy engineering equipment in the mid 19th century, through an era of specialized batch production of electrical and agricultural products in the early 20th century, to a modified form of mass production in the 1930s. In many respects the experience of Allis-Chalmers and its ambiguous relationship with mass production was probably more typical of (even large) American companies, than the experience of Ford (see also Raff, 1991; Amberg, 1989). In this case, mass production did not suddenly capture the company, but emerged initially as an adjunct of batch production and then gradually spread until mass and batch production co-existed within the company. Even though by the 1930s Allis-Chalmers was the third largest US tractor manufacturer (and as we have seen Ford successfully applied its mass production techniques to tractors), it did not follow assembly-line methods or sequential work organization and continued to use much skilled and specialized labour. Even after 1937 when the company substantially modernized its production, it retained a bifurcated internal structure – part craft and part modified mass production. Even on the modern side of the firm, management was often remarkably ignorant of how mass production was supposed to work and implemented only a 'pale imitation' of Detroit methods.

As Daniel Nelson shows in Chapter 14, the tire industry was pulled more closely along in the wake of auto mass production, being driven through 'compulsive sequences' of mechanization, reorganization, and investment in the inter-war years. Yet mass production in the tire industry remained persistently different from the auto industry. Nelson highlights two important areas of difference. Firstly, although mechanized handling eliminated swathes of manual labourers, mass production did not eliminate the need for skills, judgement and dexterity in the production process, though it did reduce physical effort. Secondly, machine pacing did not spread widely. In fact, employers preferred to eliminate machine-paced tasks which, they believed, limited the output of the fastest workers, and placed as many as possible on piece-rates (thus generating a crucial focus for heightened union–management conflict over rate cutting in the 1930s and 1940s). Nelson, therefore, rejects any simple 'de-skilling' hypothesis, though he does link the new techniques strongly to important trends in the professionalization of management.

Many writers have identified war and military demand as a powerful driving force for mass production (Smith, 1987), and we have already noted the role of military demand in the rise of the American System. However, the next two articles suggest the need for a more complex picture. David Hounshell (Chapter 15) focuses on Ford's disastrous attempt during World War I to apply his system to the mass production of Eagle submarine patrol boats. Ford's effort foundered on hubris and miscalculation. Ford had unbridled confidence in the wholesale applicability of his methods to boats and failed to understand that marine engineering involved quite different design problems and construction techniques to auto making. As a result, Ford failed to revolutionize shipbuilding techniques and highlighted the problems of transferring mass production techniques more widely.

Jonathan Zeitlin (Chapter 6) places these issues in a wider comparative context of war production. He stresses that the central characteristics of military demand in war include not only uniformity and order, but also 'baroque' needs for maximum performance, speedy redesign and response to unpredictable shortages. In other words, they present in a heightened form many central ongoing dilemmas of manufacturing.

Britain, the United States and Germany all reacted in different ways to the need to balance qualitative gains from design changes against quantitative losses arising from interruptions to continuous production, reflecting contrasts in national traditions, resources and civil–military relations. The British solution was to pursue volume production but to accept continuous incremental improvements and modifications spliced into existing products and designs. By drip feeding in improvements they were able to achieve dramatic design advances over time while output still rose steadily, as in the case of the Spitfire. The American solution (notably at the giant Willow Run bomber plant) was to go all out for mass production by freezing designs and then spend freely on retrospective modification of the results to incorporate best design practice. The results (compared to Britain) were remarkable achievements in output, but also significant negatives in terms of long delays in getting into production, poorer quality, and huge but largely hidden costs in the 'modification centers' which put a big question mark over apparent productivity successes. In Germany, until 1944 the military exercised free rein over incessant design changes, relying on its traditions of technical excellence and skilled labour. Their need to catch up with British designs perhaps gave them little choice about constant innovation but the co-ordination and administration of the changes were erratic and unrealistic, reflecting the decision-making structures of Nazi Germany. Only after 1944, when Speer reorganized around mass production of a few standard types, was a remarkable quantum leap in output achieved. But this was a matter of desperation, and sheer quantity was no panacea for inadequate designs and poor quality (see also, Holley, 1987).

Zeitlin questions simple readings of this comparison which argue (as Corelli Barnett (1986) has done) for the powerful superiority of American mass production, an image which played a major role in shaping post-war reconstruction planning. The chief lesson drawn by many officials and industrialists was that extraordinary productivity could be achieved only if continuous output and standard models were not disturbed by 'abnormal' disruptions to large-scale production programmes. In fact, however, these 'abnormal' conditions of fluctuations and changing demand are as much a part of normal manufacturing conditions as a condition of stable and predictable markets, and in the post-war world it was the countries like Japan (that most effectively grasped this and built such factors into their production systems) that were to enjoy most dramatic success. As Zeitlin underlines, all manufacturers need to find ways to balance countervailing objectives: quality and quantity, design change and production continuity; or standardization and customization. In many respects, British innovations in handling incremental change during the war anticipated many later Japanese methods of dealing with these issues such as continuous improvement, Just-in-Time or complex mechanisms to orchestrate collaboration with suppliers and subcontractors in product development and process improvement, that formed part of the Japanese reworking of mass production in the 1970s and 1980s (see Chapters 22–8).

The preceding articles have all concentrated on the rise and functioning of mass production or hybrid mass production systems. Michael Storper, however (Chapter 17), looks at a case of transition from an established mass production regime to a system of flexible specialization. The case he analyses is the Hollywood film industry. Here as in many other industries, the origins of the film industry lay in artisanal small-scale production, but in the 1920s the industry became industrialized, concentrated in California and focused on the production of standardized products for mass audiences, at times, literally 'sold by the foot'. The factory-like 'studio system' which dominated Hollywood from the 1920s to the 1940s was set up to fabricate and assemble 'formula' pictures produced to tight specifications according to mass production principles, emphasizing routinization, capacity utilization, flow techiques and task fragmentation. Until the 1940s, oligopolistic control of the industry was cemented by forward integration into closely controlled distribution channels (the cinemas).

But in the 1950s, antitrust action forced the studios to divest their cinema chains. This made it harder to predict and control audiences, especially since it coincided with the rising challenge of television. As profits fell, the studios responded by seeking to lower their fixed costs. They did so by differentiating their products and reducing their fixed costs by externalizing functions and vertically disintegrating. Both of these actions brought into existence conditions for independent producers, screenwriters and specialist technical services to flourish. Once started, the process of externalization took on a dynamic of its own. The specialized subcontractors started to compete with the studios that had helped to create them, while externalized labour markets freed 'stars' from monopoly studio control. Thus studios found their costs still rising and pushed the process of externalization still further to offset this. Despite attempts to restabilize the situation in their favour, the studios were not able to reassert control and stability in the industry, and ultimately opted for partnership with a now heavily externalized industry, spreading their own risks by diversifying into related entertainment services. The result has been the creation of an 'entertainment industrial complex' around Los Angeles using flexible methods to produce, and to collectivize risk.

Storper draws a number of general conclusions out of this (considered further in Part VI on Industrial Divides (Chapters 29–32)). Why are certain selections made from a range of possibilities of industrial structure and organization? Why for instance did Hollywood produce flexible specialization rather than industrial dualism? He develops a theory based on contingent choices among alternatives creating 'branching points'. Drawing on contingency and path dependency theories, Storper argues that alternatives once chosen, for whatever reason, are not easily reversible and large outcomes may be driven by small events. Once Hollywood turned to seeking external economies it created a chain of events where successive crises meant that the externalized services further destabilized the old internal scale economies and triggered still further vertical disintegration. The outcome did not necessarily depend on the long-term relative efficiencies of the alternative structures, but on the outcomes of specific choices at strategic moments.

## 4. Labour under Mass Production

The transformation of work under mass production has been the subject of an enormous literature, which is covered in Part IV. Most work in this area has been quite narrowly focused. For the auto industry, there are several detailed studies of the early impact of mass production organization on work and life in the factories (Gartman, 1986; Peterson, 1987; Raff, 1988; Raff and Summers, 1987) which also discuss the impact of de-skilling, the break up of old forms of union organization, and the rationale of high wages in the industry.

Stephen Meyer (Chapter 18) analyses Ford's Five-Dollar Day, and places it in the context of wider American experiments in 'Americanization' and welfare capitalism (see also

Meyer, 1981). By 1914, some 70 per cent of Ford's workforce consisted of raw immigrants from some of the least industrialized regions of Italy and Central and Eastern Europe. The plant was plagued with absenteeism and high levels of labour turnover, and Ford was concerned about the debilitating effects of alien cultures and the 'racial inefficiency' of pre-industrial work habits in a factory that required a new intensity of labour discipline. His response was the Five-Dollar Day and the intrusive investigations of the Sociological Department to combine the carrot of high wages and the stick of an inquisition into personal life to drive the Americanization of the workforce and its acclimatization to new work disciplines. Meyer's study highlights the manipulative and coercive elements of these systems, and shows also that they were very short-lived. Neither high wages nor welfare survived the changed economic conditions of the war and early 1920s, and Ford soon turned to a tougher and more conflict-oriented approach.

By focusing on the inter-war history of shopfloor work in American auto factories, Ronald Edsforth and Robert Asher (Chapter 19) highlight the devastating physical and psychological effects of mass production on auto workers and show how resistance to speed up became a focal point of grievance and contributed to the rise of the United Automobile Workers union (UAW) (see also Gartman, 1986; Peterson, 1987; Friedlander, 1975; Gabin, 1990; Fine, 1963; Lichtenstein and Meyer, 1989; Lichtenstein, 1980; Babson, 1991; Asher and Edsforth, 1995). The reaction against a coercive work environment meant that the UAW from the outset focused on issues such as security and regularity, which they sought to enforce via contract language and seniority systems, controlling workloads and channelling anger into orderly procedures. These constraints on employers were not the result of the radicalism of labour, but of a drive for fairness and order. These restraints became well established in the post-war years, and some have argued that they made deep inroads into managerial prerogatives (MacDonald, 1963), though other accounts have argued that they never amounted to a fundamental challenge for control (Tolliday and Zeitlin, 1986; Harbison and Dubin, 1947).

In a wider spectrum of industries, the impact of mass production was more varied, though as yet relatively little attention has been paid to these sectors in comparison to automobiles (Schatz, 1983; Harbison and Dubin, 1947). Nelson (Chapter 14) stresses the relatively limited role of de-skilling in the tire industry, and Meyer (Chapter 13) shows how patterns of labour reorganization could be complex and contradictory even within a single firm. At Allis-Chalmers for example, management persistently used mechanization and new technology to destroy old craft unions before World War I. From the 1920s, the workers did not generally resist the new technology – its spread corresponded largely to management plans – but they contested the terms of its introduction. As we have seen, Allis-Chalmers production technology was bifurcated, and so too was trade unionism in the plant. Craft unions remained the unionism of the batch production side, while the mass production side became a bastion of militant AFL-CIO (American Federation of Labour-Congress of Industrial Organizations) encompassing unionism, developing a reputation as a 'red' local before World War II. In parallel to the 'difficult coexistence' of batch and mass production in the factory came an equally fraught co-existence of craft and mass unionism.

A substantial literature has also emerged attempting to understand the predicament of assembly line workers as well as their dreams and aspirations (Walker and Guest, 1952; Chinoy, 1965). The assembly line has also produced a long line of first-hand accounts of life and work on the shopfloor. Many have been written by academics or journalists working in the plants, and some of these are very fine, notably Watkins at Ford in the early 1920s, Linhart at Renault in the 1970s and Kamata at Toyota in the 1970s. Together they present a fascinating cultural/social critique of labour conditions under mass production (Watkins, 1920; Linhart, 1981; Kamata, 1984). But none of these accounts is as first hand or as devastating as the account by Ben Hamper, a riveter at General Motors' Truck and Bus Plant in Flint, Michigan in the late 1970s and early 1980s (Chapter 20). Hamper's powerful writing captures the boredom and anger, the frustration, creativity, drunkenness and nihilism that become the lot of a line worker in a dehumanizing system. In particular, it describes the waste of people's talents on the shopfloor and the grim daily battle against authority and the line to carve out space for individual identity and avoid becoming a zombie. Although the line between truth and fiction may sometimes be blurred in Hamper's work, it is nonetheless, in many senses, the most authentic voice from the assembly line that we have had.

By the 1980s, the post-war collective bargaining system in American mass production industries was coming under stress, and Michael Piore (Chapter 21) links the decline of organized labour in this period to a broader crisis in mass production. When industrial unionism emerged in the 1930s, he argues, it rose on the back of mass production. The particular form that it took enjoyed a powerful resonance because it linked the interests of industrial workers with a 'social vision' of more general relevance (Boyle, 1995; Lichtenstein, 1986). It supported high wages which seemed to underpin consumer incomes and offset risks of the return of Depression. It linked protective collective bargaining and labour legislation with a wider programme of social welfare. It also provided a set of rules and constraints that were widely seen as legitimate countervailing forces against the immense technically determined powers of mass production. In this capacity it broadly accepted management's right to manage (and the exclusion of trade unions from any role in the operation of business) in return for a role in enforcing rules and procedures to ensure that these rights were exercised in a fair and non-discriminatory fashion (Tolliday and Zeitlin, 1986).

But by the 1970s both the classic mass production system and the union 'vision' related to it were in decline. Driven by Reaganite politicians and neoclassical economics, unions increasingly came to be seen as narrow sectional interests, and their defence of rules seemed to damage competitiveness, productivity and, ultimately, prosperity. Unions were seen as obstacles to the functioning of efficient free markets, and new needs for flexibility and co-operation were widely interpreted as requiring non-union modes. But, as Piore notes, effective flexibility also requires skills and trust, which in turn require that competitive pressures should not be allowed to drive self-defeating short-term cost cutting. Under mass production, the necessity for complex relations of mutual dependence between employer and employed were obscured by the technical power of the system. They were more transparent in earlier craft systems. But there is no way to go back to those. The need in the 1980s, he concludes, is to marry older craftist traditions with the best aspects of industrial unionism, notably its encompassing and social aspects, to build a more creative union tradition.

## 5. The Japanese Challenge

The success of the Japanese challenge to Western production methods in the 1970s and 1980s has been interpreted in many ways. For some, it is the result of quite new productive principles of 'lean production' which reject the fundamentals of mass production. For others it is a barely disguised, and intensified form of mass production. These issues provide the focus of the articles in Part V.

Kōichi Shimokawa (Chapter 22) argues that the Japanese Just-In-Time system (JIT) is closely related to the Ford system, with similar starting point and many similar principles. But is has been developed in different ways, resulting ultimately in two quite contrasting patterns. The early Ford system, as we have seen (Chapters 3–8), initially focused on synchronization and low inventories. But there was always a tension in mass production between imperatives to maximize economies of scale through extreme specialization and continuous production, and needs to synchronize processes to avoid stockpiling. Western mass production systems prioritized one aspect, Japan prioritized the other. Shimokawa argues that the 'Fordist' variant emerged in the 1920s with an emphasis on scale and continuous-output in large batches. Both work processes and jobs were closely defined and routinized and workers' jobs minutely divided. The JIT approach, in contrast, emphasized process synchronization and small lot production, relying heavily on the flexible handling of tasks by versatile workers. Shimokawa analyses the conditions that led to the rigidification of mass production systems in the USA, and the very different conditions that gave birth in Japan to JIT and the Toyota Production System (TPS) (see also Cusumano, 1985; Tolliday, 1998; Udagawa, 1995).

Shimokawa's analysis of the slow emergence of an alternative approach to mass production is complemented by the first-hand account of Taiichi Ohno (Chapter 23), widely recognized as the founding father of the Toyota Production System, of the origins of the process. Ohno is clear about his debts to Henry Ford and the American supermarket model, but also emphasizes how the principles of synchronization can lead to very different conclusions in productive practice. The two articles by Yasuhiro Monden (Chapters 24 and 25; see also Monden, 1993) are detailed engineering descriptions of the principles and practices of the Toyota Motor Company at the end of the 1970s. As such, they provide a fascinating counterpart to the classic descriptions of the Ford system by Colvin, Arnold and Faurote, and Van Deventer (Chapters 3, 4 and 5). Monden, a close associate of Ohno, succinctly outlines the key features of JIT and provides concrete examples of their application. These two articles provide easily understandable expositions of the workings of key features of the system such as kanban, production levelling, rapid setups and automation that are often discussed rather abstractly by social scientists (for other analyses see, Coriat, 1979; Sayer, 1986; Schonberger, 1982; Williams *et al.*, 1991).

By the 1980s these features had become the basis for what was widely regarded as a new system of production organization that many argued was superior to the traditional patterns of Fordist mass production. This critique of mass production was perhaps most fully formulated in the influential work of the International Motor Vehicle Project (IMVP) at the Massachusetts Institute of Technology in the late 1980s (Womack, Jones and Roos, 1990; see also Cusumano, 1994). John Krafcik's research (Chapter 26) was a fundamental building block of this work. He compared global manufacturing practice and sought to explain wide variations in productivity, quality and flexibility across companies and national industries. Krafcik is concerned to define the key differences between different production systems: 'craft', 'Fordist' and 'lean production' (or the Toyota Production System).

The crucial distinction developed here, and supported by extensive statistical material, is the difference between Fordist and lean systems. The former, Krafcik argues, are high volume 'buffered' systems in which it is accepted that cost has to be traded off against quality and where risks are controlled by accepting the relatively higher costs of buffers at all levels of production. In contrast, lean systems are based on the assumption that high quality and high productivity are inseparable and that low costs can only be achieved by making the system 'fragile', so that problems are quickly made visible and resolved. Lean systems, he argues, share many common principles with Fordist mass production, but have developed in a quite contrasting way. They do not depend on national or cultural factors to succeed, but primarily on corporate factors under the control of management. Interestingly, he also notes that well-managed 'buffered' systems can at times produce results comparable to those of some lean systems.

Not all have shared the IMVP's optimistic interpretation of the Toyota Production System. The article by Knuth Dohse, Ulrich Jürgens and Thomas Malsch (Chapter 27) provides a classic and much-cited statement of the view that Toyotism departs much less from Fordism than is often claimed (see also Parker and Slaughter, 1988; Babson, 1995; Dassbach, 1996; Berggren, 1993). The writers stress that, at its core, the system relies on the intensification and exploitation of labour, through systematic undermanning and individualized wage systems. Despite its appearance of flexibility, work in Japanese factories is intensely standardized and Taylorized, and the absence of buffers in the production system is used to deliberately 'stress' production and force the workforce to rapidly solve problems. The flexibility of the system also depends heavily on an ability to exploit low wage suppliers and peripheral workers. The success of the system, therefore, derives not so much from innovations in productive method (though there are some of these), as from a system of management–labour relations that allows almost unlimited managerial prerogative. This in turn is the result in part of cultural factors, but also of a history of defeat for independent worker organization, and of worker dependence on powerful firms. As such, they conclude, the system relies on many exclusively 'Japanese' features, and cannot easily be transferred to other national contexts.

In contrast, Martin Kenney and Richard Florida (Chapter 28; see also Kenney and Florida, 1993), reject views that the Japanese system is a sort of hyper-Fordism, pumping more surplus value out of workers. Instead they argue that Japan has given birth to a 'postfordist' or 'Fujitsuist' form of productive organization embodying 'structured flexibility' and moving away from the extreme functional specialization and de-skilling of Fordism. 'Fujitsuism' is not a more advanced version of Fordism but a qualitatively superior alternative to it, and the beginning of another phase of capitalist development. Fordist mass production, they argue is based on task fragmentation, functional specialization, mechanization and assembly-line principles. But Fujitsuism is based on the integration of innovation and production, the abolition of the strict dichotomy between markets and vertical integration, the use of polyvalency and learning in the factory, and the operation of the factory as an information system. As such, the system is not culturally specific to Japan, but is both imitable and transplantable, and can be rapidly diffused across national contexts.

## 6. Industrial Divides?

Part VI considers a variety of other recent challenges to the conventional practice and theorizations of mass production. The end of the long post-war boom and the profound changes affecting the advanced capitalist economies since then, have precipitated widespread attempts to make sense of these shifts. Perhaps the most prominent of these have been theories of postfordism and theories of flexible specialization. Both approaches argue that the crisis of the advanced economies can be traced to an underlying crisis in the established basic pattern of manufacturing, which they generally refer to as Fordism or mass production. For the Regulation School and variants of this approach in Marxist political economy, the central issue is the breakdown of a coherent social formation based around the general principles of Fordism. According to these arguments, the social organization of production has shaped both technology and the labour process, and has required supportive organizational forms at corporate, governmental and social levels. In the case of mass production, these forms are the functionally specialized and vertically integrated corporation, Keynesian state economic intervention and high wages, rising living standards and mass consumption (see Noel, 1987; Lash and Urry, 1987; for critiques see Hirst and Zeitlin, 1991). These 'totalizing' approaches have aimed to develop a general theory of the form and timing of changes in modes of production.

Equally influential have been theories of flexible specialization. The crucial text here is Michael Piore and Charles Sabel's *The Second Industrial Divide* (1984), which argues that the present crisis 'results from the limits of the model of industrial development that is founded on mass production', and presents flexible specialization as an alternative model of industrial development. Piore and Sabel's work makes a fundamental distinction between mass production and flexible specialization as principles of organization, technological paradigms or ideal–typical models. They counterpose the 'mass production' model of standardized production in high volumes, using special-purpose machines and predominantly unskilled labour to the craft or flexible specialization model which is based on a wide and changing array of customized products, produced with flexible general-purpose equipment and skilled adaptable workers. The two modes have existed in rivalry since the 19th century. But early in the 20th century the force of mass production created an 'industrial divide'. It was able to impose a technological trajectory and gain an ideological hold that largely blocked the path of alternative flexible systems. The triumph of mass production was not determined by sheer efficiency, but rather by the exercise of power, and the commitment of political and financial resources. However, in the 1970s and 1980s internal and external shocks to the mass production system created the possibility of a new branching point and revealed again the potential of networks, industrial districts and flexibly specialized industrial systems. The combination of external shocks, notably the breakdown of Keynesian regulation, and of internal shocks, particularly the fragmentation of mass markets, created the conditions in which islands of flexible specialization could potentially move again to the mainstream. In a complementary wide-ranging and seminal study (Chapter 29) Charles Sabel and Jonathan Zeitlin explored the long-run history of alternatives to mass production, particularly in European industrial districts, and examined both their productive successes and the reasons for their decline or marginalization.

These views were from the first highly influential but also very controversial. One avenue of criticism (Chapter 30) by David Landes, for example, was fiercely to question Sabel and Zeitlin's narrative and to reassert the power and continuing primacy of mass production systems. 'Mass production', Landes asserts, 'is *the* mode of production best suited to the needs and means of the great majority of consumers. Mass production is cheaper production ... and the logic of the machine pushes it in the direction of uniformity and standardization ... [T]he logic of industrialization has pushed all economies towards mass production, some faster than others'. A second line of critique came from Karel Williams, Tony Cutler, John Williams and Colin Haslam (Chapter 31) who rejected the conceptual framework of the flexible specialization argument. They argued that the sharp distinction between mass production and flexible specialization was both conceptually and empirically unsustainable. In fact, most productive systems can be seen as blurred or hybrid versions of the two. Moreover, the ideal type of mass production is unrealistic. From its early days, mass production involved significant product differentiation, and it has been adapted and modified over time to deal with lower volumes and greater product variety. In some senses the Model T was the exception and Sloanism the norm. Any simple notion of a generic system of mass production should be treated with care. Williams, Cutler, Williams and Haslam also questioned Piore and Sabel's diagnosis of the crisis of the 1970s and 1980s, doubting the extent of market fragmentation or market saturation, and stressing the compatibility of many forms of mass production with product differentiation and innovation. Rather than a crisis of mass production, they argued, there is a permanent problem of enterprise calculation which requires different production strategies for different market conditions. In a related vein, Andrew Sayer (Chapter 32) also stressed the dangers of 'binary histories' and questioned the concept of a 'crisis of Fordism', insisting that 'mass production is thriving' and criticizing notions of the inherent limits of mass production (compare also Hyman, 1988; Elam, 1990; Badham and Matthews, 1989; Mahon, 1987).

## Conclusion

The articles in this collection and the preceding review of the literature suggest that the use of narrowly defined and restrictive archetypes of mass production may be a mistake. The classic Fordist archetype of mass production was a singular historical event, and most of the history of mass production has been one of change, adaptation and innovation to deal with changing process, product and market constraints. The rigidity and standardization of the system, though undoubtedly significant, have often been overstated. In the automobile industry, Sloanism with its emphasis on 'keeping the customer dissatisfied' soon took centre stage and produced a much more complex situation for production and factory organization (which has yet to receive adequate historical attention) (see Kuhn, 1986). In core American 'mass production' industries, as we have seen (Chapters 12–14), mass production took various hybrid forms. In Europe in the inter-war years and in Japan after the World War II, selective and creative adaptation and innovation opened up new ways of interpreting key elements and themes of mass production (Chapters 9, 16 and 23–6). And in a wide variety of industries different ways of combining standardization and variety, or economies of scale and scope have emerged (Chapters 17 and 28). Polarized concepts of mass and flexible production have perhaps impeded the development of more subtle languages of analysis to understand these productive forms.

At a broader level, this analysis suggests that it is also dangerous to erect abstract superstructures on an oversimplified concept of the productive system. The Regulation School, Flexible Specialization theories and theories of postfordism all use mass production as a central organizing concept. But it is doubtful whether the concept of mass production can bear the weight ascribed to it. Clearly, 'mass production' describes certain elements of a productive transformation, but it is a concept that needs to be handled in a careful and realistic fashion with close attention to historical periodization, continuity and change, adaptation and hybridization. 'Mass production' has been frequently deployed in powerful economic, cultural and political discourses. And these discourses themselves have undeniably been powerful forces in the world (Chapters 10 and 29–32). But the relationship between the concepts and the reality requires constant careful study and uncritical use of the term may conceal more than it reveals.

## Further Reading

### *Part I. Antecedents*

Battison, Edwin A. (1966), 'Eli Whitney and the milling machine', *Smithsonian Journal of History*, **1**.

Burn, D.L. (1931), 'The genesis of American engineering competition, 1850–1870', *Economic History Review*, **2**; reprinted in Sidney Pollard (ed.), *The Metal Fabrication and Engineering Industries*, Oxford: Basil Blackwell (1994).

Davies, Alun C. (1993), 'British watchmaking and the American System', *Business History*, **35** (1), January.

Floud, R.C. (1974), 'The adolesence of American Engineering Competition, 1860–1890', *Economic History Review*, **27** (1), February.

Fries, Russell I. (1975), 'British response to the American System: The case of the small-arms industry after 1869', *Technology and Culture*, **16** (3), July.

Hoke, Donald R. (1990), *Ingenious Yankees. The Rise of the American System of Manufactures in the Private Sector*, New York: Columbia University Press.

Hounshell, David (1984), *From the American System to Mass Production, 1800–1932. The Development of Manufacturing Technology in the United States*, Baltimore: Johns Hopkins University Press.

Howard, Robert A. (1978), 'Interchangeable parts re-examined: the private sector of the American arms industry on the eve of the Civil War', *Technology and Culture*, **19** (4), October, 633–49.

Lumley, Roger (1989), '"The American System of Manufactures" in Birmingham: Production Methods at the Birmingham Small Arms Co. in the Nineteenth Century', *Business History*, **31** (1), January.

Mayr, Otto and Robert C. Post (eds) (1982), *Yankee Enterprise: The Rise of the American System of Manufactures*, Washington DC: Smithsonian Institution Press.

Musson, A.E. (1969), 'The growth of mass-production engineering' in A.E. Musson and E. Robinson (eds), *Science and Technology in the Industrial Revolution*; reprinted in Sidney Pollard (ed.), *The Metal Fabrication and Engineering Industries*, Oxford: Basil Blackwell (1994).

Page, Brian and Richard Walker (1991), 'From Settlement to Fordism: The agroindustrial revolution in the American MidWest', *Economic Geography*, **67**, 281–315.

Rosenberg, Nathan (1963), 'Technological change in the machine tool industry, 1840–1910', *Journal of Economic History*, **23**, 414–43.

Rosenberg, Nathan (ed.) (1969), *The American System of Manufactures: The Report of the Committee on the Machinery of the United States 1855, and the Special Reports of George Wallis and Joseph Whitworth 1854*, Edinburgh: Edinburgh University Press.

Rosenbloom, Joshua L. (1993), 'Anglo-American technological differences in small arms manufacturing', *Journal of Interdisciplinary History*, **23** (4), Spring.
Saul, S.B. (1960), 'The American impact on British industry, 1895–1914', *Business History*, **3** (1), December; reprinted in R.P.T. Davenport-Hines (ed.), *Capital, Entrepreneurs and Profits*, Frank Cass, 1990, London.
Smith, Merritt Roe (1977), *Harpers Ferry Armory and the New Technology. The Challenge of Change*, Ithaca, NY: Cornell University Press.
Smith, Merritt Roe (1987), 'Army ordnance and the "American System" of manufacturing, 1815–1861', in Merritt Roe Smith (ed.), *Military Enterprise and Technological Change. Perspectives on the American Experience*, Cambridge, MA: MIT Press.
Thomson, Ross (1989), *The Path to Mechanized Shoe Production in the United States*, Chapel Hill: North Carolina University Press.
Uselding, Paul (1974), 'Elisha K. Root, forging and the "American System"', *Technology and Culture*, **15**; reprinted in Sidney Pollard (ed.), *The Metal Fabrication and Engineering Industries*, Oxford: Basil Blackwell (1994).
Woodbury, Robert S. (1960), 'The legend of Eli Whitney and interchangeable parts', *Technology and Culture*, **1**.
Zeitlin, Jonathan (1997), 'Between flexibility and mass production: strategic ambiguity and selective adaptation in the British engineering industry, 1830–1914', in Charles F. Sabel and Jonathan Zeitlin (eds), *World of Possibilities, Flexibility and Mass Production in Western Industrialization* (Cambridge: Cambridge University Press).

## *Part II. The 'Moment' of Fordism*

[See also Hounshell (1984) above]

Biggs, Lindy (1995), 'Building for mass production: factory design and work process at the Ford Motor Company', in Robert F. Asher and Ronald Edsforth (eds), *Autowork*, Albany: State University of New York Press.
Dassbach, Carl (1991), 'The origins of Fordism: The introduction of mass production and the Five-Dollar wage', *Critical Sociology*, **18** (1).
Filene, Edward A. (1924), *The Way Out: A Forecast of Coming Changes in American Business and Industry*, New York: Doubleday Page & Company.
Ford, Henry (1922), *My Life and Work*, New York: Doubleday Page and Company.
Gartman, David (1986), *Auto Slavery. The Labor Process in the American Automobile Industry, 1897–1950*, New Brunswick: Rutgers University Press.
Gramsci, Antonio (1971), 'Americanism and Fordism', in Q. Hoare and G. Smith (eds), *Selections from the Prison Notebooks*, New York: International Publishers.
Lewchuk, Wayne (1987), *American Technology and the British Vehicle Industry*, Cambridge and New York: Cambridge University Press.
Lyddon, Dave (1996), 'The myth of mass production and the mass production of myth', *Historical Studies in Industrial Relations*, **1**, March.
Meyer, Stephen (1981), *The Five-Dollar Day: Labor Management and Social Control in the Ford Motor Co., 1908–21*, New York: State University of New York Press.
Nevins, Allan (1954), *Ford. The Times, The Man, The Company*, New York: Charles Scribner & Sons.
Nevins, Allan and Frank Hill (1957), *Ford: Expansion and Challenge, 1915–1933*, New York: Charles Scribner & Sons.
Russell, J. (1978), 'The Coming of the Line: The Ford Highland Park Plant, 1910–14; *Radical America*, **12**.
Williams, Karel, Colin Haslam and John Williams, with Andy Adcroft and Sukhdev Johal (1993), 'The Myth of the Line: Ford's production of the Model T at Highland Park, 1909–1916', *Business History*, **35** (3), July.

## *Part III. Diffusion and Variations*

[See also, above: Lewchuk, 1987; Lyddon, 1996; Smith, 1987]


Abernathy, W.J. (1978), *The Productivity Dilemma: Roadblock to Innovation in the Auto Industry*, Baltimore: Johns Hopkins University Press.

Amberg, Stephen (1989), 'The Triumph of Industrial Orthodoxy: The Collapse of Studebaker-Packard', in Nelson Lichtenstein and Stephen Meyer (eds), *On the Line: Essays in the History of Auto Work*, Urbana: University of Illinois Press.

Barnett, Corelli (1986), *The Audit of War: The Illusion and Reality of Britain as a Great Nation*, London: Macmillan.

Batchelor, Ray (1994), *Henry Ford. Mass Production, Modernism and Design*, Manchester: Manchester University Press.

Chandler, Alfred D. (1964), *Giant Enterprise: Ford, General Motors and the Automobile Industry. Sources and Readings*, New York: Harcourt, Brace and World.

Coriat, Benjamin (1992), 'The Revitalization of Mass Production in the Computer Age', in Michael Storper and Allen Scott (eds), *Pathways to Industrialization and Regional Development*, London: Routledge.

Egle, Walter (1938), 'The progress of mass production and the German small-scale industries', *Journal of Political Economy*, **46**, February–December.

Fridenson, Patrick (1978), 'The Coming of the Assembly Line to Europe', in W. Krohn (ed.), *The Dynamics of Science and Technology*, Volume 2. Dordrecht.

Giedion, Siegfried (1948), *Mechanization Takes Command: A Contribution to Anonymous History*, Oxford: Oxford University Press.

Greenhalgh, Paul (ed.) (1990), *Modernism in Design*, London:

Holley, Irving Brinton (1987), 'A Detroit dream of mass-produced fighter aircraft: The XP-75 fiasco', *Technology and Culture*, **28** (3), July.

Hughes, Thomas P. (1989), *American Genesis. A century of invention and technological enthusiasm*, London: Viking Penguin.

Kaplinsky, Raphael (1994), 'From mass production to flexible specialization: a case study of micro-economic change in a semi-industrialized economy', *World Development*, March.

Kogut, Bruce and David Parkinson (1993), 'The Diffusion of American Organizing Principles to Europe', in Bruce Kogut (ed.), *Country Competitiveness. Technology and the Organizing of Work*, Oxford: Oxford University Press.

Kuhn, Arthur J. (1986), *GM passes Ford, 1918–1938: Designing the General Motors Performance-Control System*, Pennsylvania State University Press.

Langlois, Richard N. and Paul L. Robertson (1989), 'Explaining vertical integration: Lessons from the American Automobile Industry', *Journal of Economic History*, **49** (2), June.

Layton, Edwin T. (1974), 'The diffusion of scientific management and mass production from the US in the 20th century', *Proceedings of the XIVth International Congress in the History of Science*, 4, Tokyo.

Meyer, Stephen (1989), 'The persistence of Fordism: Workers and technology in the American automobile industry, 1900–1960', in Nelson Lichtenstein and Stephen Meyer (eds), *On the Line: Essays in the History of Auto Work*, Urbana: University of Illinois Press.

Nolan, Mary (1994), *Visions of Modernity. American Business and the Modernization of Germany*, Oxford: Oxford University Press.

Mumford, Lewis (1946), *Technics and Civilisation*, London:

Raff, Daniel M.G. (1991), 'Making cars and making money in the interwar automobile industry: Economies of scale and scope and the manufacturing behind the marketing', *Business History Review*, **65** (4), Winter.

Scott, Allen (1991), 'The aerospace-electronics industrial complex of Southern California: the formative years, 1940–1960', *Research Policy*, **20**.

Scranton, Philip (1991), 'Diversity in diversity: Flexible production and American industrialization, 1880–1930', *Business History Review*, **65** (1), Spring.

Scranton, Philip (1994), 'Manufacturing diversity: production systems, markets, and an American consumer society, 1870–1930', *Technology and Culture*, **35** (3), July.
Shiomi, Haruhito and Kazuo Wada (eds) (1995), *Fordism Transformed. The Development of Production Methods in the Automobile Industry*, Oxford: Oxford University Press.
Sloan, Alfred P. (1965), *My Years With General Motors*, New York: Doubleday.
Snell, B. (1970), 'Annual style change in the automobile industry as an unfair method of competition', *Antitrust Law and Economics Review*, **4** (1 and 2).
Sutton, Anthony C. (1968), *Western Technology and Soviet Economic Development, 1917–30*, Stanford, CA: Stanford University Press.
Thompson, G. (1954), 'Intercompany technical standardization in the early American automobile industry', *Journal of Economic History*, **14**, Winter, 1–20.
Tolliday, Steven (1987), 'The failure of mass production unionism in the motor industry, 1914–1939', in Chris Wrigley (ed.), *A History of British Industrial Relations. Vol. 2. 1914–1939*, Brighton: Harvester.
Tolliday, Steven (1998), 'The diffusion and transformation of Fordism: Britain and Japan compared', in Robert Boyer, Elsie Charron, Ulrich Jurgens and Steven Tolliday (eds), *Between Imitation and Innovation: The Transfer and Hybridization of Productive Models in the International Automobile Industry*, Oxford: Oxford University Press.
Tolliday, Steven and Jonathan Zeitlin (eds) (1986), *The Automobile Industry and Its Workers. Between Fordism and Flexibility*, Cambridge: Polity Press.
White, Christine (1986), 'Ford in Russia: in pursuit of the chimeral market', *Business History*, **28** (4), October.
White, Lawrence J. (1972), 'The American automobile industry and the small car, 1945–70', *Journal of Industrial Economics*, **20** (2), April.
Womack, James P., Daniel T. Jones and Daniel Roos (1990), *The Machine That Changed the World*, New York: Macmillan.

## *Part IV. Labour Under Mass Production*

[See also, above: Gartman, 1986; Lewchuk, 1987; Meyer, 1981; Tolliday, 1987; Tolliday and Zeitlin, 1986]

Asher, Robert F. and Ronald Edsforth (eds) (1995), *Autowork*, Albany: State University of New York Press.
Babson, Steve (1991), *Building the Union. Skilled Workers and Anglo-Gaelic Immigrants in the Rise of the UAW*, New Brunswick: Rutgers University Press.
Boyle, Kevin (1995), *The UAW and the heyday of American Liberalism, 1945–68*, Ithaca, NY: Cornell University Press.
Braverman, Harry (1974), *Labor and Monopoly Capital: The Degradation of Work in the Twentieth Century*, New York: Monthly Review Press.
Chinoy, Eli (1965), *Auto Workers and the American Dream*, Boston, MA: Random House.
Fine, Sidney (1963), *The Auto Under the Blue Eagle*, Ann Arbor: University of Michigan Press.
Friedlander, Peter (1975), *The Emergence of a UAW Local, 1936–1939: A Study in Class and Culture*, Pittsburgh: University of Pittsburgh Press.
Gabin, Nancy (1990), *Feminism in the Labor Movement. Women and the United Auto Workers, 1935–1975*, Ithaca, NY: Cornell University Press.
Harbison, Frederick H. and Robert Dubin (1947), *Patterns of Union-Management Relations. UAW, GM. Studebaker*, Chicago, IL: Chicago University Press.
Kamata, Satoshi (1984), *Japan in the Passing Lane*, London: Allen & Unwin.
Lichtenstein, Nelson (1980), 'Auto Worker Militancy and the Structure of Factory Life, 1937–55', *Journal of American History*, **67** (2), September.
Lichtenstein, Nelson (1986), 'Reutherism on the Shopfloor', in Steven Tolliday and Jonathan Zeitlin (eds), *The Automobile Industry and Its Workers. Between Fordism and Flexibility*, Cambridge: Polity Press.

Lichtenstein, Nelson and Stephen Meyer (eds) (1989), *On the Line: Essays in the History of Auto Work*, Urbana: University of Illinois Press.
Linhart, Robert (1981), *The Assembly Line*, London: John Calder.
MacDonald, Robert M. (1963), *Collective Bargaining in the Automobile Industry: a study of wage structure and competitive relations*, New Haven, CT: Yale University Press.
Moberg, David (1978), 'No More Junk: Lordstown Workers and the Demand for Quality', *Insurgent Sociologist*, **8**.
Parker, Mike and Jane Slaughter (1988), 'Management by Stress', *Technology Review*, October.
Peterson, Joyce (1987), *American Automobile Workers, 1900–1933*, Albany: State University of New York Press.
Daniel M.G. Raff (1988), 'Wage Determination Theory and the Five-Dollar Day at Ford', *Journal of Economic History*, **48** (2), June.
Raff, Daniel M.G. and Lawrence H. Summers (1987), 'Did Henry Ford pay efficiency wages?', *Journal of Labor Economics*, **5** (4).
Schatz, Ronald W. (1983), *The Electrical Workers. A History of Labor at General Electric and Westinghouse, 1923–60*, Urbana, IL: University of Illinois Press.
Tolliday, Steven and Jonathan Zeitlin (eds) (1986), *The Automobile Industry and Its Workers. Between Fordism and Flexibility*, Cambridge: Polity Press.
Tolliday, Steven and Jonathan Zeitlin (1986), 'Shopfloor bargaining, contract unionism and job control: An Anglo-American comparison', in Tolliday and Zeitlin (eds), *The Automobile Industry and Its Workers*, Cambridge: Polity Press.
Walker, Charles R. and Robert H. Guest (1952), *The Man on the Assembly Line*, Cambridge MA: Harvard University Press.
Watkins, Myron W. (1920), 'The labor situation in Detroit', *Journal of Political Economy*, **28**, December.

## *Part V. The Japanese Challenge*

[See also, above: Kamata, 1984; Parker and Slaughter, 1988; Shiomi and Wade, 1995; Womack, Jones and Roos, 1990]

Babson, Steve (ed.) (1995), *Lean Work: Empowerment and Exploitation in the Global Auto Industry*, Detroit: Wayne State University Press.
Berggren, Christian (1993), 'Lean Production – the End of History?', *Work, Employment and Society*, **7** (2), June.
Conti, Robert F. and Malcolm Warner (1993), 'Taylorism, new technology and Just-in-Time systems in Japanese manufacturing', *New Technology, Work and Employment*, **8** (1).
Coriat, Benjamin (1979), *L'atelier et le chronomêtre, essais sur le taylorisme, le fordisme et la production de masse*, Paris: Bourgeois.
Cusumano, Michael A. (1985), *The Japanese Automobile Industry. Technology and Management at Nissan and Toyota*, Cambridge MA: Harvard University Press.
Cusumano, Michael (1994), 'The Limits of Lean', *Sloan Management Review*, **18**, Summer.
Dassbach, Carl (1996), 'Lean production, labor control and post-Fordism in the Japanese automobile industry', in William C. Green and Ernest J. Yanarella (eds), *North American Auto Unions in Crisis: Lean Production as Contested Terrain*, Albany: State University of New York Press.
Friedman, David (1983), 'Beyond the Age of Ford: The strategic basis of the Japanese success in automobiles', in John Zysman and Laura Tyson (eds), *American Industry in International Competition: Government Policies and Corporate Strategies*, Ithaca, NY: Cornell University Press.
Jacot, Jacques-Henri (ed.) (1990), *Du fordisme au toyotisme, les voies de la modernisation du systeme automobile en France et au Japon*, La Documentation Francaise, Etudes et Recherches du CGP, No. 7–8, Paris.
Kenney, Martin and Richard Florida (1993), *Beyond Mass Production. The Japanese System and Its Transfer to the US*, Oxford: Oxford University Press.

Monden, Yasuhiro (1993), *Toyota Production System: An Integrated Approach to Just-in-Time*, Georgia Atlanta: Industrial Engineering and Management Press, 2nd ed.
Sayer, Andrew (1986), 'New developments in manufacturing: The Just-in-Time system', *Capital and Class*, **30**.
Schonberger, Richard J. (1982), *Japanese Manufacturing Techniques: Nine Hidden Lessons in Simplicity*, New York: Free Press.
Tolliday, Steven (1998), 'The diffusion and transformation of Fordism: Britain and Japan compared', in Robert Boyer, Elsie Charron, Ulrich Jurgens and Steven Tolliday (eds), *Between Imitation and Innovation: The Transfer and Hybridization of Productive Models in the International Automobile Industry*, Oxford: Oxford University Press.
Udagawa, Masaru (1995), 'The development of production management at the Toyota Motor Corporation', *Business History*, **37** (2), April.
Williams, Karel, Itsutomo Mitsui and Colin Haslam (1991), 'How far from Japan: A case study of Japanese Press Shop practice and management calculation', *Critical Perspectives on Accounting*, **2**.

## *Part VI. Industrial Divides?*

[See also Abernathy, 1978, above]

Abernathy, William J., Kim Clark and Alan Kantrow (1983), *Industrial Renaissance. Producing a Competitive Future for America*, New York: Basic Books.
Badham, Richard and John Matthews (1989), 'The new production systems debate', *Labour and Industry*, **2** (2).
Elam, Mark (1990), 'Puzzling out the post-Fordist debate: Technology, markets and institutions', *Economic and Industrial Democracy*, **11** (1).
Gilbert, Nigel, Roger Burrows and Anna Pollert (eds) (1992), *Fordism and Flexibility. Divisions and Change*, New York: St Martin's Press.
Hirst, Paul and Jonathan Zeitlin (1991), 'Flexible specialization versus post-Fordism: Theory, evidence and policy implications', *Economy and Society*, **20** (1), February.
Hyman, Richard (1988), 'Flexible specialization: Miracle or myth?', in R. Hyman and W. Streeck (eds), *New Technology and Industrial Relations*, Oxford: Blackwell.
Lash, Scott and John Urry (1987), *The End of Organized Capitalism*, Cambridge: Polity Press.
Mahon, R. (1987), 'From Fordism to ?: New technology, labour markets and unions', *Economic and Industrial Democracy*, **8**.
Noel, Alain (1987), 'Accumulation, regulation and social change: an essay on French political economy', *International Organization*, **41** (2), Spring.
Piore, Michael J. and Charles F. Sabel (1984), *The Second Industrial Divide: Possibilities for Prosperity*, New York: Basic Books.
Sabel, Charles (1982), *Work and Politics: The Division of Labor in Industry*, Cambridge and New York: Cambridge University Press.
Storper, Michael and Alan Scott (eds) (1992), *Pathways to Industrialization and Regional Development*, London: Routledge.
Williams, Karel (1984), 'Made in the USA', *Economy and Society*, **13** (4), November.

# Part I
# Antecedents

# [1]

# ELI WHITNEY AND THE AMERICAN SYSTEM OF MANUFACTURING*

**Merritt Roe Smith**

Commonly acknowledged as the "father of American technology," Eli Whitney (1765–1825) towers above other early 19th-century inventors as a technological folk hero. Best known is his invention of the cotton gin, a device which revolutionized agriculture and made cotton king in the plantation South. Of equal technical significance is his association with developing methods of interchangeable production, a seminal part of the Industrial Revolution, which became known during the 1850's as the "American System" of manufacturing. These contributions made Whitney a legend in his time and one of the great men of the modern era.

One thing Whitney never lacked was good publicity. Biographies, popular histories, and textbooks abound with glittering accounts of how he almost singlehandedly inaugurated a "dual revolution" in agriculture and industry, a revolution which purportedly made Americans a "people of plenty" and defined the character of their national life. To generations of schoolchildren Whitney's name became synonymous with the new machine tech-

*Parts of this essay are adopted from my book, *Harpers Ferry Armory and the New Technology: The Challenge of Change*. I am grateful to Cornell University Press for permission to publish this information.—*Merritt Roe Smith*

nology, a democratizing force which helped to stabilize America's shaky economy and legitimize the new republic in the eyes of the rest of the world. Yet, as often happens in the building of heroes, publicists created a myth larger than the man. The purpose of this essay is to re-evaluate Whitney's role as inventor of the American System and place him in the larger context of the 19th-century technological development.

* * *

The origins of the legend date back to the spring of 1798 when the debt-ridden Whitney, his cotton gin business on the verge of collapse, approached Secretary of Treasury Oliver Wolcott about manufacturing muskets for the United States. "I am persuaded," he wrote Wolcott, "that machinery moved by water (and) adapted to this business would greatly reduce the labor and facilitate the manufacture of this article." This suggestion, coupled with the promise of rapid delivery, evidently appealed to the government hovering on the brink of war because Wolcott offered Whitney a contract for 10,000 regulation muskets. Signed on June 21, 1798, the pact called for final delivery by September 30, 1800. The government, on its part, promised to pay Whitney $13.40 a musket and make generous monetary advances at various stages of the contract. Whitney considered the latter provision particularly important. "Bankruptcy and ruin were constantly staring me in the face," he confided to a friend. "By this contract I obtained some thousands of Dollars in advance which saved me from ruin."

Whitney's entry into the arms business undoubtedly was an act of desperation; the threat of "bankruptcy and ruin" drove him into making a very rash proposal. The scale of the undertaking was unprecedented. No American armory, not even the large national establishments at Springfield, Massachusetts and Harpers Ferry, Virginia had ever produced 5,000 arms annually. What is more, Whitney knew little or nothing about arms making. He soon learned that manufacturing firearms was a good deal more complex than building cotton gins. Before a single musket could be produced he had to outfit a factory, recruit a labor force, and procure the necessary raw materials. Just how rash his proposal was became evident when he failed to meet his first contract deadline, the delivery of 4,000 muskets, on September 30, 1799.

With less than a year remaining on the contract, he still had not equipped his armory. Even so, through some astute maneuvering he managed to convince federal authorities to grant one extension after another until he finally completed the contract in January 1809—more than eight years beyond the original deadline.

Whitney successfully forestalled cancellation of his contract by cultivating an idea that captured the imagination of government officials. In July 1799 he responded to expressions of concern about not fulfilling his obligation by announcing that he was developing a "new principle" of manufacturing that would not only revolutionize the arms industry but also vastly improve the quality of arms delivered to the government. "One of my primary objects," he wrote Wolcott, "is to form the tools so the tools themselves shall fashion the work and give to every part its just proportion—which when once accomplished, will give expedition, uniformity, and exactness to the whole.... In short, the tools which I contemplate are similar to an engraving on a copper plate from which may be taken a great number of impressions perceptibly alike." Here was the first clear intimation that Whitney planned to make arms with uniform, even interchangeable, parts. The copper plate analogy not only made sense; it convinced Wolcott and his staff that Whitney was indeed on the brink of an important technical discovery.

Having announced his plan, Whitney continued to champion the uniformity concept before members of Congress and other government officials. His most famous demonstration took place at the nation's capitol early in January 1801, before an audience that included President John Adams and President-elect Thomas Jefferson. Acting with typical aplomb, he showed how he could fit ten different lock mechanisms to the same musket with the use of a simple screwdriver. The intimation was clear: Whitney was making firearms with uniform parts. Those present were astonished. Even Jefferson, who was familiar with the earlier work of Honoré Blanc in France, marvelled at the display. "Mr. Whitney," he later wrote James Monroe, "has invented moulds and machines for making all the pieces of his locks so exactly equal, that take 100 locks to pieces and mingle their parts and the hundred locks may be put together as well by taking the first pieces which come to hand." "This is of importance," he continued, "because out of 10 locks, e.g. disabled for want of different pieces, 9 good locks may be put together without employing a smith." Jefferson clearly recognized the military significance of uniformity.

The 1801 demonstration made a deep impression on America's power elite. Thereafter Whitney's name became closely identified with the development of interchangeable parts, an association that eliminated further difficulties in getting extensions on his contract and made him one of the most respected men of his era. Suddenly he found himself lionized by public leaders and hailed as the "Artist of his Country." With recognition came respect, not only as an arms maker but also as a man of affairs whose opinion and advice was sought on a broad range of social and economic issues. A measure of the man's growing popularity could be seen at his Mill Rock armory near New Haven, Connecticut, where scores of curious sightseers flocked to see the inventor and his system at work. Diarists and gazetteers also began to enshrine "Whitneyville" as a monument to American technological genius. Typical was Edward Kendall's *Travels Through the Northern Parts of the United States*, a three-volume work published in 1809 that briefly described Whitney's operations and echoed Jefferson's praise. During the next four decades other writers such as Timothy Dwight (1823), Denison Olmstead (1832), and Henry Howe (1844) embellished the story. From this literature emerged the full-blown Whitney myth incorporated into hundreds of 19th and 20th-century textbooks and taught to millions of schoolchildren.

* * *

Whitney's reputation as inventor of the American System stood unchallenged until the 1960's when researchers began to detect serious discrepancies between the written record and extant artifacts. Most damaging was the discovery that Whitney muskets, including those made under his first contract, did not have interchangeable parts. The physical evidence was unmistakable. "In fact," one scholar exclaimed, "in some respects they are not even approximately interchangeable!" This, coupled with the fact that individual lock components of Whitney muskets carry special identifying marks—something which is unnecessary for truly standardized parts—indicates that Whitney must have staged his famous 1801 demonstration with specimens specially prepared for the occasion. In short, it appears that Whitney purposely duped government authorities in 1801 and afterwards encouraged the notion that he had successfully developed a system for producing uniform parts.

Is the Whitney legend completely fabricated? If Whitney did not make arms with interchangeable parts, should he be written off as a charlatan and dismissed as having played no part in the development of the American System? The answer to both questions is emphatically *no*. Clearly part of the legend is salvageable, but an accurate understanding of Whitney and his work requires a reassessment of the man as innovator, advocate, and symbol. Such an analysis not only addresses problems of technical creativity but also sheds light on the complex social and ideological character of 19th-century American industrial development.

Mechanization was a vital part of the American System, and Whitney recognized its importance at an early date. Indeed, his initial emphasis on machine production, alluded to earlier, was instrumental in gaining his first contract in 1798. Yet, curiously enough, a survey of equipment at Whitney's Mill Rock armory reveals a very modest attempt at mechanization. The articles listed are not nearly as sophisticated as one might expect to find in a pioneering "best practice" establishment.

In September 1801, a few weeks before Whitney made his first delivery of 500 muskets under the 1798 contract, young Philos Blake wrote home about the machinery he had seen at his uncle's factory. The letter mentioned a drilling machine, a boring machine, a screw machine, and a trip hammer. All four devices were well known to Whitney's contemporaries. The most interesting unit was the screw machine, a device for hollow-milling screw forgings which Whitney may have copied from the Frenchman Blanc. By 1825, the year of his death, Whitney had added to his mechanical repertoire a stamping machine, a "nitching" machine for slotting screw heads, a pair of cast iron shears, and two polishing machines. In terms of design, none of these machines represented new concepts. Even the famous "Whitney" milling machine, a self-acting device employing toothed cutters capable of producing both plain and irregular surfaces, is now known to have been built not by the master but by his nephew, Eli Whitney Blake, around 1827 and probably copied from a similar "straight cutting engine" designed by John H. Hall of the Harpers Ferry armory.

Mill Rock was a transitional establishment, a halfway point

between the traditional craft shop and the fully mechanized factory. The physical plant and labor force were never large (the latter averaging around 50 workers), and production rarely exceeded 1,500 muskets a year. Compared with other New England armories circa 1820, Mill Rock's machinery was neither plentiful nor particularly distinctive. Whether built on the premises or acquired from others, the designs reflected simplicity and cheapness. In fact, the conspicuous absence of machinery for performing the weightier tasks of gunmaking reveals that Whitney purchased a large number of components from outside sources. From the very beginning he had subcontracted for finished gunstocks and barrels, and he continued the practice into the 1820's. He intentionally confined his operations to the on-site manufacture of lock mechanisms, unquestionably the most complicated part of the musket but also the one part that could be produced without a large investment in heavy machinery. This strategy fully comported with Whitney's view of Mill Rock as an assembly shop rather than a fully integrated manufacturing establishment. He made his muskets almost entirely by hand methods, and his limited stable of machinery accentuated this orientation.

That labor-intensive methods dominated production at Mill Rock is clearly shown by the large number of files carried on the armory's inventory. Indeed, hand filing represented the touchstone of Whitney's manufacturing system. The filing shop accordingly became the focal point of activity. Visitors invariably commented on the "moulds" used by workers at Mill Rock, attributing much of Whitney's success to their widespread adoption. Commonly known as "jigs," these implements consisted of tempered steel patterns which guided armorers in filing down components to their proper dimensions. Their use was not new, even though Whitney claimed to have invented the technique and extended its applications far beyond anything achieved by his contemporaries. Ironically, physical limitations in the jigs themselves prevented him from achieving what he had promised government authorities in 1801—the production of muskets with interchangeable parts. As it was, jig filing allowed Whitney to substitute unskilled for skilled labor in the productive process and thereby effect substantial cost savings. Herein lay the true significance of his manufacturing system.

A distinguishing feature of industrial development is work

specialization through the division of labor. When labor is divided, as in jig filing, individual work assignments become more simplified while the overall production process becomes more complex. The need for coordination and control is therefore more urgent, and this, in turn, intensifies managerial responsibilities. Further demands are placed on managers when the eventual growth of an enterprise increases the speed and scale of production. Whitney, to his credit, recognized this problem and began to establish procedures for monitoring the work process at Mill Rock as early as 1798. His efforts took two forms: cost analysis and labor regimentation. Both aimed at improving the efficiency and flow of production, and both constituted important managerial advances.

A shrewd businessman, Whitney knew that proper attention to costs could mean the difference between profit or loss, success or failure. Unlike other private arms makers who simply estimated costs by adding a dollar or so to the going rates for labor and raw materials, Whitney sought to specify costs with precision. During his 1807 contract negotiations with the War Department, for example, he figured interest as well as insurance charges in calculating the price of his product. By 1812 he was beginning to add depreciation allowances for "wear and tear" on his plant and machinery. Coupled with the use of piece-rate accounting, his method of estimating costs placed him in the vanguard of factory masters who were trying to rationalize production through systematic procedures and regularized practice. As a result, Whitney enjoyed lucrative profits from the firearms business while many of his less advanced contemporaries found themselves operating at the margin. This aspect of Whitney's career has received little scholarly attention, but preliminary research indicates that his contributions to management are quite significant and certainly overshadow the mechanical inventions for which he is best but undeservedly remembered.

Whitney's desire for order and discipline applied especially to the people he employed. Like other early factory masters, he styled himself as a "steward of the Lord" whose special calling was spiritual as well as worldly leadership. Seen in this context, Mill Rock became not only a place of production but also a moral gymnasium where "correct habits" of diligence, thrift, and sobriety were inculcated through the systematic control of everyday life. Such an orientation allowed Whitney to amalgamate the sacred and the profane, thereby identifying godliness with productivity and dealing a serious blow to agrarian critics who railed against the moral decadence of industrialization as a threat to republican virtue.

***The Eli Whitney Gun Manufactory at Mill Rock, Connecticut, attracted many visitors to see Whitney's manufacturing and management innovations (Painting by William Giles Munson).***

***Source: The Mabel Brady Garvan Collection, Yale University Art Gallery***

William Giles Munson's painting of the Mill Rock "Gun Manufactory" reflects Whitney's penchant for social control. Everything in the bucolic setting centers around the neat and orderly armory. Even elements in the landscape—the mill dam, the covered bridge, the symmetrical arrangement of worker housing, the fenced pastures, the rich foliage—convey a sense of balance and stability. Life within the village paralleled this orderly view of the universe. The factory bell called armorers to and from work as clocked time gradually supplanted irregular patterns of toil. Rules punished indolence and encouraged diligence while the strict regimen itself systematically eliminated individuality from the productive process. Above all, prohibitions against drinking, loitering, swearing, and other "improprieties" during working hours eroded pre-industrial traditions that had formerly mixed life and labor. Outside the shops armorers and their families were expected to observe the Sabbath and lead moral lives. Those who transgressed established norms were chastized, fined, and in some cases deprived of credit and housing. Repeat offenders were fired and blacklisted, their chances of finding employment elsewhere in the region virtually eliminated. These and other sanctions set a pattern of paternalism that would be repeated time and again in 19th and 20th-century America.

Despite his public charisma, Whitney was not popular with his employees. Armorers chafed under the pressures of working in a closely regimented environment. Constant supervision accompanied by carrot-and-stick methods gave rise to anxieties that vented themselves in labor disturbances and expressions of discontent. Not surprisingly, Mill Rock had one of the highest labor turnover rates in the arms industry. On several occasions Whitney posted bills for the return of runaway apprentices. Far more frequently, dissatisfied armorers simply left their jobs for more appealing positions elsewhere. Those who remained at Mill Rock continually grumbled about low wages and the rigidity with which foremen enforced the rules. Serious morale problems existed at Mill Rock, and Whitney's inability to attract and retain competent workmen severely hampered his efforts to attain the engineering ideal of uniformity.

Even though Whitney sought to develop new manufacturing and management methods, his efforts fell short of success. Except for a promising attempt at cost analysis, he introduced little that was startlingly new at Mill Rock. From filing jigs to hollow mills, the techniques he used originated with others. Yet his zealous advocacy of uniform production made him one of the key progenitors of the American System. From the start he not only popularized the uniformity idea but also persuaded politicians to support policies aimed at standardizing the manufacture of military arms. Time and again he conferred with ordnance and armory officials regarding the preparation of model arms and the introduction of new techniques. A frequent correspondent was Roswell Lee, a former protégé whose innovative management made the Springfield armory the hub of the American firearms industry. Because of his reputation and influence, Whitney participated in virtually every important government decision bearing on the manufacture of firearms between 1801 and 1825. In June 1815, for instance, he hosted a meeting at New Haven, Connecticut that devised a formal strategy for standardizing the manufacture of military muskets at the national armories. Present were Lee, Benjamin Prescott (Lee's predecessor at Springfield), superintendent James Stubblefield of the Harpers Ferry armory, and Colonel Decius Wadsworth, chief of the U.S. Ordnance Department and one of Whitney's closest friends. Out of this meeting came the decision to introduce a new Model 1816 musket as well as special inspection gauges at the national armories and eventually to distribute them to private contractors. The adoption in 1823 of these gauges, though primitive by modern standards, signalled the beginning of a sustained effort toward the production of truly interchangeable parts, a goal that would eventually be achieved during the 1840's.

In the final analysis, Whitney's most important contribution to the American System stemmed not from any technical innovations he made but from his active promotion of the engineering ideal of uniformity. He was, to be sure, an entrepreneur who saw opportunities for profit and took them; but he was also an *advocate* who perceived the larger benefits of precision manufacture and worked assiduousiy to keep the concept before the public eye. Beyond this, his conscious cultivation of the uniformity system had a larger cultural purpose. At a time when the fledgling United States badly needed psychological anchors and a collective social vision, Whitney helped to identify technical progress with political and economic stability. In his mind technology meant something more than a means of controlling nature and producing goods; it represented a barometer of national development, an accomplishment in which Americans could take pride and from which republican institutions could draw strength and vitality. Through technology, then, everything associated with the American Dream—material abundance, democracy, freedom, perfectability, destiny—could be achieved. In time Whitney came to symbolize this union. Along with Samuel Slater, Oliver Evans, and other lesser known individuals, his name not only stood for material progress through technology but also for republican virtue in an era of developing industrialism.

* * *

At least three major streams of inventive activity converged during the 1820's and 1830's to produce the mechanical synthesis known as the American System. Two centers of this activity —Middletown, Connecticut and Springfield, Massachusetts— were located along the Connecticut River within 40 miles of each other and shared a common heritage going back to the earliest Puritan settlements of the 17th century. The third center stood near the junction of the Potomac and Shenandoah rivers at Harpers Ferry, Virginia and reflected mores and traditions quite different from those of New England. Yet, despite these cultural contrasts, all three communities developed exciting new technologies, built on one another's advances, and eventually succeeded in manufacturing firearms with interchangeable parts. Together they exercised a cumulative impact on American metalworking and shop practice that prevails to this very day.

Middletown enjoyed a widespread reputation as the home of a number of distinguished arms makers. Among the most prominent were Henry Aston, Robert and J. D. Johnson, Edward Savage, and Nathan Starr and son. Yet, as able as they were, none of these men displayed the mechanical ingenuity and skill of Simeon North (1765–1852).

A quiet, unassuming person, North had a very prolific career as an arms maker. He secured his first government contract in 1799 and continued as a private contractor, manufacturing horsemen's pistols, common rifles, and Hall breechloaders, until his death in 1852. Apart from his remarkable longevity, North possessed a talent for technical creativity that made him a trend setter. At a time when most armorers relied almost exclusively on hand tools to execute their work, he sought to mechanize his operations and adopt precision standards. Success came early. By 1816 he had not only equipped his "Staddle Hill" factory in Middletown with advanced gun-stocking and barrel-turning machinery but had also succeeded in manufacturing pistol locks with unmarked interchangeable parts. Equally significant, he had begun to form some of the small lock components with a plain milling machine—the first of its kind in America—thereby effecting significant cost savings in file consumption as well as higher standards of accuracy. Ordinarily these accomplishments should have catapulted North to fame and fortune. But lacking Whitney's flamboyance and charm, he failed to attract public notice. While recognized by peers as an ingenious inventor, he died as he had lived—in relative obscurity.

What North initiated at Middletown, John H. Hall extended at Harpers Ferry. A native of Portland, Maine, Hall (1781–1841) received a government contract and moved to Virginia in 1819 to manufacture a breechloading rifle bearing his name. At the outset he asserted that he would make his rifles with interchangeable parts, a point scoffed at by skeptics but strongly supported by ordnance officials who emphasized the experimental nature of the work. Provided with makeshift facilities at the Harpers Ferry armory, between 1821 and 1827 Hall built an impressive stable of machinery that included water-powered drop forges as well as several types of milling, drilling, and profiling machines. Featuring automatic stop mechanisms, formed cutters, and total metal construction, this equipment represented a significant step forward in mechanical design. Attention to small details, such as the use of balanced pulleys, not only increased the speed and efficiency of operation but also extended engine life as well. Beyond this, Hall sought to insure precision by executing work from special bearing points, adopting standard screw threads, and using over 63 gauges during the manufacturing process. These innovations enabled him to achieve tolerances of one-thousandth of an inch and in 1826 successfully complete the first fully interchangeable weapons ever made in the United States.

Although Hall conducted his operations in a small pilot plant and failed to achieve significant economies of scale, he is nonetheless a pivotal figure in the annals of industry. What was so startlingly new about his work was the extent to which he mechanized his operations and the impressive results he actually achieved. No one at the time Hall went to Harpers Ferry had been able to master the problem of attaining complete interchangeability in firearms. Much of the excitement generated by his work can be traced directly to his success in combining men, machines, and precision-measurement methods into a practical system of production. In 1819 no one—not even the chief of ordnance—was absolutely certain that mechanized processes could be successfully applied to every aspect of gun making. Hall showed that they could and, in doing so, bolstered confidence among arms makers that one day they would achieve in a larger, more efficient manner what he had done on a limited scale. In this sense, Hall's work represented an important extension of the Industrial Revolution in America, a mechanical synthesis so different in degree as to constitute a difference in kind.

Like Hall and North, armorers at the Springfield armory played an indispensable role in fashioning the American System of manufacturing. Under the dynamic and resourceful leadership of Roswell Lee (1815–1833), Springfield not only became the "Grand National Armory" but also emerged as one of the most progressive manufacturing establishments in the United States. Lee recognized that, no matter how refined manual skills became, muskets made by hand simply could not compare in accuracy and workmanship with those finished by machinery. Process refinement consequently became the keystone of his policy, and from his appointment as superintendent in 1815 to his death in 1833, he doggedly pursued the mechanized production of uniform parts. Although complete success eluded Lee during his lifetime, he nonetheless established a managerial strategy at Springfield which culminated during the 1840's with the design and manufacture of the U.S. Model 1842 musket. This was the first regular-issue weapon ever to be mass-produced with interchangeable parts and, as such, constitutes a milestone in the history of technology.

Springfield's leadership in the mechanical arts depended less on in-house inventive activity than on an uncanny ability to monitor and assimilate work being done elsewhere. Lee and his successors, John Robb and Major James W. Ripley, excelled at ferreting out technical novelties and incorporating them into the productive process at Springfield. Because the most important private arms makers—like Simeon North and Asa Waters—depended almost exclusively on the federal government for contracts and because ordnance officials insisted that contractors share new technological information with the national armories, Lee and his staff had access to virtually every important workshop in New England. Inspectors and machinists from Springfield were continually being dispatched to private armories during the 1820's and 1830's to examine new machinery and make drawings so that the devices could be replicated at Springfield. Through this intelligence network Lee procured invaluable information on the latest techniques of milling, forging, and gauging. Through the same network he learned about Thomas Blanchard, a young mechanic from Millbury, Massachusetts who in 1818 devised an ingenious cam-operated lathe for turning the irregular surface of gun barrels. After conducting a demonstration at Springfield, Blanchard not only installed the machine at the national armories but went on to apply the principle of eccentric turning—tracing a master pattern and reproducing it—to the manufacture of gunstocks. Lee deemed this invention so important that he subsidized Blanchard by bringing him to Springfield in 1822 as an inside contractor. During the next five years the Millbury mechanic designed and perfected 14 different woodworking machines which completely mechanized the process of gunstocking, insured uniformity, and eliminated the need for skilled labor in one of the three major divisions of armory production. Rarely have the contributions of one person effected such a sweeping change in manufacturing practice in so short a time.

The presence of Blanchard and other gifted mechanics at Springfield helps to explain the armory's surge to international prominence during the pre-Civil War period. As early as 1825 Springfield had become the linchpin of the American arms industry and a pivotal clearinghouse for the acquisition and dissemination of technical know-how. If an arms maker, or any other manufacturer for that matter, wished to learn about the latest developments in metalworking, he invariably visited the national armory. There he could examine the most advanced tools and machinery, compare notes, and discuss common problems with such master machinists as Adonijah Foot, Cyrus Buckland, and Thomas Warner. After satisfying his curiosity and determining his needs, he might then set out for other establishments to gain further knowledge about new and interesting mechanical ideas. The ability of arms makers to move freely from one shop to another contrasted with more closely guarded practices in Europe. Such an "open-door" policy doubtlessly helps to explain the highly integrated character of the American arms industry as well as the relative speed with which manufacturers assimilated the new machine technology. The same policy also accentuates the public-service orientation of the U.S. military service and armory system during the antebellum period.

By the 1840's, manufacturing techniques aggregated and synthesized at Springfield began to spread to other armories and machine shops throughout the United States. Frequently this influence manifested itself through workmen who had received their early training at Springfield and subsequently moved to new positions as machinists and supervisors at other manufacturing establishments. These "graduates" of the armory became key disseminators of the new technology and could be found in most of the important metalworking centers in America. In 1842, for instance, Thomas Warner, one of the architects of the system at Springfield, resigned his position as master armorer and moved to the Whitney armory at New Haven. Several years later the Ames Manufacturing Company lured a skilled machinist named Jacob Corey MacFarland to its factory at Chicopee, Massachusetts. Like Warner, MacFarland possessed intimate knowledge of the techniques used at Springfield and as foreman of the Ames machine shop introduced "armory practice" at Chicopee. Other manufacturers followed similar strategies, thereby acquiring the latest armory know-how and relieving themselves of lengthy time lags and costly expenditures associated with learning-by-doing. Such methods of transfer and convergence carried mechanized production to a heightened level of maturity during the 1840's and 1850's, paving the way for the rise of the American machine tool industry and its entry into European machine markets prior to the Civil War.

The experience of the Ames Manufacturing Company aptly illustrates how private firms availed themselves of the new technology. Established in 1834 by the brothers Nathan P. and James T. Ames, the company became one of the first businesses in the United States to manufacture and market a standard line of machine tools. The firm also undertook special jobbing contracts and made a wide variety of millwork, mining equipment, cutlery, small arms, cannon, statuary, and other metal castings. Yet, despite their obvious technical skill and versatility, the Ames brothers were basically copyists rather than innovators. And for good reason. Chicopee was only a few miles from Springfield and the brothers had ready access to patterns and drawings owned by the national armory. Aided by a number of former Springfield armorers, their stock of machinery clearly reflected this influence.

Through the marketing activities of Ames and other early machine tool companies, armory practice began to spread to technically related industries and by the 1850's could be found in factories making sewing machines, pocket watches, padlocks, and hand tools. From these modest beginnings it was only a matter of time before the new technology would be adapted to the production of typewriters, agricultural implements, bicycles, cameras, automobiles, and a whole host of products associated with the mass-production industries of the 20th century. Interestingly enough, mechanized techniques rather than precision processes constituted the most important transfer from the arms industry. Since precision production was expensive, most businessmen contented themselves with manufacturing highly uniform but not necessarily interchangeable parts. Only the government could afford the luxury of complete interchangeability.

The capstone of two generations of creativity in the American firearms industry came in 1851 at the London Crystal Palace Exhibition. For the first time little-known Yankee manufacturers impressed visitors with the quality and precision of their machine-made products. Exciting considerable comment and admiration were the unpickable padlocks of Alfred C. Hobbs of New York, Samuel Colt's revolving pistols, and six completely interchangeable rifles made by Robbins & Lawrence, a small "armory practice" company from Windsor, Vermont. All three firms won medals at the exhibition, but more significantly the excellence of their wares prompted the British government to send an investigatory commission, headed by Joseph Whitworth and George Wallis, to New York's Crystal Palace Exhibition and, from there, on a fact-finding tour of the northeastern section of the country during the summer of 1853. The following spring Her Majesty's Ordnance Board, eager to take advantage of the American System, dispatched another three-man "Committee on Machinery" to the United States for the express purpose of introducing similar improvements at the Enfield armory near London. By the time the investigators returned home in August 1854, they had placed orders for over $105,000 worth of machinery, chiefly with Robbins & Lawrence and the Ames Company. Within four years several other governments sent similar investigatory teams to the United States. Although little noticed at the time, these events signaled America's coming of age as an industrial power.

Eli Whitney doubtlessly would have been pleased with the outcome. Although he failed to achieve interchangeable standards during his lifetime, he could take considerable pride in the fact that he had been instrumental in popularizing the engineering ideal and fixing the concept in public consciousness. Through Whitney the uniformity system became a cornerstone of Ordnance Department policy. Through the Ordnance Department, creative mechanics like Hall and North received government contracts enabling them to pursue the elusive goal of interchangeability. Through Hall, North, and other contractors the new techniques coalesced at the Springfield armory; from Springfield they filtered through the nascent machine tool industry to permeate American and European markets. Out of this complex tapestry of invention, cooperation, and diffusion can be discerned a mechanical genealogy that directly links Whitney and other early arms makers with the mass production industries of the 20th century. The achievement sealed the fate of pre-industrial culture in America.

## References

Hindle, Brooke, and Steven Lubar. *Engines of Change: The American Industrial Revolution, 1790–1860*. Washington, DC, 1986.

Hughes, Thomas P. *American Genesis: A Century of Technological Enthusiasm, 1870–1970*. New York, 1989.

Marx, Leo. *The Pilot and the Passenger: Essays on Literature, Technology, and Culture in the United States*. New York, 1988.

Morison, Elting E. *From Know-how to Nowhere: The Development of American Technology*. New York, 1974.

Smith, Merritt Roe, ed. *Military Enterprise and Technological Change: Perspectives on the American Experience*. Cambridge, MA, 1985.

# [2]

# Product Design and Cost Considerations: Clock, Watch, and Typewriter Manufacturing in the 19th Century

**Donald Hoke**
*Outagamie Museum, Appleton, Wisconsin*

This paper examines three products-- wooden movement clocks, watches, and typewriters-- to illustrate the heretofore overlooked importance of product design in 19th century manufacturing. It does this within the context of the scholarly debate on the rise of the American System of Manufactures.[1]

Nineteenth century mechanics and entrepreneurs were acutely aware of cost considerations in manufacturing. They designed products to be adjustable as an integral part of the manufacturing process due to the nature of the product itself and the materials with which they worked. They were driven by a technological imperative-- the product and its materials-- to lower cost through product design.

Mechanics and entrepreneurs used new designs to enter existing markets and compete with existing designs, thus creating such new industries as typewriter and watch manufacturing. These new industries followed similar development patterns, characterized by the appearance of a host of new firms with new products, intense competition, and the evolution of the industry from a highly competitive one to an oligopoly dominated by large firms with successfully designed products.

## The American System Debate

Merritt Roe Smith fired the first shot in the debate on the origin of the American System with his prize-winning book, *Harpers Ferry Armory and the New Technology* [3]. Smith's study followed the evolution of fire arms manufacturing at a federal armory during the first half of the 19th century. According to Smith, the American System originated in the federal armories-- the public sector.

David Hounshell takes up the "Armory Practice" banner in his prize-winning book, *From the American System to Mass Production* [1]. Hounshell also believes the American System originated in the federal armories (hence the term "Armory Practice") and then spread through the migration of armory-trained mechanics to the private sector, notably in sewing machines, reapers, bicycles, and automobiles. His survey of the literature is exemplary and the technical accounts of manufacturing processes are excellent.

---

[1] The data in this paper is drawn directly from my *Ingenious Yankees; The Rise of the American System of Manufactures in the Private Sector*, (New York: Columbia University Press, 1989). The reader is directed to that work for primary source citations.

*BUSINESS AND ECONOMIC HISTORY*, Second Series, Volume Eighteen, 1989.
 ISSN 0849-6825.

Smith and Hounshell have generated a paradigm for interpreting the rise of the American System-- government subsidized R&D that created a body of knowledge called "Armory Practice" followed by the appropriation of that technology by the private sector for its own enrichment. As David Noble has summarized [2, p. 337]:

> As is well known, the uniformity system developed in the armories became the basis of the so-called American system of manufactures, characterized by special machinery, precise gauges, and interchangeability of parts. Men left the arms business to set up the machine tool industry and went on from there to carry the principle of uniformity into the manufacture of railroad equipment, sewing machines, pocket watches, typewriters, agricultural implements, bicycles, and so on. The rest, as they say, is history, the history of progress.
>
> But there was another side to this story, which we have not heard much about.

There certainly is! My own book on the American System focuses on developments in the private sector. My interpretation of the rise of the American System differs quite sharply from that of Smith and Hounshell.

The American System is primarily and overwhelmingly a *private sector phenomenon*. American private sector manufacturers held the technological lead in America throughout the 19th century and shared it only briefly (between 1820 and 1840) with the federal armories. While private sector manufacturers certainly derived some benefit from the technical developments in the federal armories, they developed new products, new methods, new materials, new sales and promotion techniques, and new designs without any federal subsidy.

One especially important aspect of the American System's development in the private sector is product design, a topic notably absent from my good friend David Hounshell's work. For example, he gives us the clearest, most concise description of Ford's production technology, but never once considered how the Model T's design changed between 1908 and 1927 and how those design changes influenced manufacturing technology and costs.

This paper examines two aspects of product design as it influenced the American System of Manufactures:

a) at the microeconomic level-- cost cutting.

b) at the industry level-- competition and patterns of industry development.

**Cost Cutting**

Cleverly designed products, particularly products *designed to be adjusted as an integral part of the manufacturing process*, cut final product costs by increasing the range of acceptable tolerances on many parts and reducing the assembly time required by including adjustable features that permitted the use of those parts with wider tolerances and eliminating the need to alter the parts themselves. During assembly, only the relationship between the parts changed, not the parts. Three 19th century products, wooden movement clocks from the antebellum period, and typewriters and watches from the postbellum period, illuminate this concept.

**Wooden Movement Clock Design**

The manufacture of hang-up or wag-on-the-wall clocks began in 1807 with the pioneering work of Eli Terry. In 1814 Terry invented a shelf clock that he perfected over the next few years. It was eventually manufactured in great quantities by over twenty different makers. The new shelf clock differed in many ways from its hang-up clock predecessor, most notably in its escapement, the device that regulates the clock's speed. The escapement is the most critical part of a wooden movement clock, the only part requiring a truly close fit. The escapement wheel and the verge must be properly "depthed" (located in relation to each other) in order to run properly. Depthing was a problem with Terry's 1807 hang-up clock, but not with the new shelf clock-- it featured an *adjustable escapement*.

Terry's new shelf clock design brought the escapement outside the clock plates. The escape wheel arbor was carried through a hole in the top plate and was supported by a bridge. Beneath it, the verge was mounted on an iron pin set off-center in a brass plug. To depth the escapement, the assembler had only to apply some pressure to the escape wheel, then turn the plug carrying the verge until the verge was in the proper position with respect to the escape wheel. Then the brass plug was nailed into place. In the late 1830s brass clock makers adopted this design. Indeed, the adjustable escapement remained a feature of American clocks through the 1920s.

The economic implications of this design are important. This design allowed clock makers to produce parts within a wider range of tolerances and to bring each clock "into beat" quickly and easily by adjusting the relative positions of the verge and the escape wheel, not altering the parts themselves but their relative positions. The same is true with watches and typewriters.

**Watch Design**

Like wooden movement clocks, watches required properly adjusted escapements to keep time. However, they were more complex and precise, requiring a more sophisticated approach than the clumsy wooden movement clocks.

Watches were routinely adjusted as an integral part of their manufacture. This is particularly true with the escapement and balance. A discussion of this mechanism, its assembly, and its adjustment will illuminate the extent to which all American System manufacturers and watch manufacturers in particular were driven by the nature of their products and the materials with which they were made to adopt particular technologies.

Waltham and most American watch manufacturers used the detached straight line lever escapement with a club-tooth escape wheel and an escape lever with jeweled pallets. Similarly, most manufacturers used a temperature compensating (cut), bimetallic balance with timing and poising screws in its rim and a blued steel (hardened and tempered) hairspring.

The entire mechanism consisted of the escape wheel and its arbor, the escape lever and its arbor, two pallet stones, two banking pins, a guard pin, the roller (and sometimes a double roller), roller jewel, balance staff, hairspring, hairspring collet, hairspring stud, hair spring stud screw, hair spring regulator, balance wheel, and up to twenty-two timing and poising screws (and perhaps timing washers), the cocks and screws to hold the escapement, and the end stones and hole jewels and jewel screws-- as many as fifty-six individual parts depending on the model. This was a complex precision mechanism.

"Matching the escapement" was a particularly skilled job, as the watch's performance depended greatly on the relationship between the escape wheel and the escape lever with its jewels. "Matching the escapement" consisted of cementing the pallet jewels into the pallet fork using heated shellac with special hand tools in an alcohol lamp flame. Despite the production of escape wheels and forks and jewels to precise standards, each had to be matched independently during the process of assembly. The pallet fork and its jewels were designed to be adjusted as an integral part of that assembly.

Watch manufacturers never automated the assembly and adjusting phases of production. Depending on the model and grade, adjusting could take up to five months. In the higher grades, adjusting consisted of timing the watch in six positions-- pendant up, pendant down, pendant sides, dial up, dial down-- in temperatures ranging from 38° to 95°, and adjusting for isochronism.

The adjusters could make as many as eleven adjustments in the process of escapement assembly. These include:

1) placement of the two pallet stones in the escape lever,
2) adjusting the two escape lever banking pins,
3) placement of the roller jewel in the roller table,
4) placement of the roller table on the balance staff,
5) placement of the hairspring collet on the balance staff,
6) adjusting the effective length of the hairspring with the regulator lever,
7) adjusting the placement of the balance jewels (hole jewels and end stones),
8) adjusting the timing and poising screws (may include adding timing washers or undercutting screws),
9) bending the guard pin on the escape lever,
10) repositioning the hairspring in its stud, and
11) raising & lowering the hairspring stud.

The watch escapement and balance was undoubtedly the most precise assembly produced by American System manufacturers in the nineteenth century. Typewriters, however, were more complex.

**Typewriter Design**

The most important performance criterion for typewriters was "perfect alignment," the degree to which the lower edges of the letters line up when typed. Typewriter manufacturers realized that their extremely complex product required adjustment to accomplish "perfect alignment," and they designed adjustability into their machines.

The design ingenuity in the Remington typewriter is evident in the typebar hanger, which was secured with a machine screw. Another clever design aspect of the typebar hanger was its reversibility. The same hanger suspended all the typebars, but, to provide clearance for the typebar arms, every other hanger was installed at 180° from the hanger on either side of it.

This is not the only Remington typebar adjustment. Another adjustment was the steel wire connecting the typebar to the key lever. This adjustment assured that the keys on the keyboard could be adjusted to the same height without affecting the typebars themselves. The Remingtons advertised their earliest machines, the *Sholes & Glidden*, as being adjustable by the user.

> If any type should get a trifle out of alignment, a gentle pressure against the inner end of the type-bar, one way or the other, as may be needed, will put all right again. If a type should get radically out of place, it can be adjusted by loosening the screw of its hanger-bearing, but this should not be attempted till one is fully familiar with the machine.

The Remingtons developed special tools to insure the proper alignment of the type, notably a special typebar adjusting fixture, which insured that when the type was hung in its hanger, each type would strike a common point. The typebar ring was placed in a frame supported on four legs, similar to the typewriter's top plate. The typebars hung as they would when finally assembled. Across the back of the fixture was an adjustable iron bar on which was mounted an indicating arm which marks the center point of the typebars.

Despite specialized machinery, a gauging system, and its dedication to the interchangeable system, the American Writing Machine Company also found it necessary to build adjustability into its writing machine, the *Caligraph*. The examination of a *Caligraph* reveals four separate adjustment points between the steel type face and the wooden type lever. First, there was a turn buckle joining the "Long Connecting Rod" with the "Short Connecting Rod" between the typebar and the wooden type lever, allowing each bank of keys to be adjusted to a single height. Second, there was the typebar hanger, held in place by the "Hanger Washer" and a machine screw. Third, the typebar and its hanger featured adjustable conical bearings to take up wear. Fourth, each type was forced into a tapered hole in the steel block brazed onto the end of the typebar. Additional *Caligraph* adjustments included the rack, dog, carriage tension, paper feed, ribbon feed, and finger-key tension adjustments.

The complexity of the typewriter and its thousands of moving parts forced typewriter mechanics to develop "exercising machines" to manipulate these mechanisms mechanically before final adjustment and aligning. This idea of making parts work together during the manufacturing process appears not to have been unique to the typewriter industry, but apparently was not widely practiced. Only a sewing machine manufacturer, Wheeler & Wilson, is known to have adopted the practice of "breaking in" its machines with a machine as a step in the manufacturing process [1, pp. 68, 75].

The American Writing Machine Company exercised its *Caligraph* typebars in 1886 before final assembly. The typebar and hanger assembly was screwed into place on a "Working Jack" as it would be later in the machine itself. A reciprocating rack and pinion arrangement rapidly moved the typebars, forcing the bearings to wear into each other. The result was to "obtain an accurate and easy movement of the type bars when . . . inserted in the machine." In 1903 the "Oliver Exerciser" worked the completed machine rather than a particular sub-assembly. After it was "exercised," each Oliver typewriter was again subjected to another rigid inspection and alignment. In 1924 the L. C. Smith & Bros. "Typebar Exerciser" worked each ball bearing typebar for two hours, the equivalent of 36,000 keystrokes, before the typebar was assembled into the machine. Even the Hall Type Writer Company, which produced an inexpensive index or indicator machine, used a special "device for easing the 'motions,' that they may run smoothly."

The complexity of the typewriter forced typewriter mechanics to organize their factories to accommodate the assembly, adjustment, and alignment of the machine. Each sub-assembly took place in a separate department. As early as 1886 the Remington factory had at least three different departments, an assembly department for "Putting in Connecting-rods and Levers," an "Aligning Room," and an "Adjusting Room."

In 1906 shortly after its "recent enlargement," the Remington Typewriter Company's new factory was divided into production and assembly sections. Production required 70 percent of the factory complex. The remaining 30 percent was used to assemble, adjust, and inspect the individual machines. In Remington's "great machine hall" some 3,000 machines were in process of assembly at any one time by "several hundreds of skilled assembling experts." "After receiving a registered number," the machine "rapidly [grew] . . . to a frame . . ." The various components of the machine (the type basket, carriage, ribbon mechanism, etc.) were added at various stages of the assembly process as the increasingly complete machine progressed through the factory. There were several sub-assembly areas in which the various components were assembled, while some minor assemblies were put together in the production wing. After its assembly, the machine was ready for its first adjusting, followed by its second or "touching up" alignment, and then its "ordeal of final inspection and adjustment." "Seldom [was a machine] passed without criticism."

Significantly, in 1906 the Remingtons illustrated their promotional literature with only three production machine illustrations and six assembly and adjusting illustrations. Assembly and adjusting and aligning were the highly skilled, labor intensive aspects of typewriter manufacturing at

Remington in the early 20th century. Interestingly, they seem to have approached the concept of an assembly line-- the typewriters moved down the length of the building as they gradually grew from frame into completed machine -- and perhaps William K. Jenne (who designed the factory and much of its machinery) considered such an idea only to discard it due to the problems of intricate assembly. Perhaps in the interest of quality, assembly was not rushed.

The Oliver Typewriter Company also developed an extensive assembly, adjustment, and alignment organization in its Woodstock, Illinois factory. Their organization consisted of at least six separate departments: Type Bar Department, Carriage Department, Assembly Department, Tabulators and Adjustment, Inspection Department, and Aligning Room.

L. C. Smith & Bros. divided their factory into production and assembly departments. They divided the assembly process according to the parts and mechanisms of the machine itself. Each department employed many operatives for each assembly. At least forty men assembled ball bearings into the typebars. Each machine passed through a series of ten assembling departments and "in each department certain parts [were] added until finally the machine [was] complete."

The need for adjusting assembled and aligned typewriters continued at least through the early 1950s. In the Royal Typewriter Company's factory in 1954 the final adjuster was an especially skilled person who adjusted only three or four typewriters per day.

Like other private sector American System manufacturers, the typewriter industry faced a technological imperative in producing its machines. The typewriter had special assembly problems, problems that were related to the nature of the typewriter itself, not the process of manufacturing its various parts.

All the manufacturers faced the problem of assembling, aligning, and adjusting a very complex mechanism. All the manufacturers responded by subdividing the assembly process into the various components of the typewriter and hiring vast numbers of highly skilled people to do the complex assembly, adjusting, and aligning work.

### Competition and Patterns of Industry Development

Product design played an important part in bringing competitors into an industry. In the case studies of wooden movement clocks, and particularly watches and typewriters, a familiar pattern repeats itself. Once the pioneering firm had created a market by manufacturing and selling its product, other firms began to compete, usually with newly designed products. The typewriter industry provides the best illustration of this phenomenon.

The variety of new typewriter designs that appeared following the Remingtons' pioneering work is astounding, not to mention the firms that simply copied the Remington designs. Between 1873 and 1890 some twenty-four new firms appeared. Between 1890 and 1900 another twelve tried their luck in the typewriter market. In 1900 there were thirty-seven typewriter companies in addition to the twenty-five that had already failed. Most firms were based on new typewriter designs, and most failed after a few years, often due to unworkable designs. Yet out of the chaos of newly invented typewriters came the new designs (visible writing) that were to change the typewriter industry and provide new corporate leadership. By 1910 such firms and L. C. Smith & Bros., and especially Underwood, had wrested design leadership from Remington.

Product design quickly became an advertising tool used to compete with existing manufacturers. New firms often advertised their newly designed typewriters as superior to the old designs. For example, its multiple adjustments were an important advertising point for the American Writing Machine Company and its *Caligraph*. The firm took special pride in promoting its adjustable features, notably the typebars, both in its advertising and its instruction manuals.

> DURABILITY. This is an important consideration, as writing machines are expensive and subject to continuous use and in this respect we claim the *Caligraph* is far ahead of all competing Machines. Its type-bars are adjustable, and in event of any lost motion in the journals, it can be taken up easily, no other machine possesses this unquestioned advantage. The paper feed bands on the *Caligraph* are of tempered steel and always adjustable, a great improvement over machines using rubber bands for this purpose [the *Sholes & Glidden* and the *Remington Nos. 1 & 2*].
>
> ALIGNMENT. The following cut, which represents the new type-bar hanger, has an adjusting screw and shows how the wear can be taken up, from time to time, by the operator. Remember this is the only machine that can be aligned by users, and shows the best work under hard strain and rapid manipulation. The parts which move, in any kind of machinery, will wear. The faster they move the faster they wear. This is common to all; and the *Caligraph* alone is adjustable. Look out for durability!

There were other factors in the rise and fall of many of the new firms, including capital resources, marketing expertise, management quality, and patent conditions. Nevertheless, the workability of new typewriter designs played perhaps the most important roll. Those firms with exceptional product design-- Royal, Underwood, and L. C. Smith & Bros.-- succeeded, while no firm with a poor design survived.

**Conclusion**

The American System is primarily and overwhelmingly a *private sector phenomenon*. Although private sector firms briefly shared the technological lead with federal armories in the 1820-1840 period, the armories soon stagnated and the technological action quickly shifted to the private sector. Many private sector industries shared a common technological ground with the armories, but they quickly expanded to include areas far outside "Armory Practice." This paper stresses the importance of looking at 19th century objects to interpret business and technological history. Product design, notably adjustability as an integral part of the manufacturing process, is an important idea that must enter the debate on the rise of the American System.

Just for fun, lets take this concept of product design and adjustability and suggest a reinterpretation of a small piece of David Hounshell's work, his bicycle chapter in *From the American System to Mass Production*. Our sharply different perceptions lead David and me to interpret the same data in very different ways. David's outstanding chapter provides historians with the first detailed history of bicycle manufacturing. "Clearly," says Hounshell, "the bicycle industry as a staging ground for the diffusion of armory practice cannot be overemphasized" [1, p. 8]. But what Hounshell describes instead is a private sector industry that initially used *some Armory Practice* but which then quickly departed radically from it to develop and adopt new technology. Such techniques as electric resistance welding for wheels, the manufacture and use of seamless steel tubing for bicycle frames, and the development of presswork are all outside the "Armory Practice" experience. Furthermore, the bicycle's frame and wheels and chain were specifically designed to be adjusted as an integral part of the manufacturing process. Even such assembly tools as Pope's wheel truing stand were described as "adjustable all over." One could argue convincingly that Pope's manufacturing process parallelled other private sector manufacturers (such as typewriters and watches), particularly in responding to the technological imperative of the bicycle. Pope was forced for technical reasons related directly to the nature of the bicycle to develop and adopt particular (non-"Armory Practice") technologies.

Following Pope's creation of the bicycle industry in 1878 and the expiration of his patent protection in 1886, competing firms rapidly entered the bicycle market. Predictably, many chose to produce cycles that were, by Pope's definition, "awfully cheap looking." Yet these new cycle manufacturers with their innovative press work and stamping techniques captured a greater share of the cycle market. Notice particularly that press work and stamping are not "Armory Practice." They were new techniques developed in the private sector that enabled their users to compete very successfully with the so called "Armory Practice" producers.

The firms like the Western Wheel Works not only brought new techniques to the industry, but broadened the bicycle's public appeal by producing cycles over a range of quality, precisely what Pope had sneered at. Thus the bicycle industry represents not so much a transfer of "Armory Practice" into the private sector as it is the appearance of a new industry with *some-- but very little-- Armory Practice*, noted particularly for developing new materials, technologies, and designs, and attracting new innovative firms which broadened the range of available quality. This is precisely what happened in the watch and typewriter industries, both of which used *some-- but very little-- Armory Practice*, but neither of which relied heavily on it.

In conclusion, I recognize the hazards of drawing such sharp distinctions between my work and my friend David Hounshell's. Time marches on and as certainly as Eli Terry manufactured fully interchangeable clock parts in 1814, there is some hot shot graduate student out there ready to challenge my thesis and *clean my clock*. When he does (to continue the horological analogy), I hope he'll *ream my pivot holes and polish my pivots* with zest. I hope he'll *adjust my rate to six positions, temperature, and isochronism* with the same exhilaration and pleasure as I had in *adjusting David's escapement*.

As the late Lynn White, Jr., wrote, "The most important thing that can be said about any scholarly pursuit is that it is fun. The history of technology is, emphatically, fun" [4, pp. 349-51]. Let the fun continue!

## References

1. David A. Hounshell, *From the American System to Mass Production* (Baltimore, 1984).
2. David F. Noble, "Command Performance: A Perspective on the Social and Economic Consequences of Military Enterprise," in Merritt Roe Smith, ed., *Military Enterprise and Technological Change: Perspectives on the American Experience* (Cambridge, MA, 1985).
3. Merritt Roe Smith, *Harpers Ferry Armory and the New Technology* (Ithaca, 1977).
4. Lynn White, Jr., "The Discipline of the History of Technology," *Journal of Engineering Education*, 54 (January 1964), 349-351.

# Part II
# The 'Moment' of Fordism

# [3]

# Building an Automobile Every 40 Seconds

**Fred H. Colvin**

**Synopsis**

*An introduction to the problem which led to the marvelous growth of the plant of the Ford Motor Company, Detroit, Michigan, and some of the methods used. The development of vanadium alloy steel; its heat treatment; handling a million pounds of material per day and other problems. The manufacturing methods will appear in later articles.*

Before attempting to intelligently describe the methods used in building the Ford automobile, it is necessary to look back a few years. Only in this way can we appreciate its development, which we must call marvelous, as very few adjectives fit the case. Nor can we easily compare either the shop or its methods with any other; for, when we realize that in spite of the multitude of automobile builders, this one plant is now producing *over half of the entire output* of this country, we can see that there is nothing with which to compare.

We can also get a better idea of the mechanics of the whole proposition if we glance back briefly over its development. In spite of the trend of automobile design toward heavier and heavier cars, with added complications and equipment, Henry Ford grimly adhered to his ideal of a car that would be light, yet stand the strain of hard usage and could at the same time be built and sold at what we are pleased to call a popular price.

This necessitated simplicity of design, and this simplicity has made possible its manufacture in large quantities and at low cost. The spirit of simplicity will be noted all through the manufacturing operations, which are to follow. And while some features of the design and of these methods may perhaps strike some as being too simple, we must always bear in mind that they accomplish the result desired – a satisfactory, serviceable product, at an almost unheard-of cost.

## Trying to Realize What 200,000 Means

When we get up into numbers with over four figures, we find that mere numbers mean very little to the average person. We think of 200,000 automobiles in a single season as being unheard of, if not impossible, as we can hardly imagine such an output. And so, when we see the huge piles of front axles ready to be machined; the stacks of rear axles ready for the assemblers; the motors, the wheels, tires, tops, windshields, steering handles, etc., we lose all sense of proportion, and we get to the point where we are quite as ready to accept a million as the proper figure as the paltry (?) 200,000.

As we are all familiar with the locomotive, I have used that as a basis of comparison. The total output of locomotives in the United States last year was 4513, according to published statistics, or say, 15 per day. Comparing this with the daily output of Ford cars, by reducing the 15 to one giant locomotive and doing the same with the 800 automobiles, we have Figure 1 with the locomotive looking rather small, though having about the same horsepower. Incidentally the money value is about in the same ratio as the sizes. The battleship gives an idea of the actual sizes compared with a familiar object.

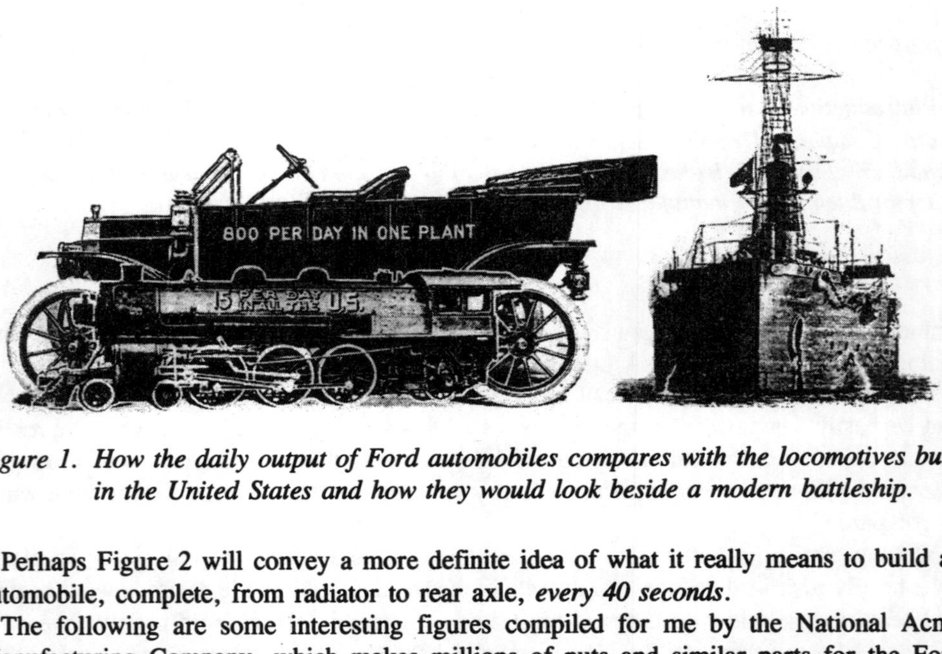


*Figure 1. How the daily output of Ford automobiles compares with the locomotives built in the United States and how they would look beside a modern battleship.*

Perhaps Figure 2 will convey a more definite idea of what it really means to build an automobile, complete, from radiator to rear axle, *every 40 seconds*.

The following are some interesting figures compiled for me by the National Acme Manufacturing Company, which makes millions of nuts and similar parts for the Ford company. It took 22 carloads of steel bars to make the ⅜-in. castellated nuts on one order of 4,500,000. The holes drilled in these nuts, if the nuts were placed end to end, would reach from the National Acme plant in Cleveland, clear across Lake Erie to the Ford plant in Detroit, and extend 25 miles beyond, 80 miles in all. The dies on two screw machines would travel a straight distance from Cleveland to Pittsburgh (140 miles), in cutting the threads on just one size of capscrew. There are 4200 miles of thread on these screws. Over a million and a quarter pieces a week are shipped to Ford, or on average five tons a day of small parts. On some of the Ford pieces, several machines have been run continuously for more than three years.

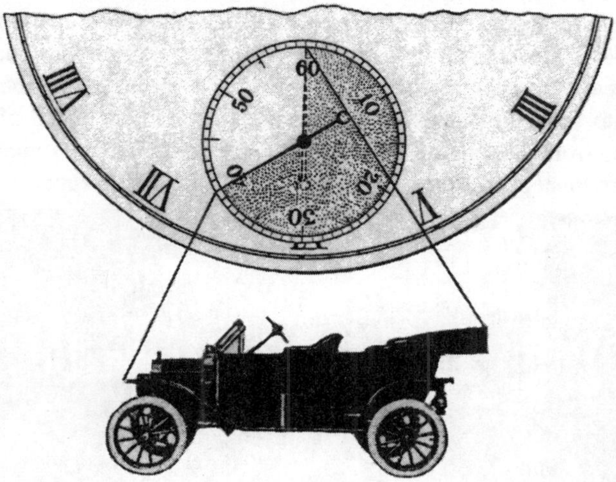


*Figure 2. A complete automobile every 40 seconds.*

**More Interesting Comparisons**

A few figures of the necessary equipment of the year's output are also of interest. These include 1,000,000 lamps; 800,000 wheels and tires, 90,000 tons of steel; leather from 400,000 cattle and 6,000,000 lb. of hair for the seats and backs, nearly 2,000,000 sq. ft. of glass for windshields, and so on. Table 1 compares Ford cars with locomotives built in the United States every day.

*Table 1. Comparisons between the Ford automobiles and the locomotives built in the United States every day.*

| | Ford cars | Locomotives |
|---|---|---|
| Number | 800 | 15 |
| Weight | 960,000 | 2,400,000 |
| Selling price | $480,000 | $240,000 |
| Horsepower | 16,000 | 15,000 |

On one day in January, 1336 cars were finished and shipped, taking over 200 freight cars or 5 trains of 40 cars each, and the shipments for 1913 will require 35,000 freight cars. This one day's shipment alone, leaving the factory a half mile apart, would reach from Detroit to New York City. Having impressed ourselves with the output necessary, we are in a better position to appreciate the methods which have led to such possibilities.

**Developing Vanadium Steel**

The question of lightness is dependent on the available materials. But so vitally did this phase of the question appeal to Henry Ford, that he established a testing laboratory, and aided by C.H. Wills, developed the vanadium alloy steel, which has since become so well known. Much of the success is due to this alloy, and the heat treatment which has also been developed to secure the best results from its use. Every piece of steel used in the Ford automobile, from the engine crankshaft to the irons which support the mud guards or fenders is heat treated in a large building devoted to that purpose.

*Figure 3. The portable baths for heat treating the drop forgings.*

A view of this is shown in Figure 3 as it is one of the keynotes of the whole process. This shows two of the heat-treating furnaces, numbers 7 and 8, and a large steam hammer in the front row, with other furnaces behind, giving some idea of the size of the heat-treating building, which is 72 × 363 ft. The cars shown in front of the furnaces carry the quenching or cooling baths and are run along the track into position in front of the furnace, which is ready to unload. The doors are then opened, and the heated forgings pulled out with a hoe, after which they slide down the inclined front and into the heavy wire basket, shown in the car carrying the bath. This basket is afterward lifted out with a crane. Nor is it a small part in actual production, for over 50 tons or 100,000 lb. of forgings are heat treated every day. This requires about 5000 gallons of crude oil per day.

How well Henry Ford's object of producing a light, serviceable car at a moderate price has been attained, can be judged from the plant covering 65 acres, and employing 15,000 men. And the end is not yet, for a new shop, 1700 ft. long and two storeys high, is now being contracted for.

The present machine shop is 862.5 ft. long by 643 ft. wide, having been enlarged several times, to over treble its original capacity in three years. Such rapid growth has made it impossible to always keep departments in just the proper relative positions, but the results have always been forthcoming, the production steadily increased and the cost lowered. What more could the greatest high priest of efficiency expect?

## Handing a Million Pounds a Day

One of the vital problems of such a shop is the handling of material. For with an average of 1,000,000 lb. or 500 tons going through the plant each day, this cost might easily assume gigantic proportions. As a matter of fact, however, the plan of manufacture carries the parts along from machine to machine, with comparatively little labor; far less than would be possible if the departments were arranged by operations instead of by work.

The heavier handling of work to machines is done by the overhead monorail system, as can be seen in Figure 4. Platforms or trays of convenient sizes for the work to be handled, are used in transporting what work is necessary. These trays have large rings at the corners and are easily picked up by the hoist.

In this way cylinder castings are brought from the foundry and similar transportation afforded where necessary. After the castings reach the cylinder department, however, they move steadily from machine to machine without mechanical appliances. When advantageous to do so, they are handled by a light hand hoist, but this is seldom necessary. Their light weight adds materially to the output as they are very easily handled to and from the machines.

Two regular traveling cranes are used, one in the long craneway in the center of the shop, as in Figure 5. This is for temporary storage of parts, as can be seen, a few frames and rear axles being shown at the left, with storage bins for parts at the right. Another and shorter craneway runs in a cross-bay, which is used as the heavy stamping department. This has a somewhat novel arrangement, as will be seen later. The cross-bay crane is so arranged that loads can be transferred from one crane to the other.

So thoroughly is the sequence of operations followed that we not only find drilling machines sandwiched in between heavy millers and even punch presses, but also carbonizing furnaces and babbitting equipment in the midst of the machines. This reduces the handling of work to the minimum; for, when a piece has reached the carbonizing stage, it has also arrived at the furnace which carbonizes it, and, in the case of work to be finished by grinding, the grinders are within easy reach when it comes from the carbonizing treatment. Exhaust fans carry off the fumes of the cyanide, and the furnaces are now housed in with partitions, although the general plan is to avoid all obstacles. Apart from these, and a few tool-grinding rooms at convenient points and offices for shop executives and time clerks, there are no partitions or obstructions in the whole shop.

## Fifty-Car Units the Rule

Small pieces are kept in metal barrels having trunnions, so as to be easily moved by trucks for this purpose or carried by the monorail. After going to stock, however, they are put up in lots sufficient for 50 automobiles. In the case of valves, for example, 400

*Figure 4. The monorail for shop transportation.*

are held in a rack, this being a shallow box with the bottom open, the stems hanging through. Bolts, nuts and similar parts are put up in boxes in the same way. The sizes of boxes and cartons vary widely, but the system does away with the use of bins and makes it easy to draw stock and keep an account of it.

The toilets are elevated as shown in Figure 6, the space beneath being used according

*Figure 5. Temporary storage in main craneway.*

*Figure 6. Elevated toilets with exhaust fans for ventilation*

to requirements. Heating is by overhead hot-air ducts in the old part, but in the new portion it comes up around the steel columns as can be seen in Figure 7. Great attention has been paid to this feature, which also includes an air washing and humidifying plant to add to the comfort of the men. This is said to reduce the temperature ten degrees or more in summer and make even the hottest day fairly comfortable inside the shop.

*Figure 7. Heat and air outlet.*

**Dividing the Working Force**

The handling of 15,000 men presents many serious problems. No car line provides transportation for such conditions. So the working hours are varied to divide the traffic. The first day gang starts to work at 6 a.m.; the second at 6.30 and a third at 7 o'clock. They also leave a half-hour apart at night. In the same way, they divide the work of the paymaster. Every day is pay day except Saturday, the daily payroll being between $30,000 and $35,000, the work being lessened by the plan of paying a part of the force each day.

The night gangs come to work in the same order, the first beginning about 4 in the afternoon and coming in a half-hour apart. They work 9 hours a day or 54 hours a week, the night gang getting in its half day on Saturday afternoon. The changing of gangs is an interesting study. It is done far more quietly and with far less confusion than seems possible. You notice a man leave his place at a machine or bench and apparently coming back to work in a few minutes. You notice men donning hats and coats and leaving the shop. And, looking again, you discover that the man who came back is another individual entirely. And so the work goes on with hardly a break. There is no whistle or gong, so that you barely notice the change made. Then, too, the cleanliness of the shop deserves comment. Floors are cleaned with hot soda ash every Sunday and are kept very clean during the week, no dirt or chips being allowed to accumulate around the machine.

**Ford Testing Methods**

The Ford testing method is unique and simple, but thoroughly practical, and secures satisfactory results. It is the unit method carried to its logical conclusion. Each unit is built in standard fixtures and inspected by standard gages, before being assembled. Every unit is tested before being assembled to any other unit. It is, therefore, perfectly logical to assume that the units, having run before assembly, will also run after they are put together.

Having careful inspection of motor parts before assembly, with testing of cylinders and heads for leakages, etc., it is taken for granted that the motor cannot help running when given proper conditions. It is simply a case of locomotive building over again. So, instead of the usual block test, with its noise and headachy odors, the engine is simply run in by electric motor, with oil in the crank case to limber up all bearings, to know positively that the parts run well together, and that the transmission is right in every way. This being so, the motor simply must run when given proper gas and spark.

This is proved after the whole automobile is assembled, and that is all the running the motor does before leaving the factory. But no user ever seems to complain that the cars do not run, and that is the real test of any machine. There is no road test before shipment, any more than there is in locomotive building. Why should it be necessary if parts have been made and assembled correctly? And imagine the congestion on the streets with all the cars in Detroit being tried out as in the old days.

## The Largest Gas Engines

Power is obtained from large gas engines, built by the Ford company because no builder had nerve enough to build as large units as it desired. The first unit is 3000 hp., and a new one of 5000 hp. is being installed, making the largest gas-power units in this country, if not in the world. In this connection it is only fair to mention Edward Gray, the gas engineer, who designed the engines and supervised their construction.

Power distribution is by line shaft and belting, divided into groups, of course, and driven by electric motors with current from direct-connected generators on the gas-engine shaft. The line shafting runs either lengthwise or crosswise of the shop, according to the layout of tools desired. And this layout has been made to economize space as much as possible in spite of the large proportions of the shop. I do not know of a place where more men are worked in the same area, although the Baldwin Locomotive Works are in the same class. While a shop with lots of room around machines presents a fine appearance, there is more opportunity for work to accumulate instead of being finished and moved to the next machine. Where men work close together, it becomes absolutely necessary for them to get the work out of each other's way, thus increasing the output.

## Scheduling the Output of Machines

Another interesting feature of the plant is the way work is scheduled. Every machine has an allotted daily output, and the work is planned accordingly. Each foreman receives a daily schedule of work to be turned out by his department, and it is up to him to see that it is forthcoming. As the men are all paid by the day – rather an unusual method for a large shop and on such an essentially manufacturing basis – the output is largely a matter of foremanship. And it is interesting to note how carefully each workman keeps tab on his output, just as is the custom where piecework is employed.

This scheduling of machine output tends to produce maximum results from machines, as it pins everything down to the actual cutting of metal as is being advocated under the name of the 'machine rate'. Everything is done to reduce the idle time of machines. Assume, for example, that there are four machines with a scheduled capacity of 200 pieces each, or a total output of 800 per day. Should the schedule only call for 600 per day, one machine would be shut down, instead of reducing the output of all four.

In order to prevent the output from falling behind, it is checked up every hour by special time-keepers. In this way reports from all departments go to the office of the shop superintendent every hour. This shows him immediately when a department falls behind, and he can at once see what the trouble is, and so prevent holding up another department and the daily output of 800 automobiles.

## Assembling 800 Cars per Day

Sixty cars in all stages, grow to completion before your eyes. There are no elaborate assembly stands. A truck goes down the aisle and leaves a front axle assembly. The rear axles, which include the drive shaft, follow suit. Then a frame is placed in position by a third gang and the real work of assembling begins. This is divided into several gangs, each doing its allotted work, and no more, and then moving to the next frame or stand. The method is similar to that used in assembling locomotives at the Baldwin and other plants, and works out well.

Everything is delivered in proper sequence to the assembly floor by gangs of truckers. Springs are fastened to axles and to frames, motors go into place and are coupled to the drive shaft. Dashboard assemblies seem to come automatically and then the steering gear can be connected up. In the meantime, the exhaust pipes and mufflers have been placed on the floor in handy positions. Wheels have appeared, already tired. The hand brakes go into place; then the felt washers; the wheel hubs are greased and go into position. Differential housings are shot full of dope with a grease gun; the gasoline tank is placed and connected; the body, which has mysteriously appeared, as do all the parts in their turn, is quickly fastened in place; and lastly, the footboard and fenders. A special truck, shown in Figure 8, is run under the rear wheels, and off they go to the testing room – to be tried out under their own power for the first time.

*Figure 8. Truck for rear wheels.*

And, while we have been watching this car go together, 60 others have also been assembled. Each man or gang does his own part and goes to the next stand to repeat the operation there. Needless to say, they become expert and put on dashboards, fenders or what not, in rapid time. And if we stand by the door leading to the testing room we can see that the production of 800 cars a day is not merely guesswork. For a car averages to go through the door about every 40 seconds of the 9-hour day, into the testing department.

# [4]

## [a]

## THE FORD STOCK SYSTEM AND EMPLOYMENT METHODS

IT is absolutely correct to say that the Ford Highland Park shops are unique. The place certainly stands alone in three vital particulars, which dominate the practice at every point:—

First, the company has but one solitary item of commercial product, the Ford motor car, made in one form only, because the differing body forms do not in any way modify the single chassis.

Second, the entire human race desires to ride, and to ride fast and cheaply, and the Ford car meets these conditions of human desire more suitably than any other automobile meets them, and hence is truly and in every sense a certain money maker.

Third, the Ford car holds so closely to the one unchanged model that it becomes commercially possible to equip the shops with every special tool, great or small, simple or complex, cheap or costly, which can be made to reduce production labor-costs.

The Ford shop production has never yet met the purchase demand for cars. There has never yet been a year in the history of the Ford Motor Company in which it did not refuse "immediate delivery" spring and summer orders. The Ford factory product is always over-sold; no car is ever stored at the Highland Park plant for one single day, because this plant has absolutely no place to store its product save in railway cars, and 90 per cent of the automobiles shipped from Highland Park carry with them cash-on-delivery bills of lading. The remaining 10 per cent of the Highland production is shipped to the various Ford "branches," which are, in fact, merely removed members of the Highland Park plant, owned by the Ford Motor Company, and operated by men paid by the Ford Motor Company, who sell this 10 per cent of the plant production at retail, to supply the branch local demand. This branch handling capacity, equaling 10 per cent of the total production, is the reservoir which takes care of the difference between day-by-day actual sales and the total output. The fact that 10 per cent of the total production provides a sufficient storage volume for cash-sales fluctuations shows the unparalleled steadiness of the increase in the demand for the Ford Motor Company's one and only automobile.

These three points of generic distinction—first, that the factory produces but a single article for sale; second, that it is an article of universal human desire, sold for cash to within 10 per cent of the total volume of production the moment it is completed; third, that being exempt from commercial demand for change of form, this article can be built through special machine-tool equipment of enormous cost—these three extraordinary characteristics place the Ford shops in a class by themselves for their certainty of money-making and exemption from rivalry with their product or competition for their gains.

These are not merely unconsidered assertions made in the interests of the Ford Motor Company. On the contrary they are the carefully matured conclusions of a widely inclusive search for the ultimate causes of the profits which compel a steady flow of vast sums of money through the Ford treasury, day in and day out, the whole year round.

The motor car had its epoch of opportunity, and Henry Ford was fortunate enough to make the car best fitted of all cars of its date to seize this opportunity, and to give this car sufficient vitality and stamina to enable it to hold what it grasped. He who best supplies a universally existent human desire has Fortune at his feet, as all students of commercial success are well aware. Mankind at large will surely purchase whatever meets its desires, and the lower the price the wider the field of sale. That is the foundation of trade at large, and what is here said of the true cause of the Ford car success cannot possibly be controverted.

The Ford factory holding thus, as it certainly does, a position commercially unique, may be safely expected to show very marked features of factory practice, wholly unlike the practice of any other existing factory, because the practice of every factory must be dominated by commercial conditions solely, if it is to continue in active existence, and the Ford factory does not disappoint those who expect to find its practice unique at vital points.

## Maximum Production the Objective

The urgent demand for maximum production is the dominant condition which governs every activity of the Highland Park shops. It is true that the automobile trade at large has its dull trade season of the year, that the spring and summer sales of motor cars vastly exceed those of the fall and winter, so far as the purchases of actual car-users are concerned; but nothing of this season-of-the-year-governed call for cars is ever felt in the Ford shops to the extent of giving the factory a slack period. The Ford system of branches and local sales agents acts as a reservoir of production; so that, because the maximum yearly production of the Ford car-building shops has never yet equaled the prospective demand for cars in the near future, the Highland Park shops are always worked to full capacity.

The Highland Park shops are very far from manufacturing the entire Ford car. A long list of components, both rough and finished, are purchased from outside suppliers, who produce and sell to the Ford Company the bodies, wheels, tires, coil-box units, carburetors, lamps, 90 per cent of the car-body painting, all drop forgings, all roller and ball bearings, grease cups, spark-plugs, electrical conductors, gaskets, hose connections and clips, horn, fan-belt, muffler pipe, and part of the bolts. It is the policy of the company to deal with several suppliers of the same component, to ensure supply by drawing from several sources (which certainty of supply could not be had from any single producer) and also to make possible avail of competitive advantages not to be had if the total orders were placed with a single producer only.

Because the Ford Motor Company has but one production unit, the Ford car, and because the commercial demand for these cars has been, from the very first up to now, in excess of the plant production supply, the Ford production order is largely a matter of form and anticipatory surmise, based, as to quantity, on the "best judgment" of the sales department.

This commercial condition of being always over-sold must, of course, dominate the Ford factory activities in all directions, and does, in point of fact, make the fixing of a production limit by an inflexible production order an impossibility.

The Ford fiscal year begins and ends October 1, and inventory is taken at that time, by actual count and survey. The Ford shops are minutely systematized, and all accounts are carefully and accurately kept; no purchase order is issued without what amounts to a complete inventory of the component under replenishment, so that, save to discover theft, the Ford shops might dispense with the costly function of "inventory by count and survey" altogether, so far as fixing purchase order quantities is concerned.

Prior to October 1 of each year a consultation between the sales-department heads and the manufacturing department is held, and the estimate of the number of cars to be turned out during each individual month of the year to come is marked in the twelve month-spaces on a large sheet of section-lined paper, ten spaces to the inch. This monthly car-production number is communicated to the factory heads, and is used as a basis for the purchase-order department; but the original numbers of cars to be produced in any month to come may be changed on advices from the sales department.

As the fiscal year goes on, the factory superintendent plots a "production curve" on the large sheet of section-ruled paper headed by the twelve spaces filled with the car-production quantities of the individual months, and thus obtains a diagram which shows the factory car output in detail and in total, at a glance, as the months progress.

Meantime, the factory departments are producing finished components to meet the requirements of the production-order schedule, and it would seem that, with an accurately compiled record of finished stores produced, checked by an accurate record of complete cars, complete assembled units, and finished components shipped from the Highland Park plant, there should never be a moment's doubt or question as to the individual finished-component stores on hand and ready for use by the assemblers, so that all prospective shortages of components would be accurately known and replenished to the safety mark long before the prospective shortages became a present actual condition; but such is not the fact.

## Location and Arrangement of Offices

The offices of those departmental heads directly instrumental in machine-shop supply and working are located on the west side of the machine shop, begin at an east-and-west line through the middle of the Administration building, and extend northward, in the following order:

First and second, anteroom and office of the machine-shop superintendent, who has one assistant, the two working together in such absolute harmony that I was informed that they were of equal rank, which appears to be the truth in practice, though not in accordance with the register of officials; third, inspector-in-chief; fourth, buildings maintenance; fifth, the clearing house, which is also the stock-superintendent's office, and sixth, the shortage-chaser's desk, shared with the clearing-house chief clerk. Close at hand the clerks working under orders from the clearing-house chief clerk, who keeps board and book records of all components produced, are stationed, all in an unenclosed space.

**Interior of Machine-Shop Superintendent's Office and Group of the Principal Officials of the Machine Shop**

## The Stock Superintendent

The Ford Company's purchase routine is very full and elaborate, so elaborate, indeed, that no short story of it can be made adequate and it must be reserved for a supplementary chapter.

Requisitions are based upon the production volume determined upon at the first of the year, in connection with the data of a remarkable continual inventory of components on hand in the main and branch establishments. Contracts are not generally made for the entire season's need, but in larger or smaller lots, following the indications of the market.

When a delivery has been ordered on a purchase contract, the purchasing agent turns over the particulars of the transaction to a "follow-up" assistant, who keeps after it continuously until shipment is made. As soon as he has details of shipment date, car numbers, and routing, the supervision of the movement passes to the traffic department, who keep in touch with the consignment until the goods arrive at the unloading platform. They then enter into the custody of the material department,

**Shop Transportation. Sprague Monorail System Passing through the Ford Plant**

by whom they are routed to the exact point of use in the shops. A system of sectional letters and numbers, conspicuously painted on columns and girders, guides even the uneducated foreign labor used for handling and trucking to the correct destination. In case of over-accumulation of stock at any point, the routing clerk is notified by telephone and directs a new place of delivery until the stock has been reduced at the point of congestion. Wherever possible, purchase orders provide for shipment in packages of exact count size, thus greatly facilitating inspection for quantity. All material is strictly charged and credited from department to department, according to the quantity that passes inspection. Rejections are charged, billed, and shipped back to sellers.

The stock receiving is in charge of one individual, the stock superintendent, who has an office between that of the machine superintendent and that of the shortage chaser, with one principal office assistant and six office clerks. He has besides 1,285 men outside of the office—department stock keepers, checkers, counters, and weighmasters, with a horde of stock handlers and truckers, who go where he says go, and come as bidden. Besides this considerable manual force, the stock superintendent has control of all traveling cranes and the electric monorail service.

The machine-shop traveling crane-way extends through the shop on north and south lines and covers a railway siding. The monorail lines surround the machine-shop floor and extend from the factory building line on John R Street to the eastern end of the foundry. besides having various other lines of travel inside the machine shop.

The Ford shops receive by rail about seventy carloads per day of freight from outside suppliers, and send out about one hundred and seventy-five cars loaded with Ford automobiles, daily.

Test pieces from rail shipments of pig-iron and bar-steel go to the chemist for analysis before unloading begins, and if the analysis is not to specifications the car is side-tracked; such instances have been known but are, of course, extremely rare. The analysis is satisfactory as a rule, and the checkers and counters and weighers are set to unloading the car as soon as may be after the chemist's report is in hand. Sheet steel is taken in without analysis and the stock received may go anywhere on the premises, but is usually sent to one of the many regular stock-receiving departments. Drop forgings, of which the Ford Company uses about $5,000,000 value yearly, are unloaded on the platform south of the smithy and foundry, which presents a very busy scene during work hours.

## The Ford Shop Transportation

One of the principal problems placed before the Ford engineers is that of transportation within the factory walls. The lightest complete Ford car weighs about 1,400 pounds, and if a thousand cars a day are turned out at least 1,400,000 pounds of materials must be moved daily, not once only, but many, many times over, each day, in the course of construction. Hence the traveling cranes, hence the many monorail lines, and hence the great numbers of truckers, and pullers and pushers and roustabouts in general, everywhere visible in the Ford shops during working hours. The unrelenting push of sales orders has forced the management to work for constantly expanding production capacity, regardless of everything else. At the time of this writing the first two of a series of new six-story buildings, each floor having seventeen landing stages served by an overhead traveling crane, are approaching completion; and with this great increase of floor area, supplemented by efficient traveling-crane service, much of the present hand trucking and roustabout work will be eliminated.

The Ford pay-roll is split about evenly between direct and indirect production labor, and the stock general superintendent's item of 1,285 men outside of his own office gives something of an idea of what the mere handling and placing of the 1,000-cars-per-day construction-materials cost. Not only is the fullest possible use made of the traveling cranes and the monorail system, but endless-belt and endless-chain transportation is established in many locations. Most of all, however, the Ford engineers have taxed the convolutions of their brain surfaces to shorten the lines of natural work-travel on the factory floors, first by crowding machine-tools together far closer than I have elsewhere seen machine-tools placed, and next by first finding the shortest possible lines of production travel of every car component, integral or assembled, and then placing every production agent needed either directly in that shortest line, or as near that line as possible, to the extent of placing even the brazing fires where most travel-saving advantage demands.

It is of record that in the old Piquette Street days, previous to the time when any attempts at Ford shops systematizing were made and chaos reigned supreme, the first systematizer found that the Ford *en-bloc* four-cylinders casting traveled no less than 4,000 feet in course of finishing, a distance now reduced to about 334 feet.

At first sight the Ford close tool-placing seems to be carried even

Shop Transportation. Hand-Truckers Moving Filled Component Receptacles

**Illustrating the Close Placing of Machine Tools in the Ford Machine Shop**

beyond the limit of economical application in labor-cost reducing; but a somewhat extended observation of the more congested areas of Ford machine-shop floor space failed to find a single instance in which the workmen did not have all the room necessary for economical action. It is true that the wanderer in the Ford shops is forced to keep to main lines of travel, where much care is needful to avoid a mishap through the monorail trains overhead or by way of the never-pausing hand-truck traffic on the floor level, but I am forced to the conclusion that even the closest spacing of machine tools in the New England manufacturing machine shops is still prodigal of floor space. Every factory economist well knows that every square foot of floor space carries the same tax of overhead costs, which cannot be recouped save by placing a profit-earning load thereupon. Yet the traditions of the Elders yet carry weight in spite of the missionary labors of Mr. Taylor and his disciples, and perhaps the Ford shops are doing an unguessed cost-reducing service in showing how closely even the larger of the small machine tools may be placed with no loss of per-hour efficiency.

Of course, after what has been here said, the visitor will not expect to find in the Ford shops any examples of orthodox machine-tool placing in generic groups, lathes together in one place, drilling machines, milling machines and planing machines each in a group by themselves.

The automatic screw machines, which are great floor-space consumers unless placed diagonally, force segregation, and obtain it. The large drawing-presses which produce the pressed-steel crank-box are also in a line by themselves, since labor saving and transportation saving both demand the grouping, and two long lines of Bliss presses producing many different articles in sheet metal are placed back to back to enable one long horizontal belt traveling between them to carry all the work

**National-Acme Automatic Screw Machines, Showing Diagonal Mode of Placement**

produced to the eastern end, where two men stand and sort the products, each into a receptacle of its own kind.

The gear cutters, both spur and bevel, are placed in bunches where their finished product can be built into the appropriate assembly with least transportation, and a long line of Potter and Johnson automatic lathes are the first machine tools to meet the eye of one entering the machine floor from the Administration building, because these tools work on single pieces and can economize floor space by close end-to-end placing.

But when it comes to brazing furnaces, which are usually regarded as unthinkable on a machine-shop floor, and are relegated to separate and despised quarters, often far out of the natural line of component travel, the Ford engineers have no hesitation in placing them as nearly as may be in the natural production line of travel, and brazing fires are seen in many unexpected places in the Ford shops.

### The Ford Workers

Having shown the factory buildings, and stocked the department, rather hurriedly it must be confessed, the place is now ready to hire in men. Molders? Core makers? Smiths? Machinists? Not in numbers to notice. The foundry superintendent asserts that if an immigrant, who has never even seen the inside of a foundry before, cannot be made a first-class molder of one piece only in three days, he can never be any use on the floor; and two days is held to be ample time to make a first-class core maker of a man who has never before seen a core-molding bench in his life.

### Tool-Makers, Experimental-Room Hands and Draftsmen

These constitute the aristocracy of every shop. They must be good, they must be experienced, and the better they are and the bigger wage they can earn the more valuable they are to the shop. The Ford shops work about 250 tool makers and 20 men or so in the experimental room, and a large force of metal-pattern makers in the foundry pattern shop, all of whom must be first-class mechanics. The Ford management asserts that its tool makers are as good as can be found in the world, although they are nearly all of western growth, and what tool-making jobs I have noticed in the tool-making department were certainly being well done. The Ford Company has a large quantity of shop special-tool work constantly in progress, and all special tools at work in the machine shop so far as noticed showed as good design and construction as need be. Nothing is scamped or hurried in the Ford shops tool-making, as might be assumed without writing the assertion in so many words, because economy in tool-making is always rank extravagance on the balance sheet. The general appearance of the tool room and its equipment is shown in the illustration on page 6 of the preceding chapter, in which the large installation of Hendey lathes is especially noticeable.

As to machinists, old-time, all-round men, perish the thought! The Ford Company has no use for experience, in the working ranks, anyway. It desires and prefers machine-tool operators who have nothing to unlearn, who have no theories of correct surface speeds for metal finishing, and will simply do what they are told to do, over and over again, from bell-time to bell-time. The Ford help need not even be able-bodied.

I had been told that all applicants for Ford jobs must be up to grade under a "military examination" by the Ford surgeons. Nothing could be farther from the truth. The employment agent, who looks applicants over before sending them to the surgeon for examination, said to me that so long as it seemed to him that a man could do work enough to pay overhead charges on the floor space he would occupy, he sent him to the examining surgeon. He quoted Henry Ford as saying "We must all live. If a man can make himself of any use at all, put him on, give him his chance, and if he tries to do the right thing we can find a living for him, anyway."

This Ford labor policy must have given hope, most valuable of all human possessions, to many and many a despairing brain.

Pursuing the work-applicants-examination story, Dr. Meade said: "No, the examination is not 'military' in any degree. We might accept a man with but one eye, or one hand, or one foot. The applicant must have no contagious disease, or if he has such a disease he may receive hospital treatment gratis if he shows a disposition to take care of himself until he has passed the condition of possible disease-communication. Chronic contagious disease is cause for peremptory rejection."

The Ford Company does not demand physical perfection in its workers; it simply takes such precautionary steps in hiring workmen as are, from the standpoint of the wise factory manager, necessary to give a fair prospect of that mutually profitable interdependence of the employer and the employed, which is the only satisfactory and altogether just embodiment of the working relations of capital and labor.

It is of vital importance to successful factory management that workmen should be steady in their habits, and dependable in the point of continuous service. Of course, there are occasions which fully justify the worker's absence from his place during work hours, and by making statement to his foreman the Ford shop hand can obtain leave of absence for such time as may be needful. Absence without previous notice is rigidly disapproved, and upon return from unpermitted absence the worker is at once taken to the watchman's office, a form is filled out, with excuse for absence fully detailed, and signed by the workman, who is then taken to the physician's office and "investigated." Should the medical staff find the excuse valid, the workman is permitted to resume his duties, in good standing. If the workman's excuses are not deemed satisfactory he may be either discharged outright, suspended, or "admonished" by his superiors with a view to his own good, before beginning work anew.

And in general, the Ford Company claims that it is best for all concerned that it should stand in an advisory position as to the conduct of each of its workmen, and should assist, lead, and guide him in the right direction to make such use of his natural powers and abilities as will best conduce to his own value to himself as a man, and to the Ford Company as an employee.

Of course, the man who begins to act as conscience to another man has a long road before him, as such a course can easily be made to end in paternalism carried to a degree of inclusiveness commonly looked for only in the most rigidly self-righteous of New England households.

Let not the reader hasten to conclusions based on meager information. All economists are agreed that the only reason why any one man works for another man is because the hired man does not know enough to be the director of his own labor. And, incontrovertibly, the employer being wiser than the employed, the wisdom of the employer should be applied to the benefit of the employed, to some extent at least.

So far, the Ford Company's efforts to be of real service to its toilers has been productive of absolutely nothing but good, with no qualifying addendum whatever.

## [b]

## HOW THE WORK IS DONE

A DETROIT daily newspaper not long ago printed a story anent a new farm tractor, said soon to be placed on the market by the Ford Company, and added, in the fulness of reportorial wisdom, that while the Ford factory was at present equipped for building Ford cars only, yet all of the machine-tools had been designed with such cunning foresight that "by merely changing the dies" they would produce farm tractors just as well as they had before produced Ford cars.

Probably the average Ford factory visitor has the same general broad conception of the "How" of making 1,000 automobiles (it is now more than 1,100 automobiles) per day.

The man who really knows anything of the entrails of a machine shop, who knows that nothing, of advantage, at least, ever "happens" in a machine shop, has plenty to think about as he stands at the "start-to-run" end of the Ford chassis assembling lines. This man of experience asks himself how the component production is evened up to assembling requirements. Here are, say, from 1,000 to 4,000 separate pieces of each chassis component to be supplied daily, infallibly, and constantly. How is this done?

In brief, first, by unremitting record-keeping of every finished component produced. Hour by hour, with endless toil and pains, an absolutely correct record is kept of the Ford component production and of the Ford factory out-put.

The factor of safety of component supply is placed by the official production head at a sufficiency for 25,000 cars, a month's assembling supply at the production rate of 1,000 cars per day. This is official, but, as will be seen from the "shortage-chaser" story, the factory practice does not follow the production-head schedule, but quite to the contrary, places a maximum component supply at a sufficiency for 5,000 cars, three-days' assembling, with a danger line at components enough for 3,000 cars.

Is it easy to hold these figures by accurate factory superintendence and accurate production accounting?

Easy or difficult, the accurate factory accounting records are made, but are not used for emergency decision in directing component production.

Seeing is believing. When a Ford car assembled-component assembler suddenly discovers that his requisition on finished stores for components wanted at once is not filled because there are no such components in finished stores, no grand conclave of factory accountants is summoned.

Quite to the contrary, a brisk and intelligent young man, styled the "Shortage Chaser," whose face already begins to show fine lines etched by the stress of concentrated attention, boards the department drifting within sound of breakers, seizes the helm of component production, and pilots the department into smooth water again—sometimes but barely escaping the surf-line, it is true, but yet always managing to escape disaster.

## The Shortage Chasers

In point of fact there are two shortage chasers—the day shortage chaser, and the night shortage chaser who combines shortage chasing with other duties from the time he goes on duty at 3:30 p. m.; he stops at 7:30 p. m. to eat 10 minutes on his own time, begins again at 7:40, and works to 11:40 p. m.

At 3:30 p. m. when he goes on duty, the night shortage chaser first inspects the day shortage-chaser's final report, made at 2:30 p. m., and is thereby informed as to the condition of all assembling departments which are working through his shift, the day shortage chaser specifying in his 2:30 p. m. report what component production is most needed and specifically calling the night shortage-chaser's attention thereto. Guided by the day shortage-chaser's report, the night shortage chaser urges the most needed component production, has checkers and counters under him, and fills tags authorizing the removal of finished components from their production department to the assembling department which needs them. Before leaving at 11:40 p. m. the night shortage chaser makes a record of components which he has moved from a production department to an assembling department, for information of the day shortage chaser, who comes on duty at 6:30 a. m.

The usual number of reports made by the night shortage chaser to the day shortage chaser is about six.

**The Shortage-Chaser's Desk. Shortage Chaser at the Left and Clearing-House Chief Clerk at the Right**

## The Day Shortage Chaser

The story of the shortage chaser must be told as briefly as is consistent with intelligibility. His desk is about 50 feet north of the machine-shop superintendent's office; he has a desk and one assistant.

At 6:30 the day shortage chaser, henceforth specified as "the shortage chaser" only, first inspects the night reports, making notes of impending shortage; next he notifies the "checkers" who direct the movements of components reported at or below the 3,000 danger limit; this is done at 7:00 a. m., the shortage chaser making a personal delivery of his own pencil memoranda to the proper "checkers." Next the busy shortage chaser goes in person to each assembling station reporting shortages and informs himself by personal observation of shortages—this because the final reports were made to component-production officials at 2:30 p. m. of the previous day. Being then advised of actual conditions by personal observation, the shortage chaser goes to the foreman of the machine department producing the most needed component, informs

**Office at the Clearing House with Shortage-Chaser's Blackboards**

The same office is used by the general stock superintendent (standing at the right of the picture) and the clearing-house head (seated at the large desk in the center).

him of conditions, and from this department foreman to such other department foremen as he may deem needful to see in person, say from three to ten individuals. Next the shortage chaser makes his 8:30 a. m. report at the checker's office by writing the same on his checker's office blackboard, 40 inches wide by 40 inches high, with 59 lines for writing in chalk. On this blackboard the chaser makes his reports by first inscribing each symbol and following the symbol with numbers of this individual component finished since 6:30 a. m. of the current day.

This blackboard record is then typed in manifold in the clearing house; one copy is sent to each department foreman interested, and one copy is sent to the shortage-chaser's desk, these typed report copies being distributed at about 8:40 a. m.

The next succeeding morning the shortage chaser compares his shortage reports for that day with the typed record of his shortage reports made on the blackboard at 8:30 of the previous day. He adds, in pencil, to the typed report of the day before such shortages as may be revealed by the reports of the current day, and delivers this added-to-in-pencil typed list of yesterday's shortages to the clearing house tag-blank filler, who is a clearing house official, but is placed at the shortage-chaser's desk for speedy communication. His duties are explained below.

The day shortage chaser makes three reports daily to the clearing house, at 8:30 and 11:30 a. m. and 2:30 p. m.

From 6:30 a. m. to 11:30 the shortage chaser covers all factory production departments and all assembling departments, and at 11:40 a. m. he chalks all production results from 6:30 a. m. to 11:30 a. m. on the clearing house blackboard. The shortage-chaser's clearing-house blackboard reports are thus made the guide for the activities of the checkers and counters and the checker's truckmen in moving components first to such departments as are most in need of replenishments.

Having made his 11:30 a. m. reports on the clearing-house blackboards in the stock superintendent's office, the day shortage chaser then again turns his attention to such assembling departments as most need his presence, and at 2:30 p. m. makes his final blackboard records in the stock-superintendent's office, which is also the "clearing-house" head's office.

### The Clearing House

The function of the "Clearing House" is to initiate and control the movements of foundry finished product, in the form of snagged and tumbled, patched and sand-blasted castings (forming the entire finished product of the foundry), the completed machine-shop departments product, and the products of the snagging department, pressed-steel department, bushings department, magneto department, fender department, gasoline-tank department, coil-box department, brake-band department, muffler department, and painting department, which finishes certain components checked into and out of the painting department by the clearing-house checkers.

### Clearing-House Officials

The Clearing House has one head, one chief clerk, stationed at the shortage-chaser's desk, and above referred to as the tag blank filler, thirteen "checkers," and fifty "counters," under the checker's orders.

### The Checkers

When any component production department has completed a production order the department foreman reports to a clearing-house checker orally, and the checker then takes one or more counters, as may be needed, with him to the production department and there proceeds to count the product into suitable receptacles, to enter the quantity on a proper tag, and to wire the body of the tag, or all of the tag, to the component receptacle, to serve as a direction for delivering the components to the proper department by crane, mono rail or otherwise, and he also takes such other steps as are needful to give the clearing house a record of the components transfer, all as fully detailed in the description of the use and functions of the clearing-house tags, presently following.

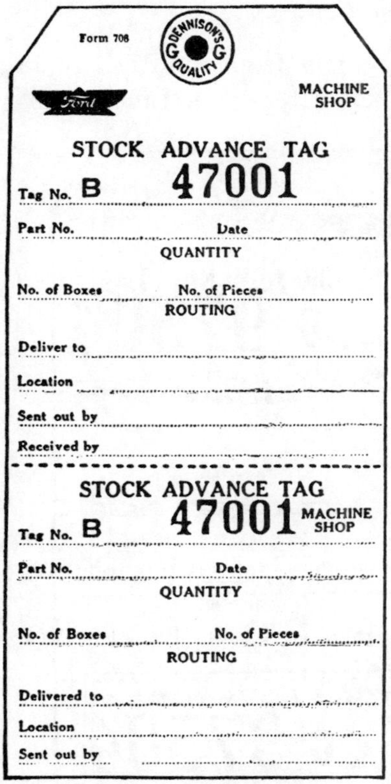

Form 708

DENNISON'S G G QUALITY

MACHINE SHOP

STOCK ADVANCE TAG

Tag No. B 47001

Part No. Date

QUANTITY

No. of Boxes No. of Pieces

ROUTING

Deliver to

Location

Sent out by

Received by

STOCK ADVANCE TAG

Tag No. B 47001 MACHINE SHOP

Part No. Date

QUANTITY

No. of Boxes No. of Pieces

ROUTING

Delivered to

Location

Sent out by

**Form 708. Machine-Shop Components**

## Clearing-House Original Records

These are all made on wired tags and the coupons of these tags, presently to be described as to forms, colors, and functions.

## The Clearing-House Chief Clerk

The chief clerk is the responsible official of the clearing house, who collects, compiles, and separates into appropriate groups and divisions all the tag records which comprise the entire original records of the clearing-house transactions.

The clearing-house chief clerk is stationed at the shortage-chaser's desk and has three clerks and one messenger. The clerks are located close by the chief clerk, and are employed solely in transferring production records, taken from tags filled by checkers, to certain "record boards" and books.

Each day's component-production records for the entire factory are completed at the end of each day, but the chief clerk consults the shortage-chaser's clearing-house blackboard reports, and also the shortage chaser himself, to aid in producing completed records of components which are short, first, before completing the records of components which are in good supply, so that early attention will be given to shortage production.

## Clearing-House Tags

The only form blanks used by the clearing house are wired tags, suitable for attaching to component receptacles. Their size, color, and use are as follows:

### Form 708, Blue, With Coupon

Filled by checker in a production department from counter's report to checker, body and coupon both; checker separates coupon and sends it

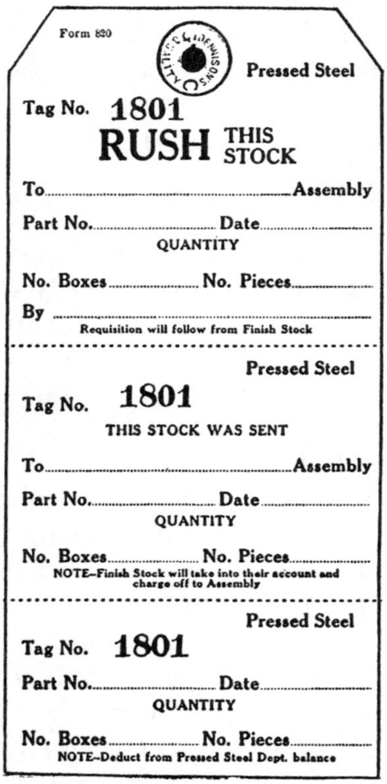

Form 820

Pressed Steel

Tag No. 1801

RUSH THIS STOCK

To ........................ Assembly

Part No. ............ Date ............

QUANTITY

No. Boxes ............ No. Pieces ............

By ........................

Requisition will follow from Finish Stock

Pressed Steel

Tag No. 1801

THIS STOCK WAS SENT

To ........................ Assembly

Part No. ............ Date ............

QUANTITY

No. Boxes ............ No. Pieces ............

NOTE—Finish Stock will take into their account and charge off to Assembly

Pressed Steel

Tag No. 1801

Part No. ............ Date ............

QUANTITY

No. Boxes ............ No. Pieces ............

NOTE—Deduct from Pressed Steel Dept. balance

Form 820. Rush Tag

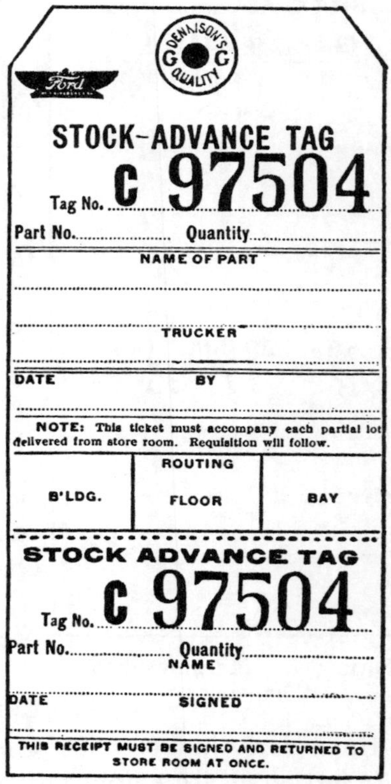

STOCK-ADVANCE TAG

Tag No. C 97504

Part No. ............ Quantity ............

NAME OF PART

TRUCKER

DATE BY

NOTE: This ticket must accompany each partial lot delivered from store room. Requisition will follow.

ROUTING

| B'LDG. | FLOOR | BAY |
|---|---|---|

STOCK ADVANCE TAG

Tag No. C 97504

Part No. ............ Quantity ............

NAME

DATE SIGNED

THIS RECEIPT MUST BE SIGNED AND RETURNED TO STORE ROOM AT ONCE.

Partial Quantity Tag

Form 705 is similar to Form 820, but is red in color and is used for machine-shop components

to the clearing-house chief clerk, and then wires blue tag body to the component receptacle, which then passes into control of the shop transportation system. Used for components produced in the machine shop, only.

Form 708, wired blue tag, $3\frac{1}{8}$ inches wide by $6\frac{3}{8}$ inches high, coupon $2\frac{9}{16}$ inches high; stiff card, printed in black on one side only, with duplicate serial numbers in red on body and coupon. Used for components produced in the machine shop only, which may be delivered either to the finished-stock department or to an assembling department, "Stock Advance" Tag.

### Form 709, Pressed-Steel Components

Form 709 is same as 708 in all particulars save color, which is bright green, serial numbers in red. Its use is the same as that of Form 708, save that the component originates in the pressed-steel department.

### Form 705, Deep Red, Rush from Machine Shop

Used for components produced in the machine shop to be delivered directly to an assembling department.

Two coupons are filled by the checker from the counter's report. The checker removes the end coupon and sends it to the clearing-house chief clerk, who transfers the record to his production records. The checker also removes the second coupon and sends that to the finished-stock department. He then wires the body to the component receptacle which is transported to the specified assembling department. The finished-stock department then fills an assembling-department requisition and sends same with the second coupon to the assembling department, where the department foreman compares the amount, and if alike, signs the requisition and returns it to the finished-stock office.

All the clearing-house tags are the same over-all dimensions—a stiff tag deep red, with two coupons; the first is $1^7/_{16}$ inches high, the second 2 inches high, printed in black, on one side only, body and both coupons bearing same serial number.

### Form 820, Rush for Pressed-Steel Components

Form 820 is the same in all particulars as 705, for the same use, but is yellow in color, and is the rush tag for pressed-steel department production, printed in black on a deep yellow tag, serial numbers printed in red.

### Partial-Quantity Delivery Tag

Form (no number) "Stock Advance Tag," "Series C" "Partial-Quantity" delivery tag is filled by the checker, both body and the one coupon. The coupon is detached by the checker and held by him until the quantity for the day is known, when he makes out a department requisition for the total coupons quantity, and sends same with the coupons to the department which verifies the quantity, signs the requisition, and returns the requisition with the verifying coupons to the checker, who retains the coupons and delivers the requisition to the clearing-house chief clerk. This form is used for one or more partial deliveries from the same department, all on one day, in all departments save and except the machine-shop departments and the pressed-steel department. Color, vivid salmon; coupon $1\frac{3}{4}$ inches high, printed all in black, on one side only.

This story has told how workmen are taken into the Ford shops, and enough of the finished-components handling is given to show how the men are kept busy, and to show the duties of the shortage chaser, who is a result of the closely limited maximum production working of the Ford factory, which has never yet been able to produce cars enough to enable the acceptance of immediate delivery orders during the spring and summer.

If the shops at Highland Park were not worked to maximum capacity continuously, the shortage chaser would be superfluous. With the factory production capacity in excess of the demand for cars, the finished stock (that is to say, the finished components in the finished-stock department), would always be kept sufficiently in advance of the requisitions of the assembling departments to meet all requisitions promptly and certainly—a condition of affairs which would at once eliminate the shortage chaser, because there would then be no shortage possibility. This story shows how the Ford shops manage to face a constant shortage probability—perhaps it might be said that the real condition is that of actual shortage of components constantly—and yet escape serious delays.

If the factory production capacity were in excess of the Ford car demand, and the component production were in excess of assembling demand, there would then be an overhead interest and storage charge against the factor-of-safety excess of finished components held in the finished-stock department, which does not now exist.

On the other hand, the shortage chaser now avoids disastrous shortages only by the use of extraordinary care and exertions, together with sudden changes, which are always costly factory expedients and conditions. The entire situation at the Ford shops is novel, and is met by novel methods, and is of much interest to factory managers in general. The ideal factory condition is undoubtedly that of perpetual tranquillity, all operations balanced and co-ordinated, so that there is no "hurry up" pressure required in any direction whatever. Yet this condition of perpetual tranquillity necessitates the carrying of large values of finished components, with an inevitable increase of fixed charges, greater or less as the margin of safe component-excess supply is fixed at a higher or lower level. There is a point, of course, where abundant component supply carried in the finished-stock department becomes an extravagance, and is hence reprehensible; so that it is not easy to say with certainty that the hand-to-mouth component production, even when coupled with the anxieties and make-shifts of the shortage chaser, really costs more than it would cost to carry an ample supply of finished components of all kinds. Leaving these ultimate considerations out of view, this story simply shows the practice of a factory which is so fortunately situated that its market is ready-made, so that the selling of the factory product is of negligible importance, while the constant maintenance of the factory production-capacity at maximum becomes the one objective of vital importance and all absorbing interest to the management.

Six Views of the Ford *en bloc* Cylinder Casting

## [C]

## INSPECTION AND ASSEMBLING

FORD shop inspection is, of course, thorough, and begins before the unloading of incoming rail shipments, by analysis of samples of bar steel and forgings, and pig-iron.

The inspection of material and components in process of finishing in the various shop departments, is in charge of one chief inspector, who began as machinist with the first Ford car model built in the shop at 81 Park Place, 1901–2. It is worthy of note that when Henry Ford began at 81 Park Place, 1901, he worked as a machinist on the first Ford car with his own hands and had with him one young machinist, one draughtsman, and one boy. The draughtsman is now Ford factory superintendent, the young machinist is now chief inspector, and the former boy is now the metallurgist-in-chief of the Ford Highland Park plant, which may be taken as evidence that Henry Ford was unusually fortunate in the selection of his first three men, or, again, as evidence that he is willing to aid in advancing individuals who once begin to work for him.

The head inspector has three assistants, two for day work and one for night work, and about six hundred others under him, variously employed but all classed under the general title of "inspectors."

All inspectors take orders from the head inspector only, but of course respect the convenience of the foreman of the department in which they are stationed.

The inspectors are divided into classes having special duties and designations.

The "incoming-material inspectors," one hundred and twenty in number, deal with all material received partly finished from outside suppliers, inspecting to Ford Company's specifications and drawings.

The fifty foundry inspectors inspect every piece of casting made by the foundry, first in the rough, then as to tumbling, snagging, and visi-

Inspector-in-Chief in His Office. Office Assistant Seated at Typewriter

ble defects, and see that all work sent from the foundry to the machine shop is to all appearance satisfactory.

The incoming-material inspectors and the foundry inspectors pass upon materials entering the factory departments where they are to be finished and completed.

The factory departmental inspectors are divided into three classes, as follows:

(1) "Machine inspectors," who are inspectors of machine work in progress.

(2) "Operation inspectors," who inspect work after the completion of individual operations.

(3) "Floor" or "final" inspectors, who inspect completed components only. These are day inspectors.

In addition there are about one hundred night inspectors, mostly acting as machine inspectors, some six or eight "rejected-components" inspectors, and finally, about twenty "scrap" inspectors, whose duties will be presently specified.

The machine inspectors, one or more in each machine-shop component-production department, move from one machine to another and note work in progress. There are about one hundred and twenty day machine inspectors and about one hundred night machine inspectors. The machine inspector notes any fault in any operation in progress, and may either correct faulty tool-setting himself, or may call the department-foreman's attention to the fault, or may order a change of tools or may call a tool-setter to remedy a fault. Machine inspectors enough are placed in each department to cover all operations in that department at frequent intervals, so that no faulty operation shall proceed for any great length of time. The office of the machine inspector is highly important and his powers are large and are exercised at discretion.

The operation inspectors inspect the work at certain periods in progress of finishing.

The floor or final inspectors deal with components in completed form.

Each department is cleared daily of wasters or components rejected by the final inspectors, and all wasters from all machine-shop departments are taken at once to the same room, close to the machine-shop superintendent's office, where each waster is carefully examined by one of the six or eight wasters inspectors to see if the rejected component can be made good.

All wasters finally condemned by these wasters inspectors then go to the "scrap" inspection department, and there undergo individual examination by the twenty scrap inspectors who place the blame of the scrap-making where it belongs.

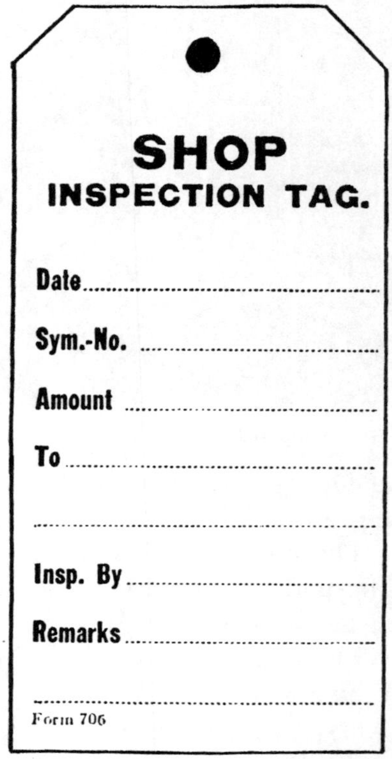

**SHOP**
**INSPECTION TAG.**

Date ........................................

Sym.-No. ........................................

Amount ........................................

To ........................................

........................................

Insp. By ........................................

Remarks ........................................

........................................

Form 706

Form 706. Shop Inspection Tag

The wasters inspectors may call on the head inspector, and the head inspector may summon the foreman of the production department where the waster was originated, and the machine-shop superintendent may be notified, so that the waster makes plenty of trouble for those whose faults assisted in its production. The same procedure may be followed in the scrap-inspection quarters, so that all in fault are very likely to be made fully aware of vigorous disapproval of scrap production even as a rare performance.

Every effort is made to save everything which can be profitably saved, and to place blame where it belongs, with a view to curing faulty practice of every description.

### The Traveling Inspector

It often happens that matters under discussion with an outside supplier can be soonest brought to a mutual agreement by a face-to-face talk and in such intance a traveling inspector is sent to the outside supplying factory.

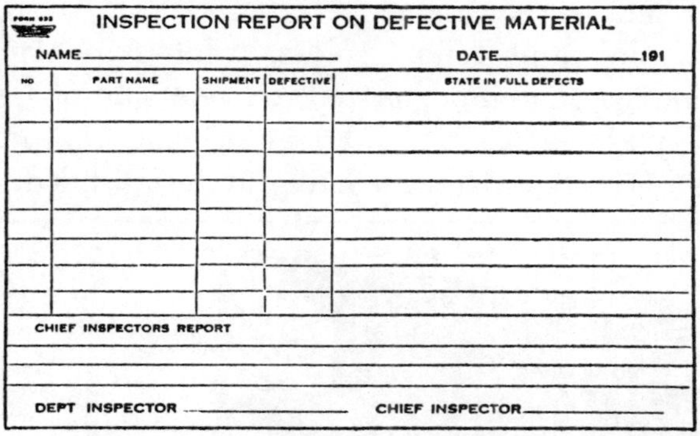

INSPECTION REPORT ON DEFECTIVE MATERIAL

NAME ________ DATE ________ 191

| NO | PART NAME | SHIPMENT | DEFECTIVE | STATE IN FULL DEFECTS |
|---|---|---|---|---|
| | | | | |

CHIEF INSPECTORS REPORT

DEPT INSPECTOR ________ CHIEF INSPECTOR ________

Form 933. Inspection Report on Defective Material

In cases of sufficient importance the head inspector may go to an outside factory, though this official has plenty to do at home, and leaves the Ford shops in cases of emergency only.

### Inspection Form Blanks

Form 706, "Shop Inspection Tag," size $2^1/_{16}$ wide by $4^3/_8$ inches long, is wired to the receptacle containing finished components by a floor, or final inspector, only. No other inspection official is authorized to use Tag 706. It is printed in black on stiff white paper, filled by the final inspector, who makes the inspection and wires the tag to the components receptacle, where it remains until the components are used in an assembling department. The inspector dates the tag, fills in the component symbol space, directs the finished components to destination, signs his initials, and certifies accuracy of dimensions and finish.

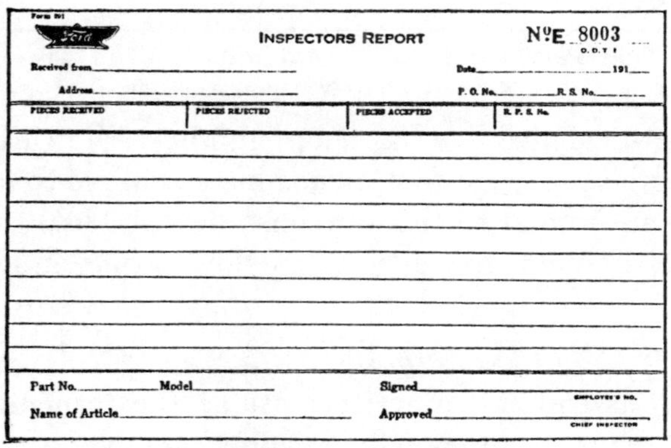

INSPECTORS REPORT

Nº E 8003

Received from ________ Date ________ 191__

Address ________ P. O. No. ________ R. S. No. ________

| PIECES RECEIVED | PIECES REJECTED | PIECES ACCEPTED | R. P. S. No. |
|---|---|---|---|
| | | | |

Part No. ________ Model ________ Signed ________

Name of Article ________ Approved ________

Inspector's Report

Form 933, Inspection Report on Defective Material. Printed in black on one side only of substantial white-paper form, $8\frac{1}{4}$ wide by $5\frac{1}{4}$

inches high. Filled by the department inspector with carbon duplicate, sent to the head inspector, who signs his name to both original and carbon, sends the carbon to the purchase agent, and files the original; when final adjustment is made it is recorded by the head inspector on his filed original, completing the entire record.

### Inspector's Report

Triplicate form, green, pink and yellow colors, printed in black, serial letter black and same serial number in red, on one side of thin paper, only, 8 inches long by 5¼ inches high. Green original and pink and yellow carbons. This is the report on incoming material, filled by

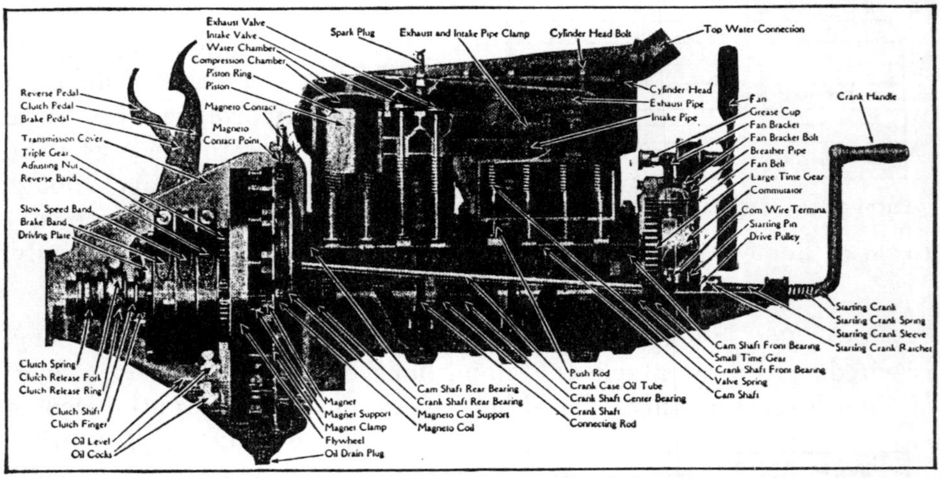


**"Ghost" View and Partial Section of Ford Motor**

Giving a good idea of the problems and the work involved in its assembly.

incoming-materials inspector, signed by his shop number; the three forms are sent to the head inspector, who appends his signature to all three forms, sends the green original to the purchase department, the pink carbon to the stock department, and files the yellow carbon in his own department.

### Ford Shops Assembling

Ordinary machine-shop assembling practice stations the principal component in a convenient place on the shop floor, sometimes over a pit, or on a suitable platform having more or less elevation above the shop floor, on horses or on the bench, as may seem best, and proceeds with the assembling by bringing the other components to the principal component and applying and fixing them to the principal component which remains in the one place until the assembly is completed.

It is also ordinary practice to place all the components of a given assembly, as of a lock, or a dozen locks, in a box, or in a box with twelve compartments, and to pass this box to an assembler, who may wholly assemble each lock, or perhaps may partly assemble each lock and pass the box on to one or more assemblers who complete the assembling; or to place the components of (say) a watch movement in each of a series of boxes and pass these boxes to the first one of a line of assemblers, who does his part of the watch assembling and passes the box on to the next man, and so on, until the watch is completely assembled. All of this well-known practice may divide the assembling operations needful to produce the completed unit assembly more or less minutely, and, up to a certain limit the greater the number of the assemblers in the line, the less the total assembling time will be and the better the work of assembling will be done.

These methods are used for small unit assemblies only, and the Ford motor and chassis assembling methods are believed to show the very first example of minutely dividing the assembling operations of so large and heavy a unit assembly as an automobile. These Ford motor and chassis assembling lines are believed also to show the very first examples of chain-driving an assembly in progress of assembling, and hence are worthy of the closest study by all builders of comparatively small assemblies of any description.

The Ford shops assembling practice is to place the most suitable component on elevated ways or rails, and to carry it past successive stationary sources of component supply, and past successive groups of workmen who fix the various components to the principal component, until the assembly is completed and ready to leave the assembling line.

In some cases, where the shape of the component is unsuited to travel on rails, the principal component is pushed along on a finished iron table from one man or group of men to another man or group of men, past sources of component supply, each workman or group of workmen completing the placing, or the placing and fixing, of one component before moving the assembly in progress to its next station.

In case the assembly in progress moves on elevated rails or ways, it is common Ford practice to drive the assembly in progress by means of a slow-moving chain, and if the components are perfectly to gauge, so that all operations can be performed in predetermined times, it is better to drive the assembly in progress at a fixed suitable speed by chain, at a uniform rate, than to move it on the ways by pushing.

This Ford method of moving the assembly in progress has effected remarkable labor-saving gains over stationary assembling with all components brought to the one point for each assembly, the labor-saving gains being in all cases accompanied by great reductions in floor space required for the assembling operations.

Thus, up to September, 1913, the Ford car chassis assembling occupied 600 feet length of floor space, and required 14 hours of one man's time to assemble one chassis, standing still in one place while being assembled.

April 29, 1914, with the chassis chain-driven while assembling, 1,212 Ford chassis were assembled on three parallel elevated-rail assembling lines, by 2,080 hours of labor, giving one chassis assembled for each 93 minutes of labor, as compared with 840 minutes of labor in September, 1913.

The stationary chassis assembling in 1913 took 600 feet in length of floor space, while on April 29, 1914, the assembling lines were only 300 feet long.

As for Ford motor assembling, in October, 1913, 9,900 labor hours were required to assemble 1,000 motors in one day, which gives 9 hours 54 minutes = 594 minutes for each motor assembled; May 4, 1914, 1,003 motors, chain-driven on rails, were assembled with 3,976 labor hours, or 238,560 minutes = 237 minutes and 52 seconds time for each motor-assembly completed, a saving of 356 minutes 8 seconds = 5 hours and 56 minutes, per motor. In other words, more than 2½ motors were assembled on May 4, 1914, in the time it took to assemble 1 motor in the month of October, 1913, when the motor assembly was made by first-class American mechanics, working in what was believed by the Ford engineers in the month of October, 1913, to be the very best manner possible.

Besides these almost unbelievable reductions in assembling time, the Ford shops are now making equally surprising gains by the installation of component-carrying slides, or ways, on which components in process of finishing slide by gravity from the hand of one operation-performing workman to the hand of the next operator, this use of work slides being in some instances combined with operation divisions.

All of this Ford practice is of great importance to manufacturers at large, because the Ford engineers assert that these improved methods of handling work by slides, of moving assemblies in progress, and of minutely dividing assembling operations, can be applied to any and all small-machine manufacturing, with very large reductions of labor-cost.

The writer of these Ford stories has had sixty years of machine-shop

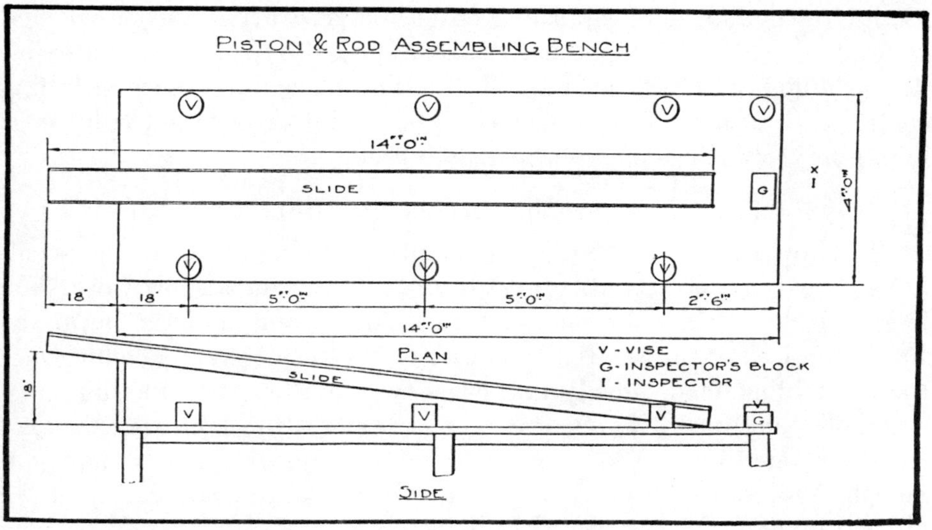


**Plan and Elevation of Piston-and-Rod-Assembling Bench and Slide, Showing Dimensions**

experience, ranging from cleaning castings and helping in the blacksmith shop to holding the position of chief draftsman and superintendent, and he believes that these assertions of the Ford engineers as to the extended scope and applicability of these new methods of work-in-progress handling and assembly moving, are broadly true.

I should say, in addition, that it was only at the end of three weeks of study and photograph-taking in preparation of this chapter that I became aware of the underlying principles by the application of which these incredible labor-savings of the Ford shops have but very recently been obtained.

What is more, if the reader experiences but a very small part of the surprise and admiration the writer experienced upon being made aware of this new Ford practice, he will certainly approve of the space devoted to full illustrations and descriptions of the Ford new practice in the directions specified. The Ford engineers are now moving over 500 machine tools in the Highland Park shops, and are having a large number of new machine tools constructed, many of them showing striking novelties of design, in order to take full advantage of the new things they themselves have learned in the last ten or twelve months.

The new Ford method of finishing the cylinder bores of small gas-engines by a rolling process, illustrated on page 126, gives an excellent interior cylinder surface, and is here first illustrated and described. This method of cylinder-bore finishing costs but very little, and is believed to produce results far superior to those obtained by the very best cylinder grinding practice, at a small fraction of cylinder-bore grinding costs, and this first disclosure of the Ford cylinder-bore rolled finish is highly important gas-engine construction news.

### The Ford Piston and Connecting-Rod Assembling

To show what may be done by simply dividing an operation seemingly already reduced to its lowest terms, and placing a short work-slide lengthwise of the assembling bench; the first example of the improved Ford practice here illustrated and described is the piston and connecting-rod assembling, changed within the last two months, so that now 14 men assemble 4,000 pistons and connecting-rods in one 8-hour day, instead of the 28 men employed to do exactly the same work less than two months ago, and with no change whatever in the tools used, nor in the ultimate operations performed.

In addition to the labor-time saving, the present practice of piston and rod assembling includes an inspector, who gauges and inspects each piston and rod assembly, with the result of no rejections from the motor-assembling line. With the former practice, where one man did the

**Piston-and-Rod-Assembling Bench with Men at Work**
South bench, looking east; inspector at left front.

whole job of piston and rod assembling, numerous returns were made from the motor-assembling line, causing costly delays in the motor assembling, to say nothing of the costs of pulling down and reassembling the faulty piston and rod assemblies.

### Ford Pistons and Pins

The finished weights of Ford pistons vary, maximum, about six ounces. Each piston is weighed and marked on the head by a heavy center-punch used without a hammer, with one, two, three, or four center marks, dividing the pistons into four weight classes, maximum weight variations in each class three-quarters of an ounce. After inspection, the inspector places the assemblies on one or another of four shelves, according to center marks on the piston head, and the pistons are paired for weight on opposed crank-throws by the motor assembler.

The pistons and pins come to the piston-and-rod-assembling bench with the pins in the pistons. The rods come to the bench by themselves.

### Piston-and-Rod-Assembling Bench

The work bench is covered with sheet metal on top. In the old style, where each man did the entire job (average time about 3 minutes), each bench had 7 piston-holding special vises on each side, with no inspector, and no inclined work-slide over the bench—14 men to each bench, 2 benches, 28 men in all, who assembled 175 pistons and rods, average, in 9 hours of each man's time, or about 3 minutes 5 seconds time, each. Operations, tools, and benches were the same as now used in working the new methods, save that 7 of the 14 vises are now removed from each bench. No inspection, and many returns of faulty piston-and-rod-assembling from the motor-assembling line attended the older practice. The flat-top sheet-metal covered benches are 14 feet long and 4 feet wide.

Here was a 3-minute operation, very simple: push pin out of piston, oil pin, slip rod in place, slip pin through rod and piston and tighten the pin-pinching screw in the rod top-end, and place and open the pinching-screw split pins; and, although the time was not very small and the work not faultless, no one had studied the job carefully, or held a stop-watch on the operations to find how the 3 minutes were actually expended. Finally the motor-assembling foreman analyzed the time with the stop-watch and found that 4 hours out of the 9-hour day were spent in walking—that is to say, in body movements of each assembler made by moving his feet.

In a day or two the foreman had split the single man into 3 men, and reported to the machine-shop superintendent that he had no use for 14 of

**North Bench for Piston and Rod Assembling**

Looking from the southwest corner; inspector at the left front

**Another View of the North Bench for Piston and Rod Assembling**

Showing four assembling shelves at the left; the inspector, also at the left, is trying the stiffness of the rod movement

the 28 men on the piston-and-rod-assembling job, and the superintendent laughed at him.

Seeing convinced the superintendent that the laugh was misplaced, and he then said that it was surprising that the job had not been changed before.

It is of no use whatever to tell this story without detailing it as minutely as a split second-hand details operation-motion time-losses; therefore, one diagram and three photographs are shown, together with a fairly complete operation-time analysis so that the reader can see for himself much more than he could learn as a mere uninstructed spectator in the Ford shops, watching this piston-and-rod-assembling job in actual work.

### Old-Style, One Man Performing Six Operations

Operation 1. Drive out pin, with special hand-hammer.

Operation 2. Oil pin, by dipping end in box of oil. Hand.

Operation 3. Slip pin in rod-eye. Hand.

Operation 4. Turn pin to take screw. Screw-driver.

Operation 5. Turn in pinch-screw. Hand brace.

Operation 6. Tighten screw, with open-end wrench, and put in cotter-pin; spread pin-end with special tool.

Time 3 minutes and 5 seconds; no inspection; 14 men on one bench. Average production per man, 175 pistons and rods assembled in 9 hours working time.

### New-Style, Operation Split Into Three Divisions

Bench provided with slide, 3 men on each side of bench and inspector at one end of bench.

Operation 1. Drive out pin, oil pin, enter pin-end in piston, average time, 10 seconds.

Operation 2. Place rod in piston, pass pin through rod and piston, with screw-driver turn pin to position to take screw, turn screw in with brace. Time 10 seconds.

Operation 3. Tighten screw with open wrench, place cotter-pin, by hand, spread cotter-pin ends with special tool. Time 10 seconds.

Operation 4. Inspection. Inspector gauges piston with flat steel gauge, places piston in pin-holding jig, tries rod to see if rod is pinched tight on pin, then holds piston horizontal in both hands and vibrates it slowly in vertical plane to see that the weight of rod free-end will barely rock pin in piston-pin bushes, and that pin has friction enough in bushes to keep rod from moving freely, a delicate test for pin-fit in piston-bushes. If rod works either too stiffly or too freely the assembly is rejected, goes back to assemblers, and has a larger or smaller diameter pin put in, as case may demand.

Actual inspecting time about 8 seconds, leaving inspector 2 seconds time to place the assembly on its proper shelf, according to the weight as shown by the center-punch mark or the marks on the piston head.

The best time record for 7 men, 6 assemblers and 1 inspector, is 2,600 piston and rod assemblies turned out in 8 hours, equal to one assembly in $77^1/_{13}$ seconds of one man's time. Average time, 2,400 assemblies in 8 hours, with 7 men, gives one assembly in 84 seconds of one man's time, or better than double the work of one man doing the entire job with no inspector, and with a saving of 101 seconds of time of assembling.

With inspection, under new style, as said before, there are no returns of pistons from the motor-assembling line.

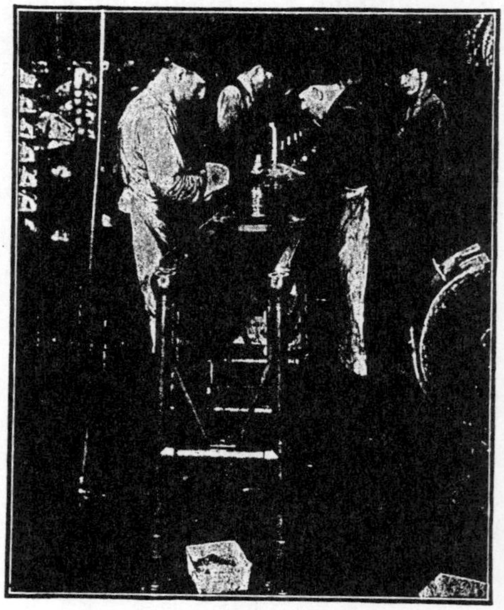

**West End of Flywheel-Magneto Assembling Line**

This is the first of the Ford sliding assembly lines, and at this end the old pipe construction yet remains in place.

This piston-assembling job teaches two lessons of first importance. The first is that there are great savings in labor to be made by splitting operations to such an extent that the workman does not need to change the position of his feet, and the second lesson is that a work-slide so located that the workman can drop his completed operation out of his hand in a certain place, without any search for a place of deposit, and also can reach to a certain place and there find his next job under his hand, is also a very important time-saver.

The vises are 60 inches apart, so that there is only 30 inches reach required for the pistons and pins, which are placed on the bench after operation 1, ready to hand to the man who performs operation 2, who in turn places the pistons where they are readily reached for operation 3. The slide is used for the completed assembly only, and delivers the com-

**Middle of Flywheel-Magneto Assembling Line**
Showing the gray-iron and angle-bar construction now adopted in the Ford assembling lines, and the bottom line of the driving chain.

pleted assemblies close in front of the inspector so that not a movement need be wasted anywhere.

In commenting on the late change from the old routine of piston and rod assembling to the new method by which 14 men are made to do more and better work than 28 men did before, the foreman of assemblers said, "We were asleep over that job, asleep and dreaming. I don't see how we came to overlook the possibilities the way we did."

**Eastern End of Flywheel-Magneto Assembling Line, Showing Part of Chain Drive**
The chain is belt-driven to a worm shaft with worm gear on the outside (rear) end of sprocket shaft.

### Height of Moving Assembly Lines

It was in this same assembling department that the first moving assembly line, that for assembling the Ford fly-wheel magneto, was installed. Of course, every one had everything to learn, and this first Ford assembling rail-line was built 8 inches lower than it should have been. The correct height for the magneto assembling, three illustrations given herewith, is 35 inches above the floor for this job. There were the same uncertainties as to the best height in the case of the chassis-assembling ways. In all instances it is of first importance that the

**Assembling Transmission Casings**

Two gray-iron finished flat-top benches are used, that at the right rear being higher than the one at the left front.

workman should stand upright. A stooping posture very soon tires the workman, and greatly reduces his efficiency. When the chassis ran on its own wheels on the floor it brought things about "work-high"; some of the operations were too high for convenience, and platforms were placed on the floor where needful. When the first high line was placed for chassis assembling, with the chassis sliding on its axles on top of the rails, it was made 26¾ inches high; two other chassis lines were installed each 24½ inches high, one on each side of the middle 26¾-inch rail-line. These two heights, 26¾ and 24½ inches, are retained with much satisfaction, the tall men being worked on the high line and the shorter men placed on the two low lines. The Ford engineers attach so much importance to this "work-high" condition that they are now placing a great number of gray-iron raising bases under various machine tools, particularly under presses, to bring the work at such a height that the workman can either stand or sit erect, any stoop being now well known to cause a marked reduction in the worker's output.

The first fly-wheel magneto moving assembling line was installed, ready for work, about May 1, 1913, but the desirability of general application of the moving assembly line to the Ford motor assembling and the chassis assembling was not at once fully conceded by all the Ford engineers.

### Fly-Wheel Magneto Moving Assembly

This moving-assembly line is of historical importance as being the first moving assembly placed in work anywhere, so far as revealed by information to date. It is, of course, possible, or perhaps probable is the better word, that the moving-assembly line has been used somewhere in the world, but it is new to the Ford engineers and entirely novel to me. If the moving-assembly line has been used elsewhere, probably this publication will bring the previous use to public knowledge.

The fly-wheel-magneto-assembling story will not be told in full at this time. The Ford motor is the only one used for automobile driving which fires the charge by current generated by a magneto built directly on the fly-wheel, and hence sure to run as long as the fly-wheel revolves —which, of course, gives a more direct and certain magneto drive than can be had with a separate magneto, gear-driven from the motor crankshaft, after the usual practice. Besides this, the Ford fly-wheel magneto is a novel construction, and it is hoped that it may be fully described at a future date, but at this time such description is outside of the line of thought.

Previous to the installation of this moving magneto-assembling line, the Ford fly-wheel magneto had been a one-man assembly, each workman on this job doing all the assembling of one fly-wheel magneto and turning out from 35 to 40 completed assemblies per 9-hour day. The work was done by experienced men, but was not so uniformly satisfactory as was desired, and was costly as a matter of course, as all one-man assembling must of necessity be forever.

Forty assemblies per 9-hour day, best time for one-man work, gives nearly 20 minutes time to each one.

When the moving-assembly line was placed in work with 29 men, splitting the one-man operations into 29 operations, the 29 men began turning out 132 magneto assemblies per hour, or 1,188 per 9-hour day, one man's time producing one fly-wheel magneto assembly in 13 minutes 10 seconds, a saving of nearly 7 minutes time on each assembly, or more than one-third of the best one-man time.

**First Operations in Motor Assembling**

Pouring crank-shaft boxes with two babbitt ladles, one in each hand; babbitt-melting furnace at the right. Cylinders slide over hot plate before babbitting and are heated so that thin babbitt will pour well. The man in the middle puts on and takes off the babbitting jig; the man at the left chops off the babbitt risers.

**Operation 10—Cleansing the Cylinders**

The cylinder castings (which weigh 101 pounds rough as sent from the foundry) are placed on pendulum carriers pivoted to a slowly revolving gray-iron wheel, turning top to left, which carries the cylinders down through a hot cleansing bath. The operation requires two men, one to place the cylinders on the shelves and the other, standing at the right, to take them off and place them on the assembling line.

A new high line with chain-drive was installed for magneto assembling about March 1, 1914, when the Ford work day had been shortened to 8 hours. At that time the magneto-assembling force had been improved by substitutions and experience, and 18 men were assembling 1,175 magnetos in 8 hours, or a little more than 7 minutes of one man's time to assembling one magneto.

Operation 14—Finish Valve Seats on Foote-Burt 8-Spindle Drilling Machine

The chain-drive speed was a matter of trial; it was first made 5 feet per minute, which was much too fast; then 18 inches per minute was tried, and found much too slow; the third trial was 44 inches per minute (3 feet 8 inches), and is yet in use, though the foreman believes it could now be increased to advantage. The chain drive proved to be a very great improvement, hurrying the slow men, holding the fast men back from pushing work on to those in advance, and acting as an all-round adjuster and equalizer.

Operation 15—Grinding in the Valves
The work of two men was saved by moving this Foote-Burt machine from its original place on the floor into the assembling line.

As soon as the men became accustomed to the automatically moving assembly 4 men were taken out of the line and the production was 160 in excess of previous performance; 14 men working 8 hours assembled 1,335 magnetos, making 5 minutes of one man's time assemble a magneto, as against 20 minutes when one man assembled the entire job.

The next attempt at moving the assembly in progress was made with the job of placing the crank-shafts in the *en-bloc* cylinders. An accident occurred which resulted in personal injury and put a stop to new installations of the moving assembly for a time, but in June, 1913, the foreman of the assembling room took courage and split the transmission-cover assembling into 23 operations, not on rails but on flat-top metal tables, the shape of the transmission cover making rail-sliding impracticable.

One man assembling the entire transmission cover produced from 20 to 30 assemblies per 9-hour day, 18 minutes for each one.

At the present time 23 men, each working one of the 23 operations for 8 hours, complete 1,200 transmission assemblies, which gives 9 minutes 12 seconds for each assembly, or a little more than one-half the best time with one man doing the whole job.

These great labor savings led to the first trial of full-length assem-

**Operation 17—Cleaning Cylinders by Live Steam**

They are lifted into a steam box, which is closed tight by toggle levers, and cleaned by 72 live-steam jets. The washing-wheel hot bath (Operation 10) stands at the left. The machine is of Ford design and construction.

**Operation 19½—Hydraulic Testing of Cylinders for Leaks**

It was impossible to place this machine in the line. At the left is the long inclined gravity cylinder-carrier, down which the cylinders run upon rollers. The machine is of Ford design and construction.

bling ways for motor assembling, in November, 1913. Motor assembling on separate benches gave, in October, 1913, 1,100 men working 9 hours to assemble 1,000 motors.

By installing full-length motor-assembling lines and building some new tools to go into the motor-assembling line, along which the motors in process of assembling are moved by hand, now, May 8, 1914, 472 men working 8 hours assemble 1,000 motors.

In November, 1913, it took nearly 594 minutes of one man's time to assemble one motor.

Now, one motor is assembled in 226 minutes of one man's time, as against that 594 minutes in November, 1913.

This very great saving is due to the continuous line assembling, to the installation of work-carrying chutes, and to holding the motor as nearly as may be at one level during the entire course of motor assembling.

## [d]

# MOTOR-TEST BLOCKS AND CHASSIS-ASSEMBLING LINES

A GRAY-IRON foundry is not an automobile-body upholstering shop, and an automobile-upholstering shop is not a gas-engine assembling nor a chassis-assembling department. These assertions will be accepted by all without argument or demonstration.

How, then, can the Ford engineers be expected to reduce labor-costs enormously in the gray-iron foundry, in the motor-assembling department, in the chassis-assembling department, and in the body-upholstering department, by use of one and the same mechanical element, a slowly-moving, endless chain?

The Ford engineers certainly do make great labor reductions in the foundry, the machine shop, and in body-upholstering by simple endless-chain installations, because in all assembling operations it costs much less in labor-time expenditure to move an assembly in progress past fixed points of component supply than it costs to hold the assembly in progress stationary and bring the components to the one assembling location, and because the moving assembly gives more floor space for assemblers and hence permits the more minute subdividing of assembling operations.

In the foundry, in ordinary practice, a poured mould is neither more nor less than an assembled unit, all assembling operations being performed in the one fixed location where the flask is first placed on the foundry floor. The sand is brought to the flask by hand and shovel, the pattern, gaggers, parting-sand, and cores are brought to the mould by hand and assembled, and, as the last operation, the melted metal, final component of this foundry assembled unit, is brought to the finished mould and poured into it.

The Ford four-cylinder *en-bloc* casting, the rough casting weighing 101 pounds, is made in this way at this time. The Westinghouse air-brake-cylinder mould is made and poured and shaken out on an intermittingly-moved endless train of cars (240 in number, if memory serves correctly), drags and copes on separate cars moving to stations under the overhead sand-chute, past the groups of moulders and core-setters and flask-closers, to the cupolas where the moulds are poured; then over shake-out grates where the sand falls into low troughs, provided with spiral conveyors, wherein the sand is moistened, cut-over, and led to an elevator which lifts it to supply the sand chutes filling the flasks. Undoubtedly the Ford *en-bloc* cylinders might be moulded and poured on a Westinghouse endless chain of cars running on two parallel tracks, connected by semi-circles of track at the ends; but the Ford cylinders are not so moulded and poured; they are moulded on the floor and poured from ladles hung from overhead rails.

All the Ford-foundry small work is machine-moulded, copes and drags separately in most instances; then the moulds are placed on endless-chain-driven mould-carriers which take them past the pouring

**Single Motor-Testing Block, with Motor in Place for Running in**

**From the Test Block the Motors Are Taken to within Reach of the Chassis-Assembling Line in Truck Loads of 5 Motors Each, 200 Truck Loads per Day of 8 Hours for 1,000 Autos Assembled**

zone and over the shake-out gratings, as in the Westinghouse method. The Ford moulding machines stand under sand chutes which are served by an endless chain of sand-distributing blades or pushers, slowly moving in troughs placed above the heads of the machine moulders.

The Ford gray-iron foundry is now installing a new core oven, served by two endless-chain lines, all as will be fully described in the chapter on the foundry, later in this book. The main themes of this chapter are the motor testing and chassis assembling; but the spectacle of the body-and-top chain-driven assembling led to a general survey of the Ford endless-chain assembling lines, which naturally recalled the Westinghouse endless-train air-brake-cylinder moulding and pouring, and so suggested the close analogy between machine-shop assembling and foundry moulding operations, notwithstanding the fact that foundry operations and machine-shop operations are so universally regarded as widely dissimilar.

Before following the chassis assembling in detail, however, further attention must be given to the motor block-test and the details of motor "running-in" on the block. These important operations conclude the motor assembly (which was described in the preceding chapter) and precede the delivery of the motor, with its motor-assembler's record, to the automobile-assembly line.

## The Motor-Test Blocks

The motor blocks are nineteen in number at the date of this writing (June 1, 1914) and the day's motor production, whatever it may be, must be handled on the nineteen blocks. This number will soon be increased.

Each test-block consists of a Westinghouse direct-current shunt-wound motor, constant speed, showing 20 horse power at 230 volts, 73 amperes and 750 revolutions per minute, equipped with Weston instruments and supported on a stand adapted to take the motor and hold the transmission shaft in line with the motor-armature shaft. The motor under test is placed on the block and driven at 750 revolutions per minute until the needle drops to a 20-ampere reading, this operation taking, say, from 3 to 5 minutes; each motor is given as much time on the block as may be, but the floor must be kept clear of untested motors. No motor is ever passed until it makes 750 revolutions per minute with 20 amperes of 230-volt current.

The head block-tester has one assistant, one tester to every test-block, and enough helpers to handle the motors on and off the test-blocks. Each block-tester has an alphabetical steel stamp with his own individual letter, and marks a certain screw-head with hammer and letter stamp at the end of each test, and also paints one other certain screw-head red when a motor is passed from the test-block for use in the chassis-assembling line.

The block-tester examines the motor while on the block for all the possible imperfections enumerated in the list on the following page.

Imperfections discovered are noted in pencil, on small, white, stiff cards, using list terms, and a straight vertical pencil mark is made in the corresponding blank space at the left of the specified fault. One card is filled for each fault found, and all cards are fixed by one particular screw to the faulty motor, which is then sent to the motor-repairs department, working four men only, who are all-round machinists, and who take care of all the motor repairs when as many as 1,100 motors per day are being turned out.

This list of motor repairs is long because it includes every fault which has been so far discovered in any Ford "Model T" motor.

Each day's report of motor repairs is sent by the head motor assembler to the machine-shop superintendent. After repairing, the motor goes back to the tester, with its fault-cards, is placed on the block and given another trial, and so on until the motor shows no faults. By applying test certification marks to two specified bolt heads, finishing record-spots on motor castings is avoided; also, if the two specified screws should be removed, no one could be sure the motor had been tested, as it would then be without test certification marks.

## Motor-Assembly Record

Triplicate sheet, Form No. 386, printed in black, on one side only, as shown on the next page, thin paper, three colors, size 9¾ inches wide by 8¼ inches high. Filled by head of motor assembling, one original and two carbon copies. The green original is placed by the head assembler in a heavy envelope, wired to the pedal, and goes

LIST OF POSSIBLE IMPERFECTIONS FOR WHICH EVERY FORD MOTOR IS TESTED

Oil leaks in transmission covers.
Oil leaks around pedals.
Oil leaks between transmission cover and pan.
Oil leaks at door of transmission cover.
Oil leaks between transmission cover and crank case.
Oil leaks at magneto contact.
Oil leaks between crank case and pan.
Oil leaks at end of cam shaft.
Oil leaks at end of crank shaft.
Bottom door bolts leaking.
Loose rods of gasket knocks.
Rods striking pan.
Manifolds, cracked, reset, etc.
Defective cylinder heads and cylinder-head gaskets.
Tight piston rings.
Bent cam shaft.
Valve trouble, push rods, bent valve stems.
Change water connection.
Metal in transmission.
Hole in cylinder.
Change cam shaft, bearing frozen.
Put on water connection.
Intake pipe leaks.
Noisy transmission.
Cylinder head bolts won't go down; dirt in hole, etc.
Clutch won't release.
Adjust transmission bands.
Hole in crank case.
Crank case arms out of line.
Put in pet-cocks.
Change small gear.
Repair carburetor.
Clutch shift grinds on shaft.
Broken gear cover.
Hole in leather pipe too small.
Tight ball-cap bearing.
Loose front-cam bearing.
Threads stripped in drain-plug hole.
Tight cam-shaft bearing.
Cam-shaft brass bushing too tight.
Magnets striking spool.
Change fan pulley—wabbles.
Cracked and leaky pans.
Noisy gears.
End play in clutch-lever shaft.
Clutch adjusting-screw too tight.
Motor out of time.
Hole in cylinder base.
Hole in water jacket.
Motor pumps oil.

with the motor to the chassis-assembling line. The yellow carbon is filled by the head assembler and also placed in the envelope wired to pedal, going with the motor to the chassis-assembling line. The white carbon is filled and filed by the head motor assembler as his own record of production.

All of these motor-assembly triplicate record forms are numbered in red, in numerical sequence, as are also the automobile-assembly records described near the end of this chapter, and to be more fully explained by an examination of the Ford sales and shipping practice.

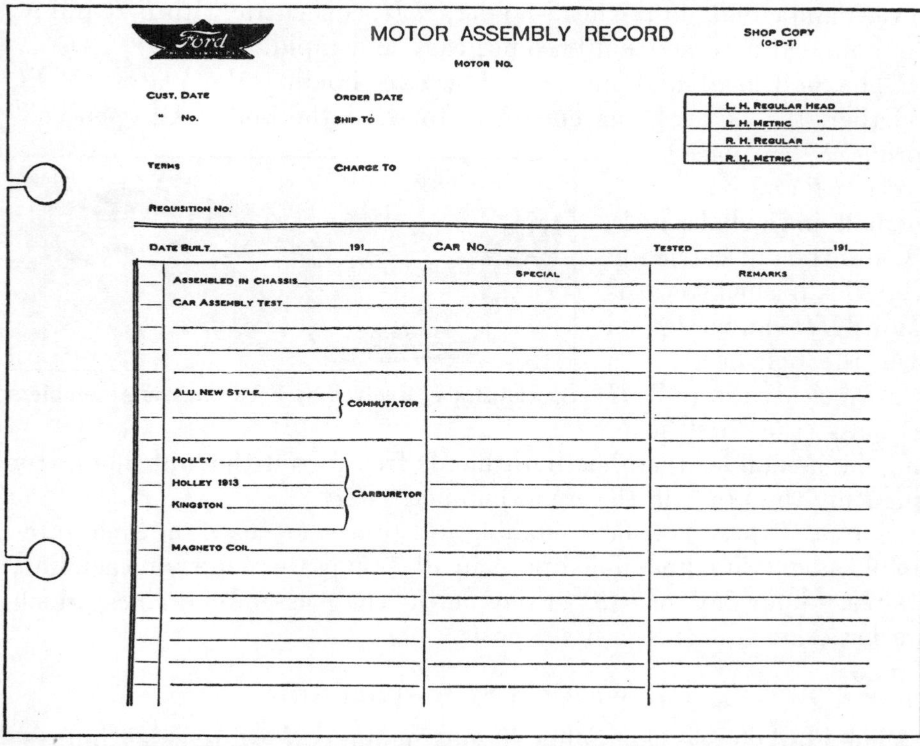
Ford

MOTOR ASSEMBLY RECORD

SHOP COPY (O-D-T)

MOTOR NO.

CUST. DATE

" NO.

ORDER DATE

SHIP TO

L. H. REGULAR HEAD

L. H. METRIC "

R. H. REGULAR "

R. H. METRIC "

TERMS

CHARGE TO

REQUISITION NO.

DATE BUILT ____ 191 ____ CAR NO. ____ TESTED ____ 191 ____

SPECIAL

REMARKS

ASSEMBLED IN CHASSIS

CAR ASSEMBLY TEST

ALU. NEW STYLE } COMMUTATOR

HOLLEY

HOLLEY 1913

KINGSTON } CARBURETOR

MAGNETO COIL

Motor-Assembly Record. Filled in Triplicate. One Copy Wired to Motor, the Others Filed as Explained on the Preceding Page

## Ford Special Nut-Brace

The Ford shops have brought out no end of original small tools and fixtures, and a considerable number of highly original machine-tools, but this is the only hand-tool involving notable invention that has so far been brought to my attention in the Ford assembling.

In case of placing a free-body-fit bolt, the head of the bolt must be held while the nut is turned on and forced to place. With ordinary wrenches, the assembler holds the head from turning with one wrench, held in one hand, and turns the nut home with another wrench, worked by his other hand—a slow operation at best, and, in many cases, very difficult because of obstructions which prevent ready access to the bolt-head while turning the nut to place.

With the special brace (page 135), which has a solid nose fitted to the nut at *N*, a block, *S B*, fitted to slide on the brace-body, *B B*, with an off-set extension, fixed to *S B*, reaching outward beyond *N* and then bent inward and ending in the head-holder, *H H*, concentric with *B B* and *N*, the bolt-seating operation becomes easy and rapid.

The bolt-head and nut are shown as holding the plates *P*, *Pl*, together, the plates being cut away to show the bolt. An open coil-spring, *S*, is placed between *S B* and *N*. With the bolt in the holes in *P*, *Pl*, and the nut carried in *N*, *S B* is pushed towards *N* until *HH* can be slipped over the bolt-head; then *S B* is released, *S* pulls *H H* onto the bolt-head, and the assembler then has both hands free to run the nut home by operating the brace in the usual manner.

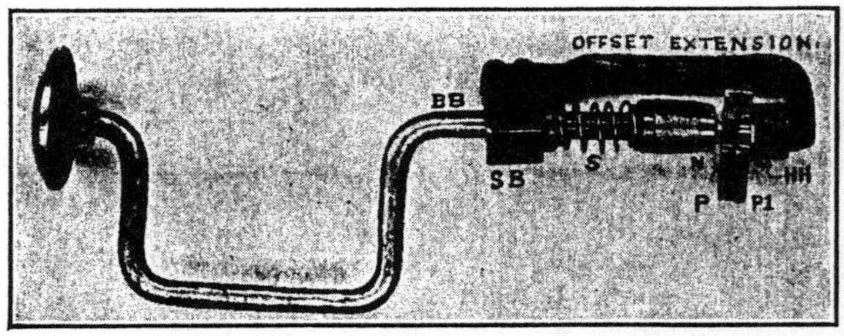

**Special Brace, Evolved by Ford Chassis Assemblers**
See pages 134, 135

Three of these bolt-head-holding nut-braces are used on each automobile assembling line, nine braces in all, saving three men on each line, $15 per 8-hour day, or $45 per day on the three assembling lines, which is a large interest on the brace cost.

## Ford Chassis Assembling

The Ford chassis assembling in moving lines affords a highly impressive spectacle to beholders of every class, technical or non-technical. Long lines of slowly moving assemblies in progress, busy groups of successive operators, the rapid growth of the chassis as component after component is added from the overhead sources of supply, and, finally the instant start into self-moving power—these excite the liveliest interest and admiration in all who witness for the first time this operation of bringing together the varied elements of the new and seemingly vivified creation, on the three Ford chassis assembling lines where over 1,200 have been put together and driven out of doors into John R Street in one single 8-hour day.

## Chassis Assembling in Chain-Driven Lines

Up to August, 1913, the Ford chassis was assembled in one location. First the front and rear axles were laid on the floor, then the chassis frame with springs in place was assembled with the axles, next the wheels were placed on the axles, and the remaining components were successively added to complete the chassis. All components needed to make up one chassis had to be brought by hand to each chassis-assembling location. This routine of stationary chassis assembling was, in September, 1913, worked with two lines of assembling-floor space, 600 feet long, 12 feet chassis-to-chassis centers, 50 assembling locations in each 600-foot line, 100 cars in process of assembling in the two lines. Working in this routine 600 men were employed, 500 being assemblers who were supplied with components by 100 men acting as component carriers.

About April 1, 1913, the first sliding assembling line, used for assembling the Ford fly-wheel magneto, was placed in work and immediately showed a large reduction in assembling labor-cost. Consequently, the possibility of lowering chassis-assembling costs by introducing the moving assembling line for chassis assembling became a matter of discussion among the Ford engineers.

In the month of August, 1913 (the dull season) 250 assemblers, with a stationary assembling location for each chassis, the assemblers being served by 80 component carriers, worked 9 hours per day for 26 days to turn out 6,182 chassis assemblies. Total labor hours $330 \times 9 \times 26 = 77{,}220$ hours, giving 12 hours and 28 minutes for labor time on each chassis, about as good as was ever done with stationary chassis assembling.

The assembling line was long—600 feet—but even at that did not give room enough, and 12½ hours of labor time seemed altogether too much for one chassis. It was in the dull season, and an experiment was made with rope and windlass traction on a moving assembly line 250 feet long. Six assemblers traveled with the chassis as it was slowly pulled along the floor by the rope and windlass past stationary means of component supply, and the chassis-assembling time was reduced to 5 hours and 50 minutes of one man's time, over 50 per cent saving.

October 7, 1913, on a moving-assembly line 150 feet long, with no helpers, components being piled at suitable locations, 140 assemblers in the line completed 435 chassis assemblies in one 9-hour day, 2 hours and 57 minutes of one man's time for each chassis assembling.

The assembling line was lengthened by degrees to 300 feet, giving the men more room, and on December 1, 1913, 177 assemblers working 9 hours turned out 606 completed chassis assemblies, about 2 hours 38 minutes of one man's time to each chassis.

December 30, 1913, working two assembling lines, 191 men completed 642 chassis assemblies in one 9-hour day, a little less than 2 hours 40 minutes of one man's time for each chassis, the cars being pushed along by hand.

January 14, 1914, one assembling line was endless-chain driven, with favorable results.

**Construction of Chassis-Assembling Line of Elevated Rails**
Showing the endless-chain idler at south end of chain. The crosswise-placed gasoline-tank platform is shown filled two-high, with tanks standing on end

January 19, four chassis-assembling lines were worked, only one line being chain-driven. The wheels were put on as soon as the axles and the chassis frames were assembled, and the assemblies in progress ran with their front wheels on the floor and their hind wheels carried in 3-wheeled cradles, used to give easy

**Another View of the Chassis-Assembling Lines, from a View Point North of the Tank-Platform**

The chain is in place, but not stretched. Pile of mufflers at left hand

placing of the rear wheels on the motor-starting drive at the end of the line.

February 27, 1914, the first high line of rails with chain drive was used. The chassis slid on its axles as pulled by the chain, and the wheels were applied only a short distance before the motor-starting was reached. This first high line was made with rails 26¾ inches above the shop floor, and at once showed great advantages, the best time for one chassis assembling being only 84 minutes, while the worst time was 2 hours. Two other high lines were soon installed, 24½ inches high, with chain drives; tall men worked on the line 26¾ inches high, and short men on the other two lines, 24½ inches high.

The Ford engineers make a point of "man-high" work placing, having learned that any stooping position greatly reduces the workman's efficiency. The differing heights of the chassis-assembling high lines are believed to be decidedly advantageous.

On these three high lines, on April 30, 1914, 1,212 chassis assemblies were completed in one 8-hour day, each chassis being assembled in 1 hour 33 minutes of one man's time, as against 12 hours 28 minutes, the best time with stationary chassis assembling, September, 1913—93 minutes as against 728 minutes—and it must be borne in mind that the September, 1913, Ford practice in chassis assembling was fully abreast of the best known in the trade. Very naturally this unbelievable reduction in chassis-assembling labor costs gave pause to the Ford engineering staff, and led to serious search for other labor-reduction opportunities in the Ford shops, regardless of precedents and traditions of the trade at large.

The chassis was not completed when it ran out into John R Street, and the first practice was to let the driver run the chassis up and down until he thought best to abandon it to the motor inspector and the rear-axle inspector, and to return to the end of the assembling lines for another chassis to drive out into John R Street. The bodies were allowed to slide down an incline from the second floor, and were then dragged along the pavement by one man and stood on end in a bunch south of the chute.

When the assembly was completed on the John R pavement, and had been inspected by the motor inspector and the rear-axle inspector, it was again boarded by a driver and taken to the bunch of bodies, where four men lifted a body into place on the chassis, and the completed automobile assembly was then driven to the shipping-clerk's office, between the railway tracks, ready for shipment.

This procedure afforded plenty of gaps and vacancies for discretionary proceedings on the part of all the men working outside under the head assembler. The next radical improvement was made by laying down the angle-iron John R street track, running southward from the exit door, under the body-chute and something more than a chassis

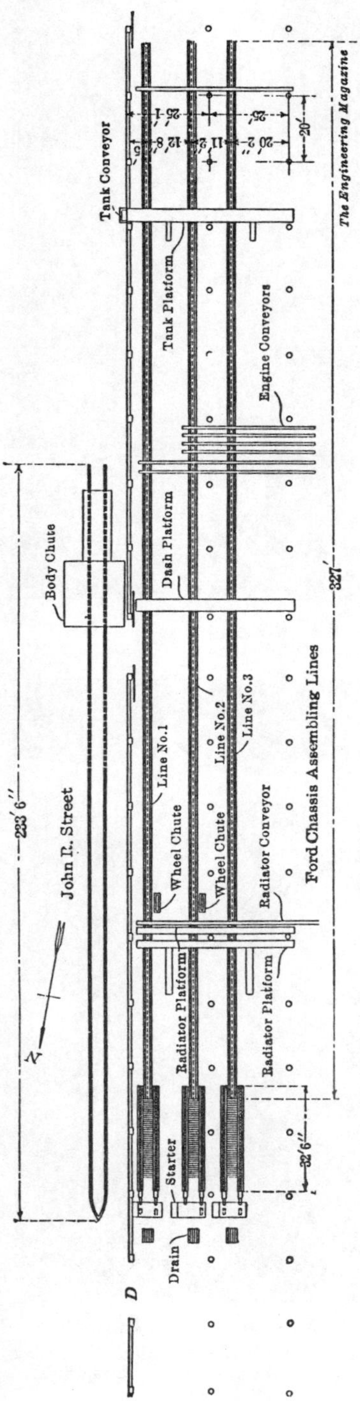


**Plan of Chassis-Assembling Lines**

The chassis assemblies begin at the south (right-hand) end, and move to the north (left-hand) end, under the overhead gasoline-tank platform, the motor-carrying chain-hoist tracks, the dash assembly platform and the radiator platform. They then take the wheels, run on the wheels on roller frames over the pit where a workman caps the front-axle bracing globe, and then down a short incline onto the motor-starting drive for the rear wheels. Then the chassis is driven under its own power, through the door, *D*, and on the John R street track to the southward

length to the south of the chute; and the chute itself was presently equipped with a car-body handling rig, not regarded as the final thing, but serving to place a body on the chassis with one handling only. The ground plan of the three chassis-assembling lines inside the shop, and of the John R street track, is given to scale on this page and the numerous illustrations in half-tone will give the reader a fair idea of the chassis-assembling procedure.

## Chassis-Assembling Operations

It must be clearly understood that the moving-assembly speed is varied to suit exactly each individual assembling job. As each operation is performed while the work passes slowly across the station occupied by each assembler or assembling gang, the time of transit past this station must be sufficient for good work and no more.

The first assembling line established in the Ford shops (for the magneto) was originally speeded at 60 inches per minute, which proved much too fast. The next speed tried, 18 inches per minute, was found to be as much too slow. The third guess, 44 inches per minute, answered so well that it is yet retained.

The dash-assembly line travels 72 inches per minute; the front-

**Mounting Chassis Frames on Springs**
View looking southeast, toward the gasoline-tank platform

**Press Operator Fixing Fender and Run-Board Brackets to Chassis Frames**
This view is at the north end of the chain-driven chassis-and-springs assembling line, on which the chassis frames are carried to the south for various operations after the frames are mounted on the springs

141

axle assembling line 189 inches per minute, and the body-and-top assembling line 144 inches per minute. The speed of the chassis-assembling line is 72 inches (6 feet) per minute.

The work is so divided among the assemblers that each operation is performed in 7 minutes and 36 seconds, turning out 300 complete chassis assemblies on each chassis-assembling line in 8 hours of working time, save in case of operations 1 and 2.

[e]

## CONVEYORS, WORK SLIDES AND ROLL-WAYS

EVERY square foot of every workshop carries a certain and inevitable overhead charge, and is hence a debtor to the plant. The only possible way in which any square foot of any factory floor-surface can be made to pay its existence debt to the plant at large, is to make this square foot of floor surface support a profit-earning load.

There must be shop-floor lines of travel which cannot be infringed upon by direct-profit-earning agencies, there must be floor-spaces which support rough stores only, other floor-surfaces which carry finished components only, and in the great majority of industrial exploitations there must be floor areas of considerable extent which carry merchantable finished product only, this merchantable product being subject to certain fixed charges for every day of its housing before it finds a purchaser.

This last form of unproductive floor-load is almost wholly avoided at the Ford Motor Company's Highland Park plant, where 90 per cent of the entire production is loaded into railway cars and shipped to cash purchasers as fast as completed, and the only finished-product stores space is the open surface of John R. Street.

Otherwise, the Ford shops are subjected to the previously specified indirectly-productive floor-space over-charges, which cannot be escaped by any manufactory whatever.

But there is another and hitherto unmentioned occupation of factory floor-space, which may be relatively large or small in direct proportion to the intelligence, ingenuity, and resourcefulness of the factory manager—that of shop-floor occupancy by components in process of finishing. This particular form of indirectly profit-earning factory-floor-space occupancy is commonly regarded by factory managers as an inevitable and unrelievable source of added overhead costs of factory product, and one which must certainly be augmented as the factory production is increased.

An increase of total production for a given time must bring more components in process of finishing into the factory; these components in progress must have floor-space to rest upon because they certainly cannot be suspended in the atmosphere between successive operations; there is no place for work in progress support save the factory floor; and consequently, since the ultimate limiting factor of production must always be the floor area available for the placing of direct-production agencies—the workmen and the benches and tools used by the workers—any increase in factory production per day must always result in decrease of factory maximum-production limit. That is to say, the more work in progress, the less the floor-space available for doing the work.

Perhaps not one factory economist in a thousand will find any fallacy in the foregoing summary of factory-floor unavoidable occupancy by components in process of finishing; yet it does contain an error, so glaringly obvious as soon as it is pointed out as to make it seem impossible that even the most conservative of factory managers could fail to perceive it instantly, to recognize its economic significance, to bear this significance constantly in mind, and to exert all his mental powers to avoid limiting his maximum-production possibilities by needless placing of non-paying loads upon his factory floor-space.

The error mentioned lies in the assumption that components in progress of finishing cannot be carried in the air between factory operations, but must, of necessity, be supported on the shop floor.

The enormous increase in the Ford-car demand very early forced the Ford factory management to place workmen and machine tools as closely together as possible on the floor; but it was not until the beginning of 1914 that it was found that, in some special instances, the convenience of the workmen could be served by the installation of gravity work-slides, usually inclined sheet-iron troughs, so placed that the workman upon completion of his own operation could drop the component into a trough, so inclined as to carry the piece by gravity to within the easy reach of the workman who was to perform the next succeeding operation on the component, instead of dropping the partly finished piece into a box or can and having this box or can carried, pulled, or pushed by a laborer to the next workman, and thus become an occupier of non-paying floor-space—with the transporting laborer's time added to floor-space overhead charges, to increase needlessly the finishing costs.

It was at once discovered that not only were the production labor costs greatly reduced by work-slide installation, but that the floor was cleaned up, making room for more tools and workmen where it was thought the limit of close placing of productive agencies had previously been reached.

Perception of these facts brought instant activities of the Ford engineers to the front in the endeavor to reduce production costs by gravity slides, endless belts, endless chains, gravity roll-ways, and overhead carriers, each one of which as soon as installed produced unbe-

**Two Lathes and Power Press for Finishing Large Thrust Washers of Babbitt Metal**
At left, inspector and two lathe hands. The lathe hands face each other, and the tail-stock ends of the lathes are set close to the power press. The lathe hands take washers as they come down the slides, with outside and inside edges finished, from the press, face the slides and cut a spiral groove in each side, and then drop the finished piece into the short slide marked 4, seen between the lathe men. The washers slide to the end next the inspector, who stacks them on his table for inspection

lievable labor-cost reductions, so that, at the date of this writing, October 6, 1914, these means of transporting components in process of finishing through the air instead of on the floor, are the most surprising feature of the Ford shops interiors.

Broadly speaking, it is safe to say that anything in the way of a work conveyor which keeps work in progress off the shop floor, will be found to effect a large saving in direct-labor costs, and to increase the factory maximum-production capacity.

## Labor-Cost Reductions Gained by Gravity Work-Slide Installation

Surprising labor-cost reductions invariably follow the placing of work-slides, although the first gravity slides were installed merely with a view to the advantages gained in the way of transportation of components in process of finishing from one operation location to the next; but, as before mentioned, it was at once made plainly apparent that gravity work-slides vastly increased each worker's efficiency. In every instance of work-slide placing there was a gain of from 30 to 100 per cent in the production volume, with the same methods, machines, small tools, and men; seemingly nothing done to decrease labor costs, yet large savings shown immediately.

The labor savings were due, as is now fully understood in the Ford shops, to giving the workman a fixed place of reception for work leaving his hand, and a fixed point within his easy reach where he could pick up the next piece, thus avoiding thought, delay, and fatigue of the workman caused by breaking into the volition routine of his movements.

**Same Job and Tools as shown on page 273, Different Workmen**

Showing work slides from press to left hand lathe on page 273, and inclined slide for finished work between the two lathe hands. The sheet iron trough slides are same for each lathe. Slide 1 delivers to slide 2, which opens to tray 3. Finished work-slide between the two lathe hands is marked "4." Press hand not visible

**Piston Gravity Slide from Four Drilling Machines to Piston, Pin, and Connecting Rod Assemblers**

The four machines are retained for emergency use, but three men now ream as much work as formerly required four, and the work of two truckers is also saved. 4,000 pistons with bushes, total weight about 12,250 pounds, come to this reaming station in cans holding 112 pistons each, every 8-hour day

Where a workman can perform absolutely similar successions of movement, he very soon gains great skill combined with great rapidity of muscular action; but if the routine of the workman's movements is broken, he must inevitably call his brain into action to find the best means of bridging his troubles, and must lose some time in devising and executing his unusual line of procedure. All of this seems too obvious to demand detailed specification, but it is certain that in very many factory operations great time losses are the rule rather than the exception, while it is also certain that a very large part of these time losses can be avoided by the placing of work-handling conveniences of insignificant cost.

These conveniences fall into the following classes, which will be shown and described in the order of mentioning:

1. Gravity work-slides.
2. Gravity roll-ways.
3. Slides on which work in progress is moved by hand.
4. Endless belts, power-driven.
5. Endless chains, power-driven.

My attention was first drawn to the labor-saving powers of gravity work-slides by the example shown on pages 273 and 274, front and side views of a group of three power-driven machines, a press and two engine lathes, employed in finishing a babbitt friction washer, now obsolete. The machines were placed as closely as possible—so closely that the work-slides placing seemed an absurdity. The three workmen, one press hand and two lathe hands, stood close together; the washers (babbitt-metal castings) were first passed through a die working against a spring plunger die and ejector in the bolster, this operation finishing the washer hole and the washer circumference, and leaving the washer about 3$^3/_4$ inches outside diameter, hole 2$^1/_4$ inches, and were then faced in the two lathes to about $^3/_{16}$ inch thickness and had spiral oil grooves cut in their sides.

Before the work-slides were placed, the regular production for the three workmen and three machines was 800 washers per 8-hour day; this rose to 1,100 washers per 8-hour day—or 37$^1/_2$ per cent increase—as soon as the work-slides were used.

On page 275 are shown gravity slides which take the pistons from four drillers on which four men ream the piston-pin holes in about 4,000 pistons per 8-hour day, doing away with two truck men.

Many more gravity work-slides might be shown, as these transportation facilities appear in every direction in the Ford shops, but one or two examples are enough to direct the attention of the intelligent and aggressive shop manager to these cost-reducing agencies, since every installation of this character is a job by itself and copying is not feasible. It is enough to say that the best placing of any gravity work-slide must be preceded by a careful motion study of the movements of the workmen who are to make use of it, that most men are wiser after the fact than before it, and that no competent workshop official will hesitate to change his own work if he thinks he sees a way to better it.

[5]

[a]

Industrial Management
The Engineering Magazine
WRITTEN FOR Industrial Executives and Mechanical Engineers
DEVOTED TO Labor Saving Machinery and Labor Saving Management

Vol. LXIV. August, 1922 No. 2

# Ford Principles and Practice At River Rouge

"RIVER ROUGE," as the latest and greatest of the Ford plants is called, is the home of the Fordson Tractor. In addition to this, it is perhaps the most significant and vitally important industrial development in the world today.

The River Rouge Plant of the Ford Company has never yet been adequately described, nor has the manufacture of the Fordson tractor been depicted, nor have the principles back of these things been written for the benefit of industry. It is extremely gratifying, therefore, to be able to announce that INDUSTRIAL MANAGEMENT, The Engineering Magazine, has been given the privilege of undertaking this work, just as it was privileged eight years ago to publish the first complete description of Mr. Ford's work at Highland Park. No bigger task could we undertake, nor one we are sure of more promise and profit to our readers, than to try to make clear the basic principles of successful quantity production and of practically frictionless management, as exemplified at River Rouge. The series as now planned, will start in September and will continue for ten or twelve months. Profusely illustrated with photographs of plant, equipment and individual operations, it will cover every important phase of activity of interest to our readers.

River Rouge is *more* than the home of the remarkable Fordson—more than a gigantic machine shop and an enormous foundry with a host of supplementary industrial units. *It is the exemplification of tested Ford principles.* Hardly five years old and most of it, in fact, of much more recent construction, it epitomizes all of the industrial principles that Ford and his engineers have found successful during twenty years of phenomenal and unparalleled success. Nowhere else have engineering genius and improved machinery been applied more effectively to secure quality production at minimum cost.

River Rouge may be called a complete exposition of the most effective use of modern production, transportation and transmission machinery applied to quantity production. Every particle of work that can possibly be done better or more quickly by machinery than by hand, is done by machinery.

Even the major problems of management, strange as this statement may seem, have been solved, at this plant, by machinery and engineering. The twenty-two thousand workers at River Rouge require an astonishingly small amount of supervision. Clerical and administrative labor has been reduced to an almost unbelievable minimum. Routing and dispatching are handled as engineering, rather than as management problems, and solved by machinery, through transportation, rather than by "management" through supervision. Stock keeping presents no unsolved problems because the various processes, from the ore pile to the finished tractor on its freight car, have been mechanically timed to such a degree that production is a continuous flow *without the need of stock reserves.* Management difficulties that force themselves into the foreground at the average plant, are non-existent or at least invisible at River Rouge. Henry Ford and his engineers have made sure that the big basic principles are right and stay right so that details that would otherwise annoy and perplex seem to arrange themselves without obvious effort.

Aside from all this, the River Rouge Plant of the Ford Company is of absorbing interest as a unique example of a substantially complete industrial economic chain. Ore from the Ford iron mines hauled on Ford's railroad is converted into pig at the Ford blast furnace and made into castings at the Ford foundry. Coke for foundry and blast furnace use is produced at the Ford coking ovens. Gas for steam-electric power production, ammonia, distillates, tar and illuminating gas are produced at the Ford by-product house and gas house. All of these units in the industrial chain are geared up as one huge and complete machine at River Rouge, presenting one of the clearest and most instructive examples of present day ultimate industrial efficiency that can be found anywhere in the world.

Mechanical transportation forms the links which unite these separate units (and the separate operations within the units) into one smooth-working, continuous process. Starting with the Hewlett and Bridge cranes that unload and transfer ore and limestone to the twelve million tons capacity concrete storage pits, through the interplant conveyors that feed product parts, fuel and supplies in continuous coördinated streams, everything is kept moving—*by machinery*. Even the molds in the foundry are poured "on the move," shaken out of their flasks "on the move," and snatched, red hot, by conveyors to platforms where they are snagged and cleaned—"on the move." Dry sand cores are baked "on the move." The Machine shops are virtually huge conveyors through which the product parts flow with but momentary stops at the necessary machines. Tractor assembly is a mechanical moving stream with feeder branches from which a complete tractor is born and comes to life each forty seconds. The spirit of the plant impresses the trained observer as an allegorical picture—a picture of Engineering Genius wrestling with Time, for the prize of Accomplishment.—The Editors.

[b]

Vol. LXIV. September, 1922 No. 3

INDUSTRIAL MANAGEMENT
The Engineering Magazine

WRITTEN FOR INDUSTRIAL EXECUTIVES AND MECHANICAL ENGINEERS

DEVOTED TO LABOR-SAVING MACHINERY AND LABOR-SAVING MANAGEMENT

# What Ford is Doing and How He Does It

## An Editorial Announcement of First Importance

***By The Editors***

ON page 131 of this issue is begun the detailed account of the most significant individually-owned industrial activity undertaken since the building of the pyramids—the description of Henry Ford's newest plant, and what is done there.

This series—the first complete relation of the astonishing activities of the Ford Motor Company at River Rouge—will deal with demonstrated principles, actual practice and definite facts rather than with economic philosophy or interesting personality. The keynote will be *what* Ford is doing and *how* he does it, rather than *why* he does it.

The first outstanding fact about Henry Ford's work is that he makes it pay—makes it show a profit. This fact alone makes the account of his work profitable reading for executives who are faced, as their first requirement of industrial survival, with the necessity of also making a profit.

There are those who state that the desire for profits is the chief impelling motive back of the Ford operations, that the principles and practice applied to Ford undertakings have been cleverly and farsightedly planned toward this objective. They who hold this opinion point to the fact that this company earns greater total profits than any other individually owned concern in the world.

Again there are those who hold that Ford is an industrial prophet who cares nothing for money and who desires profits merely in the measure that they enable him to demonstrate his economic principles, of which even the most skeptical admit profit earning power. Those of this belief point out as evidence the high wage—short working day—low cost—low price—tremendous production cycle, already demonstrated by Ford at Highland Park, and the still more advanced principles now being demonstrated at River Rouge.

There are those again who hold Ford to be a practical philanthropist whose chief object has been and is to give his consumers the most automotive value for their dollars, and his employees the best remuneration for their work.

From the standpoint of the publication of the series in this magazine it matters not which, if any, of these motives is the impelling one. As engineers and executives we are not concerned with the question of *motives*, but with the analysis of *methods* and *results*. Therefore, in this series, we shall seek to "know the man by his works" which is much safer than forming an opinion from that which is said either by friends or enemies.

The second fact that stands evident is that the Ford success is not accidental. Quite a number of people thought at one time that it *was*; that circumstances and the fortunate choice of the right men as aids had driven Henry Ford to the position that he occupies. Luck may build huge edifices over night but it never runs consistently for twenty years. Each year since 1903 (with the exception of the war years 1917-18) has seen a big but consistent increment in Ford production. Each year has seen Ford in more secure and independent position, culminating in his remarkable financial coup in 1921, when he successfully and dramatically steered his huge craft through the breakers of deflation * and tight money.

Mr. Ford's success is not accidental, nor is it impossible of duplication in large or small scale. It is based, not upon chance, nor freak of genius, nor mystic power. but upon the intelligent interpretation and application of

* See Industrial Management, September, 1921, page 129.

certain basic principles. Genius indeed is required to pioneer in the interpretation of these principles, as Mr. Ford has done, but less is required once the example has been provided.

That this is so is proven by the fact that there are today some 80,000 Ford employees in various Ford activities, in the Detroit district alone. Years ago, therefore, this business was past the "one-man management" stage. It is run by *organization,* and upon the principles interpreted by Mr. Ford but now applied by many subordinate officials to many varied branches and departments of the Ford business. If this were not so it would have ceased to function properly long since; for one man, no matter how superlative his ability or exhaustless his energy, cannot carry the task of personally and continually directing 80,000 workers. This significant fact should convince any one that the principles which Mr. Ford has developed and applied so successfully are applicable by other men and to other industries—when and where these principles are thoroughly grasped and understood. It should remove the misconception that "these things are all right for Ford and the Ford plants, but they won't work anywhere else!"

Another misconception that the reading of this series will remove is that regarding the use of machinery at the Ford plants. It is true that they are the most highly mechanicalized plants in the world and that in them the process of integrated manufacture has reached its present day apex. But contrary to somewhat general impression the bulk of mechanical equipment at the Ford plant is not complicated *special* machinery, but *standard* machinery—of best grade and carefully selected, it is true, but such as is available to any manufacturer. There is no doubt but that River Rouge contains the most impressive aggregation of high grade mechanical equipment that one could find within the boundaries of a single manufacturing plant and that the various pieces of equipment are coördinated and balanced with extreme nicety; but the great majority of individual machines are instantly recognizable by the engineer as old acquaintances. It is their *use* that is novel.

Nor is the impression correct that the class of labor attracted by the Ford wage scale and bonus plan is such as to supply a *superman* personnel either in physical quality, skill or average mentality. The labor average is probably neither above nor below that in the better factories of Detroit, and is quite representative of what may be found in the average well-conducted plant anywhere in America. It is true that the turnover is much less than that found in the average plant and that hence the employees as a whole have the advantage of a longer familiarity with the plant conditions and methods, but there is nothing in the superman idea. One sees as many physically light men as one does heavy men at River Rouge, but one also sees that the man is placed at work best suited to his physical characteristics.

HERE then we have a remarkable profit earning institution in which success has not been accidental, nor due to mystic personal power. It has been attained with the sort of men and machinery that may be found in the better classes of American plants everywhere. It is a success based upon the definite following of definite principles, which as evidenced by the continual growth of the organization, are self-building and self-prepetuating. It is a success due to Henry Ford, the man who discovered these principles, interpreted them and applied them; but made independent of Henry Ford or any other man by virtue of the fact that these basic principles are economic truths not difficult to apply when one fully understands them.

It is the function of this series to develop these economic principles which Mr. Ford has interpreted and applied with such astounding success. They will be developed gradually as the series proceeds, because to understand them one must first understand what Henry Ford and these principles have accomplished at River Rouge.

Aside from the interest which our readers will have in the delineation and explanation of these success-building principles, there are other phases of this account whch are so significant and instructive as to merit pointing out at the start. One of these phases is waste elimination. No more striking example of this could be found than River Rouge.

THE thoroughness of waste elimination at River Rouge goes as far beyond the average conception of waste elimination as Ford's conception of a complete industrial plant, as exemplified at River Rouge, goes beyond the average conception of the present day industrial plant. Mr. Ford's conception of a motor car and tractor plant is not limited to a highly efficient machine shop plus a highly efficient foundry. It goes away beyond this and embraces all of the steps back to the natural resources. Likewise his conception of preventable wastes is magnified in like proportion. Losses are eliminated through coördination of functions in this complete industrial chain—losses which cannot in their nature be either apparent or corrected in the one-step or two-step plant although they exist nevertheless and take their toll both of profits and purchasing power. Here is an economic theory being rapidly demonstrated through actual practice, a theory of immense significance—a theory which challenges the generally accepted theory of large scale independent specialization, and hence a demonstration of the utmost interest and importance to executives, economists, engineers, and even legislators. For if this theory works out successfully for such differing engineering products as the Ford car, the Lincoln car and the Fordson tractor, it will inevitably compel sooner or later a complete revision and reorganization of our great industrial undertakings and will thus vitally affect the work and future of every man who is connected with industry or who profits by it.

We must be pardoned for emphasizing and repeating so strongly the importance to our readers of this series. We honestly believe that it is a series of which no man seriously interested in industry of any kind can afford to overlook a single chapter. We hold no brief for Henry Ford or for the products of his company. But we are thoroughly alive to the significance of his work, awake to the fact that it is the most significant industrial undertaking yet recorded and appreciative of the privilege of being able to present this detailed account to our readers.

**[C]**



# Ford Principles and Practice at River Rouge

## I--Links in a Complete Industrial Chain

*By John H. Van Deventer*

Mr. John H. Van Deventer graduated from Sibley College, Cornell University in 1903, with the degree of M. E. after which he served a special apprenticeship at the Niles Works of the Niles-Bement-Pond Company. Entering the Cost Department of the Goulds Manufacturing Company of Seneca Falls, N. Y., in 1905, he soon became manager of costs and later, superintendent of production.

In 1907, he became general superintendent of the Buffalo Forge Company, remaining in this capacity until 1914, at which time he entered the field of technical journalism becoming associate editor of the American Machinist. In 1915, he was made managing editor of that publication and, in 1916, its editor-in-chief.

Called to Washington in 1917, to assist in organizing the Ordnance Department, Mr. Van Deventer was commissioned a major and served as assistant to the chief of the Engineering Division. Upon returning to his editorial chair in 1918, he organized the Army Ordnance Association.

In 1920, he became the active editor of Industrial Management and in 1921, was made president of The Engineering Magazine Company. He is the author of a number of books on management and shop practice.

It would be difficult to select any industrial plant that presents more features of absorbing interest to engineers, industrialists and executives than does the River Rouge plant of the Ford Motor Company.

River Rouge is "all things to all men."

To the transportation expert, it offers the last word in methods of material handling. The chemical engineer discovers in its coke oven and by-product plant, practice and equipment that is second to none. The metallurgist who specializes in steel and iron encounters a quality of design, construction and operation in the blast furnace and auxiliary equipment, that compels unstinted admiration. The power plant engineer finds a model of economical and reliable power production in the huge main power station and in its six operating substations. The foundryman sees at this plant, the largest and most completely mechanicalized foundry in the world. The mechanical engineer in turn, observing the principles and methods applied to machining and assembling the Fordson tractor, notes ways of conserving human energy and of utilizing machinery, that are at the forefront of present-day shop practice. And the man who handles men sees at River Rouge a maximum of sustained accomplishment with a minimum of supervision.

> The sparkling water of a great inland sea lapping playfully upon white sand which meets the foot of red hued cliffs, dotted here and there with gray rock outcroppings and crowned with forests of spruce, hickory and pine. A primeval scene created by the hand of God.
>
> A smooth white roadway winding ribbon-like between the hills; traversed by a speeding motor car. A modern scene created by the hand of man.
>
> Centuries upon centuries apart are these two pictures—the ancient cliffs and the modern motor car.
>
> Steam and power from the water, glass from the sand, ore and coal from the cliffs, limestone from the gray rocks and lumber from the trees.—and thus the motor car and tractor.
>
> Henry Ford has brought the hand of God and the hand of Man closer together at River Rouge than they have ever been brought in any other industrial undertaking.

Beyond all this, however, the experienced observer of industrial undertakings catches a glimpse of something larger and more significant. He sees these units not only in their impressive individual and astounding collective magnitude, *but he also sees each unit as part of a huge machine*—he sees each unit as a carefully designed gear which meshes with other gears and operates in synchronism with them, the whole forming one huge, perfectly-timed, smoothly operating industrial machine of almost unbelievable efficiency.

BEFORE proceedng in this introductory article to acquaint the reader with the general physical arrangements of the plant as a whole, it will be well to tell what River Rouge stands for; to state, in other words, its reason for existence. Some have the impression that it is a war plant, hurried into being in 1917 for the purpose of building Eagle boats. On the contrary, this plant is the result of the determination of Henry Ford to eventually control every step in the process of making Ford products. It was planned long prior to the war, and a goodly portion of it built before 1917. It is true that construction was hurried in order that Eagle boats might be built here as part of Ford's 100 per cent participation in industrial war work, and that the "B" building, now used for tractor assembly and other operations, was constructed and located particularly for shipbuilding. But war or no war, River Rouge was destined to come into being as the home of a large number of completing links for the Ford industrial chain. This plant, like the Ford iron and coal mines, the Ford forests, the D. T. & I. Railroad and the plate glass plant now building at Flat Rock, Mich., are all links in the same industrial chain—a chain that is being rapidly completed and that will shortly put within Ford control every step from the primary raw materials to the finished cars, trucks and tractors.

In getting a proper conception of the existing plant, it must be remembered that fixity and progress are not compatible in an organization which so continually and rapidly progresses in its ideals and aims. A description that is accurate today will be incomplete a month from today and perhaps obsolete a year from today. For new units are constantly being added and older ones modified or improved. Even as this first article is being written, construction is being rushed upon a steel converting unit of the largest and most modern type, whose electric furnaces will be in operation long before this series is completed.

Again, there is an instance which shows the almost ruthless following of the torch of progress, and the obedience to the dictates of efficiency. Due to its more recent design, construction and equipment and its particularly favorable situation with regard to fuel, the River Rouge power plant is able to produce power more efficiently and economically than it can be produced in the power plant at Highland Park, the home of the Ford cars. Therefore the great Highland Park station—one of the show places of Detroit—with its nine sets of six thousand horse-power gas-steam engines and its tiled floors and plate glass windows, is doomed. This power plant, itself of quite recent design and construction and undoubtedly one of the most beautiful power plants in the world—a plant producing and transforming a total of some 60,000 horse-power (enough to supply the needs of a city of half a million population for water, gas, electricity and ice)—this enormous engineering achievement whose cost ran into the millions must retire in favor of its younger, turbine-equipped brother at River Rouge. This latter plant when the proposed additions to it are completed, will produce and transform upwards of one hundred thousand horsepower of steam and electric energy, and will be the central power station for all of Ford's Detroit industries.

The first impression of the visitor to River Rouge is the same that will be experienced by the reader when he examines the panoramic views that accompany this issue as a folded insert. What an enormous plant, and what a multitude and diversity of buildings and facili-

**Fig. 1. The southwest end of the plant. At the left is the "High Line"; next to it the Coke Screening Building, the D. T. & I. Shops, the Coal Breaker with conveyor leading to the Coke Oven Coal House from which the coke ovens may be seen extending on both sides. At the right, beyond the Coke Oven stack, is the quenching tower where the incandescent coke is cooled.**

**Fig. 2. The Great Central Power Building at the River Rouge Plant is shown at the right. Its four stacks, eleven feet in inside diameter, rise to a height of 330 feet—twice the height of Niagara Falls! When the equipment planned for is installed there will be four more such stacks and a power production of over 100,000 kw.**

ties! Extending in length a mile and one half, and with a depth of three quarters of a mile, the 660 acres comprising the plant proper* are dotted with some 42 buildings which house the industrial activities.

This somewhat confusing first impression begins to clear when longer study and more familiarity with the purposes of the various buildings leads one to subdivide the plant into certain broad natural units. A glance at the accompanying diagram which outlines the location of the various principal buildings will be helpful at this point. (See reverse side of insert.)

STARTING with the Rouge River as a natural beginning as well as a plant boundary, (for a good share of the ore, coal and limestone arrives by water), one sees a great slip and turning basin one third of a mile long and with a depth of 25 feet, and having dockage for 12 coal or ore carriers. Railway tracks skirt this slip.

Next, and running for a full half mile along the river, one sees enormous concrete, open—top storage bins 25 feet in depth, extending back from the river more than two hundred and fifty feet. These huge bins, having a total capacity of *over twelve million tons,* form the primary storage for coal, iron ore and limestone, and assure a reserve of basic raw materials sufficient to meet any emergency. Two huge Hulett ore unloaders run along the margin of the slip traversing the entire length of the storage bins, as does a Mead-Morrison Coal Unloader. These machines reach into the holds of ore and coal carriers and transfer their cargoes in eight and ten ton bites, to the primary storage bins. Tow gigantic transfer bridges, also of Mead-Morrison construction and five hundred and fifty feet in overall length, traverse the slip and ore bins and serve not only to ballast material within the bins but in addition to transfer materials from the huge primary bins to the secondary, or "active" storage on the "High Line."

This High Line may be called the backbone of the plant. It is a concrete structure resembling a viaduct 40 feet high, of sufficient breadth to carry five standard gage railroad tracks abreast upon its upper surface, and extending for three quarters of a mile parallel to the slip and in the direction of the long axis of the plant. At one end a ramp leads down from the High Line to the normal ground level, permitting trains to be run to and from the upper surface of the High Line. The outer track, overhanging as it does the primary storage bins and being of open girder construction, permits bottom-dumping cars to be emptied directly into the primary storage bins.

The two principal functions of the High Line are active storage and distribution. By active storage is meant the short time storage of materials for current use as distinguished from their long time storage in the big concrete primary storage bins. To carry out this function of active storage, hopper bins are located in the High Line below the upper surface railroad tracks. These are filled by transfer cars which have themselves been filled by the ore bridges. Those of these bins on the Northern section of the High Line, being designed to serve the blast furnaces, empty by gravity into weigh cars which in turn fill the furnace skip buckets and which run on tracks located within the structure of the High Line but on the lower or ground level.

Hopper bins located on the Southern section of the High Line serve as points of delivery for coal and coke

* The Company owns thousands of acres beyond the river between the Rouge and the Dearborn plants, sufficient to expand the present plant ten times if desired.

of certain sizes, coal here commencing its conveyor journey to the coke ovens through breaker, crusher and pulverizer houses; and coke, smaller than foundry size, being delivered to transfer cars and thence to storage at this point.

The mechanical features serving the high line are so adequate and extensive that, supplemented by fifty miles or railroad track within the plant, it is possible to transfer or to place carloads or even trainloads of materials into or out of storage or to any part of the plant in an incredibly short time.

Aside from its principal functions of storage and distribution, the High Line, which is really an elongated concrete building of one story height, is made to house many activities. For example, one finds under it the pipe shop, flue shop, light machine shop, plate rolling shop and stock room of the D. T. & I. Railroad. Here also is the power plant stock room, the pyrometer repair and testing station, power substations, paint storage, etc. The structure is so solidly built that practically no vibration is felt when a loaded train of cars passes over it.

THERE is one building that will be mentioned, a bit out of sequence, at this point. It is one of the few buildings on the river side of the High Line. Known as the "B" building, and designed particularly for the construction of Eagle boats, it at present houses the assembling and a part of the machining of the Fordson tractor, as well as the millwork, the body stamping and partial body assembly of "tour-car"; sedan and "coop" as the coupé is familiary called.

This building, which is a four-story structure, is one third of a mile in length and contains 30 acres of floor space. A central bay which provides crane service for all floors runs lengthwise of the building, dividing it into two halves. This is perhaps the only building on the grounds that was not designed especially for its present work. While it has been made quite adaptable to its present purposes, it is not improbable that the work it now houses will some day be performed in one story buildings more suitable for the purpose and not quite so remote from the contributing units.

In addition to the Storage, High Line and "B" Building, which have been mentioned, the plant may be subdivided into the following units:

Coke and By-Product Plants
Power Plant and Sub-stations
Blast Furnace and Cast House
Foundry
D. T. & I. Shops
Electric Converter Plant (under construction).

In addition to these principal units there are others, such as a cement plant, now building, which will produce cement from blast furnace slag; a paper mill in which paper for plant and publication uses will be produced, etc. These, however, do not come within the scope of this present series.

One question which comes to the mind of the engineer who visits River Rouge is "Is this plant an example of Ford engineering and planning, or was it planned by outside engineers and consulting experts?" The answer to this is simple. The planning was done and is done by members of the Ford staff. Outside engineering talent is naturally made use of in connection with construction details, but the general plan is entirely a Ford product.

In this connection it is interesting to note the thoroughness with which a contemplated construction of any magnitude is studied before carrying it through. In the administrative office there is a wooden model, perhaps eight or ten feet in length, in which, to reduced scale, every important feature of the steel converter plant and that section of the High Line that has to do with it are reproduced. This model was built from blueprints before any construction was undertaken and was studied with great care by Mr. Ford and the Ford engineers. Every movement of material from and to one part of the unit or another was planned in miniature actuality on this model, even the slag cars being

**Fig. 3. A view showing two sides of the largest and most remarkable foundry in the world. Moving platforms and conveyors are installed throughout for both molds and cores. Fifteen thousand tons of iron must be melted daily to take care of the present requirements of Ford cars, trucks and tractors.**

**Fig. 4. Since Mr. Ford took over the D. T. & I. Railroad, its locomotives have been transformed into models of motive power. This shows one end of the general repair and assembling shop of the D. T. and I., located at River Rouge.**

reproduced in wood so that their functioning might be followed from start to finish. It is this careful and accurate preliminary planning—*in the solid*—that helps to explain the surety and speed of Ford construction work when once undertaken.

A brief description of the principal plant units is all that can be attempted in this introductory article, leaving their more detailed description for later issues.

The coke and by-products plant follow the Semet-Solvay system. There are two batteries of 60 ovens each, extending as wings from a central coal storage building. Electric charging cars run along the top of each battery, dropping a load of 16 tons of pulverized coal into each oven that requires filling. An electric "pusher" running parallel to the ovens on the ground level on one side, pushes a finished oven every eleven minutes with clocklike precision, the charge of incandescent coke falling into a huge electrically driven hopper car on the opposite side of the ovens. Coke and by-product plants have been in operation at River Rouge since October 1919. It goes without saying that the process, both from a physical and chemical standpoint has been perfected to a high degree. Particularly is this true with respect to transportation. At present, 2000 tons of coal are coked every 24 hours. The by-products secured from the coke oven gases are:

Illuminating gas
Coal tar
Motor benzol
Ammonium sulphate

One not conversant with the subject might question the policy of the Ford Motor Company in building and operating a complete coke and by-product industry instead of buying the materials needed. In the ordinary plant, even of exceptional size, it is true that such an undertaking would be out of the question. But River Rouge is no ordinary plant, particularly when considered as a part of the huge Ford industrial chain which keeps over 80,000 men at work. Of every fifteen hundred tons of coke produced from these ovens, 1270 tons are of furnace and foundry size—both of these sizes usable in the blast furnace and foundry units. Of the remaining sizes, the egg, nut and pea are sold to employees for domestic purposes, while the breeze, mixed with powdered coal, forms fuel for the boilers at Highland Park.

OF THE illuminating gas produced (between 21 and 22 million cubic feet in 24 hours) 43 per cent is needed for maintaining the temperature of the coke ovens. Three-fourths of the 57 per cent surplus remaining is used in the Ford industries for such purposes as heating core ovens, heat-treating furnaces, etc., at Highland Park, being pumped through a ten-mile pipe line. The surplus of 13 per cent is sold to the Municipal Gas Works of Detroit.

The 10,000 gallons of benzol produced every 24 hours make it unnecessary for the company to buy gasoline for trucking or testing purposes, and in addition, is used in the manufacture of artificial leather for upholstery. The only product that is not of direct use in the Ford industrial chain is the ammonium sulphate, which is sold for fertilizer.

It will be seen from the foregoing that the coke oven

**Fig. 5. At the left is one block of the coke ovens, with pipes leading therefrom to the by-product plant. An average of 1500 tons of coke is produced daily, while the by-product plant turns the waste gases into twenty-two million feet of illuminating gas, ten thousand gallons of motor benzol (gasoline) and tons of tar and fertilizer.**

and by-product plants at River Rouge form an exceedingly useful part of the complete industrial mechanism, and one that contributes materially to its operation on a low cost plane.

Strategically located adjacent to the High Line, from which it receives its supply of iron ore, furnace coke and limestone, the Blast Furnace unit is a notable part of the River Rouge plant. Two furnaces, each with a capacity of 500 tons every 24 hours form the present equipment, with space provided for a third. One furnace is operated to supply present demands which approximate 400 to 450 tons of pig metal per day. As soon as the converter plant is completed, these requirements will be materially increased.

Space in this introductory article will not permit of an adequate description of the design and operation of this unit which might be aptly termed a smokeless and dustless blast furnace plant, nor of the ingenious means by which the molten pig metal is led to the foundry and mixed with the scrap and riser remelt from the cupolas, thus eliminating entirely the cost of remelting pig-iron—a big item in foundry cost. So interesting and instructive are these methods that an entire article will be required to describe them.

It *is* in proper order, however, to speak at this time of the way in which the blast furnace unit gears up with some of the other units in the complete industrial machine. Obviously, of course, it serves the foundry as the source of pig metal. Quite as obviously, it is served by the coke-ovens in the matter of coke. In addition to this, however, the blast furnace unit serves the River Rouge power plant as the principal source of fuel, the blast furnace gases, after thorough cleaning, being pumped to the combustion chambers of the boilers and thence having their heat energy transformed into electrical horse power. An average of 36 million cubic feet of this gas is thus consumed daily.

No possible source of profits in the form of heat, of power, of basic materials or of waste or scrap is allowed to go long unharnessed at River Rouge, as is indicated by the construction, now taking place, of a cement plant which will use blast furnace slag as its principal ingredient.

The foundry at River Rouge produces all of the iron, brass, bronze, and aluminum castings consumed by the various Ford plants. It goes without saying that an enormous building is required for this purpose, even when one takes into consideration that continuous pouring and the use of moving platform and conveyor systems throughout, utilizes floor space many times more efficiently than it is used in a foundry operating on the customary stationary floor and one heat per day plan. Notwithstanding its efficiency of operation, however, so great are the demands made by a production of four thousand cars and five hundred tractors per day, that this building is the largest foundry in the world. It covers 1,119,100 sq. feet of floor space. At present an average of 1500 tons of iron are being poured daily.

The heart of any industrial undertaking is its power plant. River Rouge has a strong heart. The floor space covered by the main power building is over two hundred thousand square feet. The roof of the building is 133 feet above the ground level, and four enormous stacks, eleven feet in inside diameter and more than 330 feet high, form a landmark that identifies River Rouge from miles away.

As previously said, this power plant is to be the central power station for all of the Ford industries in the Detroit district. It is equipped with 4 Geo. T. Ladd boilers rated at 2600 horse power each, but capable of developing 10,000 horse power. The combustion system, designed by the Combustion Engineering Company and known as the "Lupulco," utilizes both powdered coal and blast furnace gas, the latter being introduced at the bottom of the combustion chamber, while the coal drops into the incandescent fire box from an elevation. The combustion is practically perfect, without ash, cinders or smoke. Fifty tons of powdered coal and 36 million cubic feet of blast furnace gas are consumed daily.

The present generating equipment consists of two 12,500 Kilowatt Curtis-General Electric units. Steps are being taken to shortly supplement this equipment with two 30,000 k. w. units. Blast for the blast furnace line is provided by three Curtis-G. E. turbine compressors with a combined capacity of 45,000 cubic feet per minute.

When finally completed, the floor of the River Rouge power plant will be covered with white tile. It will never, however, be as beautiful or impressive as the present Highland Park power house, for a turbine cannot convey the impression of power and motion that is conveyed by a huge reciprocating engine. But it is exceedingly efficient and clean and so cool, quiet and free from vibration that a banquet might be comfortably served either in its boiler room or generator room!

There is another power plant at River Rouge which is of interest as illustrating the policy of waste utilization. This plant, as will be noted on the diagram, is located near the "B" building. The sole fuel for this plant is sawdust, shavings and wood scrap, coming from the mill work and wood working operations carried on in the "B" building in connection with body manufacture for Ford cars. This wood in the first place is a Ford product, growing as timber on top of the earth beneath whose surface are the Ford iron-ore mines. Thus these forests not only supply the raw material for the millwork on Ford bodies, but also furnish fuel for all of the power necessary to operate the entire body plant!

I have said that the first impression of River Rouge is of vastness and complexity. The second impression is that of motionless quiet. Unless one enters a building, there does not seem to be much going on. One does not see many people about the yards, perhaps the reason being that space is so vast that men are less noticeable. As a matter of fact comparatively few men are required outside of the buildings because of the extent to which mechanical transportation has been developed. Covered conveyors criss-cross the huge plant in every direction, particularly at the West End where coke and coal flow in continuous streams. At first sight it seems impossible to untangle in one's mind the courses and functions of this multitude of conveyors, but after a little study they resolve themselves into logical paths arranged for coke and coal. These conveyors in the yard at River Rouge strike the keynote of the plant—*transportation*—keeping the product continually on the move—a keynote that is carried out within the buildings as well as outside of them.

This then is River Rouge as one might see it upon a hurried trip and without examining any of the buildings or processes in detail. It is an interesting plant indeed, but beyond and above its interest, a *significant* one. It is not an experiment, from the dollars and cents standpoint, but a sound, practical, money-making fact. And yet it *is* an economic experiment, or better, a demonstration of what may be done by the complete control and coördination of all of the steps in a great industrial chain. Some day, perhaps River Rouge will be the battle ground upon which is challenged the interpretation of the time-honored law of Supply and Demand. To the observer it appears, consciously or unconsciously to be moving in that direction, forming a striking object lesson of what the removal of pyramided profits upon successive steps of manufacture, and the elimination of waste, can do to the final selling price of a product and to the purchasing power of the dollar.

**Fig. 6. As an indication of the scale of operations at River Rouge, here is one section of the pump house of the by-product plant. The battery of General Electric-DeLaval centrifugal pump units, which stretches as far as the eye can see, is used just for circulating ammonia liquor for by-product extraction.**

[d]

Vol. LXV. May, 1923 No. 5

INDUSTRIAL MANAGEMENT
The Engineering Magazine
WRITTEN FOR INDUSTRIAL EXECUTIVES AND MECHANICAL ENGINEERS
DEVOTED TO LABOR-SAVING MACHINERY AND LABOR-SAVING MANAGEMENT

# Ford Principles and Practice at River Rouge

## IX—Machine Tool Arrangement and Parts Transportation

*By John H. Van Deventer*

Mr. John H. Van Deventer graduated from Sibley College, Cornell University in 1903 with the degree of M. E., after which he served a special apprenticehip at the Niles Works of the Niles-Bement-Pond Company. Entering the Cost Department of the Goulds Manufacturing Company of Seneca Falls, N. Y., in 1905, he soon became manager of costs and later, superintendent of production.

In 1907, he became general superintendent of the Buffalo Forge Company, remaining in this capacity until 1914 at which time he entered the field of technical journalism, becoming associate editor of the American Machinist. In 1915, he was made managing editor of that publication and, in 1916, its editor-in-chief.

Called to Washington in 1917, to assist in organizing the Ordinance Department, Mr. Van Deventer was commissioned a major and served as assistant to the chief of the Engineering Division. Upon returning to his editorial chair in 1918, he organized the Army Ordinance Association.

In 1920, he became the active editor of Industrial Management and in 1921, was made president of The Engineering Magazine Company. He is the author of a number of books on management and shop practice.

**The machining of castings at the River Rouge plant is a splendid example of intensive production on a large scale. Parts for five hundred Fordson tractors and eight thousand cylinder block castings for Ford cars are machined here every twenty-four hours. This remarkable production is made possible by a highly coordinated system of intra-machine conveyors, which knit the separate operations into one continuous process.—The Editors.**

TO ONE who is familiar with machine shop manment and operation the cast-iron machine department at River Rouge is of exceptional interest. One becomes accustomed at this plant to encountering the unusual and to finding reasons, practical, logical ones, for the existence of unusual practice. Nowhere about the plant does one find more departures from the customary way of doing things than in the huge department where the castings produced at the River Rouge foundry are machined.

The very location of this department is unusual. It is housed in the same building in which the moulds and cores are made and in which the castings are poured, cleaned and tumbled. In fact, at this plant the machining is considered an integral part of the making of castings. Machining starts where the tumbling barrels leave off, the whole process from core making to final inspection of machined castings being knit together by a "close coupled" system of intra-operation transportation.

This practice is brought about by strict adherence to the principle of integrated manufacture. The more continuous and uninterrupted are the product streams, the less chance there will be for delays and accumulation of stock. Hence, why place an unnecessary gap between the cleaning operations and the machining operations; why consider these things as belonging in distinctly separate families or groups and to be housed in separate buildings? Particularly so when dust removal systems for foundry and cleaning rooms have been perfected to such a stage that the danger of accelerated depreciation of expensive tools is practically done away with once for all.

Fig. 1. Fordson cylinder blocks enter the machine shop on the gravity conveyor shown above. The first operation, milling tops and bottoms, is performed on station type milling machines.

The casting machining department at River Rouge, therefore, has been placed under the same roof that houses the molding, pouring and cleaning operations. It occupies the eastern portion of the foundry building, its floor space covering a strip some 1200 feet long and 150 feet wide, running north and south. It is separated from the foundry proper by walls through which openings are provided for conveyors which bring in the cleaned castings and by doors which are used for trucking purposes.

If you will refer to the article entitled "Plant Facilities of the World's Biggest Foundry," published in September, 1922, you will see on page 134 a general floor plan of the foundry building in which the machine shop space is designated. A portion of this floor plan has been redrawn to larger scale in Fig. 2 of the present article to show the transportation connections between the tumbling barrels and cleaning mezzanines and the machine tools. The layout of machines is in groups, each group being comprised of sufficient balanced equipment to completely finish all of the operations required on one type and style of casting. There are thirty-two of these groups at present in operation, beginning at the extreme south end with the Fordson tractor cylinder blocks (which enter the machine department by means of the gravity roller conveyor shown in Fig. 1.) and terminating at the north end in the group assigned to the machining of Ford car cylinder head castings.

It will be noted from Fig. 2 that the direction of flow of the incoming castings is from west to east, at least until these castings meet the rough stock conveyor—a wood slat distributing conveyor which runs from south to north for nearly the entire length of the western border of the machine department. Castings coming from the foundry are distributed by means of this rough stock conveyor to the proper machining units, being switched into the unit by means of intra-machine conveyors of various types which carry them from one operation to the next, finally delivering them at the finished stock conveyor which marks the eastern border of

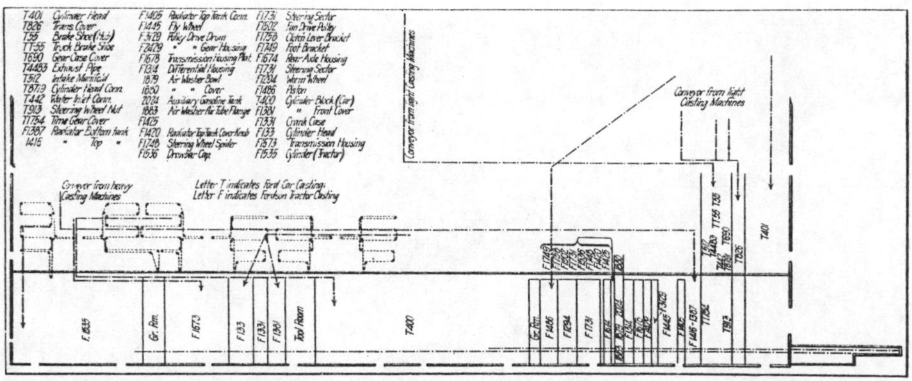


**Fig. 2. In this diagram one may visualize the arrangement of machine groups in the machining department at the River Rouge foundry. Each group of tools is properly balanced to handle all of the machining operations on the casting assigned to it.**

the machine department, and which is close to one-third of a mile in length. The intra-machine conveyors which tie the machine tools of each group or unit into one continous production process or huge coördinated machine, are so numerous that it is impossible to even indicate them in this diagram. They can be seen, however, very clearly in nearly all of the pictures which accompany this article.

Thus the direction of flow of product is in general from west to east into the department, then from south to north to the machine group or unit, then from west to east and from south to north through this unit and finally from south to north on the finished stock conveyor which leads to the car loading platform. Some idea of the perfection which has been reached in coördinating this movement of stock may be had from the fact that in the group devoted to machining Ford car cylinder blocks, and in which forty-three distinct operations are performed, it requires not more than one hour of elapsed time for the complete travel of the cylinder from the rough stock conveyor to the finished stock conveyor through all the operations involved. Some two thousand cylinders "in process" are required to comfortably "balance" this process but when one considers that the daily output is eight thousand cylinders of the T-400 or Ford car type it will be realized that this ties up a remarkably small investment in work in process.

Both the huge production quotas obtained and the cutting to the bone of this stock in process investment are due to the principle of integrated manufacture or "continuous flow of product," which in turn is only possible because of the conveying systems. Still another production economy may be attributed to these systems and that is the elimination of any possibility of loafing or soldiering on the job when each operator is faced with the necessity of keeping up with the procession or else seeing his stock piled up to a point where it becomes distinctly noticeable by the immediate management.

Transportation being the key-note in this department as in all others at River Rouge, one finds the machine tools placed strictly in accordance with the working out of this principle. A man who is accustomed to the space usually allowed between machine tools in average shop practice would say that the River Rouge departments were crowded and congested. If you will examine the photographs presented here with this point in mind your first thought as that of the visitor will be: "how do men find room to operate in this dense mass of tools and conveyors?" Fig. 10, for example, is a good illustration of this point.

The answer to this is that in the average shop a great deal of floor space is unnecessarily wasted, or, to express it better, not fully used for production. Under the usual operating conditions in the average plant a machine operator is required to take many steps that have been eliminated at River Rouge where "the work moves and the men stand still." An old time machinist might feel himself decidedly cramped if confined to the space allotted him in this machine shop. Inasmuch as the majority of operators at River Rouge, however, are specialists who perhaps have never even seen a machine tool before their employment by the Ford Company, they have no precedents or ingrained habits with respect to tool operation and they soon become accustomed to carrying on their operations in the space provided, which in all cases is carefully figured to avoid the drawback of actual crowding.

Again, in the average plant, very wide passageways must be maintained for the movement of trucks and the passing back and forth of people. At River Rouge there is no trucking because of the complete conveying systems which frequently are arranged (as in Fig. 7) to pass *over* the tops of machine tables thus providing transportation of stock free of floor space cost. Again, in the average plant, space must be left for an accumulation of work between machines. At River Rouge this is largely eliminated because the immense production permits the splitting up of operations into their barest elements, and this in turn permits of a very accurate solution of the ordinarily difficult equation of machine balance. The system of automatic and semi-automatic transportation, which is so evident in the Ford plant depends upon huge quantity production as a requisite, thus the extremely close placing of machines and the

Fig. 3. A good idea of the elaborate system of intra-machine conveyors which characterize this shop is had from the above picture. This is also part of the tractor cylinder block machining group

elimination of stock in process cannot be hoped for to this extent in a plant less fortunately situated with respect to production quotas. However, the principle of tying up successive operations wherever possible into one coördinated process, by means of some form of mechanical transportation, is an objective which, if adhered to, is bound to increase production and decrease costs whether the means utilized be conveyors, lift trucks, overhead trolleys or other forms of handling equipment.

A striking example of what a properly coördinated transportation system will do toward reducing the stock in process investment is given in Fig. 11. In this picture you see in the foreground the finished stock conveyors on the eastern margin of the machine shop. On a week day these conveyors are as crowded with finished castings as the boardwalk at Atlantic City is with pedestrians on a summer holiday. But in this picture all that you can see are a few milling cutter heads on the outer conveyor. This photograph was taken on a Sunday and the fact that the finished stock has completely disap-

Fig. 4. Starting with the milling operations in the first picture, the final work on the tractor cylinders is performed on the main bearings. This picture shows the North end of the tractor cylinder group where babbitting, studding and drilling operations are performed.

Fig. 5. A few yards to the north along the main aisle we come to the tractor transmission housing group shown above. In the foreground is a portion of the finishedstock conveyor that extends nearly the entire length of the plant.

peared from sight is striking proof of the ability of the coördinated transportation system to keep even the busiest shop "cleaned up" to the minute. Here is a department handling perhaps what is the greatest output of machined castings of any plant in the world, turning out a daily production of five hundred completed tractors and eight thousand Ford cylinder blocks, and yet a few moments after the last shift leaves on Saturday night this huge department, thanks to automatic transportation, is free from the slightest accumulation.

Individual motor drive is the rule at River Rouge in this machine department for the large and medium-sized machine tool. Here again we have a distinct production advantage, for due to the absence of overhead belts, space above the machines can be fully utilized in aiding production. In Fig. 13, for example, you will note the steel super-structure built over the planer type millers, operating on the tops and sides of the Ford cylinder blocks. Suspended on this super-structure are hoist trolleys utilized for greater facility in loading and un-

Fig. 6. Next in order is the tractor crankcase group with its heavy station type mills for the surfacing operations. At the left, also, is the tractor cylinder head unit.

Fig. 7. This is the first glimpse of the busiest machining unit of all, that devoted to a daily production of eight thousand Ford car cylinder blocks. Note the conveyor running above the miller tables.

loading castings from the work table.

The smaller tools are group driven from jack shafts driven in turn by motors located on steel structures, such as shown in Fig. 12. There are over one thousand electric motors in this one huge department, which is one of the most completely motorized machine shops in the United States.

The problem of cutting tool maintenance in a plant where production is forced to the pitch that it is at River Rouge is a serious one. The area covered by the department makes it impractical to depend upon a centralized tool grinding department. Hence, small decentralized tool grinding groups are distributed throughout the department at the points where the demand for tool renewals is greatest.

Tool repair is a function which comes under the jurisdiction of the general tool department which is located in the B building a half mile distant. Traveling gangs of repair men are dispatched from here to whereever they may be required, their chief duty being to prevent the cause of shutdowns. As an important aid to this objective the question of machine oiling is given strict attention. Oiling is not done by the machine operators but by specially designated employees also under the jurisdiction of the tool room to make regular tours throughout the plant. Almost every tool before it is installed in the Ford machine shops is fitted with gravity feed multiple oiling systems, specialists being employed for this work and kept continually busy at it.

Since the operators in this department are one-operation specialists they are not equipped with sufficient knowledge or training for tool setting, nor is this expected of them. This comes under the control of various tool setters who are responsible for quality and sustained output of the groups to which they are assigned, and of which they act as group foremen. The tool setters do not come under the jurisdiction of the tool room but report to the superintendent of the machine shop.

The conveyors used for intra machine transportation in this department are of wide variety. They vary from the specially designed cylinder block "scenic railways" with individual seats provided for each casting to simple, smooth sheet iron surfaces, lubricated with oil or cutting compound along which the castings are slid from one machine to another. Miles of inclined and horizontal gravity roller conveyors are to be found in this department. Whenever possible the shape of a casting is availed of to aid transportation as in the case of the Fordson pistons shown in Fig. 12, in which the ability

**Fig. 8. This battery of four-spindle boring machine rough bores the car cylinder blocks. Especially designed conveyors carry the castings to and away from the machines.**

of a cylinder to roll is cleverly utilized and a conveying system established thereupon with the aid of parallel angle iron tracks.

In a department where such a multitude of tools are continuously cutting metal the problem of chip disposal becomes a serious one. Wherever possible chips are handled by conveying systems as shown in Fig. 14. This is a double conveying system, the one at the top, which is driven by chain, consisting of individual cylinder block carriages which keep the battery of four-spindle cylinder boring machines supplied with work. Beneath this upper conveyor and mounted on the same frame is a continuous belt conveyor for handling chips produced on these machines. Thus, product transportation and chip handling are accomplished with the utilization of the same floor space area. These borings at present are returned to the foundry and are melted, first being packed in sheet iron cylindrical containers. Eventually they will go to the Sintering plant to be agglomerated with the blast furnace dust into chunks for remelting.

Tool cribs for the storage and issue of small tool drills, taps and reamers required in this department, are kept in inclosures as located in convenient positions. One of these cribs is seen at the extreme right in Fig. 4, being adjacent both to tractor cylinder machining unit and to the tractor transmission housing unit.

The pictures published with this article were taken from various positions along the main aisle which runs outside of the finished stock conveyor along the eastern edge of the machine department. They present this department as it would be seen by a visitor who walks from south to north along this aisle. The first picture, Fig. 1, was taken at the extreme south and looking west toward the cleaning room. Inside of the gravity roller conveyor which brings the tractor cylinders from the tumbling barrels are huge station type milling machines on which the tops and bottoms of the Fordson cylinder blocks are machined. These castings (unlike the T-400 car cylinder blocks) are too heavy for hand lifting and hence one finds in this unit overhead facilities for swinging the castings from conveyor to machine and from machine to conveyor.

As the visitor passes a few steps north along the aisle he comes to the view shown in Fig. 3. This is still a part of the tractor cylinder machining unit. In the foreground may be seen a camshaft hole boring machine, and at the extreme right multiple drills for valve tappet and valve stem hole drilling. Overhead works for lifting castings extend over this portion also.

Fig. 4 shows a further portion of the tractor cylinder

Fig. 9. Drilling operations on the tappet and valve stem holes and boring and facing of ports is performed on the battery of sixteen-spindle drills shown in the above picture.

machining unit, the space here being devoted to drilling, studding and babbitting the cylinder block main bearings. Notice that the roller conveyors are arranged to bring the work to a convenient operating height, so that there need be no stooping over on the part of the operators.

Continuing still further northward along the finishing stock conveyor, one comes to the tractor transmission housing job, shown in Fig. 5. Here the axle flange and the transmission shaft holes are bored and faced on heavy single spindle boring machines.

Next one comes to the tractor crankcase machining unit. The weight of this part is not sufficient to call for overhead lifting works. The operations are performed on station type milling machines and on drills with the exception of turning and facing the transmission flange which is performed on the single spindle boring machine shown at the extreme left of Fig. 7. The loop cylinder conveyor shown in this same fixture marks the beginning of the Ford "T-400" car cylinder block operations, the first of which after spotting, are milling the cylinder bot-

Fig. 10. How fully every inch of floor space is utilized in this machining department is indicated by the above picture which represents the group of special bcring lathes for the car cylinder camshaft hole.

Fig. 11. This is a general view from the nothern end of th e car cylinder block machining group looking toward the Southwest. The six machines in the middle foreground are hydraulic testing machines for the final water test of the cylinders.

toms and form milling the main bearings. Still further north we come to an impressive battery of four-spindle boring machines shown in Fig. 8. Paralleling this battery is another conveyor loop which brings the cylinders to the boring machines from the preceding operation.

An extensive vista of multi-spindle drills is next encountered stretching the full width of the shop. This is shown in Fig. 9 and is where the drilling operations are performed on the valve stem and tappet holes of the Ford car cylinder block. Each drill is provided with sixteen spindles in two rows of eight. One row is in operation at a time, the work being traversed in its holding fixture to bring the second set of spindles into action. One of the sheet metal work slides, previously mentioned in this article is to be seen in this picture.

I have previously referred to the battery of camshaft hole boring lathes shown in Fig. 10 which is next encountered by the visitor. The cylinders come to these machines on the conveyors shown in the left foreground. They are next slid upon the sheet iron work tables in the

Fig. 12. The tractor piston job is a good example of Ford practice in light machine work. Note the skeleton steel structures which support the jack shafts and countershafts for group drive of the lighter tools.

Fig. 13. The large and medium-sized machine tools at this plant are driven by individual electric motors. Thus the overhead space is left free for utilization by work handling devices.

rear of each machine and after the operation is completed are pushed along on the same tables to the next conveyor route seen at the right.

Passing along further the visitor turns backward and sees the battery of hydraulic testing machines just back of the finished stock conveyor in Fig. 11. From this point on to the north end of the plant the finished stock conveyor is a double line, the additional conveyor being needed to take care of the car block cylinder output which terminates its journey through the machining unit at this point.

Next to the cylinder block machining unit we come to the various groups devoted to machining the lighter tractor parts. The first of these is the tractor piston group shown in Fig. 12.

At the extreme north end of the department the finished stock conveyors pass through openings into the loading shed. This extends outside of the building between spurs of railway track, and doors are arranged on either side of the shed so that cars may be conveniently

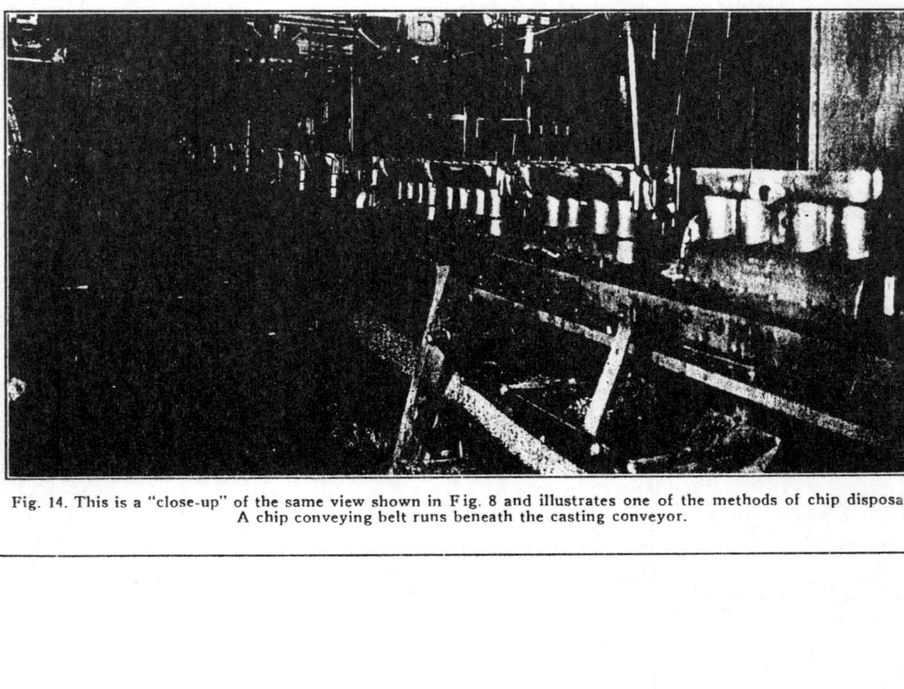

Fig. 14. This is a "close-up" of the same view shown in Fig. 8 and illustrates one of the methods of chip disposal. A chip conveying belt runs beneath the casting conveyor.

**Fig. 15. At the northern end of the building the finished stock conveyors enter a car loading shed extending between railway tracks. The castings are loaded into box cars for delivery to Highland Park or to the B building.**

loaded. All of the output of castings machined in this department are carried to their destinations in box cars, whether this be the Highland Park plant or the B building at River Rouge where tractor assembly is performed.

From the foregoing description and the photographs accompanying it, the reader will obtain a fairly clear idea of the size and general arrangement of this remarkable machining department at River Rouge. It has been impossible in this one issue to enter into the details of machine operation. To give a complete description of Ford machining methods as exemplified by the "T-400" car cylinder block and the tractor piston--typical parts which represent heavy and light casting iron machine work—will be the function of the succeeding article.

---

CONFRONTED with unusual problems of production and labor affecting the business structure of the entire country, the Nation's Industry will meet in annual conference on May 14, 15 and 16 at the Waldorf-Astoria Hotel, New York City. This will be the occasion of the twenty-eighth annual convention of the National Association of Manufacturers, whose membership is composed of great and small companies from coast to coast and engaged in almost every conceivable line of industry.

Every effort is being exerted to provide a program that will be a distinct milestone on the highway of industrial accomplishments that stand to the record of the National Association of Manufacturers in its history of more than a quarter of a century. Some of the pressing subjects to be discussed are transportation, pioneer of industry; construction, which establishes communities; production, which expands those communities; open shop, which guarantees the right to work to every citizen; immigration, upon which the country is more dependent today for its rough labor than at any time in its history; taxation, some forms of which are putting a curb on ambition and expansion; merchant marine, which every growing nation must have.

Some of the particular features already arranged are:

Industrial Production Session: This will consider every phase of the industrial situation of to-day. There will be addresses on how to improve and stabilize the present production trend and suggestions will be offered looking to maintaining sane production during the usual slack months of the summer.

Open Forum for Manufacturers: This will be conducted with the express view of giving to individual manufacturers who have constructive suggestions an opportunity to present them for the consideration of the Association. It will be the first time that the floor of the convention has ever been thrown open generally to the members, and officials of the Association are confident that it will be productive of many valuable and progressive suggestions.

[e]

# XI—Machining Tractor Parts and Assembling Fordson Tractors

*By John H. Van Deventer*

Mr. John H. Van Deventer graduated from Sibley College, Cornell University in 1903, with the degree of M. E. after which he served a special apprenticeship at the Niles Works of the Niles-Bement-Pond Company. Entering the Cost Department of the Goulds Manufacturing Company of Seneca Falls, N. Y., in 1905, he soon became manager of costs and later, superintendent of production.

In 1907, he became general superintendent of the Buffalo Forge Company, remaining in this capacity until 1914, at which time he entered the field of technical journalism, becoming associate editor of the American Machinist. In 1915, he was made managing editor of that publication and, in 1916, its editor-in-chief.

Called to Washington in 1917, to assist in organizing the Ordnance Department, Mr. Van Deventer was commissioned a major and served as assistant to the chief of the Engineering Division. Upon returning to his editorial chair in 1918, he organized the Army Ordnance Association.

In 1920, he became the active editor of Industrial Management and in 1921, was made president of The Engineering Magazine Company. He is the author of a number of books on management and shop practice.

THE two articles which have preceded in this series described the equipment and operations employed in machining cast iron parts in the huge Ford foundry at the River Rouge plant. It now becomes time to introduce the reader to the "B" building, a huge structure one-third of a mile long, situated at the northwest end of the plant. It was here that the Eagle boats were built during the war and it is here that the steel parts for the Fordson tractor are now machined and the assembling operations carried on.

**The "B" building, which houses the operations described in the accompanying article, was originally designed as an emergency building for turning out submarine chasers. The fact that this structure was not specifically planned for its present use, makes all the more interesting the details here given of the layout and operations for producing machined parts and for assembling Fordson tractors.—The Editors.**

The "B" building is perhaps the only structure in the entire River Rouge plant which was not specifically designed to serve its present purpose. It was built as a war-time emergency building and with structural details intended to facilitate the quantity production of submarine chasers. At present it houses the forged and cast steel machining operations on tractor parts, a variety of light automatic and semi-automatic screw machine work, the general tool department, the wood and metal pattern shops, the sheet metal working equipment and wood-working equipment for car bodies and the pattern storage. The building is 1700 feet in width and 308 feet in length, running North and South with a central bay of full height which extends the entire length and is served both by overhead cranes and railway tracks. On either side are mezzanine floors, three to a tier in the southern half of the building and a single mezzanine on either side in the northern half. Each of these mezzanines is provided with occasional extending platforms or bays jutting out under the crane tracks of the central bay, thus permitting heavy crane service to and from the galleries. Roughly speaking, the tractor machining and assembly is located in the southern end of the building while the body work, both sheet iron and wood, occupy the northern half. The tool room and pattern shops occupy the larger part of the entire eastern lower mezzanine.

The process of machining Fordson crankshafts is typical of the attention and care given to splitting up all operations on the Fordson tractor. The rough forg-

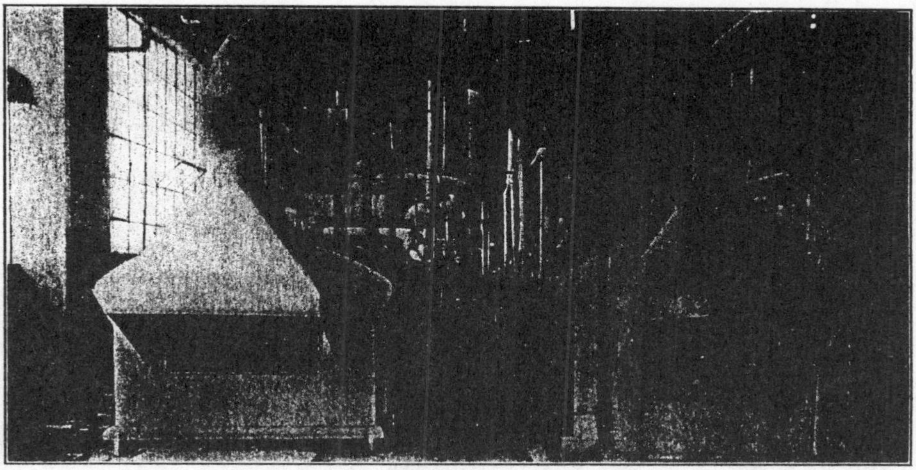

Fig. 1. Forgings for the Fordson tractor crankshaft are annealed and heat treated in the rough before machining. The furnaces shown above are of Ford construction heated by coke oven gas. Work is pushed through by ratchet conveyors.

ings are at present purchased and come to the plant as raw material. They pass through forty distinct operations before reaching completion. These are as follows:

1. Annealing in air furnace
2. Quenching
3. Reheating in air furnace
4. Air cooling on truck
5. Rough straighten on hydraulic press
6. Mill ends to length on continuous miller
7. Center ends on double end centering machine
8. Rough turn rear bearings on lathes
9. Rough turn diameters of flanges on lathes
10. Turn center bearing on lathes
11. Rough turn front bearings on lathes
12. Face center bearings on lathes
13. Rough inside face of flange lathes
14. Finish face flange and rear bearing on lathes
15. Finish bearing pulley and timer gear fit on lathes
16. Finish turn face and groove front bearing on lathes
17. Finish turn diameter of flange and rear bearing on lathes
18. Turn two outside pins on double end crankshaft lathes
19. Turn two center pins on double end crankshaft lathes
20. Recess flange end on lathes
21. Drill six holes in flange and ream two holes on drill presses
22. Restraighten crankshaft on hydraulic straightening press
23. Rough grind center and rough bearing on pin grinders
24. Rough grind rear bearing on plain grinders
25. Rough grind all pins on plain grinders
26. Finish grind two end pins on pin grinders
27. Finish grind two center pins on pin grinders
28. Finish grind center and front bearings on plain grinders
29. Finish grind thrust bearing on plain grinders
30. Grind flange on grinders
31. Finish grind pulley fit on plain grinders
32. Finish grind gear fit on plain grinders

Fig. 2. Rough turning crankshaft throws is accomplished by this simple yet rugged means. Cooling lubricant is circulated throughout the entire department through overhead pipes returning to a separating tank beneath the floor.

33. Drill and ream starting pin hole on drill presses
34. Mill keyway on hand miller
35. Balance on balancing stand
36. Chamfer gear fit on lathe
37. Inspect

Fig. 3. Fordson crankshafts are finished by grinding. The above picture gives a view of machine arrangement in the grinding department and shows the crankshaft conveyor at the left.

38. Polish end pins on double head polishing machines
39. Polish center pins on double head polishing machines.
40. Polish main bearings on double head polishing machines

Fig. 4. One of the interesting features of the gear cutting department is the multiple station type bevel gear hobber. This machine makes six cuts at a time indexing and revolving automatically.

These operations are listed in detail so that the reader may see how the Ford policy of splitting operations is carried out on even a comparatively small quantity production, the average number of tractors made daily at the present time being about five hundred—a small quantity indeed when compared with the daily output of five thousand Ford cars that is now being accomplished.

A wood slat conveyor, similar to the rough and finished stock conveyors described in the preceding article carries the finished crankshafts and other tractor parts North toward the assembling lines. Distributed through the crank machining department are gravity roller conveyors similar to that shown in Fig. 3 which convey the pieces from one operation to the next.

Operations on the crankshaft are held to extremely close limits, the crank pin diameter allowance being, for example, one-half of one one-thousandth of an inch. There are seven inspections which take place on the crankshafts at various stages and which safeguard the quality of production.

Immediately above the crankshaft machining department on the mezzanine level is the gear and pinion department. Here all of the blanks for the Ford steel gears used in the transmission are roughed, finished and cut. There are eight gears in the transmission of the Fordson tractor, giving a required total output of more than 4,000 per day, so that the machine equipment of this department is rather extensive. Heat treating furnaces and baths are located on this mezzanine, for all of the gears used in the transmission are heat treated. Inter-machine transportation is provided by means of gravity roller conveyors, such as shown in Fig. 6. The machine tools are arranged in orderly rows paralleled by these conveyors, with trucking aisles occasionally provided between rows of machines.

An interesting departure recently adopted at this plant in gear hobbing practice which will be of interest to the machine shop man is the use of a triple thread roughing hob, the installation of which practically doubled the rate of output on spur pinion gashing.

Lack of space will prevent a further description of the machine operations on the Fordson tractor parts, some

**Fig. 1. Forgings for the Fordson tractor crankshaft are annealed and heat treated in the rough before machining. The furnaces shown above are of Ford construction heated by coke oven gas. Work is pushed through by ratchet conveyors.**

ings are at present purchased and come to the plant as raw material. They pass through forty distinct operations before reaching completion. These are as follows:

1. Annealing in air furnace
2. Quenching
3. Reheating in air furnace
4. Air cooling on truck
5. Rough straighten on hydraulic press
6. Mill ends to length on continuous miller
7. Center ends on double end centering machine
8. Rough turn rear bearings on lathes
9. Rough turn diameters of flanges on lathes
10. Turn center bearing on lathes
11. Rough turn front bearings on lathes
12. Face center bearings on lathes
13. Rough inside face of flange lathes
14. Finish face flange and rear bearing on lathes
15. Finish bearing pulley and timer gear fit on lathes
16. Finish turn face and groove front bearing on lathes
17. Finish turn diameter of flange and rear bearing on lathes
18. Turn two outside pins on double end crankshaft lathes
19. Turn two center pins on double end crankshaft lathes
20. Recess flange end on lathes
21. Drill six holes in flange and ream two holes on drill presses
22. Restraighten crankshaft on hydraulic straightening press
23. Rough grind center and rough bearing on pin grinders
24. Rough grind rear bearing on plain grinders
25. Rough grind all pins on plain grinders
26. Finish grind two end pins on pin grinders
27. Finish grind two center pins on pin grinders
28. Finish grind center and front bearings on plain grinders
29. Finish grind thrust bearing on plain grinders
30. Grind flange on grinders
31. Finish grind pulley fit on plain grinders
32. Finish grind gear fit on plain grinders

**Fig. 2. Rough turning crankshaft throws is accomplished by this simple yet rugged means. Cooling lubricant is circulated throughout the entire department through overhead pipes returning to a separating tank beneath the floor.**

33. Drill and ream starting pin hole on drill presses
34. Mill keyway on hand miller
35. Balance on balancing stand
36. Chamfer gear fit on lathe
37. Inspect

Fig. 3. Fordson crankshafts are finished by grinding. The above picture gives a view of machine arrangement in the grinding department and shows the crankshaft conveyor at the left.

38. Polish end pins on double head polishing machines
39. Polish center pins on double head polishing machines.
40. Polish main bearings on double head polishing machines

Fig. 4. One of the interesting features of the gear cutting department is the multiple station type bevel gear hobber. This machine makes six cuts at a time indexing and revolving automatically.

These operations are listed in detail so that the reader may see how the Ford policy of splitting operations is carried out on even a comparatively small quantity production, the average number of tractors made daily at the present time being about five hundred—a small quantity indeed when compared with the daily output of five thousand Ford cars that is now being accomplished.

A wood slat conveyor, similar to the rough and finished stock conveyors described in the preceding article carries the finished crankshafts and other tractor parts North toward the assembling lines. Distributed through the crank machining department are gravity roller conveyors similar to that shown in Fig. 3 which convey the pieces from one operation to the next.

Operations on the crankshaft are held to extremely close limits, the crank pin diameter allowance being, for example, one-half of one one-thousandth of an inch. There are seven inspections which take place on the crankshafts at various stages and which safeguard the quality of production.

Immediately above the crankshaft machining department on the mezzanine level is the gear and pinion department. Here all of the blanks for the Ford steel gears used in the transmission are roughed, finished and cut. There are eight gears in the transmission of the Fordson tractor, giving a required total output of more than 4,000 per day, so that the machine equipment of this department is rather extensive. Heat treating furnaces and baths are located on this mezzanine, for all of the gears used in the transmission are heat treated. Inter-machine transportation is provided by means of gravity roller conveyors, such as shown in Fig. 6. The machine tools are arranged in orderly rows paralleled by these conveyors, with trucking aisles occasionally provided between rows of machines.

An interesting departure recently adopted at this plant in gear hobbing practice which will be of interest to the machine shop man is the use of a triple thread roughing hob, the installation of which practically doubled the rate of output on spur pinion gashing.

Lack of space will prevent a further description of the machine operations on the Fordson tractor parts, some

**Fig. 5. In the Fordson gear department tools are arranged in orderly rows usually paralleled by gravity roller conveyors for handling the parts. Trucking aisles are occasionally provided.**

of which are extremely interesting and unusual as, for example, the connecting rod work and the front axle bearing spindle taper turning and grinding.

So we will proceed now to the assembling department where one finds perhaps the most interesting and spectacular demonstration of the rapid growth of a complicated machine, starting with a few simple parts and ending with the familiar completed Fordson.

It requires an average of some two hours and thirty-nine minutes to completely assemble a Fordson tractor on the moving assembly lines. This is the average elapsed time from the moment that the cylinder block is placed upon the first cylinder assembling line to the time that the completed machine, fully painted and ready to run under its own power, descends from the conveyor rails of the paint line.

I have described in previous articles the principles of integrated manufacture of which the Ford Company's plants provide the foremost examples and which have proved to be miracle-working principles in their effect on cost and production. So that the reader may have his mind refreshed on this point let me state again that these principles are based upon the arrangement of work in continuous and regularly flowing streams of

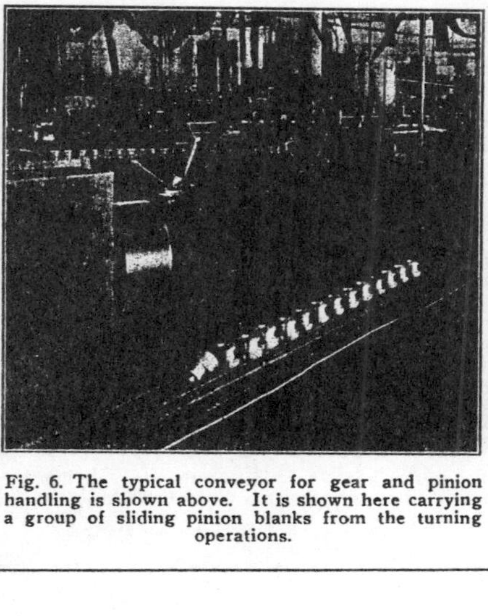

**Fig. 6. The typical conveyor for gear and pinion handling is shown above. It is shown here carrying a group of sliding pinion blanks from the turning operations.**

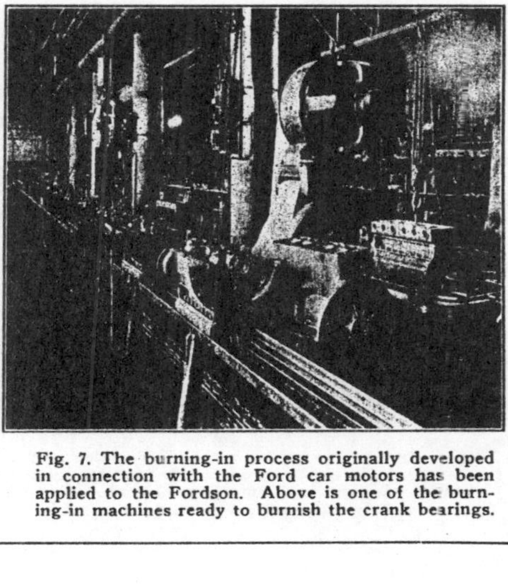

**Fig. 7. The burning-in process originally developed in connection with the Ford car motors has been applied to the Fordson. Above is one of the burning-in machines ready to burnish the crank bearings.**

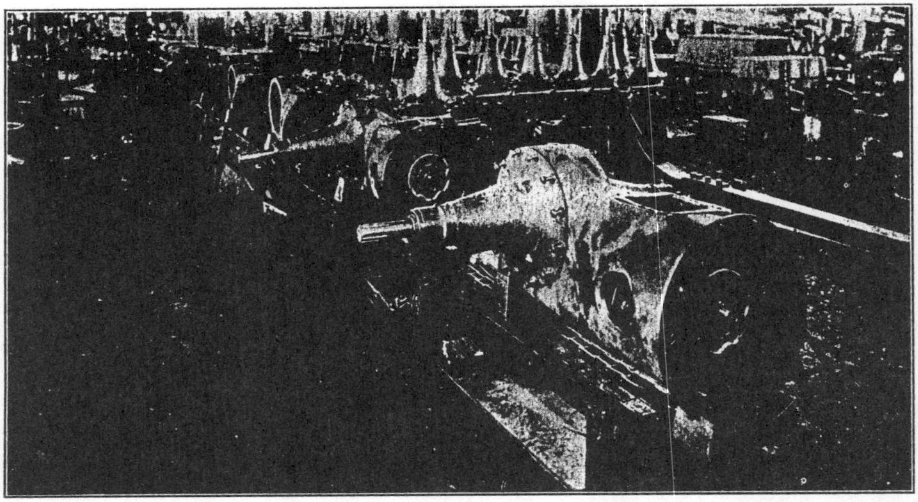

Fig. 8. This is a view of the transmission and rear axle assembly lines where the "rear end" of the Fordson is put together. This and the motor assembly comprise the two principal units.

successive operations on various parts, which come together, predeterminedly, at the right place and the right time, and which finally result in a completed product delivered in a very nearly constant but controllable rate. A system of this sort must be accompanied by a complete installation of conveyors or means of material and parts handling. The operations must be timed in accordance with the speed of these conveyors which are the mechanical pacemakers and which automatically eliminate time losses. A system of this sort requires a huge output to begin with and a large investment in conveying equipment. Once secured, however, and when applied to a suitable product made in a sufficient quantity it is the most efficient method of manufacturing yet developed, because it eliminates not only intentional "soldiering" or loafing but the even commoner and more costly time-wastes which are constantly incurred in the majority of plants through the workers not being supplied with the right materials at the right time Not only does the integrated system of manufacture largely eliminate these time losses, but it also reduces to a very minimum the amount of stock-in-process, thus minimizing the investment and releasing capital for other work. At River Rouge, for example, where

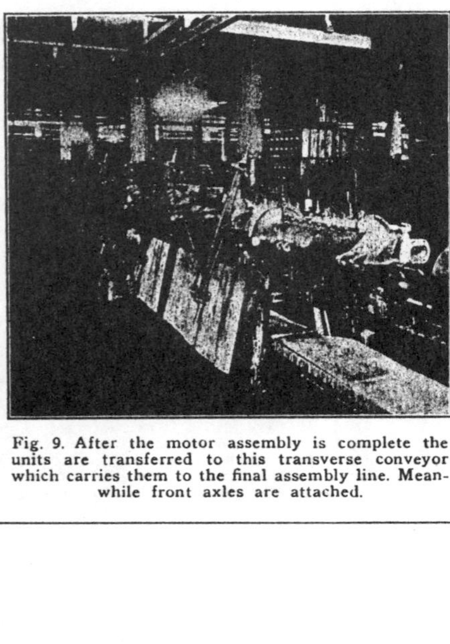

Fig. 9. After the motor assembly is complete the units are transferred to this transverse conveyor which carries them to the final assembly line. Meanwhile front axles are attached.

Fig. 10. A view of the second motor assmbly line on which piston, valve, magneto and flywheel assembly is completed. A magneto assembly line may be seen in the left background.

**Fig. 11. At the beginning of the final assembly line the two principal units, the motor and the transmission are brought together. From this point on, the Fordson rapidly progresses toward its completion.**

the integrated system of manufacture is applied from the ore and the coal piles clear through to the finished tractor, less than thirty days of stock requirements are necessary to balance the production schedules. You will immediately see the tremendous savings possible on such huge production quotas as are involved in the Ford industry when you compare this thirty-day tying up of capital with the ninety days to six months' period ordinarily necessary in plants where the production is not handled on this basis. It was through this process of integrated manufacture and its inherent ability to work on a close margin that Mr. Ford was enabled during the financial stringency to release some $30,000,000 from stock-in-process to fluid cash.

At no other point in the manufacturing cycle can one visualize the possibility of the savings due to these methods as vividly as when watching the moving assembly lines. As a matter of fact the efficiency of these lines from the mechanical standpoint cannot compare with the less spectacular but highly efficient and more completely mechanicalized flow of work through the machine shop as described in our June issue of INDUSTRIAL

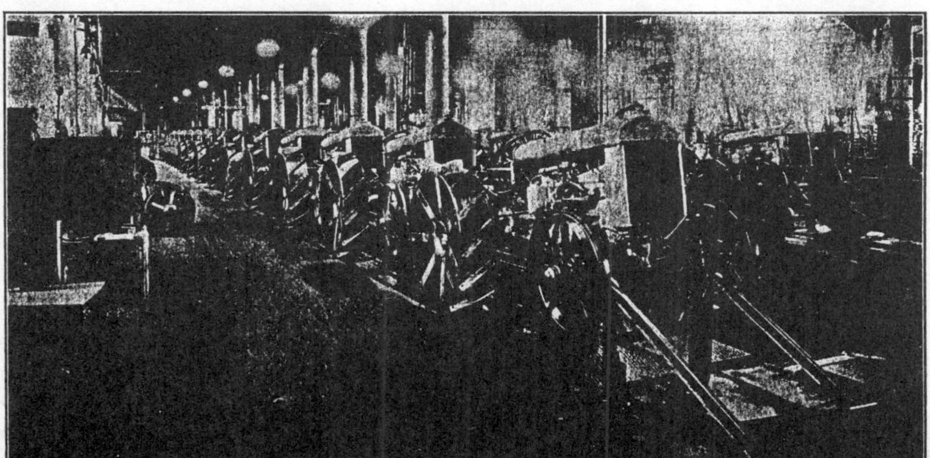

**Fig. 12. This is the end of the final assembly line. The tractor at the right is ready to descend, to be cranked by electric motor and to proceed under its own power.**

Fig. 13. The wood pattern shop covers an extensive area on one of the mezzanine floors of the "B" building. The view shown above embraces but approximately one-fifth of the total department.

MANAGEMENT. In assembly, most of the work is done by hand, the moving lines furnish merely the transportation and set the pace. Each line in assembly is bordered with intense human activity in the form of hundreds of men whose operations, each of the simplest sort, are co-ordinated by continuous conveyor movement. One can thus virtually *see* the growth of a cylinder into a motor, and of housing, gears and shafts into a complete rear end, and finally of motor and rear end units into a completed Fordson. It is this ability to visualize the rapid and continuous growth of a complicated mechanism from its merest elements that so impresses one in watching these assembly lines.

The two principal subsidiary units in the assembly of the Fordson tractor are the motor and the transmission and rear axle housing; the assembly of these units occupies separate but parallel conveyor lines running from South to North. Parallel to them also run assembly lines for such sub-units as magnetos, gear sets, piston and rod assembly, etc. The final assembling lines on which the principal units are brought together run in the opposite direction, namely, from North to South. A transverse

Fig. 14. The metal pattern shop also located on the "B" building mezzanine is shown above. All work going into production is of course plated for machine molding. In addition there are constant repairs required to running patterns.

assembly line runs from East to West, carrying the completed and run-in motors to the origin of the final assembly lines. On this transverse conveyor the front axles are pinned to the motors. The rates of movement of the various lines and the number of men working at them and the number of shifts employed are adjusted in conformity with the demands of the final assembly lines which in reality are the keys or indices to production.

Crankshaft and camshaft assembly occupies the first motor assembly line. A portion of this line midway of its length is shown in Fig. 7. All of these views, of course, have been taken on a holiday. In normal times there is an average of one man for perhaps each four feet of conveyor length on each side of each line. This distance represents the space between the pushing links of the conveyor chain which may be clearly noted in the illustration. Assembling operations are split into their most simple elements, similarly as in Ford car assembly practice. One man, for example, places the crankshaft in its bearing, another puts on the bearing caps, another starts the stud nuts, another operator runs them down with a pneumatic nut driver. Crank bearings are set up tight on this first assembly line so that it is barely possible to turn the crank over by means of a long-handled lever bar fitted with pins to engage in opposite crank flange holes. The camshaft is inserted on this same assembly line and the timing gears are fitted, by which time the block has reached the position shown in the foreground in Fig. 7. There it makes a momentary detour for the "burning-in" process, the machine employed for which is depicted in the same illustration. All Ford crank bearings both in car and tractor are "burnt-in"—a process that consists of running in by power with the bearings tight until the crank heats sufficiently to glaze the babbitt. From fifteen to twenty seconds of an actual running is required to produce a finished bearing job which is said to be even better than that which could be secured by tedious scraping.

After burning-in the motor is placed again on the assembly line and proceeds for inspection, shimming and final adjustment of bearing caps.

Reaching the end of the first motor assembly line the block with crank and camshaft in place is back-tracked to the origin of the second line, a portion of which is shown in Fig. 8. Here the piston and valve assembling operations are carried out as indicated in the foreground. Here also the magneto, flywheels, crankcase, cylinder head, etc., are progressively added, so that by the time the end of this line is reached the motor is complete. As an insurance of proper adjustment and to remove the possibility of friction completely, each assembled motor is run in at high speed by motor-operated machines.

Closely adjacent to the second motor assembly line, but not shown in our illustration, is the piston assembly line where the pistons are assembled in sets of four, the connecting rods pinned to them and the individual sets weighed to secure proper running balance.

After the motors have been sufficiently run in they are transferred to the conveyor shown in Fig. 9. This is the transverse conveyor previously mentioned which runs from east to west and which carries the motors to the final tractor assembly lines. While in transit on this conveyor the front axle assemblies are attached.

While work on the motors has been progressing as described, similar progress has been made on the rear axles and transmission housings, as shown in Fig. 10. These are assembled step by step by gangs of men stationed on either side, each man, as in the motor line, having some specific operation to perform on each unit as it comes to him. Most of these operations have been split into their simplest possible components, therefore

**Fig. 15. Ths is a partial view of the extensive tool room which is the center of repair activity for the entire Rouge plant. Here also the special jigs and fixtures are constructed.**

skill is acquired in performing them in a minimum time. Quite naturally some operations require more skill than others and it is on the more difficult ones that one finds the more experienced and higher paid operators.

Overhead, above the line in Fig. 10, can be seen some of the ingenious, pneumatically-operated gripping and lifting devices which are used to swing these heavy units to and from the line with the least possible expenditure of time and effort. Subsidiary and adjacent assembly lines feed this one with partial assemblies, for example, a procession of rear axle housings may be seen in the right background.

A view of one of the final assembly lines taken from a point near its origin is shown in Fig. 11. Here the two principal units, the motor and the transmission housing, are lifted and swung into position, and bolted together. From this point on the transformation into the finished Fordson takes place at a rapid rate.

These assembly lines are the busiest of all and the visitor is compelled to wonder how so many men can work around them without getting into each other's way. Each man or set of men, however, knows exactly what he has to do and how to do it, and also what the men next to him are doing, so that there is no confusion. Carburetors, manifolds, steering posts, starter boxes, radiators, fuel tanks, front and rear wheels, and drivers' seats seem to fly into position with the smoothness and speed of a moving picture.

EVERY step in the assembly of these tractors has been carefully thought out. As an illustration of this I may cite the method of filling the radiators and of introducing lubricating oil and sufficient motor fuel for the operating test. These acts are performed while the tractor is still "on the line," overhead pipes and hose and automatic measuring devices being provided for the convenient introduction of these fluids.

Perhaps the most fascinating part of the assembly line is the end where the completely assembled tractors descend from the conveyor rails. As each tractor comes off the line it is complete, has been mechanically inspected, and is ready to run. A tester jumps into the driver's seat and adjusts the spark and throttle control. Simultaneously, another operator brings into position in front of the tractor a swinging arm which mounts an electrical cranking device. This device engages the crankshaft extension and at the press of a button spins the motor. Almost without exception the tractors start after one or two revolutions. In driving the tractors under their own power to the paint line, the testers are able to determine if any adjustments are required and to make them.

Painting is performed on chain driven conveyor lines and of course, "on the move." A supply of both red and gray paint is contained in tanks which are built along the sides of the painting conveyors. As in assembly the operations in painting are subdivided into elements, each operator is assigned to cover a certain area or part. Red is used for the wheels of the tractor and gray enamel for the remaining portion. Strips of wrapping paper are laid on either side of the paint lines to catch any splash or drip, and are removed and burnt at regular intervals to prevent the floor from becoming paint soaked, thus minimizing the fire risk.

Descending from the paint lines the completed tractors are driven to parking spaces where the paint is allowed to dry. Thence they proceed under their own power to box cars in which they are shipped. Thus is completed the evolution of the Fordson tractor—a most interesting example of integrated manufacture.

A partial view of the wood pattern shop which occupies a portion of the lower eastern mezzanine in the "B" building is shown in Fig. 13. This view hardly does justice to the size of this department. In addition to the patterns for production castings which are of course of metal, there is so much construction work, special machine building and other "shop work" continually in progress in this enormous plant that a large wood pattern shop is required to take care of it.

A portion of the metal pattern shop is shown in Fig. 14. This takes care of new patterns for production castings and for repairs to running metal patterns. Like the wood pattern shop it is an extensive and well-equipped department, characterized by orderliness, both of arrangement and operation.

A VIEW of a portion of the central tool room located on the same mezzanine is shown in Fig. 15. In this department tool, jigs and fixtures are constructed, and machine maintenance is here centralized for the entire plant. Very few standard cutting tools are made in this department, however, as the Ford policy is to purchese these. A great deal of jig and fixture work and special machine building is done. This picture does not by any means convey an adequate idea of the tool department since it shows only a portion of the benches and a few of the shaping machines. Other groups of machines are arranged to the left but do not show in this picture. For example, there is a complete brass lathe department for machining bushings of all sorts, a complete gear cutting department for repair purposes, a complete unit for screw machine work for jig bushings, and a complete and separate unit for machine tool repairing and overhauling.

This article and the two preceding ones will have given the reader a fair idea of the machine shop methods and equipment employed at River Rouge both for general casting, machining and for the Fordson tractor. In our next article I will describe one of the busiest and most highly specialized parts of the work of the Ford Motor Company, namely, the manufacture of Ford car bodies, now centralized at the River Rouge plant.

---

The operation of the national health insurance act of Great Britain, for the period from 1911, the date of the enactment of the act, to 1921, forms the subject of Bulletin No. 312 just published by the United States Bureau of Labor Statistics.

Through amendments to the original act the insurance covers all persons 16 years of age and over, who are employed in any service for which remuneration is paid, subject to certain restriction. There is no restriction of membership in the approved societies in regard to manual workers, but persons employed at other than manual labor whose annual earnings are in excess of a stated sum are not eligible for membership.

[f]



# Ford Principles and Practice at River Rouge

## XI—The Manufacture of Ford Car Bodies

*By John H. Van Deventer*

Mr. John H. Van Deventer graduated from Sibley College, Cornell University, in 1903, with the degree of M. E., after which he served a special apprenticeship at the Niles Works of the Niles-Bement-Pond Company. Entering the Cost Department of the Goulds Manufacturing Company of Seneca Falls, N. Y., in 1905, he soon became manager of costs and later, superintendent of production.

In 1907, he became general superintendent of the Buffalo Forge Company, remaining in this capacity until 1914, at which time he entered the field of technical journalism, becoming associate editor of the American Machinist. In 1915, he was made managing editor of that publication and, in 1916, its editor-in-chief.

Called to Washington in 1917, to assist in organizing the Ordinance Department, Mr. Van Deventer was commissioned a major and served as assistant to the chief of the Engineering Division. Upon returning to his editorial chair in 1918, he organized the Army Ordnance Association.

In 1920, he became the active editor of Industrial Management and in 1921, was made president of The Engineering Magazine Company. He is the author of a number of books on management and shop practice.

**Obviously, the making of bodies for a production of 5,000 Ford cars per day is a problem of some magnitude. The solution of this problem has been achieved by decentralizing the body assembly—the parts being shipped "knocked down," to various assembling plants throughout the United States. By thus saving the space which would be needed for assembly, it becomes possible to make all of the body parts for the entire American production of Ford cars in the one plant. How these parts are produced on this immense scale at River Rouge forms the subject of the accompanying article.—The Editors.**

ALL of the body parts for the entire American output of Ford cars, with the exception of the coupe, are fabricated in the B building at River Rouge. In itself, this body work constitutes an industry employing over 6,000 men, in spite of the fact that comparatively few bodies are completely assembled in Detroit. For the Ford policy is to decentralize body assembly, to ship carload sets of "knocked down" body parts to the various assembling plants located throughout the United States. In these branch plants the bodies are assembled, painted and fitted to the chassis.

This policy of decentralization has been dictated both from production and cost standpoints. One has but to imagine the enormous floor space that would be required to house in one plant the complete assembly of bodies for a daily production of 5,000 cars, to see the resulting congestion and slowing down of production. From the standpoint of cost, two savings may be mentioned. The more obvious is that of freight charges. Under the present plan, from 120 to 200 complete sets of "knocked down" bodies are shipped in one box car as against a former shipment of from 7 to 9 per car of completely assembled bodies. A less obvious but equally important cost saving will occur to the thoughtful executive who considers the reduction of total cubic footage of storage space enabled under this plan in taking care of delivery schedules. For with decentralized assembly, each body (the bulkiest part of the entire car) is made to occupy minimum space until a few days before its delivery to the customer.

A system of this sort is made possible only through the evolution of a highly standardized body design, particularly adapted for quick and simple final assembly.

Fig. 1. A view of the central bay in the B building. Steel parts are processed under the mezzanine at the right and wood part fabrication and frame assembly take place under and on the mezzanine at the left.

This evolution has been taking place for the past twenty years at the Ford Motor Company and is stilll in progress today. In the case of the touring car and roadster, however, the ultimate would seem to have been reached. What further progress is to be made will perhaps lie in the substitution of sheet steel for wood in the making of various part assemblies which now contain both of these materials.

What one finds in the body plant at River Rouge today, therefore, is a carefully specialized and highly integrated process of manufacture whereby sheet steel and hard wood are transformed with almost confusing celerity into "knocked down" body parts.

One who has not investigated the subject of car body manufacture will be surprised at the number of individual parts entering such an apparently simple construction as a Ford touring car body.

The following list is taken from a stock shipping order and represents the individual parts and partial assemblies required in making up a set to be shipped to a branch assembling plant. Such sets are shipped out in units of 132, 142, 154 and 200 complete cars, depending upon the size of box car available. It must be kept in mind that this list does not attempt to itemize the individual *parts* since several of the items are assem-

Fig. 2. Right and left-hand panelled side frame assemblies for touring car. This is as far as body assembly proceeds in Detroit.

Fig. 3. Wood skeleton frame assembly for four-door Ford sedan.

blies which may require in their make-up anywhere from two to twenty fabricated pieces of wood and steel, not to mention screws, bolts, nuts, washers, etc.

Rear toe board reinforcement
Rear floor board assy
Rear end sill
Rear seat frame side R H
Rear seat frame side L H
Rear seat frame center
Rear seat back panel strainer
Sill panel R
Sill panel L
Rear seat trim rail connection R H (6)
Rear seat back panel flange reinforcement (2)
Rear seat back panel assy
Side and frame panel assy R H
Side and frame panel assy L H
Front seat back spring anti rattler

**Fig. 4. The touring car dash has been recently changed from wood to sheet steel construction. Above are the nine stages in the completion of an all-steel dash showing the piece after each operation.**

Tool box bottom assy
Front seat back panel
Rear seat toe board assy
Front seat frame (2)
Front seat trim strip
Rear seat trim strip side (2)
Rear seat trim strip corner (2)
Front heel board & finishing strip assy
Rear seat back cushion frame strainer (2)
Rear seat back cushion frame lower
Rear seat back cushion frame upper
Front seat trim rail side intermediate R H
Front seat trim rail side intermediate L H
Front seat trim rail side rear R H
Front seat trim rail side rear L H
Rear wheel board & finishing strip assembly
Cowl assembly
Rear seat trim rail rear
Side arm trim block cardboard (2)
Front seat trim rail rear
Rear seat frame bar
Rear seat back cushion frame side
Tool box lining

The list presented above covers the touring car body which is the simplest in the number of assembled parts. The sedan list of stock shipping parts includes 126 items as against the 38 listed above for the touring car.

Roughly speaking, and for purposes of description, the process of body part making divides itself into

**Fig. 5. The first operation on the dash is blanking. This is done from the sheet and results in the piece shown at 1 in Fig. 4.**

**Fig. 6. The second operation is drawing. Note the safety starting control buttons which require the use of both hands.**

**Fig. 7. Flanging is the third operation on the all-steel dash. Operators are not allowed to put their hands between the dies.**

four parts:

1. Sheet metal stamping and forming
2. Wood part sawing and fabricating
3. Framing, paneling, door hanging and fitting
4. Shipping

The sheet metal operations occupy possibly 25 per cent. of the total floor area devoted to body work. The heavy presses are located on the ground floor under the mezzanine on the eastern side of the northern half of the B building. The lighter presses for small sheet metal parts are located on the mezzanine immediately above. The sheet metal arrives in cars near the eastern wall of the building. From temporary sheet storage piles just inside of this wall the work flows from East to West through various progressive "line ups" of presses emerging at the central bay in finished form. Here such parts as are now complete without the addition of woodwork are made directly into sets for shipment to the branch assembling plants. Panels and other parts requiring framing are lifted by crane to the assembling mezzanine on the West side of the building.

Fig. 8. The next operation is piercing, in which the holes shown at 4 in Fig. 4 are punched.

Fig. 9. Next comes the piercing of the holes in the side flange.

Fig. 10. After this the holes in the top flange are punched.

Fig. 11. Next the corner flanges are pierced as shown above.

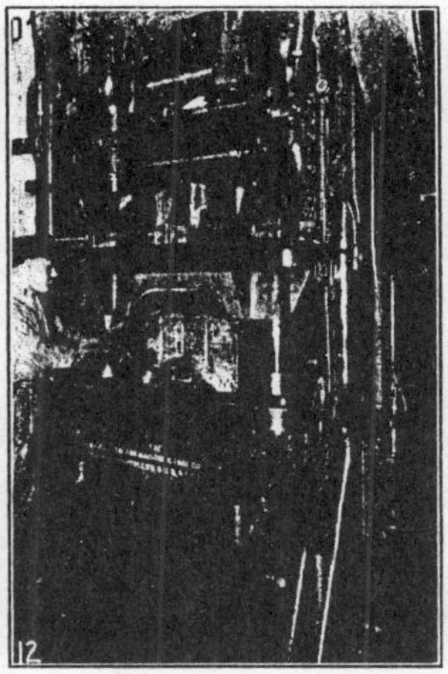

Fig. 12. After this the top flange is embossed for the rain gutter.

Fig. 13. The last operation is riveting on the hood clips.

Hardwoods arrive in various forms, kinds and sizes, both in logs and rough sawn. After kiln drying, this stock proceeds to the gang saws that are located on the ground floor under the West mezzanine at the northern extremity of the plant. Operations on the sedan woodwork progress from North to South on the ground floor level. Operations on the touring car body wood parts take place on the western mezzanine immediately above and also progress in the same direction. The framing and paneling of partial assemblies for "tour car" parts are done on this same mezzanine, the progression also being from North to South. Completed assemblies are brought by conveyor and otherwise down to the ground floor level to the center bay which at present is the car loading station. Plans are being now made to completely conveyorize this transportation. This will include the installation of finished stock conveyors running to car loading platforms outside of the building, doing away with the necessity of employing valuable crane-served floor space for car loading and thus permitting this area to be used for heavy press work.

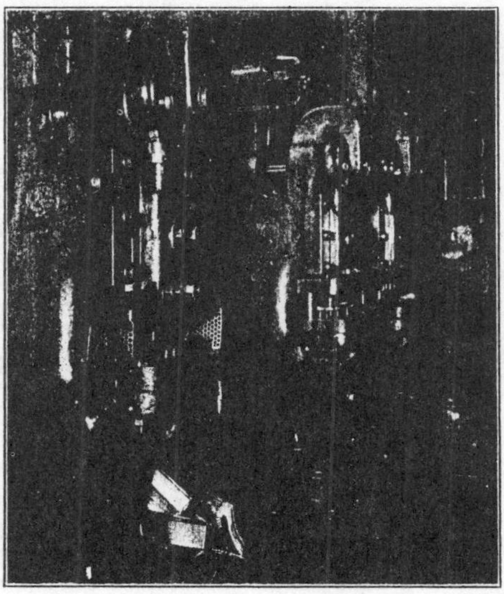

Fig. 14. This is a view of the waste disposal unit which is located across the aisle from the dash line-up.

Having presented this general description of the nature, arrangement and course of work, I will now treat in more detail of the methods employed, starting first with the pocessing of several typical sheet steel parts. Some fifty car loads of twenty gage drawing steel are consumed daily at this plant in the making of body parts. These cars are spotted alongside of various openings in the eastern wall of the B building adjacent to those press "line-ups" which are to utilize sheets of the dimensions contained in the cars. As each car load is received, fifty production parts are stamped or drawn from sheets selected at random from its contents before it is unloaded. This comprises the acceptance test. Then the sheets are unloaded upon four-wheel shop trucks and pulled into the shop by means of Fordson tractors.

Along this eastern inside wall of the building are the temporary storage piles which feed the various batteries of "line-ups" of presses which fabricate the sheet metal parts and which are arranged in distinct progressive groups running from East to West. Production on these presses is balanced so that there is practically no inter-operation storage or piling. There may, of course, be a few pieces ahead of each operator but strictly speaking they work from hand to hand, operations being so timed that it is a case of out of one press and into another without delay. Under such a system while the individual efficiency of one or two machines may suffer and their maximum output not be obtained, nevertheless,

The arrangement of these groups is of course what is termed "fractional." That is to say, each group comprises a little shop for the production of one piece. Each group, therefore, is provided with all of the equipment necessary to produce this piece in the given quantity. As a rule it is arranged to get the quota of production in two shifts, thus providing a one-shift "cushion" or surplus for contingencies.

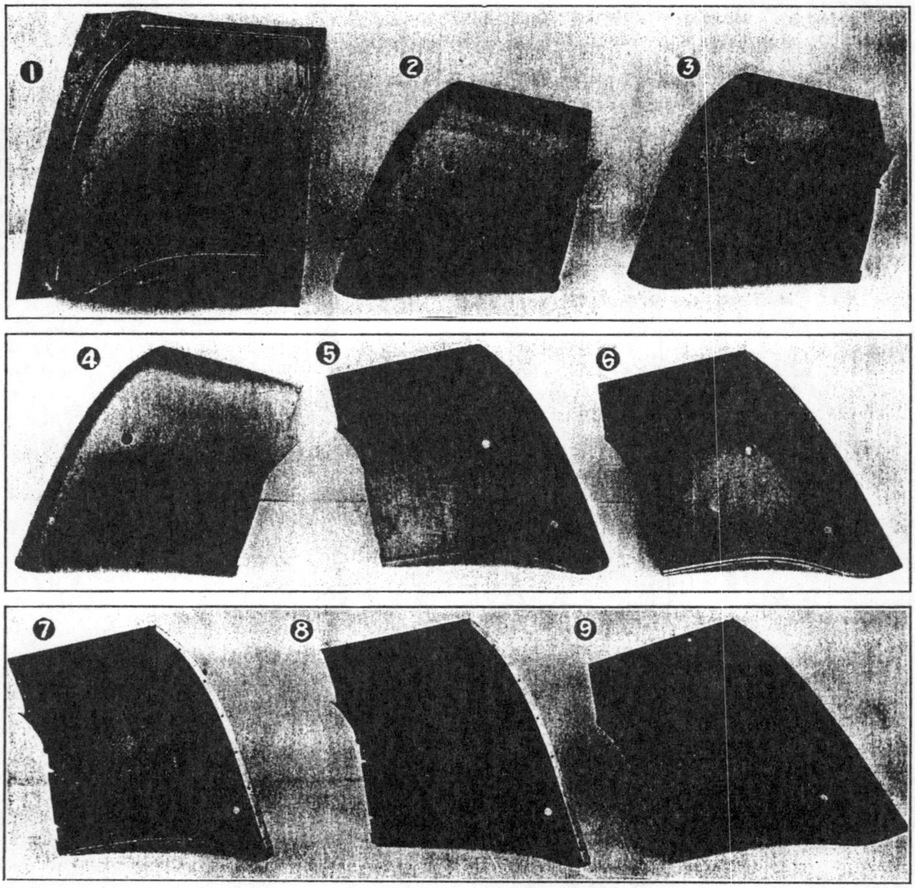


**Fig. 15. Above are shown the successive stages in the making of a rear one-quarter touring car panel.**

the efficiency of the group which represents the total efficiency or over-all efficiency of making the part is a maximum. In most of these groups the presses are located in such manner that the press operators pass the parts from one to the other, either from hand to hand or by sliding on gravity conveyors or slippery slides. Thus in the majority of "line-ups," from the time that the sheet leaves the storage pile to the time that the last operation has been completed there is no necessity for the employment of roustabouts or material handlers.

In connection with these fractional groups or little factories there is to be found a most interesting and characteristic example of waste utilization. The Ford idea is to "make something" out of every bit of scrap or waste whether this be wood or sheet iron; therefore the scrap coming from the trimming of presses and blanking presses has been carefully studied to determine what Ford or Fordson parts can be stamped from it. Once this has been decided, a *waste utilization group* of presses is included in the unit or little factory. Thus this waste is worked up immediately and with

Fig. 16. The first operation is shearing and bending from the sheet.

Fig. 17. Next comes drawing and beading the rear one-quarter panel.

Fig. 18. The third operation is not shown. The fourth, piercing four holes, appears above.

the least amount of transportation into usable parts, reducing to a minimum the amount of steel scrap to be baled and remelted.

Perhaps the best way to convey a really clear idea of all of this is to follow through one or two typical parts, operation by operation.

Let us take, for example, the all-steel car dash which has recently replaced the wooden dash. The successive operations are shown in Figs. 4 to 13, presented herewith. The operations are as follows:

| | Men | Hrs. | Prod. |
|---|---|---|---|
| 1. Blank 94 1/2 E. Toledo Double Geared 4" S | 2 | 8 | 3200 |
| 2. Draw 94 1/2 E. Toledo Double Geared 4" S | 2 | 8 | 3400 |
| 3. Flange 96 1/2 E. Toledo Triple Geared 6" S | 2 | 8 | 3200 |
| 4. Pierce front 94 1/2 E. Toledo Double Gear 4" S | 2 | 8 | 3200 |
| 5. Pierce side Fl. C. A. 4 Ferracute 2" S | 1 | 8 | 3400 |
| 6. Pierce top Fl. C. A. 4 Ferracute 2" S | 1 | 8 | 3200 |
| 7. Pierce corner Fl. P. 3 Ferracute 2" S | 1 | 8 | 3200 |
| 8. Emboss Top Fl. 92 1/2 Tol. Triple Gear 5" S | 2 | 8 | 3200 |
| 9. Rivet Hood Clip 2 Auto self feed Thompson riveter | 2 | 8 | 4000 |

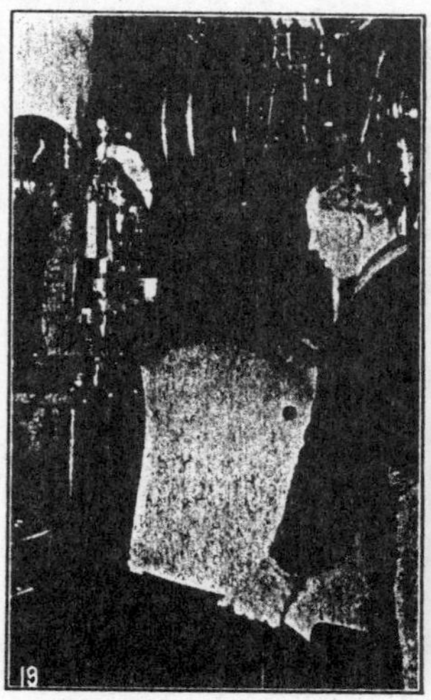

Fig. 19. Nail holes for attaching to wood frames are spaced "by eye."

Fig. 20. Flanging and embossing the rear quarter panel.

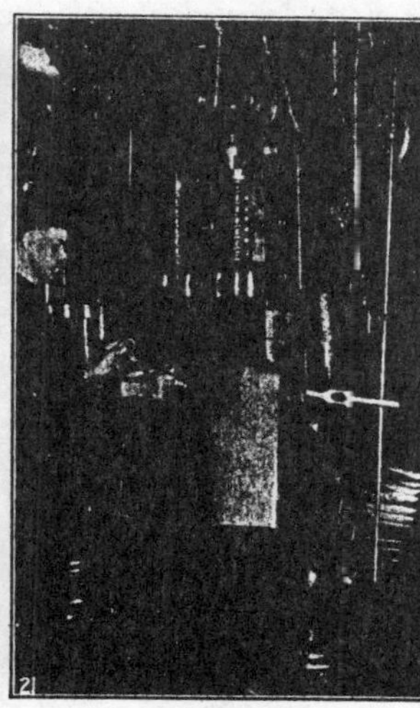

Fig. 21. Another flanging operation takes place immediately afterwards.

Fig. 22. Then this flange is pierced for the "T-mold."

Fig. 23. Recesses for the hinges are next embossed and pierced.

Fig. 24. Finally a reinforcement is spot welded to the panel.

There are fifteen men employed in the gang performing the above operations and their output is 3,500 pieces in eight hours. A marked sheet was passed from hand to hand through all nine operations in an elapsed time of one minute and fifteen seconds. The scrap from the trimming presses on this job is passed over to the adjacent scrap reclaiming unit shown in Fig. 14, and is made into door lock panels and heel board retainers. A further economy in connection with this piece is contemplated, namely, through a

Fig. 25. The sedan cowl is drawn in one operation on the heavy press shown above.

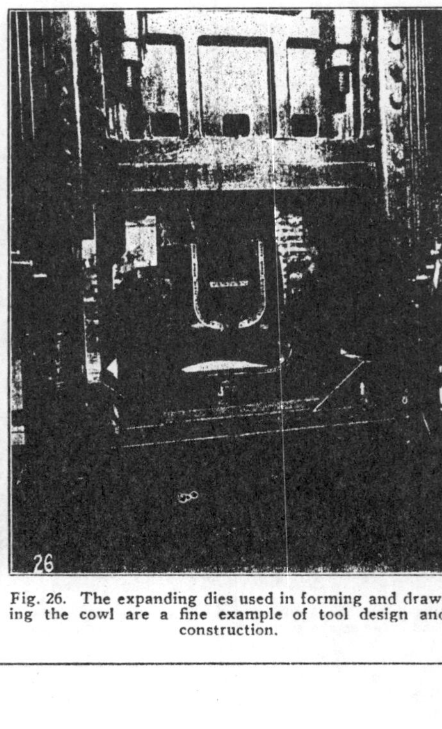


Fig. 26. The expanding dies used in forming and drawing the cowl are a fine example of tool design and construction.

Fig. 27. An ingenious apparatus is pictured above for heating the cowl corners for hot dinging.

Fig. 28. The hinge job is located on the upper mezzanine.

double punching of the holes A in Fig. 4, ten standard ⅜ inch steel washers will be reclaimed from each dash.

Another part, which will be more familiar to our readers than the dash job because a more obvious part of the body, is the rear one-quarter touring car panel. The processing of this and the changes in it after the various operations are shown in Figs. 15 to 24. Thirteen men comprise the gang in this unit and produce five thousand of these panels in eight hours. These are not complete, however, until combined with wood framing, and hence instead of going directly from the presses into the freight cars, are carried to the western mezzanine for assembly. The operations on this panel are as follows:

| Operation | Men | Hrs. | Prod. |
|---|---|---|---|
| 1. Shear, shear | 2 | 8 | 4400 |
| 2. Draw, 407 Bliss Press BT-23351 | 2 | 8 | 2500 |
| 3. Trim, 6-F straight side double crank press | 2 | 8 | 2200 |
| 4. Pierce 4 holes, P-3 press 1-½" S. BT-18984 | 1 | 8 | 2500 |
| 5. Pierce nail holes #0 Verdin bench press #23524 | 1 | 8 | 2500 |
| 6. Flange and emboss 6-D straight side double crank press Bliss BT-22880 11" stroke | 2 | 8 | 2200 |
| 7. Flange for T-mold P-4 Ford std. Ferracute press #3" stroke BT-23398 | 1 | 8 | 2200 |
| 8. Pierce for T-mold, P-4 Ford std. Ferracute press 1-½" stroke, BT-23397 | 1 | 8 | 2200 |
| 9. Pierce and emboss for hinge recess P-5 Ford std. Ferracute press 3-½" stroke BT-18412 | 1 | 8 | 2200 |
| 10. Spot weld reinforcement 3 Taylor spot welding machines | 3 | 8 | 2200 |

Unredeemable scrap goes to the baling room. After a bale is made and is elevated by the vertical plunger of

Fig. 29. This is a view of the hinge assembly table, where door hinges are put together.

Fig. 30. View in the scrap baling department showing one of the three hydraulic presses in operation.

Fig. 31. After the bales leave the press they are elevated by conveyor and dropped into waiting freight cars.

Fig. 32. The tool and die repair department is located under the mezzanine.

the press it is kicked by the quick action of the horizontal ram into an inclined chute. This leads it to the underground leg of an elevating conveyor which in turn carries it up through the side wall of the building, and drops it into an open gondola car. Train loads of this scrap are pulled to the foundry where the contents are hoisted to the back stock gallery. Some three and a half car loads of this scrap is baled every day. When compared with the consumption of fifty carloads of sheet steel per day, this gives an approximate net loss, as far as the sheet metal department is concerned, of seven per cent. Inasmuch, however, as all of this goes back into the cupola and thus into cast iron production, there is very little loss as far as the plant as a whole is concerned except that which is occasioned by remelting.

Space will not permit an extended description of even the more interesting of the various sheet metal production line-ups in this shop. The two parts which have been described, namely, the all-steel dash and the rear one-quarter panel are typical in so far as the touring car is concerned. The sedan is a somewhat different proposition inasmuch as the wood frame is assembled complete in one unit, as shown in Fig. 3. The individual sheet metal parts for the sedan are in general larger than for the touring car. The cowl, for example, is drawn in one piece in a 75 ton double action press. The dies used on this machine are a fine example of tool design and die making skill. A front and rear view of these dies is shown in Figs. 24 and 25. After the cowl is drawn and trimmed it is necessary to straighten out its upper corners by "dinging." Unlike the usual practice in cold dinging of sheet metal parts, this operation is done hot, an ingenious reciprocating work table and suspended gas torches providing

Fig. 33. Lift tier trucks are used for transporting dies and putting them into the presses.

the means employed. These gas torches heat two refractory blocks, so arranged as to drop into place on the two corners of the cowl which are to be straightened. A moment's contact brings the sheet metal to a red heat, after which the reciprocating table is slid back, thus introducing another cowl under the refractories and putting the heated one into position for dinging.

Spot welding is extensively used on the sedan bodies and some very interesting examples of its use occur in the making of the rear seat back and side panels, particularly in the stitching of these three parts together and the insertion of gusset plates for wheel clearance and rear fenders in the rear quarter panels.

Practically all of the presses in this plant are equipped with safety button control. In the case of a press which is operated by two men there are two buttons and both must be pressed to actuate the machine. In addition to this, all of the circuits are controlled by safety switches.

The dies for body production on sheet iron parts are designed and built in the huge tool department at Highland Park where 1,500 tool makers are kept busy at this and similar work. The maintenance, repair and alteration as well as the fitting up of these dies is cared for in a large die repair department adjacent to the press department in the B building. Several hundred die makers are kept busy at this work, for although the average die will turn out from several thousand to two hundred thousand pieces they are made so rapidly that, measured in days instead of pieces, the active life of the die is very short. A run of two weeks is an exceptionally long time for a set of dies to remain in condition and some must be replaced almost daily. Of course, under these circumstances, and in view of the insistence on continuous uninterrupted production which characterizes the Ford policy throughout, it is necessary to keep duplicate or even triplicate sets of dies in reserve ready for immediate change.

The method of transporting these dies, some of which are very heavy, from the tool department to the presses and of lifting them into the presses, is of interest and applicable to many plants. Fig. 33 shows the means employed for this purpose. The utilization of tier lift trucks which carry the dies from storage or from tool department direct to the press platens and which elevate the dies, forms a simple but effective procedure that eliminates one and sometimes two rehandlings

This brief description of the methods of sheet metal production for Ford car parts will give the reader an adequate impression of this part of the work. We next come to a phase of production that is, if not more efficient, at least much more interesting because more unusual, namely, the fabrication of wooden parts for the car and the framing and paneling or partial assemblies. This phase of the work, together with the utilization of wood wastes will be covered in our next installment.

[g]



# Ford Principles and Practice at River Rouge

## XIII—Manufacturing and Assembling Body Parts

***By John H. Van Deventer***

Mr. John H. Van Deventer graduated from Sibley College, Cornell University in 1903 with the degree of M. E., after which he served a special apprenticeship at the Niles Works of the Niles-Bement-Pond Company. Entering the Cost Department of the Goulds Manufacturing Company of Seneca Falls, N. Y., in 1905, he soon became manager of costs and later, superintendent of production.

In 1907, he became general superintendent of the Buffalo Forge Company, remaining in this capacity until 1914 at which time he entered the field of technical journalism becoming associate editor of the American Machinist. In 1915, he was made managing editor of that publication and, in 1916, its editor-in-chief.

Called to Washington in 1917, to assist in organizing the Ordnance Department, Mr. Van Deventer was commissioned a major and served as assistant to the chief of the Engineering Division. Upon returning to his editorial chair in 1918, he organized the Army Ordnance Association.

In 1920, he became the active editor of Industrial Management and in 1921, was made president of The Engineering Magazine Company. He is the author of a number of books on management and shop practice.

ONE of the most interesting activities of the River Rouge plant of the Ford Motor Company is the fabrication of wood parts for motor car bodies. Perhaps this is because of the skill with which these activities are carried out, perhaps it is due partly also to the ease with which wood is cut and shaped as compared with metal, and hence the speed with which the various pieces are fashioned. At any rate it is a department that the visitor to this plant should most assuredly visit and where he will receive a striking object lesson not only in planned production but also in the elimination of waste.

**The machining of eight thousand Ford car cylinder-blocks every twenty-four hours is surely an unequalled example of quantity production. And when one finds that the entire elapsed time required for 43 operations on this casting is less than one hour, the invariable question is: "How can it be done?"**

**This question is answered in detail in the accompanying article.**
**—The Editors.**

It would be interesting and instructive if statistics could be obtained and published showing the per cent. of actual utilization of the total cutting of our timber. Without doubt in the average plant by the time the wooden product is completed one would find that more than six-tenths at least of the original tree had been wasted. Modern practice in woodworking plants, such as is represented at the Ford body plant, is rapidly altering this condition, thereby materially aiding in the national conservation of our rapidly dwindling timber resources.

The raw material, *wood*, comes into this plant in various kinds, forms and sizes. A considerable portion of it is already coming from the Ford holdings at Iron Mountain and elsewhere. That received as logs is sawn into boards in a sawmill northwest of the B building, a mill having a capacity of approximately 50,000 board feet per day, power for operating which is provided through burning the bark and trash. Ultimately all of the sawing of Ford logs will be done in the forests in local mills where in fact it is intended to save considerable freight by also finishing completely many of the wood body parts.

At this early stage of the process, in the sawmill is evident one of the early utilizations of waste. The outside boards cut from the log, or the *sap wood* as it is

Fig. 1. Exterior of one of the dry kiln bays showing method of transporting lumber.

Fig. 2. Exterior of the woodworking departments of the body plant at River Rouge. Ninety tons of sawdust are caught in the dust collectors every twenty-four hours.

called is used for the making of body parts. The very first cuttings on each log are used for shipping box lumber, and the like, which purpose they serve admirably even though some of the bark may occasionally be in evidence.

By far the greater part of the hard wood, the maple, spruce, gum, oak and birch coming into this plant, arrives in the form of sawed lumber, in thicknesses running from four to twelve quarters, or in other words, from one to three inches. Arriving in freight cars at the unloading stations, this stock is inspected, graded and then stacked in temporary storage piles. Here it awaits admission to the dry kiln. The daily consumption of lumber for body work at this plant averages at present 325,000 board feet per day, equivalent to a 60-mile length of one inch by 12-inch boards. This lumber as received is both air dryed and kiln dryed but all of it nevertheless is tested for its moisture content

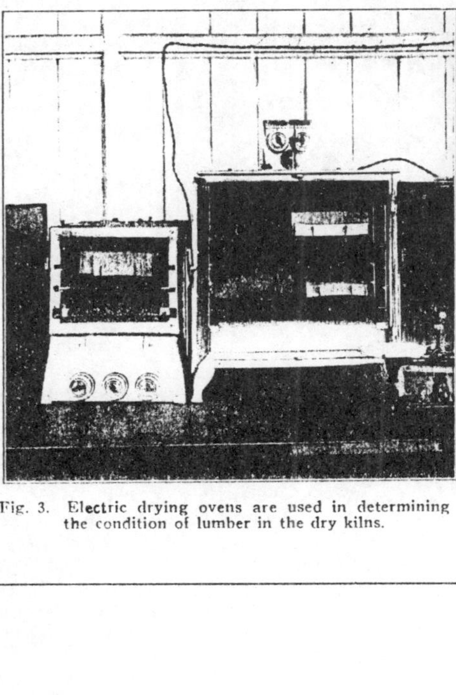

Fig. 3. Electric drying ovens are used in determining the condition of lumber in the dry kilns.

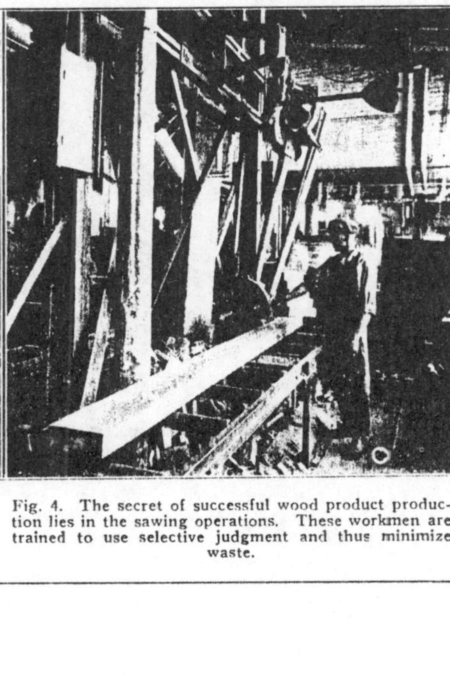

Fig. 4. The secret of successful wood product production lies in the sawing operations. These workmen are trained to use selective judgment and thus minimize waste.

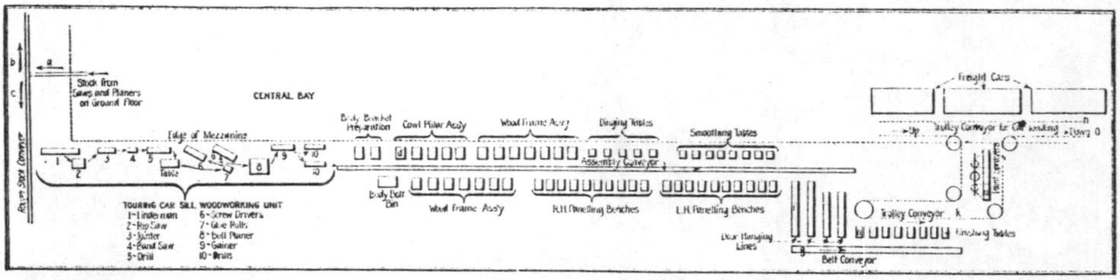


**Fig. 4a. This diagram shows course of assembly of the Ford touring car body. The sequence of woodworking operations on the sills are also shown.**

when received. This test is made from selected sections of heart wood which are weighed green, then dryed in electric ovens at a temperature of 250 degrees, and then weighed bone dry.

The maximum percentage of moisture permitted in Ford car frame lumber is seven per cent. The average moisture in the air dryed stock as received is between 30 and 45 per cent. This surplus of moisture must be removed by kiln drying. Fig. 1 shows a view of the dry kilns just back of the B building. There are 20 of these kilns with a total approximate capacity of 50,000 board feet. Lumber remains in these kilns from 8 to 28 days depending on the per cent. of moisture originally contained. Samplings are made every seven days from each batch in each kiln to determine the process of drying. In making these samplings the planks are cut at least a foot from the end to avoid false moisture percentages, and a thin section of wood taken from the center of the cut to avoid errors caused by "case hardening," as the too rapid drying of the outside fibres is called.

**Fig. 5. General view of the north end of the western mezzanine in the B building. The stock shown here is a fraction of a day's requirements for touring car wood parts.**

Each kiln is provided with steam sprays for saturating the lumber when first received and with automatic humidity control and recording instruments for registering the air condition. Starting, therefore, with 100 per cent. humidity, or better, through a gradual daily reduction, the contents of each kiln is brought to an average of between five to seven per cent. of moisture before it is ready to enter the B building and go to the saws.

As stated in the previous issue, the gang saws used for cutting up all of the body stock are situated on the ground floor of the B building under the western mezzanine at the northern end of the plant. Transfer tracks lead from the dry kiln doors to storage piles within the building immediately back of these saws. The loaded kiln wagons are drawn on these tracks by Fordson tractors.

It is truly said that the operation of the saws can make or break any woodworking plant. Boards and planks do not partake of the uniformity of dimension of other materials, such as steel sheets, beams and structural shapes, nor does wood equal these other materials in uniformity of texture. Knots and checks, for example, have a disturbing way of appearing in inconvenient places. And so in the conduct of the saws, where the scheduling of production virtually begins or where, in other words, the decision is made as to just what part shall be made out of each board, there are surprising potentialities either for waste or for economy.

In examining the wooden frame work of Ford cars of every body type, the visitor finds a complete freedom from knots and checks. This is true even on parts and in places where they would do no harm. In spite of this careful inspection and selection, however, some sixty per cent. of the total footage sawn daily is actually turned into body parts. In other words, the board foot contents of the finished parts made from a day's cutting will amount to sixty per

Fig. 6. The Linderman performs the first operation on touring car sills.

Fig. 7. Narrow boards fed into the Linderman come out automatically glued into wide panels.

Fig. 8. After sawing to width the sill stock is passed through the jointer.

cent. of the original footage sawn. The remaining forty per cent., however, is not loss as we shall see later. When it is considered, however, that in many body plants where knots are not taboo, the average utilization falls to forty per cent. or below, these figures take on their true significance.

The secret of this remarkable utilization, as said before, lies in the sawing. In many, perhaps it may be said in the average woodworking plant, if a thousand pieces of a certain dimension are to be made the sawyer will get an order for a thousand pieces of certain thickness, width, length and kind of wood and will proceed to saw up the stock straightaway as it comes, cutting out knots where too near the edge of a cut, and discarding leftovers from the end of the board. In the Ford body plant no straightaway sawing is permitted. Each swing saw takes care of a certain thickness of board and is provided with from five to seven stops permitting five to seven distinct pieces to be sawn according to the sawyer's judgment. With this leeway and with men carefully trained to use it, there is surprisingly little waste at the saws. All through the subsequent operations each workman is alert to save pieces of stock which may be successfully made into something smaller but which would normally be regarded as waste and thrown away. This practice extends to even three cornered and other odd shaped cuttings from band saws. So far does Mr. Ford's insistence on this practice go that he has these odd shapes collected in burlap bags and saved for utilization. Some of this utilization

Fig. 9. Next a band saw provides the necessary bevel.

Fig. 10. Screw holes are drilled into the sill body by means of this multispindle automatic boring machine.

Fig. 11. Each sill is made up of two pieces which are clamped in jigs for the automatic screw driving operations.

Fig. 12. The bull planer furnishes the required skew bevel surface for the touring car sills.

cannot be had by making product parts for the reason that there are not enough of these small product parts to use up all of this waste. In this case some other product must be found for which this is suitable raw material, whether it be small handles, spools, or wooden novelties.

The wisdom of this edict is apparent in the very difficulty of finding uses for all of this odd material, for since the rule has been in effect, there has been a marked increase in the degree of care exercised in sawing with a consequent marked reduction in the number of pieces of waste thus produced. The various foremen see to this since otherwise it is up to them to go to the trouble of devising ways and means of utilizing the waste produced.

It will, of course, be impossible in the scope of this article to fully describe even the major pieces of wood for one type of body. To convey, therefore, an adequate idea of the methods used in fabricating body parts after the stock leaves the saws, let us ascend to the mezzanine and follow through from start to finish the operations on a typical touring car part, namely, a frame sill. There are two of these required for each touring car, one left hand and one right. Each one forms the longitudinal base of a side frame assembly, through which the finished side with its doors is bolted to the chassis frame.

A diagram showing the line-up for the progressive operation on these sills is represented at the left in the diagram shown in Fig. 4a, the machines used for this process being numbered from one to ten. The rough sawn and plain stock, in 74-inch lengths, comes up from the ground floor

Fig. 13. A multiple gaining machine cuts the notches for the frame uprights.
Fig. 14. Body bolt holes are next drilled on the multispindle borers.
Fig. 15. The assembling conveyor on which frames are put together. Note storage of bolts and iron parts.

on an elevating conveyor A which brings it to the mezzanine level. Here it is diverted to transverse conveyor C which carries it to the first mezzanine operation on the Linderman.

A Linderman, it may be said for the benefit of those who are not acquainted with it, is an ingenious machine for making wide boards out of narrow ones. It is particularly useful where pieces are sawn to a curve, for the last piece of stock which would otherwise be wasted, is Lindermanned to fresh pieces and made up into a wide, new, sawing panel. The idea is illustrated in Fig. 6. Stock is fed into the Linderman from both ends and in going through the machine it is automatically notched or rabbitted, the edges glued and the rabbits forced together, giving a joint that is actually stronger than the original stock.

From the Linderman the panels go to a rip saw shown in Fig. 2 which saws them on a twisted bevel into double widths. Next they proceed to jointer shown in Fig. 8, and past this to the bandsaw, Fig. 9, where they are split into single widths. On the multi-spindle wood boring machine, Fig. 11, the screw holes are drilled for the screws which fasten the two component parts of each sill together. From here a gravity conveyor carries them to glue rolls, where waterproof glue is applied. The two parts are then clamped in a jig and passed to one of the screw driving machines. These are of the automatic magazine type, each stroke of the operating lever feeding a fresh screw into the chuck and driving the screw home into the upper member of the sill. From here the sills go to the bull planer, shown in Fig. 12, where the skew bevel surface and rounded nose are form-planed. Then the parts pass to the gainer where notches are milled for the upright door pillars, as the vertical members of the side frame assembly are called. Next the multi-spindle boring machines, Fig. 14, bore and countersink the nine holes in each sill required for body bolts. These operations complete the woodworking process on the chief member of the side frame assembly, which is now placed upon the assembling conveyor shown in Fig. 15, which forms the means of transportation to and from the framing and paneling operations.

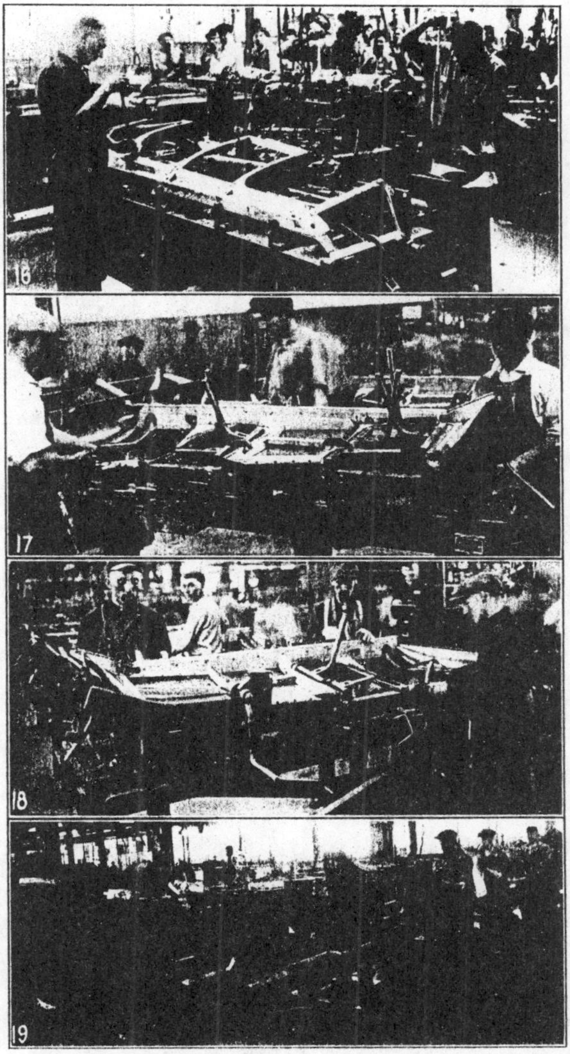

Fig. 16. Framing benches are ingeniously contrived to conserve every second of time and inch of motion.

Fig. 17. On the paneling benches the wood frames and pressed steel panels are joined together.

Fig. 18. The paneling tables are both right and left hand, one of the latter being here shown.

Fig. 19. View of one of the door fitting lines where the door hinges are attached and the doors mounted.

The first stage of side frame assembly consists in attaching the body brackets and cowl pillars at the ends of the sill. Body brackets are prepared ready for bolting to the sill on two double work benches at the origin of the conveyor, and are attached to the sill on the conveyor, on the move. For the assembly of the cowl pillar the sill

Fig. 20. One of the final fitting tables where each complete side frame is set up and inspected.

Fig. 21. As the completed sides travel on an overhead conveyor they pass spray hoods where the interiors are coated with black paint.

is removed from the conveyor and taken to one of the double work benches D. This completed, it is placed again on the conveyor and proceeds to the wood frame assembly benches, as shown in Fig. 16.

These benches are in reality extremely ingenious assembling fixtures in which every detail is the result of long experience. They provide not only the means of clamping the sill and cowl pillar securely while adding the remaining wooden parts to the frame, but they also serve as receptacles for all of the screws, nails, and tools; the glue pot, and underneath, for the wooden pillars, top members and cross braces.

One framer works at each framing table. First he takes the sill with body brackets and cowl pillar attached from the conveyor and places it on the framing table, fastening it securely with quick-acting clamps. Next he takes the brush from his electrically heated glue pot and applies glue to the gains or notches in which the

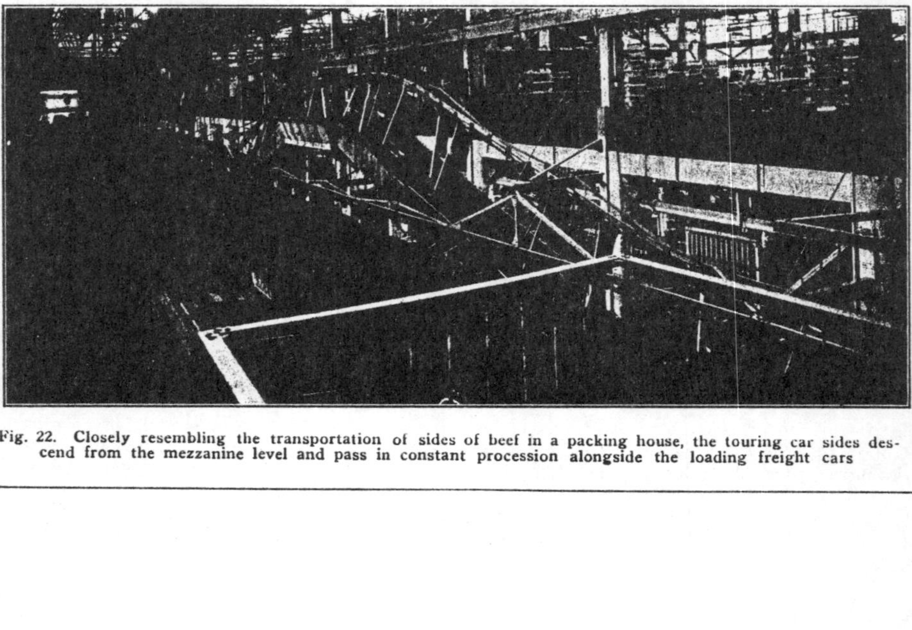

Fig. 22. Closely resembling the transportation of sides of beef in a packing house, the touring car sides descend from the mezzanine level and pass in constant procession alongside the loading freight cars

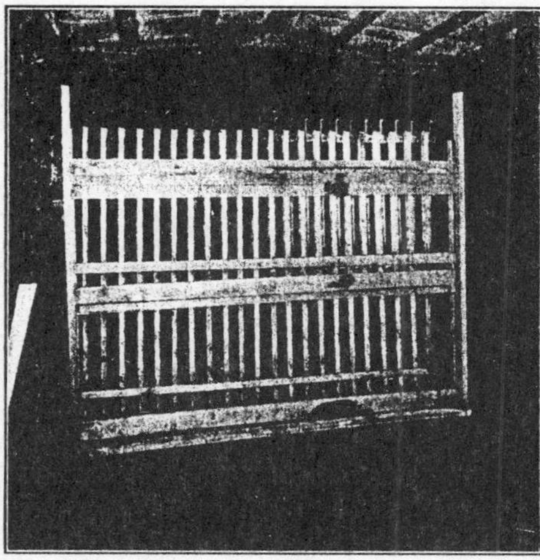

Fig. 23.. Every inch of freight car space is utilized in packing body parts. For example, note the above rack of car sides.

Fig. 24. This ingenious roll-over form jig attached to a wood planer gives the convex and concave surfaces to the Ford door frames.

upright pillars are to be fastened. While he is doing this he has already selected an upright and has prepared it for placing in the sill. Next he fits the cross members and glues them, after which he reaches for a suspended electric portable drill hanging over the table, and drills the screw holes through the wood members. Finally he inserts and tightens the screws, nails the tenons, and the frame is complete, ready for paneling. The entire operation requires about three minutes.

The completed frame, again, placed up on the assembling conveyor, proceeds to the paneling benches. There are two sets of these benches on which the metal body panels are attached to the wood frames. One set of these benches is for right-hand side assemblies and the other for left-hand, this necessitated by the fact that the two sides differ, not only in hand but in the fact of having a blind door on the left-hand or driver's side.

The paneling benches are worked out with the same

Fig. 25. Aluminum jigs rubbing against collars beneath shaper knives give the rounded corners to the door frames.

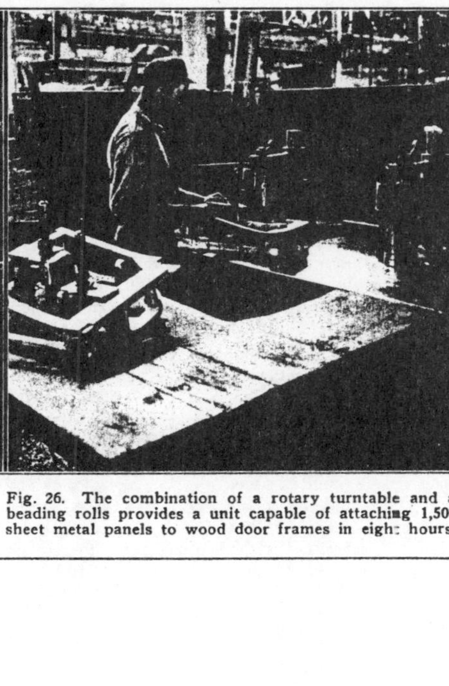

Fig. 26. The combination of a rotary turntable and a beading rolls provides a unit capable of attaching 1,500 sheet metal panels to wood door frames in eight hours.

care and ingenuity as the framing benches. They differ, however, in that three men are assigned to each bench. First the rear one-quarter and side panels are placed in the jig, after which the wooden frame is superimposed, the panel flanges then are hammered down and nailed and the paneling is completed. This operation takes approximately a minute and a half after which the assembly is again laid on the assembly conveyors, to proceed to the door hanging lines shown in the diagram at F, where the doors are hung and fitted. En route to this point various dinging and smoothing tables are passed where pieces that contain bumps or dents in the sheet metal undergo treatment.

The door hanging lines are shown in Fig 19. They are progressive lines consisting of work tables on which the paneled side frames are placed in upright position, and held by quick acting clamps while the various operations are performed. For the touring car there are four stations, each side being slid from one to the next in the direction of the arrow in the diagram, Fig. 4a, upon completion of each operation. A side frame complete with its metal paneling but without doors, is first placed on this line. It is then drilled to templet for hinge screws, after which it is slid back to the second position where the door latch catch is fitted. Progressing to the third position, the door is hung and two screws driven home, while in the fourth or last position the remaining screws are driven and the door fitted to swing and catch properly.

The doors being hung, the side frame assembly is placed on the belt conveyor G which carries it to the finishing tables H, shown also in Fig. 20. Here the assembly is gone over carefully by experts and the finishing touches which may be needed are added. This completed, the assembly is hung like a side of beef on the trolley conveyor K, which carries it to the paint spraying hoods M.

The exterior finishing of all Ford bodies is taken care of at the branch assembling plants. Therefore, the spray hoods mentioned above are used merely for coating the inside of the frames and doors to protect the metal from rust and the wood from decay. A black asphaltum mixture is sprayed over these parts, the frame then being rehung on the conveyor K which carries it to the edge of the mezzanine and down to the central shipping bay. Descending on an incline along the line O, this conveyor carries the frames to the extremity of the car loading area and then makes a 180 deg, turn, bringing the bodies in endless procession past the doors of the waiting freight cars. Thus the bulkiest parts of the entire shipment are brought directly to the cars by mechanical transportation, each shipper picking off the number of right and left hand sides that he requires to make up his car, whether this be a shipment of touring cars or roadsters.

The processing of the minor wood parts entering the side from assembly follows the same sort of progression and the same direction as the described for the touring car sills, meeting the sill at the assembly conveyor. Chief of these minor assemblies is the door frame which involves some interesting operations in templet planing and shaping to produce the curved door surfaces and the rounded door corners. In each case these curves are re-produced by the contour of the holding fixture or jig. Thus in planing the curved surfaces of the door shown in Fig. 24, one pass through the planer produces the convex surface, and, turning the fixture over, a second pass produces the concave surface. In corner rounding, as shown in Fig. 25, the holding fixture also gives form to the frame, the base of the aluminum holding fixture being a templet which is brought by manual contact against a collar mounted on the shaper spindle below the cutter.

An ingenious turntable hook-up has been devised to take care of the operation of rolling the metal door panels over the door frames. Each turntable is located adjacent to a pair of rolls as shown in Fig. 26. There are four operators to each unit. For units, or sixteen men in all, roll six thousand doors in eight hours. The first operator places the wood frame and metal door panel into a holding jig on the first station of the turntable. At the second position an operator clamps the work in the jig. At the third position a third operator lifts the jig from the turntable and passes the edge of the door through the beading rolls, then replacing the fixture on the turntable. The fourth operator loosens the fixture and removes the work, thus completing the cycle.

The conservation of wood waste which I have emphasized previously does not stop with the endeavor to utilize scrap for raw material for usable or salable wooden articles. There are two primary ways in which the waste which cannot be thus reclaimed is utilized at River Rouge. One of these is through the utilization of saw dust as fuel for generating power. The other is the use of scrap wood in the manufacture of paper.

In a preceding article in INDUSTRIAL MANAGEMENT for April, 1923, on page 213, there was shown an exterior picture of power plant number three, where electric current is generated sufficient for the entire needs of the B building. The fuel used in this power plant is saw dust and coke breeze, but primarily saw dust. Ninety tons of this excellent fuel are produced on the woodworking operations on body work every twenty-four hours. The addition of ten tons of coke breeze gives a highly effective burning mixture, which is stoker fed.

The collection and conveying of this saw dust would require an article in itself to do its description justice. Briefly, however, it is as follows: Each saw, planer and other saw dust or shaving producing tool is provided with an exhaust hood at the cutter. A veritable forest of galvanized iron ducts lead to the battery of dust collectors shown in Fig. 2. Beneath these and connected to their outlets is a chain drag saw dust conveyor 750 feets long which carries the separated saw dust and shavings to an elevating conveyor which takes it directly to power plant number three, situated adjacent to the B building. Here it is converted into some 4200 horsepower, sufficient, as said before, to operate all of the machinery in the B building.

A great quantity of heavy cardboard is used as packing for cushion stuffing in the manufacture of Ford car cushions. A large component of the paper of which such cardboard is made is hardwood. Here then, was both a need and a use which was hardly likely to be overlooked in the utilization of wood scrap. A large paper mill, therefore, has been erected within the past six months and is now in operation, where ultimately some twenty-five per cent. of the thus far unreclaimed refuse will be put to production purposes.

[6]

# Mass Production

**Henry Ford**

The term 'mass production' is used to describe the modern method by which great quantities of a standardized commodity are manufactured. As commonly employed it is made to refer to the quantity produced, but its primary reference is to method. In several particulars the term is unsatisfactory. Mass production is not merely quantity production, for this may be had with none of the requisites of mass production. Nor is it merely machine production, which also may exist without any resemblance to mass production. Mass production is the focusing upon a manufacturing project of the principles of power, accuracy, economy, system, continuity, speed, and repetition. To interpret these principles, through studies of operation and machine development and their co-ordination, is the conspicuous task of management. The normal result is a productive organization that delivers in continuous quantities a useful commodity of standard material, workmanship and design at minimum cost. The necessary, precedent condition of mass production is a capacity, latent or developed, of *mass consumption*, the ability to absorb large production. The two go together, and in the latter may be traced the reasons for the former.

## The Origins of Mass Production

In origin mass production is American and recent; its earliest notable appearance falls within the first decade of the 20th century. The mere massing of men and materials is a procedure as old as the pyramids. Basic industries, like weaving, domestic baking, house construction and wooden ship building, are carried on, with only superficial changes, much as they were in ancient Egypt. Cottage manufactures and handicrafts moulded the practices of industry until the invention of the steam-engine. With the coming of power machines the seat of industry was removed from the homes of the people and a new work centre, the factory, was established.

Harsh criticism of 'the factory system' is modified by the fact that the first effect of the factory was to emancipate the home from being a mere adjunct to the loom or bench, and its later effect was to provide the home with means to develop the dignified status which it has now attained.

### *The Factory System Giving Way*

The early factory system was uneconomical. Its beginning brought greater risk and loss of capital than had been known before, lower wages and more precarious outlook for the workers, and a decrease in quality of goods. More hours, more workers, more machines did not improve conditions; every increase did but enlarge the scale of fallacies built into business. Mere massing of men and tools was not enough; the profit motive, which dominated enterprise, was not enough. There remained the scientific motive which grew eventually into what is called mass production.

The new method followed the failure of the mercantile and financial emphasis in manufacture and the consequent resort to engineering emphasis. The advent and progress of financial control of industry were marked by two developments, the corporation and the labour revolt. Artificial combination of industrial plants into vast corporations for financial purposes proceeded on the theory that complete financial control would automatically bring complete profit advantage. The theory ignored many vital principles of business and its fallacy became apparent, but not before serious social hostility had been incurred.

However, out of the social strife thus engendered the idea began to emerge that possibly the difficulty lay in the neglect of scientific manufacturing principles. Industry was conceded to be necessary and useful; the service it rendered was regarded as of sufficient value to afford fair compensation for all engaged in it; the attention of management was, therefore, more directly focused on the actual labour processes that were employed. This led to what was known early in the 20th century as the 'efficiency movement' with its accompaniments of time-study and similar methods, although its roots began in the experiences of sound industrial observers as early as 1878. It cannot be said, however, that the efficiency experts did more than direct attention to the problem, by showing, in selected instances, how the then current methods were wasteful of men's earning power, and how their correction and improvement could lead to greater production, hence higher wages, and therefore a general betterment of labour relations. They emphasized a more intelligent management of methods then in use, but did not see that a wholly new method was needed that would simply abolish the problems of which the old method, under the most intelligent management, was inevitably prolific. For example they dealt with methods which enabled labourers whose task was to load 12½ tons of pig-iron a day, to load 47½ long tons a day for an increase in the day's pay from $1.15 to $1.85. They did not see that another and better method might be devised which would make it unnecessary for a working man to carry 106,400 lb. of pig-iron to earn $1.85. Mass production was not in their view, but only the alleviation of the worst errors of competitive factory practice.

### *The Motor Industry Leads the Way*

To the motor industry is given the credit of bringing mass production to experimental success, and by general consent the Ford Motor Company is regarded as having pioneered in the largest development of the method under a single management and for a single purpose. It may, therefore, simplify the history of mass production and the description of its principles if the experience of this company is taken as a basis. It has been already suggested that mass production is possible only through the ability of the public to absorb large quantities of the commodity thus produced. These commodities were formerly but are no longer limited to necessities and conveniences. The greatest development of mass production methods has occurred in the production of conveniences. The motor vehicle represents a basic and continuous convenience – transportation.

Mass production begins, then, in the conception of a public need of which the public may not as yet be conscious and proceeds on the principle that use-convenience must be matched by price-convenience. Under this principle the element of service remains uppermost; profit and expansion are trusted to emerge as consequences. As to which precedes the other, consumption or production, experiences will differ. But granted that the vision of the public need is correct, and the commodity adapted to meet it, the impulse to increased production may come in anticipation of demand, or in response to demand, but, in any case, the resulting consumption is always utilized to obtain such increase of quality, or such decrease of cost, or both, as shall secure still greater use-convenience and price-convenience. As these increase, consumption increases, making possible still greater production advantages, and so on to a fulfilment that is not yet in view.

The commodities that conduce to civilized living are thus far enjoyed by only a small fraction of the world's inhabitants. The experience of the Ford Motor Company has been that mass production precedes mass consumption and makes it possible, by reducing costs and thus permitting both greater use-convenience and price-convenience. If the production is increased, costs can be reduced. If production is increased 500% costs may be cut 50%, and this decrease in cost, with its accompanying decrease in selling price, will probably multiply by ten the number of people who can conveniently buy the product.

## The Principles of Mass Production

As to shop detail, the keyword to mass production is simplicity. Three plain principles underlie it: (*a*) the planned orderly and continuous progression of the commodity through the shop; (*b*) the delivery of work instead of leaving it to the workman's initiative to find it; (*c*) an analysis of operations into their constituent parts. These are distinct but not separate steps; all are involved in the first one. To plan the progress of material from the initial manufacturing operation until its emergence as a finished product involves shop planning on a large scale and the manufacture and delivery of material, tools and parts at various points along the line. To do this successfully with a progressing piece of work means a careful breaking up of the work into the sequence of its 'operations'. All three fundamentals are involved in the original act of planning a moving line of production.

This system is practised, not only on the final assembly line, but throughout the various arts and trades involved in the completed product. The motor car final assembly line offers an impressive spectacle of hundreds of parts being quickly put together into a going vehicle, but flowing into that are other assembly lines on which each of the hundreds of parts have been fashioned. It may be far down the final assembly line that the springs, for example, appear, and they may seem to be a negligible part of the whole operation. Formerly one artisan would cut, harden, bend and build a spring. In 1939 the making of one leaf of a spring was an operation of apparent complexity, yet was really the ultimate reduction to simplicity of operation.

### *A Typical Operation Described*

For its illustrative value let us trace the course of a spring leaf after it has progressed from iron ore through ingot, bloom and billet stages, and is rolled into strips. (1) Beginning as a strip of steel prepared by the steelmill, it is placed in a punch press for cutting and piercing. The workman puts the strip into press until it hits a stop, then trips the press. The cut-off and pierced piece falls on a belt conveyor which runs along the loading end of a series of heat-treating ovens. (2) A second workman takes the pieces from belt conveyor and places them on conveyor which passes through the furnace (in which temperature is automatically controlled); thence they are deposited at a certain temperature by this conveyor at the unloading end of the furnace. (3) The heated piece is lifted with tongs by a third operator and placed in a bending machine which gives the leaf its proper curve and plunges it in oil, the temperature of which is maintained at a definite degree by apparatus beyond the operator's control. (4) As the bending machine emerges from the oil bath, the same operator removes it to a conveyor that delivers it to an annealing furnace. (5) A fourth operator places the leaf on a conveyor that passes through the annealing furnace. (6) The same workman sends it by conveyor through a water-cooling bath, thence to the spring assembly line.

As a set of springs on the Ford car requires on an average 14 leaves, and 25,000 springs are a normal day's output, this operation must be visualized as employing a great battery of lines similar to the one briefly described.

As all the leaves in a spring are of different length and curve, from the bottom or master leaf to the top leaf, this operation must be visualized as one of many carried on simultaneously by different batteries of machines, each battery working on its own special size. All of these lines, with their various machines and operations, are converging on the point where the leaves are assembled into springs. The leaf whose progress has been described is the simplest one.

The operation proceeds as follows: (7) A fifth workman removes the leaf from the conveyor on its arrival at assembly line. (8) A sixth workman sprays the arriving leaves with a lubricant. (9) A seventh workman inserts a bolt through leaves assembled as a spring. (10) An eighth workman puts a nut on the bolt and tightens it. (11) A ninth workman puts in the clip bolts and tightens them. (12) A tenth workman inspects the completed spring. (13) He hangs the spring on the conveyor. (14) The conveyor carries it to the loading dock where the eleventh workman removes it.

One workman under the old system could attend the leaf through all these phases, or even make a complete spring, but his production would be limited. Where large quantities of the same article are to be made, the simplest operation may involve the whole time of one man. A one-minute operation will require one man a full day of eight hours to accomplish it on 480 pieces. Now this simple part, a spring leaf, must be identical in strength, finish and curve with millions of others designed to fulfil the same purpose, and this becomes a complicated and delicate procedure requiring automatic machinery, the most accurate of measuring devices, pyrometer controls, 'go' and 'no go' gauges – in fact, the best facilities that can be provided by modern management. The leaf described, which is a minor matter when compared with the whole great process, becomes a major matter when considered by itself; it must have its own supply of material delivered in sufficient quantities at indicated places – for example, steel at 1; heat at 2; power and oil at 3; heat at 5; water at 6; bolts at 7; lubricants at 8; bolts at 9; nuts at 10; clips at 11. In this process the secrets of many arts and trades are employed.

The story of this minor part illustrates what is meant by orderly progression of the article through the shop. It goes to meet other parts of the motor-car which have come from other parts of the plant by similar processes. The story illustrates also what is meant by delivering the work to the workman; every workman's task is prepared for him by some other workman, and delivered to his hand. The third principle also is illustrated – the analysis of a single job into its constituent operations. The simplicity of the part here described should not be permitted to exclude from view the multitude of other operations, ranging from the heaviest forgings to the lightest manipulations in bench assembly of delicate electrical instruments. Some gauge inspections involve measurements to the ten-millionth part of an inch.

The economies arising from this method are obvious. The machinery is constantly in use. It would be economically impossible to maintain all this equipment for the service of men occupied in the entire operation of making springs. Presses, furnaces, bending machines, oil baths would be idle while the workman progressed from operation to operation. Under mass production it is the work, not the man, that progresses from operation to operation. Otherwise use-convenience in the commodity would be lessened, while price-convenience would be destroyed. There is also the economy of reduced inventories, in lapse of time between raw material and finished product. Mass production justifies itself by an economy whose benefits may quickly be transmitted to the purchaser.

## The Effects of Mass Production

But it is not the history and principle of mass production which provoke the widest discussions; the *effects* of it have been placed under scrutiny. What have been the effects of mass production on society?

(1) Beginning with management, where unquestionably mass production methods take their rise, there is a notable increase in industrial control, as distinguished from financial control. The engineer's point of view has gained the ascendancy and this trend will undoubtedly continue until finance becomes the handmaid instead of the mistress of productive industry. Industrial control has been marked by a continuous improvement of the standard, for standardization does not mean stagnation, but the instant adoption of the better method to the exclusion of the old, in the interests of production. Financial control was not, in its heyday, marked by a tendency to make costly changes in the interests of the product. The economy of scrapping old equipment immediately upon the invention of the better equipment was not so well understood. Engineering control, entrenched in mass production methods, brought in this new readiness to advance. Management has been kept close to the shop and has reduced the office to a clearing-house for the shop.

Managers and men, as well as the manufacturing process, have been brought to greater singleness of purpose.

(2) The effect of mass production on the product has been to give it the highest standard of quality ever attained in output of great quantities. Conditions of mass production require

material of the best quality to pass successfully through the operations. The utmost accuracy must control all these operations. Every part must be produced to fit at once into the design for which it is made. In mass production there are no fitters. The presence of fitters indicates that the parts have been produced unfit for immediate placement in the design. In works of art and luxury this accuracy is achieved at the cost of careful handiwork. To introduce hand methods of obtaining accuracy into mass production would render mass production impossible with any reference to price-convenience. The standard quality of the product is guaranteed by the fact that machines are so constructed that a piece of work cannot go through them unless it exactly accords with specifications. If the work goes through the tools, it must be right. It will thus be seen that the burden of creation is on management in designing and selecting the material which is to be produced by the multiple processes utilized in mass production.

(3) The effect of mass production on mechanical science has been to create a wide variety of single-purpose machines which not only group similar operations and perform them in quantity, but also reproduce skill of hand to a marvellous degree. It is not so much the discovery of new principles as the new combination and application of old ones that mark this development. Under mass production the industry of machine making has increased out of all comparison with its previous history, and the constant designing of new machines is a part of the productive work of every great manufacturing institution.

(4) The effect of mass production on employees has been variously appraised. Whether the modern corporation is the destruction or salvation of arts and crafts, whether it narrows or broadens opportunity, whether it assists or retards the personal development of the worker, must be determined by observable facts. A cardinal principle of mass production is that hard work, in the old physical sense of laborious burden-bearing, is wasteful. The physical load is lifted off men and placed on machines. The recurrent mental load is shifted from men in production to men in designing. As to the contention that machines thus become the masters of men, it may be said the machines have increased men's mastery of their environment, and that a generation which is ceaselessly scrapping its machines exhibits few indications of mechanical subjection.

The need for skilled artisans and creative genius is greater under mass production than without it. The Ford Motor Company employs more than 15,000 skilled mechanics in the construction and maintenance of the machinery of production, besides other thousands engaged directly in the production of cars. It has been debated whether there is less or more skill as a consequence of mass production.

The present writer's opinion is that there is more. Practical production men assert that mass production requires an increasing degree of mass intelligence.

The common work of the world has always been done by unskilled labour, but the common work of the world in modern times is not as common as it was formerly. In almost every field of labour more knowledge and responsibility are required.

### *Some Criticisms Answered*

Mass production has also been studied with reference to what has been called the monotony of repetitive work. This monotony does not exist as much in the shops as in the minds of theorists and bookish reformers. There is no form of work without its hardness; but needless hardship has no place in the modern industrial scheme. Mass production lightens work, but increases its repetitive quality. In this it is the opposite of the mediaeval ideal of craftsmanship where the artisan performed every operation, from the preparation of the material to its final form. It is doubtful, however, if the mass of mediaeval toil was as devoid of monotony as has sometimes been pictured, but it is absolutely certain that it was less satisfactory in its results to the worker. In well-managed modern factories the tendency to monotony is combated by frequent changes of task.

The criticism of mass production as a means of reducing employment has long since been out of court. The experience of the Ford Motor Company is that wherever the number of men has been reduced on manufacturing operations, more jobs have been created. Economy in cost of production makes possible lower prices which create a larger market, and this increases employment. As to the effect of mass production on wages and shop relations, there is little need to speak. It is perhaps the mostly widely understood fact about mass production that it has resulted in higher wages than any other method of industry. The reason is at hand. The methods of mass production enable the worker to earn more and thus to have more. Moreover, the methods of mass production have thrown so much responsibility on the craftsmanship of management, that the old method of financial adjustment by reduction of wages has been abandoned by scientific manufacturers. A business that must finance by drafts out of the wage envelopes of its employees is not scientifically based. It is the problem of management so to organize production that it will pay the public, the workmen and the concern itself. Management that fails in any of these is poor management. Disturbed labour conditions, poor wages, uncertain profits indicate lapses in management. The craftsmanship of management absorbs the energies of many thousands of men who, without mass production methods, would have no creative opportunity. Here the modern method broadens instead of narrows individual opportunity.

(5) As to the effects of mass production on society, the increasing supply of human needs and the development of new standards of living are the elements to be estimated. The enlargement of leisure, the increase of human contacts, the extension of individual range, are all the result of mass production.

*See* H. Ford, *My Life and Work* (1924), and *To-day and Tomorrow* (1926); E.G. Filene, *The Way Out* (1924); and Articles in the American *Economic Review*.

# [7]

# Fordism and the Moving Assembly Line: The British and American Experience, 1895–1930

WAYNE LEWCHUK

As a production technique and industrial relations strategy Fordism is today encountering criticism of truly global proportions. In Japan, Sweden, and West Germany vehicle makers have deployed a variety of new production techniques and labor-management schemes designed to make mass production more flexible and assembly line work less dehumanizing. Likewise, in the United States, the social and economic crisis through which the automobile industry passed in the early 1980s has opened the Fordist system, with its rigid divisions of labor, hierarchical management structure, and fixation on a mass market, to a spirit of self-examination and innovation far greater than at any time since it was pioneered by Detroit's vehicle makers from 1911 to 1920.[1]

Thus, it is fitting that we reexamine the origins of this system in the early years of the century and the factors causing its enthusiastic adoption by American management but rejection by British management. In the United States a unique set of social forces and historical conditions made Fordism an effective managerial response to the labor relations crisis created by the enormous expansion of automobile industry employment in the years just prior to World War I. In particular, the Fordist system was based on the assumption that a workforce could be found, or created, which would accept higher wages in return for high-effort norms and unilateral managerial control of shop floor decision making. The hostile attitude of American employers and legislators toward unions, the relative defenselessness of a largely immigrant workforce, and the shift in American values from that of earlier "Yankee individualism" to a more consumer-oriented culture, all played a role in Fordism's American success.[2] In contrast, British managers did not believe that labor would either accept or could be forced to tolerate a system trading high wages for more effort and less shop floor autonomy. Through their unions and political channels, British labor demanded the reform of shop floor life along more democratic lines. Unable to break with the social and institutional traditions of the nineteenth-century labor movement, in which skilled workers played an organic role in the organization of shop floor production, British managers found the rigid management system embodied in Fordism both impractical and inefficient.[3]

Labor market conditions played a role in shaping acceptance or rejection of the Fordist system, but they were not, as some scholars postulate, the crucial variables in determining the outcome. Such an approach, which argues that the shortage of skilled labor in the United States forced the Ford Motor Company to experiment with new production systems, exaggerates the difference between the workforce skill profile found in the early British and American auto factories. It oversimplifies the complex process of institutional innovation by focusing our attention on the skill displacing methods employed by Ford rather than the social control mechanisms at the heart of the Fordist revolution. As we shall see, Fordism's main objective was not to fit labor demand to labor supply but rather to monitor the effort bargain of an expanding factory payroll and extract from the available workforce the maximum level of effort.[4]

Nor was Fordism simply the product of a mass market. It is true that this system emerged in the context of rapidly expanding demand for a cheap, standardized automobile. However, Ford's very high levels of output were not absolutely necessary to make his model of factory reorganization a viable management strategy. As we shall see, many elements of the Fordist system began reducing costs at annual sales levels in the low thousands, well within the reach of a number of British firms. Thus, their decision to adopt an alternative managerial system cannot be explained merely by differences in the scale of their respective national car markets.[5]

Although the changing character of labor and product markets provided the economic context in which Fordism emerged, its birth must be understood as an important chapter in the social struggle brought on by the spread of large scale factory production in the late nineteenth century. It advanced management's "frontier of control" a quantum leap, evolving out of but also moving beyond Frederick Taylor's scientific management movement. The Fordist revolution was made necessary not by the use of skill displacing machinery or the extreme division of labor but by the failure of these new machine systems, introduced between 1909 and 1913, to increase substantially managerial control of the effort bargain. To work, Fordism required a further reorganization of production departments along flow principles, a large wage increase to halt labor turnover, and a new labor relations system that curbed the independent authority of foremen and other low-level supervisors.[6] Successfully deployed by a self-confident group of American entrepreneurs, this Fordist ensemble would encounter stiff resistance on the very different social terrain existing in early twentieth-century Britain.

## Fordism and the Ford Factories in Detroit

The labor environment of Detroit was an important factor shaping the Fordist strategy. A booming, open-shop city, Detroit attracted a large number of immigrant workers and often experienced labor shortages in the years before World War I. Despite these tight labor market conditions, a strong union movement did not emerge until the 1930s. The Employers' Association of Detroit sought to regulate competition for factory labor and crush unions that tried to operate in the city. The association fought its first major battle with the International Association of Machinists in 1901, and by 1907 it had all but eliminated Detroit trade unions.[7] But the failure of unionism did not mean that Detroit employers were without a labor problem. Despite the aggressive labor recruitment drives of the Employers' Association, which attracted 20,000 workers to Detroit in 1910, local industry was unable to meet the demand for skilled and unskilled labor, forcing many firms to "pirate" the labor supply of their competition. By 1912 the situation had reached a crisis and under the direction of the association, Detroit employers called a truce. They halted competitive advertising for local labor and established a new fund to advertise for labor outside Detroit.[8]

Within the context of this tight, but union free, labor environment, the Detroit automobile producers grew and innovated. The organization of the early vehicle factories differed from Fordist factories in two important ways. First, in the manufacturing departments machines were organized functionally in contrast to the later adoption of line organization based on the kind of component being produced. Under the functional system components had to be transported between departments for successive machine operations. Second, these early auto factories assembled components and chassis on stationary trestles, and parts for each vehicle had to be delivered to each assembly stand. In the early years gangs of workers were given the responsibility for assembling an entire component or even vehicle; at a later stage the task was divided between a number of gangs, and gangs moved between assembly stations performing their tasks.[9]

The first real breakthrough into flow production in the assembly department came not at Ford, but rather at the neighboring EMF Company, whose two key managers had come from Ford. W. E. Flanders, one of the founding partners of EMF in 1908, had earlier been recruited by Ford to oversee the expansion and administration of the firm's manufacturing operations. But Flanders's stay with Ford was short, and when he left he took with him Max Wollering, who had been responsible for setting up and managing Ford's first venture into engine manufacture in 1906. EMF began production in August 1908 and in three years produced over 23,000 vehicles. Just prior to Studebaker's purchase of the firm in 1911, Flanders and Wollering installed a mechanized assembly line to boost chassis output to 100 per day. This first line had about thirty work stations, controlled by a clerk who recorded the order and completion time of each task. The entire line would then be moved forward. Wollering gave the following description of this first line: "We kept pushing them and then we thought we'd try to pull them [chassis]. We took two elevator drums, put a couple of motors on them with cables, and had proper hooks attached to the cable that would grasp the front axle of the car to tow it along."[10] This powered assembly line preceded Ford's first assembly line by at least two years. Why then did Ford delay the adoption of this system until late 1913?

Was it possible that Ford was not a sufficiently large producer to justify the capital expense of an assembly line and machine reorganization? This is unlikely as neither the cost of adopting the new system of factory organization nor the required level of output appears excessive. The first line at Highland Park in 1913 was 150 feet long and

**Table 2.1.** Output and Employment, Ford Motor Company, 1906–1913

| Year | Output | Employment |
|---|---|---|
| 1906 | 8,729 | 700 |
| 1907 | 14,887 | 575 |
| 1908 | 10,202 | 450 |
| 1909 | 17,771 | 1,655 |
| 1910 | 32,054 | 2,773 |
| 1911 | 69,762 | 3,976 |
| 1912 | 170,068 | 6,867 |
| 1913 | 195,954 | 14,366 |

*Sources:* A. Nevins, *Ford: The Times, the Man, the Company* (New York: Charles Scribner's Sons, 1954), 451; M. Wilkins, *American Business Abroad* (Detroit: Wayne State University Press, 1964), 436.

employed about 140 assemblers. It was a primitive affair designed to hold but five or six chassis.[11] The actual capital outlay in assembly departments was very small. As late as 1919 the Ford chassis assembly department accounted for only $3,490 worth of machinery, and in 1922 the entire capital outlay in the chassis department was less than $40,000.[12] Moreover, the net effect on the firm's capital requirements was probably negative because the assembly line permitted a reduction both in the number of partially finished vehicles one needed to hold in stock and in floor space.

Estimates of the scale of operation needed to make Fordism economically efficient suggest that, in manufacturing departments, a machine could be fully employed on a single piece with a yearly output of about 6,000. Machine specialization was necessary to allow machine departments to move to line organization and flow production. In the assembly areas an output of about fifty per day appears necessary to take advantage of assembly line production. In both cases, smaller savings could be had at lower levels of output.[13] Ford had easily exceeded these scale requirements by 1909 when annual output reached 17,771.

To understand why and when Ford moved to adopt assembly line methods one must look at the rapid growth in both output and employment. The relevant, and one might say spectacular figures, are presented in Table 2.1.

Ford's rapid expansion created two related problems. First, as the demand for output increased, the existing stationary method of assembly reached an upper bound beyond which production costs increased.[14] Second, as the workforce expanded, the existing institutions for monitoring and enforcing effort norms broke down. Ford faced a labor problem that was insoluble without a radical reorganization of production. The Fordist system, introduced in 1913, would successfully deal with both problems.

Under the stationary system of production, mobile trucks carried the stock along the rows of assembly stands dropping off necessary items. Eventually Ford required over 100 chassis assembly stations spread over two rows 600 feet long to meet production.[15] In machine departments, components had to be trucked between departments for successive machine operations. By 1913 congestion created by the transportation of parts stymied further expansion.[16]

The problems generated by the growth of employment are rather more complex but directly related to our thesis regarding the different behaviors of American and British management. Because Ford did little actual manufacturing in the early years, his workforce was surprisingly small. Total employment did not reach 1,000 until 1909, after which it grew rapidly. Meanwhile, other area vehicle producers employed significantly larger workforces: in 1910 EMF/Studebaker employed 5,700, Packard 4,640, Buick 4,000, and Cadillac 3,500.[17] Thus at Ford the rapid expansion of the workforce after 1910, in conjunction with the extreme division of labor, created a labor problem more keenly felt than by other local employers.

Until about 1905 or 1906, Ford shops had been run within a paternalistic framework, feasible given the relatively small size of the workforce but also necessary given Ford's dependence on skilled workers whose value to the firm gave them an implicit voice regarding how tasks were to be performed. Ford could be found regularly in the shops where he knew all the workers and the workers knew him as Henry or Hank. Work attitudes were described as cooperative, and all were enthusiastic about mounting production levels.[18] A movement away from this friendly paternalistic environment took place in 1907 and 1908 when employment grew too large for Ford to maintain personal contact with each worker. He was forced to delegate most responsibility for labor relations to low-level supervisors, now responsible for hiring, firing, and setting pay rates.[19] Labor problems were aggravated by the stationary system of assembly and the functional organization of machine departments, thus making it difficult for foremen to monitor individual workers or enforce effort norms.[20] In an attempt to keep productivity high, Ford introduced a profit sharing program in 1908 and resorted to crude time studies of labor tasks. Wollering described the system at Ford Manufacturing in 1907: "I had studies made on the various manufacturing operations. . . . We would get a man whom we had confidence in and who knew what he was doing as to whether it was a lathe or a screw machine or a grinder. He knew the fundamentals of it and he would take a stop-watch and operate the machine himself to get a fair idea of what could be done."[21] Work study was also begun in the foundry: "We [management] had to go out and make them to prove our point. We worked hard to do it and we did it. Of course, after we did it the other boys had to do it. This is essentially what I was doing back in 1908. I was setting work standards."[22]

By 1913 the rapid growth in employment levels introduced new strains on the system, although further division of tasks and the resulting reduction in average skill requirements opened up new avenues for monitoring and enforcing the effort bargain. Labor turnover reached massive proportions: 400 percent per annum at Ford, at least twice the rate of other vehicle plants at this time. Absenteeism was also on the rise with an average 10 percent of the workforce missing each day. In the words of one writer, "They are conducting a continuous, unorganized strike."[23] The tight labor market, high levels of worker mobility (due to the low level of skill needed to perform most jobs), and autocratic foremen all contributed to Ford's labor problem. Ford responded by creating a centralized employment office, establishing a new wage grade system, and curbing the authority of foremen to hire, fire, or promote the workers they supervised. Low-level supervisors could refuse to employ a worker in their department, but the employment office now decided whether the worker should be fired or simply transferred.[24]

By weakening the control low-level supervisors exercised over hiring and firing, top management also reduced their ability to enforce effort norms. Figures supplied by Stephen Meyer point to a drastic fall in the rate of unit productivity improvement. In 1911 labor productivity increased 41 percent, but in 1912 it grew only 15 percent and in 1913 a mere 4.7 percent.[25] The simplification and routinization of work, a process completed at Ford by early 1913, had thus failed to resolve the company's labor problem. The stationary system of production proved increasingly less efficient with the higher levels of output, while the existing management infrastructure could not effectively discipline the soaring number of workers.

Fordism resolved these problems. The company introduced the assembly line and flow production within months of the adoption of the new labor relations system, and the five-dollar day was adopted just three months later. Over a period of less than twelve months Ford had altered the entire basis of production in his factories. A contemporary Ford manager observed, "We are trying to make you do things with a mechanical pace setter; that is my own version of this thing . . . I never thought that would take too kindly with the average working man. He didn't like to be put on a tread mill, you know, that was the idea."[26] The success of the new strategy depended upon labor's willingness to accept regimentation, unilateral managerial control of shop floor decisions, and higher effort norms in exchange for the chance to earn relatively high wages. It is beyond the scope of this paper to explain in detail why American workers were willing to accept this bargain. In part they had little choice; the lack of trade unions and the vulnerability of immigrant workers forced many to accept Ford's offer. The rise of a consumer-oriented society certainly played a role. Moreover, many members of the Detroit workforce held the characteristically "peasant" dream of one day returning to Poland, Italy, Ontario, or Kentucky with a nest egg large enough to rebuild their shattered agrarian lives on native soil.[27]

## History of British Motor Vehicle Production, 1895–1913

The British motor vehicle industry has been criticized, as have most British industries in this period, for slowness to adopt American mass production techniques. Several scholars have argued that the vehicle makers were slow to expand, concentrated exclusively on luxury products, and continued to employ inefficient craft methods of production.[28] The evidence concerning the slow rate of expansion is clear. In 1907 Britain was the only major producer of motor vehicles unable to supply domestic demand.[29] The evidence concerning the industry's product mix is less conclusive. The luxury producers—Rolls Royce, Daimler, and Napier—were leading firms. But careful scrutiny of the models offered by other firms in the industry suggests that at least until 1906, and again after 1912, the British industry produced a large volume of less expensive models destined for the utility market. Firms such as Sunbeam, Singer, Vulcan, and Vauxhall produced mainly for this utility market before 1906, and after 1912 Humber, Rover, and Standard expanded rapidly on the strength of their small car lines. Further evidence of the extent to which British producers were supplying the small car market can be found in Daimler's turn-of-the-century decision not to produce an inexpensive model. The firm argued that there were already too many British small car producers in the market. The years 1906 to 1911 did witness a shift from small cars toward the luxury models. Former small car firms, such as Vauxhall and Sunbeam, concentrated on the larger models, but this trend was reversed in 1912 when low priced American producers threatened an invasion of the British market.[30]

The British auto industry made a larger shift to skill displacing, mechanized production methods than has usually been recognized. The failure of Britain to develop a mass market did limit the length of production runs and made some forms of mechanization less economical. However, the need for accurately machined components encouraged mechanization, particularly in the engine and chassis departments where the tastes of the British consumer were less of a factor. British vehicle producers' close ties with British cycle makers aided their shift to mechanization. The cycle trade had been slow to adopt the repetition methods pioneered by the American firms in the 1880s, but the next decade witnessed a rapid conversion. The professional journals reported a large demand by the cycle makers in 1896 and 1897 for repetition machinery, and contemporary reports suggest that by 1900 firms such as Rudge, Rover, Humber, and Northfleet had made the conversion to repetition production with unskilled labor. In his study of the British cycle trade, A. E. Harrison concluded, "it cannot be confidently asserted that the British techniques of cycle manufacture were markedly inferior, though nearly all the major developments of the 1890s in automatic and semi-automatic, and labour saving cycle-making machinery emanated from the U.S.A."[31]

The cycle trade's experience with repetition production was quickly absorbed by the young vehicle industry. As early as 1897 contemporary reports suggest that in Midland, where there was a concentration of vehicle makers, rapid progress was being made in the use of new machine techniques and the employment of unskilled labor.[32] Reports of the first Daimler factory indicate that a large stock of American machines was installed and careful attention was paid to machine organization to minimize transportation. By 1899 Daimler further added to the machine stock, including self-sequencing lathes, multiple head boring machines, and milling machines. The latter were particularly important because they replaced lathes as the work horses in repetition machine shops, making it possible for the firm to employ a high percentage of boy labor. In 1899, when the adult workers called in the Amalgamated Society of Engineers over the issue, one in four machine operators was a boy, and in the turning department over half the operators were boys.[33]

Daimler was not the only plant moving in the direction of repetition production with less skilled labor. At Humber, "Advantage has been taken of the long experience which Humber have had in the employment of labour saving tools for the production of bicycle parts, and it would be difficult to find a better regulated machine tool shop, or one with a greater output for its size. The machines are all worked at their maximum capacity, high speed steel being employed throughout . . . limit gauges and micrometer working are universal."[34] Advances were also made at the Belsize works in Manchester, described by the *Motor Trader* as the Anglicized embodiment of American methods and organization.[35]

By 1914 the lessons of American repetition production had been widely absorbed by the British vehicle producers. In that year, a management consultant argued that "in the motor trade . . . a large portion of the workers were either turret hands who do not do much more than pull certain handles, or milling machine hands who only put work in a fixture and let it go, having the speed and feed set for them, and jig drillers."[36] Similarly L. A. Legros, president of the Institute of Automobile Engineers, argued that by 1911, "the tendency in the bigger factories is to diminish the amount of responsibility left to the individual worker in respect to the employment of . . . shop knowledge."[37]

Most scholars have underestimated the extent to which repetition methods and less skilled labor were employed because the rate of change from craft to machine production differed so markedly throughout the factory. Most body production and the final assembly of the luxury models remained a craft operation until after World War I. But the majority of manufacturing operations on the engine and other components had been converted to repetition methods by a much earlier date. The *Automobile Engineer's* description of the Daimler works in 1907 reveals the uneven rate of technical progress:

> It is a curious fact that although the firm under discussion wasted many pounds in hand fitting parts, yet the engine erecting was carried out on a totally different system. . . . There was exceedingly little which was hand fitted on any of the engines . . . any part, which during the operation was found to be of a size other than that necessary, had to be returned to the machine shop and replaced with one which would fit . . . very few of the engine gangs possessed a file or other instrument which could be used for fitting in the ordinary sense of the word.[38]

Another factor receiving little attention in studies of the British vehicle industry is size of the workforce. It is certainly true that the production scale of the British firms lagged well behind those of the leading American producers. Output levels of 1,000 were considered large in the pre-1914 years, and Humber's output of 2,500 was considered exceptional.[39] But these low production runs are deceiving. Many American producers concentrated on assembling components purchased from outside suppliers, and even when these firms began to increase the level of in-house manufacturing the large component suppliers continued to play an important role. When Ford began manufacturing engines in 1906, he still purchased the more complex components from external suppliers. As late as 1913, 84 percent of the cost of the Ford Model T represented the cost of material and components purchased from outside suppliers, while less than 7 percent of costs represented labor employed by Ford.[40] The extensive use of component suppliers allowed Ford to produce over 17,000 vehicles in 1909 with less than 1,700 workers. On the other hand the early British producers both manufactured and assembled their products and hence required large workforces. By 1906 Argyll employed 3,000; Daimler had 2,000 in 1906, jumping to 4,500 in 1910 and 5,000 in 1913; and Humber employed 5,000 by 1908, and Wolseley 4,000 in 1913.[41] Thus in the critical pre-1914 period, when car producers were formulating strategies for the new industry, British managers had to oversee a far more complex production process than did the American assemblers. The latter had merely to create institutions to coordinate a much simpler production process and then gradually adapt them as they moved into manufacturing areas.

The vigor of the prewar union movement in Britain made it more difficult for management to monitor these large factory workforces relative to the American scene where heavy industry remained virtually unorganized until the late 1930s. Between 1910 and 1926 British unionism was more clearly on the offensive than in any other period of its development.[42] Membership advanced from 2.5 million in 1910 to almost 8.5 million a decade later. Officially recorded strikes, which had averaged 500 a year in the previous decade, escalated to three times that number in 1913; and, after a lull caused by the wartime "industrial truce," rose to a new peak in 1920.[43] Moreover, a radical, syndicalist mood suffused this growing movement; as British workers became increasingly disenchanted with parliamentary politics they turned with greater enthusiasm to direct action, shop steward organizations, and a demand for "workers' control."

Such an aggressive posture deeply affected British management, especially in the engineering industries, where a powerful shop stewards movement had arisen during the First World War. As labor became more vocal in its demand for a democratic form of industrial organization, British management came to realize that such radical views were widely and deeply held. In Birmingham, a group of employers suggested that the end of the war would require, "the surrender by Capital of its supposed right to dictate to Labour the conditions under which work shall be carried on."[44] As late as 1921 the chairman of the board of directors of the Daimler Company complained to his stockholders that labor has been "convinced that the conditions of which they complain result from the greed and fraud of the wealth owning section of the community. They have been taught, and now sincerely believe, that because labour is necessary to the creation of all wealth that, therefore, all wealth is created solely by labour."[45]

Faced with the specter of a highly politicized, potentially revolutionary working-class movement, a section of the British managerial elite recognized that a simple reassertion of capital's "right to manage" would prove unnecessarily confrontational and instead sought to channel and deflect this movement onto safer terrain. The so-called Whitley Councils, which proposed the postwar establishment in each industry of permanent joint bodies of employers and union officials, represented British capital's most well publicized effort to restore "orderly" industrial relations to the shop floor, while at the same time accommodating and deradicalizing the pervasive demand for a working-class voice in the management of industry.[46] Whitleyism left little mark once the postwar industrial crisis had passed, but this did not mean that British managers now felt free to reorganize the factory work regime in any fashion they chose. At the shop floor level especially, the managerial innovations adopted by their American counterparts seemed seriously at odds with both the craft tradition of the engineering workers and the syndicalist impulse so recently manifest throughout the British labor movement. Instead, British employers opted for an industrial relations strategy which provided indirect control of the labor process via an incentive wage payment system. Labor would be allowed to retain a degree of autonomy over the pace of work and technique at the actual point of production, but only because this very limited element of "workers' control" would occur within a piece rate system in which management alone set the rate schedule. Thus the chairman of the British Higher Productivity Council argued in 1919, "The whole point is that workmen now say that they want a share of the control of business and this scheme [incentive payment systems] gives them the share that they want."[47]

Incentive pay also proved popular in the United States, but the limited information we have suggests that it played a different role there, largely because bonus levels rarely reached the British norm of 50 to 100 percent.[48] The relatively high bonus earnings possible in Britain meant that incentive payment systems not only checked labor effort but also encouraged labor to perform some minor tasks of factory coordination needed to insure steady output. Under such schemes, where the bonus represented up to half of all earnings, workers had a direct monetary incentive to look after their own parts supply problems and ensure that basic maintenance was performed on the machines being used.[49] As Steven Tolliday and Jonathan Zeitlin argue in their contribution to this volume, "the line between job control and self-supervision was thin."

Daimler is an interesting case in point. Despite the extensive use of American machines, factory coordination and productivity were very poor in the early years. After much investigation the firm formalized the managerial hierarchy, reduced the number of low-level supervisors, and adopted first straight piecework and then the Premium Bonus system. P. Martin became factory manager in 1901; he wanted to replace the "driving" system of supervision with a system he called "induction." The Premium Bonus system was the key to this strategy. Under this scheme, management set minimum work norms, and workers earned a bonus once these production criteria were met. The critical factor though, and the one Martin stressed, was that the worker would not receive the entire savings in labor cost generated by higher output levels. Instead the savings would be divided, usually fifty-fifty, between firm and worker. Martin argued that since unit labor costs would fall automatically as output rose, management would have less incentive to cut piecework prices and labor could increase output and earnings without any fear of a future price reduction.

Martin's managerial reforms operated at two levels. On the one hand, he strengthened management by introducing a more comprehensive costing and planning system and the beginnings of a work study department. On the other hand, he weakened managerial control on the shop floor by having workers perform some of the basic coordination tasks and by giving labor some say in setting effort norms. As Martin told the 1906 meeting of the Institute of Automobile Engineers:

> Instead of the staff driving the workers for output, the workers drive the staff to supply them with material, and jobs are finished off by virtue of fresh ones pushing them out. . . . The usual driving system is absolutely reversed. The staff, and everybody connected with the company is kept on the jump all the time to keep the man supplied with material, so that he can earn more wages; and I think the greatest advantage of the bonus system is to be found in that very fact.[50]

This system of induction was only beginning to emerge in the pre-1914 period. It is easy to see how such a system might be attractive to managers operating factories more complex than those in America and where the labor force was less willing to accept autocratic managerial control of the labor process. This trend toward granting workers limited control over effort norms while management controlled the effort bargain through incentive pay stands in direct contrast to the Fordist system. Thus, it was not predominantly the skill content of their jobs that distinguished British workers from their American brothers and sisters. Rather, the British workforce was unique in its ability to ward off direct managerial control of effort norms and its ability to retain a voice in how work was to be done within the relatively narrow management constraints. In the final section of this paper I will examine how the induction system spread through the British vehicle industry and provided an alternative to American managerial strategies.

## Fordism in Britain, 1913–1930

Fordism spread to Britain in 1911 when Ford began producing cars at its Manchester plant. Initially, knocked-down kits were shipped from Detroit and reassembled; however, the cost of shipping bulky components such as bodies and gasoline tanks led Ford to begin manufacturing in Britain a year later. Ford proposed to manufacture these components under conditions similar to those found in its plants in Detroit. New employees hired as handymen were expected to perform any task management requested. A 1915 description of the plant confirmed that many of the jobs were designed for unskilled labor: "All parts are machine cut to templates and jigs, and little indeed is left to be done by hand, except actual putting together. The processes in the department of this factory are almost automatic, little being left for hand work beyond the insertion of screws."[51]

Ford's unwillingness to recognize existing labor customs and practices or grant any measure of control over the production process to individual workers or to the unions generated a rash of small strikes as early as 1912. Toward the end of that year, the United Kingdom Society of Coachmakers shut down the body plant for twenty-two weeks in opposition to the "American System." "We are now engaged in a struggle with the Ford Motor Company," declared the leadership of the union, "who are hoping to Americanise our Trade here, by claiming the right to employ who they like upon our work and the work of other skilled Trades, and to entirely ignore Trade Unions."[52] This strike received little more than moral support from the other unions in Manchester, and chassis assembly continued during the stoppage using bodies imported from Detroit.

Despite the strike's failure, wages rose rapidly at Ford's Manchester plant. When P. Perry, the British managing director, first arrived on the scene, he viewed the existing hourly wage of four and one-half pence per hour a starvation rate. When the coachmakers strike began, the minimum hourly rate was 5 pence per hour, but this soon rose to 6 pence and over the next few months was allowed to drift up to 10 pence per hour. These increases all took place before Ford introduced the five-dollar day in Detroit, after which a further 5 pence per hour profit-sharing bonus was added to the Manchester wage.[53] The combination of high wages, employer resistance to trade unions, and the reduced skill demands, which made workers easily replaceable, all but eliminated labor resistance to the new system of production and allowed Ford to press forward with the Americanization of the Manchester plant. In September 1914, the firm installed a mechanized assembly line, less than twelve months after Ford had adopted a similar line in Detroit.[54]

Although British workers and their unions were initially hostile to the American system, the tripling of wages at the Manchester plant between 1912 and 1914 may well have altered their viewpoint. Both the reminiscences of former workers and the lack of organized agitation in the nonunion Ford plants over the next thirty years suggest that British labor was at least sympathetic to the deal Ford offered. On a number of occasions union officials indicated they were willing to make concessions in return for Ford style wages. For example during the 1925 negotiations between the Engineering Employers Federation and the unions, the following exchange took place:

> Chairman EEP: "Do you think for one moment your men would submit themselves to the principle of Henry Ford?"
> Union reply: "They do it . . . our people seem quite well satisfied. It is true that for a period they did not take to it, but today they are quite satisfied with their employment."[55]

Moreover, British unionists, perhaps not fully understanding the contradiction between the American system of management and their own desire for democratic shop floor decision making, advocated the Americanization of their country's industry. J. T. Brownlie, the leader of the Amalgamated Society of Engineers, the strongest union in the vehicle industry, pushed for the modernization of British industry on the American model amidst the labor turmoil of 1919.[56] And two years later the Trades Union Congress also condemned British factory organization and pointed to American factories as a model: "The British engineering industry is, with few exceptions, badly organised from one end to the other. Therefore compared with many American establishments, and even with the best in this country, the great majority of British engineering workshops are badly laid out, often badly lighted, inadequately equipped, and, many of them, wastefully managed."[57]

But leading managers of British vehicle factories viewed American managerial innovations differently. Arrol Johnston's T. C. Pullinger argued against American methods and the rigid managerial control they entailed and in favor of "kindly and sympathetic" treatment of labor.[58] A. W. Reeves from Crossley argued:

> An important factor to the author's mind, and one which appears to be entirely ignored in the wonderful systems on the other side of the Atlantic, and among many idealists on this side, is that of the personal or human element. Anyone with any knowledge of the independent and, it must be confessed, awkward spirit, characterising the workers of say the Northern Midlands, would hesitate before applying the extreme methods of the latest American Scientific Management, well knowing the futility of the task.[59]

Managers of this viewpoint emphasized cooperation between labor and management and self-regulation of effort norms rather than the direct managerial control through machine pacing and close supervision used under Fordism.[60]

A detailed examination of the Austin managerial strategy reveals that it carried the pre-1914 Daimler system of "induction" into the interwar period; much evidence suggests that British managers saw this strategy as an alternative to the Fordist system. By the 1920s Austin was a mass producer and had achieved the level of output needed to adopt full-scale Fordism. Production reached 9,500 in 1924, rose to 25,000 in 1926, and exceeded 40,000 after that.[61] Some changes in the production process were made, but the system adopted differed significantly from American Fordism. Assembly lines were installed after 1924, but they remained relatively simple and unmechanized until the late 1920s. More important, Austin's system of labor relations was vastly different from Ford's. In the tradition of the Daimler induction system, the Austin workers were placed on incentive payment systems and were allowed to earn bonuses, often exceeding 100 percent. Self-regulation rather than direct control seemed to be the organizing principle at the Austin shops.[62]

Austin's observation of Fordist methods during visits to the United States certainly influenced his labor relations strategy, but not on terms that American automobile manufacturers might have approved. What impressed Austin about the Ford factories was not the use of sophisticated machinery but rather the amount of labor effort forthcoming from the workforce. He was impressed that, "everybody in the establishment seemed to be trying to do their best."[63] This led him to argue that if Britain were to compete with the United States it needed an improved spirit among labor, not new machine methods. The Austin focus on organization and shop floor social relations, rather than changes in production technology, was also evident among the leading managers of his firm. The works director held the view that "good organization and an indifferent plant achieve better results than a good plant and indifferent organization," and the head of the costing department argued that "the obvious difficulty at the moment is the lack of confidence as between employer and employed."[64] Evidently the Austin factory was facing labor and coordination problems similar to those found in the Ford works in 1913, but Austin proposed resolving his problems in a very different way.

Control over labor and the coordination of shop floor activity were to be achieved by a novel payment system called "Bonus on Time." Under this scheme prices were set in units of work time rather than money. At the end of each week workers tallied their time credits and multiplied this by their base rate to determine their final pay. Costing department head P. Keene suggested that the system worked because with "such a basis, many economic problems become common to both employers and employed, and interests flow in one direction. . . . The reason why the system of control became really efficient was that they inculcated into the whole staff a maximum idea of personal responsibility to the firm itself whereby they and the firm were likely to prosper."[65] Indirect managerial control over labor was enhanced by the custom of paying workers a base rate that was on average 25 percent lower than they could expect in other Midland shops and then allowing them to earn above average bonus rates. To Keene, the efficiency of the payment system was obvious. "The remuneration he [the worker] is able to obtain through savings is a sufficient incentive to the worker to make large output effective with the minimum of supervision."[66] Indeed, Austin managers argued that their system was an alternative to American production methods:

> There are still a few employers who object to piecework on principle. Their stand-point is that an efficient management ought to be able to get the same results at an agreed rate of wage without having to pay more money to encourage the men to work harder. . . . Some form of extra wage must be paid to a man if he is expected to work harder. The only alternative is to pay a high wage similar to the Ford system, and insist upon task achievement. The obtaining of results by this system could not be regarded with favour by an Employers Federation, as the advantage would be to one particular firm only. When every other concern came into line, the status quo would be again obtained, and a circle

> of rising wage competition would begin. . . . The daily task system at fixed wages may perhaps, be workable in American, or even Continental factories, but the necessary . . . driving works policy would not be acceptable either to English Labour or Management.[67]

Those British managers who favored the Fordist system were severely criticized by their colleagues. Thus Mr. Cole of the Employers Association ridiculed one Rover executive: "Captain Wilks, to my mind is suffering from some rather ill-digested views with regard to Capital and Labour. He is a great admirer of Mr. Ford and American methods. His idea is that everybody should receive a high day rate and then be compelled to work as hard as possible and if they do not they are to be fired out."[68] Likewise, G. Rushton, of the Associated Equipment Company, was diagnosed as having "Forditis" when he put elements of the Ford system into practice. After Associated Equipment installed moving assembly lines and high fixed day wages in the late 1920s, the London Engineering Employers Federation threatened to expel the firm from its ranks unless Associated Equipment made changes in its production system. The Employers Federation argued that "the essential difference between his [Rushton] Southall scheme, and schemes in operation at Walthamstow and at other federation firms was a payment in anticipation [of output], whereas the scheme approved by the Association were payments made after the results had been assured."[69] Thus, British management could not envision a production system without a direct link between wages and output as a check on labor behavior and as an aid to factory coordination. The Fordist notion that both could be embodied in the system of machine operation did not suit the particular needs or the general world view of British managers. Even when British workers voiced an interest in a Fordist effort bargain, British management remained suspicious. The managerial debacle associated with the industry's shift to measured day work in the late 1960s, which contributed to the collapse of the last of the British mass producers, BLMC in 1975, suggests that management's reluctance to directly control shop floor organization may have been well founded.

## Conclusions

The failure of the Fordist system to gain more than a foothold in the British vehicle industry confirms the overriding importance of social and political relations in determining the organization and technology of factory work. In the United States Ford's ability and willingness to rationalize production and increase supervision were predicated not so much upon the existence of an automobile market markedly larger than that in Britain nor relative wage rates favoring the use of less skilled workers; rather they were based upon the availability of a fragmented workforce with little tradition of organized resistance to management authority, a workforce willing to tolerate rigid managerial control of effort norms in return for relatively high wages. In contrast British managers, whose complex manufacturing operations necessitated careful shop floor coordination, were confronted with a labor movement whose traditions were hostile to labor control methods inherent in American Fordism.

A species of worker self-management, largely keyed to a wage incentive system, proved the durable solution to this managerial dilemma. Initially, such incentive wage schemes were designed to control effort norms, but in the British context they came to have a larger social function. They partially accommodated the powerful working-class desire for autonomy and control at the point of production, and from a management viewpoint they served as a substitute mechanism for improving factory coordination and fulfilling tasks which in American firms were performed by foremen or building superintendents. For the British workers, their desire for job control had been converted by management into a potent system of self-regulation. Thus, despite transatlantic similarities in production technology and product, this unique managerial outlook created a twentieth-century factory regime in the British vehicle industry which differed substantially from that of their counterparts in the United States.

## NOTES

1. See E. Rothschild, *Paradise Lost: The Decline of the Auto-Industrial Age* (New York: Vintage, 1973); R. B. Reich, *The Next American Frontier* (New York: Times Books, 1983); W. J. Abernathy, K. B. Clark, and A. M. Kantrow, *Industrial Renaissance* (New York: Basic Books, 1983); M. J. Piore and C. F. Sabel, *The Second Industrial Divide, Possibilities for Prosperity* (New York: Basic Books, 1984).

2. The production methods Ford employed have been extensively stud-

ied. See S. Meyer III, *The Five Dollar Day: Labor Management and Social Control in the Ford Motor Company, 1908–1921* (Albany: State University of New York Press, 1981); D. A. Hounshell, *From the American System to Mass Production: 1800–1932* (Baltimore: Johns Hopkins University Press, 1984); W. Gartman, "Auto Slavery: The Development of the Labor Process in the Automobile Industry of the United States: 1897–1950" (Ph.D diss., University of California, San Diego, 1980); J. Russell, "The Coming of the Line: The Ford Highland Park Plant, 1910–1914," *Radical America* 12 (1978): 29–45; A. Nevins, *Ford: The Times, the Man, the Company* (New York: Scribner, 1954). On the shift to a consumer-oriented society see D. T. Rodgers, *The Work Ethic in Industrial America, 1850–1920* (Chicago: University of Chicago Press, 1974). See also J. O'Connor, *Accumulation Crisis* (New York: Blackwell, 1984), 68–79.

3. On attitudes to Fordism in Europe see C. Maier, "Between Taylorism and Technocracy: European Ideologies and the Vision of Industrial Productivity in the 1920's," *Journal of Contemporary History* 5 (1970): 27–51. See also P. Fridenson, "The Coming of the Assembly Line to Europe," in W. Krohn, E. T. Layton, Jr., and P. Weingart, eds., *The Dynamics of Science and Technology* (Dordrecht: D. Reidel, 1978). On the history of British labor see K. Burgess, *The Challenge of Labour* (London: Croom Helm, 1980); R. Charles, *The Development of Industrial Relations in Britain, 1911–13* (London: Hutchinson, 1973); G. Brown, *Sabotage* (Nottingham: Spokesman Books, 1977); B. Holton, *British Syndicalism, 1900–1914* (London: Pluto Press, 1976); and J. Hinton, *The First Shop Stewards Movement* (London: Allen and Unwin, 1973).

4. For arguments concerning the possible impact of American social factors on American technology see E. S. Ferguson, "The Americanness of American Technology," *Technology & Culture* 20 (1979): 3–24.

5. See C. F. Sabel, *Work and Politics: The Division of Labor in Industry* (Cambridge: Cambridge University Press, 1982). On levels of output employed see L. V. Spencer, "Conveyor Systems Aids Big Production," *Automobile* 35 (July 20, 1916); *Ford Times* 8 (April 1915): 299; W. Klann, Rough Draft Reminiscences, 112, Ford Motor Company Archives, Henry Ford Museum, Dearborn, Mich.; Nevins Selected Research Papers, Acc. 572, Box 21, Ford Archives. For an interesting British view that the principles of Fordism were applicable at relatively low levels of output see F. G. Woollard, *Principles of Mass Production* (London: Iliffe and Sons, 1954), 15–48.

6. For a history of American managerial methods see A. Chandler Jr., *The Visible Hand: The Management Revolution in American Business* (Cambridge, Mass.: Harvard University Press, 1977); D. Nelson, *Managers and Workers: Origins of the Factory System in the United States, 1880–1920* (Madison: University of Wisconsin Press, 1975); D. Noble, *America by Design* (Oxford: Oxford University Press, 1977). For a survey of the evolution of American managerial throught see M. Rose, *Industrial Behaviour, Theoretical Development since Taylor* (Harmondsworth: Allen Lane, 1975).

7. G. Heliker, "Detroit Labor: 1890–1910," Acc. 958, Ford Archives; D. Montgomery, *Workers Control in America* (Cambridge: Cambridge University Press, 1979), 55.

8. Nevins, *Ford*, 513–18; Heliker, "Detroit Labor: 1890–1910," 35; G. Heliker, "Detroit Labor, 1900–1916," 8, Acc. 958, Ford Archives.

9. H. L. Arnold and F. L. Faurote, "Ford Methods," *Engineering Magazine* 47 (1914): 673; Litogot, "Reminiscences," 7, Ford Archives; Dickett, "Reminiscences," 13–14, Ford Archives.

10. Max Wollering, "Reminiscences," 43, Ford Archives; E. K. H., "Factory Transportation II," *Machinery* 11 (November 1917): 117–19; A. R. Erskine, *History of the Studebaker Corporation* (Studebaker Corp., 1924), 33.

11. Litogot, "Reminiscences," 7; Arnold and Faurote, "Ford Methods," 673.

12. Department Appraisals 1919, Acc. 73, Ford Archives; Plant Accounts, Highland Park, Acc. 571, Ford Archives.

13. Spencer, "Conveyor System Aids Big Production"; G. D. Babcock, *The Taylor System in Franklin Management* (New York: Engineering Magazine Company, 1917), 125. In 1927, A.E.C. operated an assembly line in England with a planned output of eighty to one hundred chassis per week. See EEF Membership Files, A.E.C., Interview, March 21, 1927, EEF Archives; and "The Work of the Associated Daimler Company," *Automobile Engineer* (January 1928): 9–11.

14. For a similar thesis arguing that machine innovation might be caused by the need to increase output levels rather than productivity see E. P. Duggan, "Machines, Markets and Labor: The Carriage and Wagon Industry in Late Nineteenth Century Cincinnati," *Business History Review* 51 (1977): 308–25.

15. H. L. Arnold and F. L. Faurote, "Ford Methods," *Engineering Magazine* 47 (1914): 673; Litogot, "Reminiscences," 7; Dickett, "Reminiscences," 13–14.

16. Litogot, "Reminiscences," 7; Klann, "Reminiscences," 10.

17. Nevins, *Ford*, 532.

18. A. M. Wibel, "Reminiscences," 52, Ford Archives; Nevins, *Ford*, 271.

19. G. Heliker, "Detroit Labor: 1900–1916," 25.

20. See Gartman, "Auto Slavery," 149.

21. Wollering, "Reminiscences," 26.

22. Klann, "Reminiscences," 7.

23. Meyer, *Five Dollar Day*, 82–83, 85.

24. Ibid., ch. 5. See also G. Heliker, "Detroit Labor: 1900–1916."

25. Meyer, *Five Dollar Day*, 72.

26. Wibel, "Reminiscences," 18.

27. Russell, *Coming of the Line*, 30–31. See also Sabel, *Work and Politics*.

28. Most generalizations about vehicle producers during this period stem from S. Saul, "The Motor Industry in Britain to 1914," *Business History* 5 (1962): 22–44. For a more sympathetic treatment of mass production in Britain in the nineteenth century see A. E. Musson, "J. Whitworth and the Growth of Mass Production Engineering," *Business History* 17 (1975):

109–49. For a more recent study of vehicle builders during this period see R. Church, *Herbert Austin* (London: European Publications Ltd., 1979).

29. *Motor Trader* (October 7, 1908).

30. Data on the model lines of various firms were obtained from various issues of *Motor Trader* and W. C. Bersey, *The Motor Car Red Book* (n.d.). On Daimler see, Simms Papers, University of London, files 16/10 and 16/119/i; Daimler Board of Directors, Minute Books, February 2 to March 14, 1899, February 11, 1901, June 11 and January 29, 1902. Sixth Annual General Meeting, Daimler Company, November 27, 1902. On the trend toward larger models after 1906 see J. D. Siddley's column in the *Times Engineering Supplement* (1905 and 1906).

31. A. E. Harrison, "The Competitiveness of the British Cycle Industry," *Economic History Review* 22 (1969): 294. See also *Engineer* (June 18, 1897): 620; J. Newton, "Looking Backward," *Rudge Record* (1909); "Repetition Bicycle Plant," *The Cycle Referee* (February 16, 1899, supplement); "The Firms You Do Business With," *The Cycle and Motor Trades Review* (June 7, 1906): 541. See also *Rudge Record* (October and December 1908).

32. "The Engineering Strike," *Engineer* (August 27, 1897): 207.

33. ASE Minute Books, Coventry, July 12–26 and August 15, 1899. See also Simms Papers, University of London, file 16/33; "The English Motor Industry: Description of the Plant and Practice of the Daimler Company Works," *The Cycle Referee* (January 19, 1899, supplement): ii–xiv.

34. "The Humber Works (Beeston)," *Engineer* (September 4, 1903): 232–35.

35. *Motor Trader* (October 8, 1913): 101, (September 7, 1910): 1036, (August 7, 1912) and (December 10, 1911).

36. Owen Linley, "Manufacturing on a Medium Scale," *Motor Trader* (July 8, 1914).

37. L. A. Legros, "Influence of Detail in the Development of the Automobile," *Proceedings Institute of Automobile Engineers* (October 11, 1911): 5–6.

38. "Erecting Shop Methods," *Automobile Engineer* (July 1912): 216.

39. Saul, "Motor Industry"; *Motor Trader* (March 7, 1906).

40. Model T cost books, Acc. 125, Ford Archives. See also Wollering, "Reminiscences," 7–13.

41. *Royal Commission on Motor Cars*, 1906, cmd. 3081, 153–54; *Motor Trader* (July 4 and August 15, 1906, and March 18, 1908); St. John Nixon, *Daimler 1896–1946* (London: G.T. Foulis, 1946), 112. On the reluctance of British firms to supply components to British motor vehicle producers see R. Church, *Herbert Austin*, 186; G. Maxcy, "The Motor Industry," in P. L. Cook and R. Cohen, eds., *Effects of Mergers* (London: Allen and Unwin, 1958), 360.

42. See S. Meacham, "The Sense of an Impending Clash: English Working-Class Unrest before the First World War," *American Historical Review* 77 (1972): 1343–63; J. White, "1910–1914 Reconsidered," in J. E. Cronin and J. Schneer, eds., *Social Conflict and the Political Order in Modern Britain*

(London: Croom Helm, 1982); J. E. Cronin, "Coping with Labour: 1918–1926," in Cronin and Schneer, eds., *Social Conflict*. On the history of trade unionism in Britain see K. Burgess, *The Challenge of Labour* (London: Croom Helm, 1980). On the history of unskilled unions in manufacturing see R. Hyman, *The Workers Union* (Oxford: Clarendon Press, 1971); J. Hinton, "The Rise of a Mass Labour Movement: Growth and Limits, " in C. J. Wrigley, ed., *A History of British Industrial Relations: 1875–1914* (Brighton: Harvester, 1982).

43. Statistics taken from R. Hyman's foreword to C. Goodrich, *The Frontier of Control* (London: Pluto Press, 1975), vii.

44. Cited in Cronin, *Coping with Labour*, 113.

45. Coventry Record Office, Acc. 594, box file 8, B.S.A., General Meeting, April 11, 1921, 9.

46. For details on the impact of Whitleyism see R. Charles, *The Development of Industrial Relations in Britain, 1911–35* (London: Hutchinson, 1973).

47. Letter from Howe to EEF, October 29, 1919, P(13)5, EEF Archives. On the response of labor and capital to early twentieth-century political unrest in other countries see C. S. Maier, "Between Taylorism and Technocracy: European Ideologies and the Vision of Industrial Productivity in the 1920s," *Journal of Contemporary History* 5 (1970): 27–61.

48. On British bonus levels see Coventry District Engineering Employers Federation, Board Minutes, February 27, 1922, July 27, 1931; Local Conference, December 22, 1919, P(5)49, EEF Archives.

49. On the impact incentive payment systems can have on labor see M. Burawoy, *Manufacturing Consent* (Chicago: University of Chicago Press, 1979).

50. P. Martin, "Works Organisation," *Proceedings Institute of Automobile Engineers* 1 (1906): 126.

51. *Automobile Engineer* (July 1915): 189; Ford Motor Company, "Historical Notes," 1912, Ford Archives, Warley.

52. *UKS Quarterly Journal* (July 1913). See also M. Wilkins and F. Hill, *American Business Abroad, Ford on Six Continents* (Detroit: Wayne State University Press, 1964).

53. Wage rates supplied by the Ford Motor Company, Dagenham. See also *Special Edition of the Ford Times* (1914) and the *Automobile Engineer* (1915).

54. Despite Ford's success in Americanizing his British plants, he had much more trouble producing a product acceptable to British consumers. Austin and Morris were more successful, and it was not until the late 1930s that Ford began to erode their market share. On the reasons for this see R. Church, "The Big Three: Competition, Management and Marketing in the British Motor Industry, 1922–1939," in B. Supple, ed., *Essays in British Business History* (Oxford: Oxford University Press, 1977).

55. Special Conference, May 1, 1925, EEF Archives. On British labor attitude toward the Ford operation see *Manchester Evening News* (June 1978).

56. Conference EEF and ASE, July 24, 1919, 29, EEF Archives.

57. "Comments on the Present Economic Position of the Engineering and Allied Industries," 23–24, TUC Archives.

58. T. C. Pullinger, "Opening Address," *Proceedings Institute of Automobile Engineers* 12 (1917/1918): 432.

59. A. W. Reeves and C. Kimber, "Works Organisation," *Proceedings Institute of Automobile Engineers* 11 (1916/1917): 375.

60. See the piece by Tolliday and Zeitlin in this volume where they argue that after World War II British workers were somewhat freer than American workers regarding direct managerial control but found themselves unable to turn this control in their own favor due to constraints imposed by the piece-work wage system and the lack of solidarity among workers.

61. Church, *Herbert Austin*, 84.

62. See C. R. F. Engelbach, "Some Notes on Re-Organisation of a Works to Increase Production," *Proceedings Institute of Automobile Engineers* 23 (1928). Mechanized lines were adopted in 1928.

63. Third Annual Meeting of the Institute of Automobile Engineers, as reported in *Proceedings Institute of Automobile Engineers* (1924/1925): 7.

64. For statements by Engelbach and Keene see Ward Papers, MRG1, Organisation Section, W/8/29–34/13/476, 2–14, housed at the Business History Unit, University of London.

65. Ibid.

66. Ward Papers, W/8/29–34/476, P. Keene to MRG 1, December 2, 1930.

67. "Piece Work in the Toolroom," W(3)129, February 1, 1934, 26–28, EEF Archives.

68. Cole Memo, September 10, 1930, P(20)5, EEF Archives; Membership Files AEC, Memo W. L. Bayley to A. C. Bayley, March 18, 1927, EEF Archives.

69. Membership Files AEC, "Failure to Obey Rules," 4, EEF Archives.

# [8]

*Work, Employment & Society*, Vol. 6, No. 4, pp. 517–555 December 1992

***Abstract:*** This article questions the stereotypes of Fordism and mass production. It does so by demonstrating that there is a contradiction between the stereotypes and the reality of Henry Ford's manufacturing practice in production of the Model T at the Highland Park factory between 1909 and 1919. Highland Park was not an inflexible factory which combined dedicated equipment, Taylorised semi-skilled workers and a standardised product. More positively, the article quantifies Ford's heroic achievement in taking two-thirds of the labour hours out of the product at the same time as he built more of each car. Ford used productive intervention to realise manufacturing flow through proto-Japanese manufacturing techniques which involved a commitment to continuous improvement.

# FORD VERSUS 'FORDISM': THE BEGINNING OF MASS PRODUCTION?

Karel Williams, Colin Haslam and John Williams

## Introduction

> To give a name to a thing is gratifying ... but it is also dangerous: the danger consists in one's becoming convinced that all is taken care of and that once named the phenomenon has also been explained (Primo Levi *Other People's Trades*).

So it is with 'mass production' where we have the name and the illusion that the phenomenon has been explained although the phenomenon has not even been described. But this problem is invisible in social science where the concept of mass production is increasingly invoked in texts which confidently assume that it has considerable explanatory power. With the decline of interpretive meta-theories like Marxism, organising concepts have colonised the discourse of social science and mass production is one of the more influential of these concepts for (post) modern times. It is alluring because it promises to explain so much. American and British manufacturing failure can be attributed to persistence with this obsolete system, whereas more successful manufacturing countries like Italy (and maybe Japan) have supposedly built their success on new systems of flexible specialisation or post-Fordism.

**Karel and John Williams** teach economic and social history in the University of Wales, Aberystwyth. **Colin Haslam** teaches business studies at East London Polytechnic. They have worked together on many projects and their books include *The Breakdown of Austin Rover* (1986) and (with Tony Cutler) *1992 – The Struggle for Europe* (1989). Recent work includes reports on Hanson versus ICI, and on Japanese transplant manufacturing in Britain and the United States.

Throughout this paper we will use Piore and Sabel's (1984) familiar and influential usage as a point of reference; mass production means long runs of standardised products made on dedicated special purpose equipment by Taylorised semi-skilled workers. Mass production in this sense is often used interchangeably with the concept of 'Fordism' which is a kind of historical shorthand for the manufacturing system that was discovered by Ford and then, supposedly, widely imitated. All the overlapping variant versions of the concept of mass production have their supporters and detractors and some authors, such as Hirst and Zeitlin (1991), would challenge the identification of mass production with Fordism. But the authors who dispute whether Ford should be the eponymous essence of mass production are all agreed on one point: Ford is the classic example because, as everybody knows, Ford invented the moving assembly line and built 15 million Model Ts all the same and 'any colour so long as it is black'.

Sceptics and critics, like ourselves, have questioned the usefulness of the concept (Williams *et al.* 1987), arguing that there is no reason why the defining characteristics should be bundled together as either mass production or flexible specialisation. Critics have also been exasperated by the empirical opportunism of the supporters of mass production who see confirming evidence everywhere and find ways of dismissing counter evidence. But, so far, none of the critics has questioned the classic example of Ford and the T, or even registered the point that the classic example is an unexamined stereotype which is not securely based on recent research.

The standard history of the Ford Motor Company, by Nevins (1954), is nearly forty years old and only deals incidentally with production as part of a larger story. Recent popular accounts of Ford (e.g. Womack 1990) draw heavily on the brief treatment of Ford's production methods in Hounshell's (1984) history of American manufacturing technology and Lewchuk's (1987) history of labour control in the British car industry. Both centre on the introduction of the moving assembly line in 1913–14; for Hounshell, the addition of the line constitutes mass production and for Lewchuk the line is an effective way of controlling worker effort. In a recent article (Williams *et al* 1992), we criticise this historiography, construct the first continuous series of cars per man year from 1909 to 1916 and demonstrate that there is no justification for the privilege of the line.

If the history of Ford production methods has never been written, that is not because the sources are limited and inadequate. On the contrary, the available sources are voluminous, rich and diverse. Contemporary technical journalists took full advantage of the company's openness about production methods and produced detailed accounts of Ford methods around 1914: Arnold and Faurote's (1915) book provides a process by process description of the factory, while Abell and Colvin's (1913–1915) *American Machinist* articles analyse machine shop practice in depth. *Ford Times*, the company's house magazine, provides an invaluable commentary and chronology after 1908. The archive at Henry Ford Museum in Detroit holds the written records of the company, oral reminiscences taped in the early 1950s and more than 10,000 photographs of Ford factories. The archaeological evidence includes the Highland Park site where many of the factory buildings survive in semi-derelict form.

Because Highland Park (1909–1919) is the best documented factory in the history of mass manufacture, it was possible for us to go back to the sources and analyse in detail what Ford did. The results of our research are striking: our conclusion is that Ford does not fit the stereotype of mass production because there is a huge discrepancy between the stereotype and Ford's heroic achievements at Highland Park. The mass production label is particularly unilluminating because it obscures the extent to which Ford's organisation of production at Highland Park was proto-Japanese. We do not wish to substitute one misleading label for another and therefore will put some emphasis on the differences between Ford and Japan, but, for researchers like us, who have worked in Japan on companies like Toyota and Nissan (Williams *et al* 1991a, 1991b mimeo), there are uncanny similarities between Highland Park and the current practice of the Japanese. Perhaps this is not surprising because, though eighty years separate Ford and Toyota, the structure of costs and the activity of multi-process manufacture is still basically the same. In the rest of this article we aim to illustrate the character of Ford's production methods, to demonstrate the discrepancy with the stereotype and to draw out the implications of our analysis for the concept of mass production.

## Analysing Ford's Achievements

Ford got into mass manufacture by cutting the price of the Model T car which was introduced in the autumn of 1908. A fully equipped touring Model T cost around $950 in 1908 and 1909; just seven years later, in 1916, the comparable Model T cost $360. Our own generation has seen dramatically larger price cuts on electrical and electronic products like calculators and computers. But these price reductions are mainly achieved through the regular introduction of new models which incorporate fewer, cheaper, higher performance components. In mechanical engineering products the scope for price reduction is very much smaller because there has never been a mechanical innovation comparable to the integrated circuit. The Ford company was also peculiarly handicapped after 1908 because its marketing policy was to resist the introduction of new models. Furthermore a production policy of building more of the car in-house plus detail re-design undoubtedly increased the number of components in each T. In 1914, the touring T had 7,467 separate numbered T parts; by April 1917 this had increased to 10,102 T parts (Ford Archive: accession 125 boxes 1–20). So how did Ford achieve his price cuts?

The original design of the T plus the efficiency of Ford's existing manufacturing operation provided the basis for cost reduction. In terms of design, the T was unique because the car was light yet strong. These two contradictory qualities were reconciled through careful choice of materials: pressed steel covers and cases were used in engine and transmission, vanadium alloy steel was extensively used for stressed parts of the engine and chassis and every metal component in the car was heat treated (*American Machinist* June 1913). Early adverts for the T emphasised that this 20 horsepower, 1,200 pound car was, 'the equal of a 30 horsepower, 2,000 pound car' and would bring benefits of lower running costs to the owner (*Ford Times* October 1908). Light weight design also brought the benefits of potentially lower manufacturing costs to an organisation which was well placed to exploit them. Manufacturing is physically about the conversion of materials and financially about adding value; the problem of the activity is that converting materials usually requires a substantial labour input and adding value incurs substantial labour costs. The product cannot be sold cheaply if, as in a modern American manufacturing company, labour costs account for 70 per cent of each $1 of value added. In 1908–9 Ford started with the advantage that he was running a much leaner conversion operation; from 1909 to 1916 labour's share of value added in the Ford company averaged 31 per cent (Appendix: table 1 col. iii). The share of labour in value added is generally lower in the early stages of industrialisation, but Ford's performance was superior to that of manufacturing in the United States as a whole and in Michigan State, where labour's share was almost exactly 50 per cent in 1904 and 1908 (Lee 1957: tables M6 and M8).

If the new design of 1908 and the efficiency of in-house conversion established a basis for cost reduction, it took some time to realise this potential. Like all other manufacturers in the infant American auto industry, Ford in 1908–9 bought-in most of each car from less efficient outside suppliers. When the Model T cost $950 in 1909, $590 of the car was bought-in materials and components (Appendix: table 1 col. ii); at this time, the bought-in components included chassis frames, wheels, springs, finished bodies, soft tops and some of the engines. The $590 bill represented several tiers of embodied supplier labour and conversion profit. If Ford built more of the car in-house and maintained conversion efficiency, the potential for cost reduction was quite spectacular. Insofar as Ford turned $1 of bought-in materials into $1 of internal output and incurred only 30 cents of labour cost in conversion, most of the remaining 70 cents could be given away to the consumer in the form of a price cut. The catch was that these cost reductions could only be realised if if the Ford company built and operated a large, efficient manufacturing plant.

The new factory at Highland Park was designed to meet this need. Production of Model T components at the 60 acre site began when the first phase of building was completed in spring 1909 (*Ford Times* April 1909). The new buildings represented an explicit commitment to build more cars and to build more of each car in-house; *Ford Times* (July 1908) claimed that, 'in this plant everything from screws to upholstery that enters into Ford cars will be manufactured', and this objective was partly achieved. On our calculations, in 1909 68 per cent of the value of each T was bought-in and by 1915 the comparable percentage was 52 per cent (Appendix: table 1 col. iv). The physical change in the amount of the car built in-house was certainly larger than this because, in almost every case, Ford succeeded in replacing expensive bought-in components with cheap in-house conversion and this value change damps the shift in percentage terms. In most cases, the cost of the component was quickly halved when production was brought in-house and, significantly, Ford's suppliers were generally unable to match these reductions on components which continued to be out-sourced. These points can be illustrated by some random comparisons drawn from the April 1914 and April 1916 Model T cost books which list bought-in prices and estimate in-house costs. Amongst other components, top iron, (soft) top, hood and coil set were brought in-house during this period and their 1916 cost of production was estimated at 54 per cent of their 1914 purchase price. By way of contrast, the wheel set, carburettor and springs, which were still out-sourced in 1916, shows a purchase price reduction of just under 20 per cent over two years.

At the same time as Ford was bringing production in-house, output of the Model T was increasing rapidly: the company which shipped 21,000 cars in 1910, shipped 199,000 cars in 1913 and no less than 585,000 cars in 1916 (Appendix: table 2 col. i). Factory space for this output could not be obtained simply by extending Highland Park. In Autumn 1912, the company announced a new policy of decentralising final assembly to 'branches' so that Highland Park could concentrate on the vital work of component production (*Ford Times* September 1912). By spring 1915, the company had opened 24 branch assembly plants in all 'the principal cities' and most of Highland Park's output left the factory as knocked down kits (*Ford Times* April 1915). When Ford policy was to cut the price to the consumers spread across a continental market, the policy of branch assembly had its own powerful logic. Up to 1912, the consumer paid a freight charge on top of the Detroit list price; West Coast consumers paid a substantial $75 premium for delivery to their local dealer. Branch assembly dramatically cut these freight costs; the freight rate per pound weight on kits was significantly lower than on finished cars (Jones 1916) and it was possible to pack more than 20 kits in a freight car which would only hold 3 finished cars (*Ford Times* February 1913).

In production engineering terms, this was heroic stuff: by 1915, Ford's company was building more of the car in a chain of new and extended factories that spanned the United States. Given these circumstances it would have been understandable if the company had incurred some penalty in terms of increased labour hours per car. Miraculously, however, as Ford moved into mass manufacture of the T, labour hours dropped out of the product at a rate which has probably never been equalled in the history of mechanical engineering. We have constructed a continuous series of labour hours per car from 1909 to 1916 by relating total number of employees (at Highland Park and branches) to cars shipped ex Highland Park in kit and built up form. The series based on employment totals from Nevins shows that from 1910 to 1916 total labour hours per car fell from over 400 to 130 hours (Appendix: table 2 col. iv); in just five to six years Ford took two-thirds of the labour hours out of the car. A variant series, based on *Ford Times* employment totals, shows an even more spectacular decline (Appendix: table 3 col. ii). Either way, Ford's early twentieth century American achievement makes the late twentieth century Japanese look like sluggards. Womack (1990) claims that, after 40 years of taking labour out, when the Japanese build a car they still require half the labour hours of their competitors in the West.

This reference to Japan is useful because it alerts us to the point that the preconditions of labour hour reduction were much the same at Highland Park as in Toyota City. The main labour savings were made by taking out indirect labour and operating with a very favourable ratio of indirect to direct worker hours. Reductions in direct (in-process) labour were not enough because the T, just like a modern motor car, contained only a modest amount of direct labour. The Model T Cost Books (Ford Archive: accession 125) show that in December 1913 there were only 61 direct labour hours in a Model T and by February 1916 the direct labour hours were down to 37. Ford's labour hour reduction was mainly achieved by *pari passu* reductions in the much larger input of indirect labour which maintained an extraordinarily favourable ratio of indirect to direct labour hours. After 1913 this ratio at Ford was 2.5:1. By way of comparison, the comparable ratio in Japanese manufacturing is now 4:1 (with Toyota standing at 1:1) and in modern American manufacturing it is now nearly 8 to 1. From our work on Japanese manufacturing (Williams *et al* 1989), we know that the number of indirects in any company is positively related to the level of stocks in manufacturing operations: in a high stocks factory, many indirects are employed as warehousemen, truckers and stock chasers. Thus, we would predict that Ford ran a low stocks operation.

Research confirms this supposition about Ford's stock levels: these were maintained at nearly Japanese levels which are much lower than those of modern American manufacturing firms. At company level, Ford's best ever achievement was stock equal to 6.5 weeks sales in 1914 (Appendix: table 4 col. i); this is more or less exactly the level of stock cover which Japanese manufacturing as a whole has been operating for the past twenty years and something like half the level of stock cover in modern American manufacturing. The Ford company does not steadily reduce stock cover and never achieves Toyota style stock cover equal to less than 2 weeks sales. But this 'failure' is hardly surprising given the nature of Ford's manufacturing tasks. Other things being equal, building more of the car in-house would raise stock levels and branch assembly inevitably led to long supply lines. These problems were aggravated by pronounced seasonal fluctuations in the demand for Ford cars; parts were stock piled in the branches 'during late fall and winter' so that the company could meet the spring sales peak (*Ford Times* April 1915). It is thus significant that, at plant level, Ford's Highland Park performance stands comparison with anything since achieved by the Japanese. By 1915, Highland Park was running with a stock cover of 3–5 days for major components like chassis frames or engines (Arnold and Faurote 1915:63) and, as early as 1913, buffer stocks between individual production departments were down to a 'few hours' with zero stocks inside the departments (*Ford Times* January 1913). By way of comparison, when we visited Toyota in 1988, panels were being made in batches equal to one and a half days requirement and buffer stocks between assembly and the satellite factories of Toyota and its suppliers were down to a few hours.

The implications are clear: Ford in Detroit in the 1910s discovered the benefits of so-called 'lean production', which is falsely assumed to have been first invented in Toyota City in the 1950s. The proto-Japanese character of Ford's achievement must cast doubt on the orthodox explanation of that achievement in terms of economies of scale. When asked to explain the dramatic reduction in labour hours, an economist would observe that a thirty fold increase in output, over seven years, inevitably creates new technical opportunities for any manufacturing company: as output rises, unit labour hours and costs can be reduced by substituting capital for labour and/or extending the Smithian division of labour. If you want to know the time, it may be wise to ask a policeman but if you want to understand manufacturing it is foolish to ask an orthodox economist. This is because the discourse of economics operates with a cake mix factorial concept of production that does not specify the problems and possibilities of multi-process manufacturing. This is certainly so in the case of Ford, where the two orthodox economic explanations (substitution of capital for labour and division of labour) have some force but identify necessary, rather than sufficient, conditions of labour hour reduction. At the same time, the orthodox explanations fail to identify the real dynamic at Highland Park which was continuous improvement in workflow via changes in layout.

Ford did substitute capital for labour. The Highland Park plant was, in its time, the best equipped car factory in the world. In some processes, like the foundry, high labour productivity was achieved by putting capital equipment at the elbow of semi-skilled workers (Arnold and Faurote 1915:327). As a result, the amount of capital per Ford worker increased steadily in the heroic period. The gross value of plant and equipment (excluding buildings) more or less doubled from $870 per worker to $1606 in 1917 (Appendix: table 4 col. ii). Nevertheless it was not feasible to eliminate a large amount of labour by buying in a line of machines. Automatic process technology for building cars without (direct) labour was not available; the engine transfer line and the robot body line had yet to be invented. The machine shop at Highland Park was full of bought-in machine tools whose manufacturers assumed one direct operative per machine plus back-up maintenance and hand trucking of work-in-progress. If Ford was to do better the company would have to learn how to use its equipment in a better way. Similarly, the elimination of indirect workers required a new approach to materials handling where investment in mechanical handling was not the only option.

If Ford's efforts were partly directed by the limits on what investment could do for the firm, much the same point could be made about the Smithian division of labour which again plays a positive and limiting role. What Smith described in pin manufacturer, Ford applied to car assembly. In piston and connecting rod assembly, for example, assembly time was halved when the operations which one man had carried out were sub-divided between three operators and an inspector (Arnold and Faurote 1915:103). Even so the division of labour was far from the governing principle of the Ford factory because it was applied only to hand assembly work. In machine work on the engine, gearbox and chassis parts, Ford engineers pursued with tenacity and ruthlessness, the opposite principle of combining the operations performed at one setting or chucking of the materials. Thus, the Ingersoll milling machine used on the front axle body was modified to take two axle bodies at a time and made twenty four cuts for one clamping (Arnold and Faurote 1915:166).

More fundamentally, in multi-process manufacture the principle of division of labour is a source of productivity loss as well as gain. This was particularly so in the case of a complex product like the Model T. By 1915–16, Ford was manufacturing around 5,000 T parts in-house and the work of fabrication and assembly required several hundred thousand separate process operations. To take just one example, a pressed steel cover at the bottom of the engine carried the crankshaft and doubled up as an oil sump; no less than 78 separate operations were required to make this pressed steel cover (Arnold and Faurote 1915:82). Furthermore, although the touring T was a relatively light car, it did weigh 1200 lbs: by way of comparison, the lightest European car in current production is the three door Citroen AX which weighs in at 1400 lbs (*Autocar* 18 August 1988). When the Ford company was producing 1000 cars a day by early 1914, it was shipping 625 tons of metal out of the factory each day and that weight had to be handled many times in between production stages. With a complex heavy product like the T, disintegration of production and unnecessary labour expense was the corollary of multi-process production. The smooth flow of production could not be sustained if in-process labour time was wasted on walking or waiting because the work was not immediately to hand and extra indirect labour would be required to handle any stocks that accumulated between processes.

In itself, the minute division of labour only increased the disintegration of production. It was therefore necessary for Ford to back up the division of labour and mechanisation with layout changes which improved in-process layout and inter-process workflow. Significantly, the reduction of walking and waiting time in-process was an explicit company objective. Thus in October 1912 *Ford Times* announced,

> every extra motion, every trivial waste of time on the part of any workman must be eliminated. It is the solution to this problem of "waste motion" both with men and machines, which to a very large degree makes it possible for Ford to put on the market such a high grade car at such an extremely low price.

The elimination of indirect stock handling was another explicit objective; as a senior Ford engineer observed in 1913, 'trucking is expensive and all its labour is non-productive' (Bornholt 1913). The double objective of eliminating unnecessary labour within and between processes was vigorously tackled by investing in handling equipment and by re-designing layouts so that work in progress travelled shorter, more direct, routes.

From Nevins (1954) through Braverman (1974) to Hounshell (1984) and Lewchuk (1987), scholarly accounts of Ford emphasise the moving assembly line which was first used in magneto assembly in the spring of 1913, applied to final assembly in the latter half of 1913 and then transferred to production of various sub-assemblies like dashboard, front axle and body and top. Undoubtedly, the moving line boosted productivity after 1913, but, as we argue elsewhere (Williams *et al* 1992), the moving assembly line centred view of Ford is unreasonable and should be revised. Ford had already taken 150 labour hours out of the car before the assembly line was introduced in 1913; on our calculations, 63 per cent of the 1910–1916 labour hour reduction and 70 per cent of the material cost reduction was achieved without benefit of assembly lines. Furthermore, the moving assembly line was only one device in a whole battery of mechanical handling and transfer devices which were introduced at Highland Park after 1910; these systems included craneways and an overhead monorail which played an equally important role at different stages of production.

The nature and function of the different mechanical handling systems can be pieced together from various issues of *Ford Times*. By spring 1915, moving, chain driven, lines had been applied extensively in final assembly, body and trim; at this time the factory had one and one-half miles of moving line which was roughly half manufacturing conveyor and half transfer conveyor. (*Ford Times* April 1915). But in the machine shop, the main transfer device was a one and a quarter mile long overhead monorail (*Ford Times* April 1912) which had been constructed somewhat earlier; by early 1914 this monorail system shifted 1400 tons of work in progress each day into, through and out of the machine shop to final assembly in building H. For the delivery of bought-in materials and components to the shops, strategically sited craneways played an increasingly important role. From Autumn 1910, a craneway and storage area 56 feet wide and 860 feet long bisected the main machine shop (*Ford Times* November 1910, January 1913). When new six storey buildings (W, X and Y) were constructed in 1913–1914, two narrower 40 foot craneways were placed between the buildings so that materials could be transferred directly between rail cars and projecting freight balconies which were set into the face of each building.

The preoccupation with Ford's moving lines is all the more misplaced because the company's philosophy was not only to mechanise handling in a variety of ways but also to eliminate transfer and handling wherever possible by re-designing layouts so that work in progress travelled shorter, more direct, routes. The conversion of Highland Park into a short travel factory required careful attention to three separate issues: the sequence of shops on the Highland Park site; the positioning of machines close together and in order of use within each shop; and the development of simple direct methods of inter-process transfer. The significance of the no-investment or low investment layout changes was appreciated by contemporary observers like Colvin of the *American Machinist*, but has been largely lost in later scholarly accounts which are preoccupied with the romance of the moving assembly line. In our view, layout changes are the hidden dynamic in a process of continuous reorganisation of workflow which dramatically reduces the inherent disintegration of multi-process production and generates large savings in indirect and direct labour requirements.

It is difficult to generalise about the sequence of shops at Highland Park because, over the period 1909–1916, the factory was regularly extended and, with the exception of the machine shop, the buildings were used and re-used for a variety of productive purposes. The general plan of the factory in Autumn 1912 does however indicate the company's commitment to setting shops in a logical sequence (*Ford Times* October 1912). At this time, the shops in Highland Park were laid out on a West to East axis along which work in progress travelled to final assembly. For example, in the machine shop, cutting departments were on the West side of the main craneway with engine and gearbox assembly on the East side. One step further East, on the other side of John R Street, was the four storey H building where final assembly took place on the ground floor; many of the body parts and sub-assemblies were made or stored on the upper three floors of this building before travelling down freight elevators to final assembly. If the three dimensional shop layout of the factory was logical, so was the detail of layout inside the shops.

In laying out shops, Ford's philosophy was to put the machines close together and in order of use. Visitors to the shops commented that the machines and workers were jammed together more closely than in other American factories. In 1913 Colvin wrote, 'I do not know of a place where more men are worked together in the same area, although the Baldwin locomotive works are in the same class' (*American Machinist* 5 June 1913:761). On our calculations, the 5,500 machine tools, which filled the shop in 1913, had a floor area of 55 square feet per machine including access aisles and inter-process storage space. The machine shop also included babbiting and annealing furnaces as well as enamelling tanks because the company's practice here and elsewhere was to arrange machines, regardless of function, according to 'sequences of use' (Bornholt 1913:1276–7). Arnold and Faurote (1915:223) reported that, in the production of distributors, the furnace that produced the aluminum distributor body was placed at the beginning of the line at third floor level; the foundry fed rough castings, at the rate required by the day's build schedule, directly into the machining operations from which finished cases moved straight into assembly.

When machines were placed close together and in order of use, Ford obtained spectacular reductions in work travel, stock levels and indirect labour requirement. The engine block casting, for example, originally travelled 4,000 feet between finishing operations; after putting the machines close together and in order of use, the block casting travelled just 334 feet (Arnold and Faurote 1915:38). A senior Ford engineer argued that the in-sequence, short-travel layout in engine block machining reduced stocks to nearly one-twelfth the level required by a long-travel layout with machines grouped according to function. He also calculated that, when production was running at 1000 blocks a day, the short-travel layout saved the handling of 480 tons of work in progress which would have required the full time services of 24 hand truckers (Bornholt 1913:1276–7). Furthermore, with the machines jammed close together it was possible to introduce simple direct inter-process transfer devices like gravity slides. Along with roller beds and racks of pipe and angle iron, gravity slides were used everywhere for inter-process transfer at Highland Park. These simple transfer devices not only eliminated indirect truckers but also boosted direct labour productivity and line capacity (Arnold and Faurote 1915:105).

The aim and result of layout improvement was what the Japanese call '*kaizen*' which in this context means continuous step by step improvements in workflow which progressively take labour out (Williams *et al* 1991a).

The conventional emphasis on the moving assembly line in 1913–14 is again misleading because it suggests a once and for all series of changes, when Ford's real achievement was to institute permanent progress. Any detailed examination of Ford practice shows how the Highland Park engineers were never satisfied with any given layout and expected to achieve continuing cost reductions.

The development of the moving assembly line is itself a good illustration of the company's commitment to continuous layout change, for the chain driven final assembly line was the culmination of years of progressive experiment with final assembly methods. It is mistaken to assume that, before the moving assembly line was introduced, Ford operated an inefficient and unchanging bay assembly system with one gang building the whole car. By 1911, the company was operating a moving gang system in the early stages of final assembly. Specialist gangs moved down a line of cars and fixed the same components on each station; Colvin observed that, 'each man or gang does his own part and goes to the next stand to repeat the operation' (*American Machinist* June 5 1913). In the later stages of assembly the company used a buggy system whereby part-finished cars were rolled along on their wheels between successive operations (*Ford Times* March 1914). The idea of the moving assembly line grew in an evolutionary way out of these earlier innovations. Before the chain driven line was introduced, archive pictures show a pull along line where cars ran with one rear wheel in a jockey frame that was located in a steel channel.

The Toyota-like ability of Ford to generate continuing cost reductions is best illustrated by considering the case of the 'motor' or engine plus gearbox power train of the T. Machine shop cost reduction was the main focus at Highland Park from 1909–12 because the T could not be sold cheaply as long as the engine and gearbox were expensive to manufacture. By April 1914, Ford was building each power train at a total cost of $61 for materials, labour and overhead; this was a remarkable achievement when in 1914 Ford was paying nearly as much ($55) to buy-in a light open touring body (Ford Archive: accession 125 boxes 1–20). By 1913–14, attention had shifted to final assembly which had become a bottleneck because, 'at this point we found that we were making parts a hell of a lot faster than they could put them together on cars' (Klann 1955:55). Even so the reductions in power train cost continued; the 'motor' which cost $61 in April 1914 cost only $53.15 in April 1916 and more than 80 per cent of this cost reduction was achieved by reducing the internal costs of labour and overhead (Ford Archive: accession 125 boxes 1–20). When the company started from such a low cost level in 1914 a further 13 per cent motor cost reduction in two years is eloquent testimony to the culture of *kaizen* inside the Ford company which Sorensen aptly described (1956:128) as, 'an organisation which was continuously experimenting and improvising to get better production'.

**Ford and the Stereotype of Mass Production**

After analysing Ford's productive achievements, we can now turn to measuring the difference between Ford and the stereotype of mass production. The stereotype has three elements which the supporters of the mass production concept would expect to find at Highland Park: dedicated equipment; Taylorised semi-skilled workers; and a standardised product. These elements fit together in the stereotype to sustain the working assumption that the mass production factory or enterprise is rigid and inflexible; if they were present at Highland Park, this would imply that the factory could not be easily used to produce variety and could not easily be changed over from one product line to another. In this section we will argue that the three elements cannot be found at Highland Park.

There were elements of fixity in Ford's capital equipment, most notably in the buildings at Highland Park which are a concrete disproof of the vulgar Foucauldian assumption that it is usually easy to choose appropriate architectural forms. Although the machine shop and foundry were housed in single storey sheds, most of Ford's shops were housed in multi-storey factory buildings of a type familiar in turn of the century Detroit, Berlin and Manchester. This conventional choice was hardly surprising because the original 865 feet long four storey building at Highland Park was already designed and under construction when the Model T was going into production in autumn 1908 (*Ford Times* January 1909). Ford's architect, Kahn, was responsible for the bold modernist aesthetic of glass curtain walls and for the clever detailing in the provision of ducted services like heating, compressed air and gas. He also brought to the project the Kahn family's obsession with reinforced concrete; the multi-storeys were off the form, reinforced concrete structures rather like modern multi-storey car parks. The result was a factory complex where transfer of work in progress between floors was a constant problem, even in the later 1913–14 six storey buildings, where freight elevators were backed up with craneway access to each level. As the massive floor decks could not be cut, by 1913–14 Ford's engineers were knocking out Kahn's glass curtain walls so that they could hoist and drop up and down the sides of the buildings. This is the background to a much-reproduced photograph of the H building body drop on John R Street in 1914; the bodies which were trimmed on the first floor are being ramped down outside the building to street level where finished chassis are coming off the assembly line on the ground floor (Ford Archive: PO 242). Its complement is a never-reproduced archive photograph taken in the same year, which shows bought-in, part finished bodies being ramped up from a supplier waggon on Manchester Street for trim and finishing on the first floor of building W (Ford Archive: PO 5304).

If the buildings at Highland Park did introduce an element of rigidity, the machinery inside the buildings was flexible in the sense that it was mainly re-usable. Ford did use some purpose built dedicated equipment like the machine which automatically painted rear axles (Arnold and Faurote 1915:308), but most of Ford's metal working was done on bought-in, series built lathes, drills, millers and presses. Manufacturers like Cincinnati or Ingersoll delivered a stripped down tool minus the expensive multi-speed gearboxes and universal clamps that a job shop would require. Ford engineers then added special jigs and fixtures which (temporarily) adapted a machine for a single task. As *Ford Times* (October 1912) observed,

> a trained observer is at once struck by the number of jigs throughout the big machine shops . . . . By use of jigs a standard tool can be adapted to do special work and it is, of course much cheaper to buy a standard tool than to have one built to order.

This kind of 'single purpose' equipment was generally used in other early twentieth century American machine shops which made interchangeable parts; the jig or fixture guaranteed interchangeability because it automatically set the same depth and angle of cut for each successive workpiece. Interchangeability boosted productivity because it allowed Ford to eliminate the fitter's role and to minimise inspection and rectification of finished assemblies. Single purpose tools had important additional advantages because Ford engineers could cheaply and easily re-use the equipment for a different purpose by changing the jigs and the drive sprockets which set operating speed. In the case of transfer equipment, the principle of reusability was even more important because all layouts were provisional and temporary. Ford's solution was to build transfer devices like slides, racks and roller beds out of standard sections which could be dismantled and re-assembled in a different configuration; the gravity slides, for example, were made out of lengths of pressed steel troughing which bolted together just like the parts in a child's construction kit (Arnold and Faurote 1915:274).

If the equipment at Highland Park was not dedicated, in this respect the stereotype is simply untrue. Much the same can also be said about the claim that Highland Park workers were Taylorised; a claim which has been challenged in the specialist literature (e.g. Hounshell 1984:249–52), but which is elsewhere endlessly repeated as though it was unproblematically true. Here the discrepancy between the stereotype and the reality of labour control at Highland Park is somewhat subtler but the issues raised are important. It is undoubtedly true that Ford was obsessed with labour utilisation and, more specifically, like a later generation of Japanese mass manufacturers, privileged efficient labour utilisation over capital utilisation. This is entirely understandable because the (payroll) cost of labour in mass manufacture is usually much greater than the (depreciation) cost of capital. On our calculations, this disproportion was quite spectacular in the

Ford company where, between 1910 and 1916, depreciation averages just 5 per cent of payroll (Appendix: table 4 col. iii). The privileging of labour utilisation was asserted in all the Ford company's practices about manning and machine use; if, for example, the company had excess capacity on a line, Ford preferred to shut down one machine and save the labour, rather than run all the machines slower (*American Machinist* June 5 1913:761). The issue is not *whether* Ford was preoccupied with labour utilisation but *how* the company secured this objective.

At Highland Park, Ford did not use the formal apparatus of Taylorism; time study had little influence because Ford favoured simpler more direct ways of driving the workforce. Two key company agents informally determined the workpace: output norms were routinely set and checked by 'the working foreman' who could do every job on the line and regularly covered for workers when they took a toilet break; when the company wanted to raise the output norm in a particular process, loyal company men called 'pace setters' were brought in to show that the job could be done faster (Klann 1955:10). Some contemporary observers concluded that Ford had the substance if not the form of Taylorism; Colvin (*American Machinist* 1913:149) argued that the assembly of rear axles was, 'one of the interesting propositions which show that motion study has been carefully looked into whether it is called by that name or not'. But, in our view, this interpretation misses significant differences between Ford and Taylorism at the same time as it fails to specify the nature of Ford's labour process regime.

The aim of the working foreman and the pace setters was not to define a semi-permanent scientific standard but to raise the output norm in a continuous way. In a factory of immigrant workers,

> one word every foreman had to learn in English, German, Polish and Italian was 'hurry up'. It was '*putch putch prenko*' in Polish, '*mach schnell*' in German and '*presto presto*' in Italian (Klann 1955:54).

Furthermore, although most of the workers were semi-skilled, there was never a rigid separation of execution and conception. Ford did not operate a formal suggestion scheme, but foreman or line worker suggestions about productive improvements were encouraged and workers who made useful suggestions could expect to be promoted. Klann (1955:117) recalled that this was certainly so in the machine shop which he supervised from 1912–17,

> A lot of these ideas came from the men as well as from people in supervisory positions. Lots of them came from the men. You went and asked the men for an idea and then tried it out. If it worked, you thanked him for it, or maybe made a foreman out of him.

The line between execution and conception was always permeable when Ford was a rapidly expanding company which lacked a professionally qualified management cadre. Many senior managers were recruited from the ranks and Klann's career progress, from line worker to senior manager in fourteen years, illustrates the possibility of rapid promotion. Klann was taken on as a machinist in Ford's Piquette plant in 1905; after a short stint in the tool room, in 1907 he became a roving pace setter before being promoted to assistant in charge of motor assembly in 1910; by 1919 Klann was assistant superintendent (or second in command) of the whole Highland Park factory.

More generally the label 'informal Taylorism' is misleading because it fails to specify the quasi-Japanese character of the labour control regime which Ford operated (Williams *et al* 1991a). The Highland Park regime imposed two general requirements on the workforce: they had to be infinitely flexible about the definition of the task to be performed and, at the same time, had to acquiesce in progressive de-manning of most operations. Ford workers had to accept endless changes in the labour process because workforce flexibility at this point was a crucial precondition of productive reorganisation. In assembly, tasks were increasingly subdivided to the point where, for example, different workers spot painted the fixing bolts on the left and right side of the chassis (Arnold and Faurote 1915:150). On the machine lines, tasks were combined to create multi-tasked workers who, as in modern Japan, sometimes tended more than one machine. In the machine shop, for example, some single fixture machines were placed face to face so that one worker could turn to re-load a first machine while a second ran to the end of its cycle (*American Machinist* October 1914).

With or without layout change, de-manning was a way of life in Ford's factory as it is now in Japan. The de-manning was particularly ruthless in final assembly because the company compared assembly costs at Highland Park and the branches each month and least cost assembly methods were enforced by de-manning. Klann (1955:300) recalled that when Highland Park assembly costs were above those of the Kansas City branch, the Kansas city foreman was brought up to Detroit,

> he said 'my gosh, I can't find anything wrong round here except down there I've got two men where you've got three. In putting bodies, I've got three men where you've got four'. He pulled a man off that job. By the time he got through he pulled off six or seven different men and we brought our cost down to where his cost was.

Under this regime of *kaizen* through labour intensification, management's prerogative over the labour force was infinite (and had to be so). As Arnold and Faurote (1915:328) observed,

> workmen are studied individually and changed from place to place with no cause assigned, as the bosses see fit, and not one word of protest is every spoken.

Any worker who refused instructions was sacked on the spot. Highland Park was not unionised and the company's labour control regime was designed to prevent informal resistance through practices such as 'soldiering'; for the individual Ford worker, the only available resistance option was quitting the job. As there was an external labour market in Detroit, this option was popular and, in the early years at Highland Park, Ford faced a serious problem about high rates of labour turnover which reached a phenomenal 70 per cent in 1913 (Meyer 1981:162). High wages, especially the $5 day introduced early in 1914, were an attempt to square the circle and resolve this problem. The $5 day was never simply an economic choice, it was the first of Henry Ford's social experiments which turned him into an American hero who was something more than a car maker, but high wages did have an economic function in Ford's labour control regime and, by early 1914, they also had a coherent economic basis. Ford could afford to double the hourly rate and offer shorter working hours because the company had been so successful in taking labour hours out of the car. In 1909 when hourly earnings were 28 cents and there were 400 hours of labour in a T, Ford's in-house payroll costs were $100 for each car produced. By 1915, earnings had doubled to 55 cents but, with just 125 hours of labour in a T, the in-house payroll cost was just $64 for each car produced (Appendix: table 4 col. (iv)).

The idea that the model T was a standardised product is a final misunderstanding which rests on a series of misconceptions about the changing nature of the American market for cars from 1910 to 1925 and about the possible lines of response for a manufacturer like Ford. The concept of product life cycle is slippery, but it is clear that the car was not a mature product when the T was introduced in 1908. When the T went into volume production in 1909, American cars were open, gas-lit, crank-started and artillery wheeled. By the time the T ran out in 1927, American cars were closed, electrically lit, self started and balloon tyred on demountable rims. Ford could not have maintained its position in the market for cheap cars if it had ignored these changes. The company, did however, have a choice about how to respond to changing market requirements; Ford could either incorporate the necessary changes in regularly introduced new models or develop and refine its existing model. Like most subsequent manufacturers of relatively small, cheap cars, Ford in the 1910s and the 1920s chose the second option of development to prolong the life of the existing model. It is worth nothing, in this context, that Ford of Europe's smallest model (the Fiesta) has recently been replaced after a production run of fourteen years which is not so different from the T's eighteen years. The difference is that, because the car market around World War I was changing so much more rapidly, Ford at Highland Park had to make a major commitment to continuous refinement and development (rather than occasional facelifts). As a result the model T of 1926 was, part for part, more or less completely different from the T of 1908; that is hardly surprising when by 1913 the company was making more than 100 T part specification changes every month (Ford Archive: accession 575 box 5).

The body of the T was never standardised because the T was not one car but a changing range of cars and trucks. From 1912 to 1926 the range was available in black only. This restriction of choice reflects the Ford company's low stocks productionist priority; the legend is that black paint dried more quickly and black body panels could be handled and moved on more promptly. In all other areas of body design the Ford company's policy was to extend the market for the T by offering consumers choice and change. Right from the start the company produced several different body types; the first price lists feature a two seat roadster, four seat tourer, and a closed 'town car'. When a separate load bearing chassis frame was an integral part of the T design, it was cheap and easy for Ford to develop new and existing body styles. Regular styling changes transformed the square upright town car of the 1910s into the neat soft edged sedan of the 1920s whose passengers sat fashionably in (not on) the chassis. From 1910 onwards the T range included a low volume, high style variant, the 'torpedo roadster' which was the Corrado or Calibra of its day. Utility variants like the T delivery van and the T truck followed in due course, and from 1914 onwards the company continuously listed a bare chassis so that customers could have literally any body they liked on a T. If the body variety was potentially infinite, it was also true that, before World War I, the vast majority of Ford customers chose factory-bodied open tourers and roadsters; in April 1916, for example, 97 per cent of cars sold were tourers or roadsters (Ford Archive: accession 125 box 6). This was mainly because Ford never revolutionised the building of closed bodies whose cost of manufacture was 3–4 times larger than that of an open body (Ford Archive: accession 125 box 3 and 7). Between 1909 and 1916, a closed town car always cost $235–250 more than an open tourer and, in this period, buyers of cheap cars were not prepared to pay that kind of premium for a closed body.

It is superficially more plausible to argue that the power train and chassis of the T were standardised. Like Beetles in a later period, all T's shared a distinctive identity because some increasingly idiosyncratic mechanical features were maintained throughout the model's life; in the case of the T, the identifying features included thermo-syphon cooling, a two speed planetary transmission and exiguous two wheeled brakes. Nonetheless the T power trains and chassis were not all the same. The chassis was always built in several different forms. Early customers for the car could, for example, choose between 56 and 60 inch tracks, while later purchasers of the truck were, in effect, offered a package of heavy duty modifications which involved a re-located engine in a reinforced chassis frame which carried a worm drive rear axle. Furthermore, the basic design was developed through innumerable specification changes so that, when ordering engine or chassis parts, it was always necessary to specify year of manufacture and type of body. Towards the end of the T's model life,

Fahnestock (1968:69), the leading independent technical writer on the T, concluded, 'the fundamentals are unchanged but many details have been improved', and this verdict appears to be an understatement if we set it against Fahnestock's (1968:229, 252) own account of the development of one major system, like the vehicle electrics. There were three major re-designs of the magneto ignition system in the first ten years of production during which the car also acquired electric lighting. In 1919 Ford jettisoned the magneto principle and offered the car for the first time with a modern coil and points ignition system that included battery, dynamo and starter. The final major change came in 1922 when Ford re-designed the wiring harness and all the electrical junction boxes and connectors. In design terms, the power train and the chassis of the model T was the automotive equivalent of George Washington's axe.

Ford's Model T did not succeed in the market for so long because it was a standardised product, but because it was a stretchable product. Continued success in the early 1920s, when production peaked at 2 million units a year, demonstrated Ford's competence in stretching a robust basic design. The process was nevertheless reaching its limits because like many mechanical engineering products the model T put on weight as it was re-designed and developed. Mechanical improvement raised the original 1918 weight of a touring T from 1200 to 1662 pounds by 1925, while the addition of a closed sedan body, which customers now preferred, raised the kerb weight to 1900 lbs (Van Doren Stern 1955:164–7). The extra weight spoilt the car: deterioration in the power to weight ratio could only be met by fitting lower axle ratios which increased fuel consumption, mechanical wear, noise, vibration and harshness. The necessary replacement was delayed while the ageing Henry Ford tried to develop an automatic gearbox and a new X8 engine. The mechanically mundane Model A of 1927 represented a grudging admission that such refinements could not be brought to the low cost mass market of the 1920's.

If the Model T was stretchable our argument implies that the Highland Park factory was flexible up to a point. When Highland Park combined variable layouts with re-usable equipment and a malleable workforce, the factory could have been used for turning out a wide variety or mix of volume manufactured engineering product. Ford's strategy was to develop the T range and thus the company never exploited this potential in peace time, but the flexibility of Highland Park was conclusively demonstrated in World War One when, 'practically all Ford machinery was dismantled and re-arranged to produce war material' (Ford Archive: photo caption PO 2296). In the account which follows, the facts are taken from Ford Bryan's (1990) *Beyond the Model T* which provides the only published account of this neglected but significant episode; the interpretation is, of course, our own.

For the Americans, World War I lasted only 18 months from April 1917 to November 1918. In this brief period Model T production dropped by more than one third and the spare margin of capacity at Highland Park was used to build a marvellous assortment of military equipment: 825,000 steel helmets, 5,000 state of the art Liberty V12 aero engines, around 25 sets of 2,500 hp marine boilers, steam turbines and reduction gear, and a two and a half ton tank powered by two Model T engines. This achievement is all the more remarkable when it is recalled that, during this period, Ford opened a new factory at River Rouge which initially concentrated exclusively on war production. The Ford company also had a substantial design input into several of these war time projects; the 600ft Eagle submarine chaser was designed entirely in-house and the prototype was actually built in a craneway at Highland Park before being moved to River Rouge where Ford planned to build, outfit and launch the hulls for a series of more than a hundred vessels.

The nature of wartime demand was such that the Ford company had to make a difficult shift from medium to heavy engineering. In terms of dimensions and weight, the V12 aero engine and the marine reduction gear were physically very much larger than anything that went into a T; the sudden increase in component size and weight created new problems in processing and transfer. At the same time, the Ford company faced a spiralling bill for materials, especially heavy steel plate. Both the submarine chaser and the tank were basically exercises in steel fabrication and assembly; the Eagle hull was constructed entirely by riveting cut, flat steel plate. As Ford did not make steel in-house, purchases as a percentage of sales increased from 54 per cent in 1915 and 1916 to 75 per cent in 1917 and 1918 (Appendix: table 1 col. iv). When the company faced new physical problems about conversion and was financially forced back into the old practice of buying-in, it is hardly surprising that conversion efficiency plummeted and labour's share of value added shot up to crisis levels; labour's share of value added averaged 31 per cent in the Ford company between 1909 and 1916 before jumping to 61 per cent in 1917 and 89 per cent in 1918 (Appendix: table 1 col. iii).

Highland Park was not a universal conversion apparatus; Ford could get wartime heavy metal out of the door only at the substantial cost in terms of lower conversion efficiency. In mechanical engineering, productive flexibility is never costless and infinite. This is a point which we have made elsewhere (Williams *et al* 1987) when criticising romantic accounts of the flexibility of modern robotised car body building. With old or new technology in mechanical engineering the relevant consideration is whether the production facilities can change over at moderate cost and then efficiently build a new product which is physically different but within the conversion parameters defined by the weight, size, materials and process requirements of the old product. Ford's attempt to get into the fighter aircraft business suggests that Highland Park did meet the relevant test of conversion efficiency.

In late 1917 Ford approached the US military with an offer to build 150,000 fighter aircraft for the Allies at a price of 25 cents per pound weight. The offer was immediately rejected as another one of "crazy Henry's" stunts because the military believed it was impossible to build aircraft so cheaply. A closer consideration of the facts suggests that Ford's offer was a carefully considered and genuine bid to fill his factory space with a war product which he could build efficiently and profitably. The single seater fighter aircraft was the military product which most nearly fitted the Model T conversion parameters. By 1918 the fighter to beat was the Fokker DVIII: modestly sized and relatively light weight, with a fuselage length of twenty feet and a wingspan of twenty seven feet, the 100 hp monoplane weighed around 1700 lbs (excluding pilot, two machine guns and ammunition). In conversion terms, a fighter like the Fokker was a flying T with more engine and less chassis.

The offer to build this kind of new product for a price of 25 cents per pound weight gives a real insight into Ford's conversion calculations and assumptions about flexibility. The company had already approached this level of conversion efficiency in pre-war production of the model T; in 1916, the 1200 pound touring T was listed at $360 and the selling price (including Ford's profit and dealer margin) was therefore 33 cents per pound weight, which *kaizen* would have continued to reduce. In this context, the offer to build fighters at 25 cents per pound signified Henry Ford's personal belief that his factories could change over from the T and build, with increasing efficiency, any mechanical engineering product which shared the conversion parameters of the T.

## The Failure to Emulate Ford's Achievement

If the concept of mass production has generated a series of misrepresentations about Ford's achievements, it has also promoted a misconception about the emulation of those achievements; it is generally believed that Ford's practice was a model which was widely emulated in America and other parts of the world. In this section we will argue that the reality was rather different; even in America there was only limited imitation and transference of Highland Park practice. The trajectory of American manufacturing after World War I was determined by the fact that most firms, including the later Ford company, turned their back on the achievements of Highland Park. Whatever reservations we may have about Chandler's (1962) history of American business, it does indirectly register a powerful truth: from the 1920s onwards, many American firms were increasingly preoccupied with organisational issues and forms of financial and strategic calculation which have little to do with the hard driving productionism of Highland Park. When management hierarchies and financial control displaced focused production engineering, the model of success became 1920s General Motors as a corporation rather than 1910s Highland Park as a factory.

The idea of imitation and transference is immediately plausible because by 1914 Ford's Highland Park factory was the most famous industrial site in the world whose visitors included technical journalists and public figures as well as managers and engineers from other companies. But many of the visitors to Detroit in the 1910s, like those who made the journey to Toyota City in the 1980s, left with a limited understanding of what they had seen and returned home to engage in various forms of symbolic modernisation; as with 'Japanisation' in our own time, many imitators of Ford were doomed to disappointment because they fastened on elements of his superior manufacturing practice and credited these elements with a magic efficacy. The Fiat company, for example, built a multi-storey factory at Turin in imitation of Highland Park without realising that this architectural form was one of Ford's half-solved problems. For most firms outside America, Ford's mass manufacturing practice was, in any case, of limited immediate relevance because they served relatively small national markets. American firms supplying a continental market had a different set of opportunities, but few of them faced the overwhelming demand pressure which generated and legitimised Ford's productionism in the 1910s.

Ford faced an insatiable market demand for the T, a product which in its heyday was always oversold. The aim was to supply a robust, easy to drive motor car at an ever lower price which extended the market. The company succeeded to the point where it had an unassailable position in the market for cheap new cars; in 1914 Ford made 96 per cent of all American cars sold at less than $600. When cars were open and roads were poor, there was strong seasonal variation in demand. But year on year, the policy of price cuts stimulated a phenomenal increase in demand; the company, which shipped 14,000 cars in 1909, shipped 585,000 cars in fiscal year 1915–16 (Appendix: table 2 col. i). Cost cutting through productive reorganisation was so successful that each car out of the factory still represented $100 profit for Ford, even in 1916 when a touring T sold at $360 (Appendix: table 5 col. iv). A constant profit on a falling price meant higher margins; return on sales increased from 21 to 31 per cent between 1910 and 1916 (Appendix: table 5 col. v). Seldom in the history of mass manufacture has any company ever had a stronger incentive to crank low cost output through its factories and, as we have seen, this was what Ford's engineers did. At the same time, sophisticated financial methods were used to generate extra cash flow and profit; Couzens, for example, operated a financial system under which the cars were built from materials bought on thirty days credit and always sold cash on delivery, so that the company often banked the receipts before suppliers were paid (Seltzer 1928:130). Nonetheless, between 1909 and 1916, (low cost) production was the one over-riding problem that had to be solved.

By the mid-1920's, the Ford company like others in the car business, faced a more complex hierarchy of productive, market and financial problems where the possible and necessary points of intervention were less clear cut. With some justice, observers of the late 1920s argued that the Ford company's continued productionism was an anachronism in this increasingly complex environment. As Seltzer (1928:121) wrote just after the introduction of the Model A,

> the enormous material and technical resources of the Ford Motor Company had, perhaps too long been directed at the single task of augmenting and perfecting its manufacturing organisation. Too little attention had been paid to the changing character of the automobile market in recent years.

The Ford company had always paid close attention to distribution, but, by the 1920s, General Motors was using marketing to adjust product characteristics to consumer demand: this kind of response was necessary and functional for enterprises which faced weaker and/or more complex demand patterns. The verdict of history confirms the intuitions of contemporaries about the limits of simple productionism; the lesson of the Model T, just like the VW Beetle, is that successful production of a people's car is only a passing phase in the development of a motorised society. As Kuhn (1986) argues, by the mid-1920s Ford's output was flat and his market share was being undermined by cheap second hands and the development of ever more sophisticated and powerful closed cars. Within ten years of its greatest achievements, the Ford company was no longer a shining example but a cautionary tale.

When this point has been made, it still does not explain why many of the production innovations of Highland Park should have been lost in the fifty years after 1920. In the late 1970s, when American engineers like Schonberger (1980) visited Japanese factories, they saw, for the first time, new Japanese manufacturing practices which were (at least partly) independent rediscoveries of Highland Park practice. The list of reinventions includes: cheap modification of general purpose equipment; machines close together and in order of use; pull through production; low stocks; and multi-tasked workers all operating in a Ford-like culture of *kaizen* through layout change and de-manning. A variety of external and internal factors can be invoked to explain America's loss of productive expertise. Externally, we would put considerable emphasis on institutional conditions and the pressure on public corporations to earn profit for shareholders; Ford uniquely emancipated his company from these pressures by buying out his shareholders and never borrowing from the banks, so that the family owned company was, for better or worse, free to pursue its own productionist priorities. Internally, we would argue that the transference of Highland Park manufacturing practice was never straightforward because the Ford factory's manufacturing task was artificially simplified, and because the Ford company never consolidated its achievement into a set of management techniques that could be adopted by suppliers and competitors.

Ford's manufacturing task was artificially simplified because, for nearly twenty years from 1908 to 1926, the company produced developed versions of the one power train and chassis. Quantitively, the philosophy of design stretch required many changes, possibly as many changes as if the company had produced a changing variety of non-T models. But the quality of the changes was different and distinctive as long as the company persevered with stretching the T. The necessary design changes were typically piecemeal and cumulative as old-specification T parts were replaced with new specification T parts. Of course, the company had to buy or make old-specification parts for after- market replacement sales. But, in practice, this was never a major problem because Ford's spares business was relatively modest; between 1910 and 1916, 91 per cent of the company's turnover came from selling new cars (Appendix: table 6 col. i). For nearly twenty years, Ford did not have to produce a changing family of engines and chassis with different dimensions and limited commonality; the company's engineers were able to work on the task of perfecting multi-process manufacture without regular changeover.

Cycle set up was an issue of considerable importance at Highland Park. The machining of engine and gearbox parts involved a multiplicity of sequential operations where one of the key limits on throughput efficiency was the amount of unproductive time which the company wasted on clamping the workpieces at each process stage. As Colvin explained in his 1913–15 account of machine shop practice, cycle set up time had been reduced in a variety of ways; we would add the point that most of these techniques have since been re-discovered in Japanese manufacturing. Ford used what the Japanese would now call 'one touch' quick acting clamps (*American Machinist* 1915:971) and generally favoured gravity location on steel tables (*American Machinist* 1915:1059); both cylinder head and block machining began with the drilling of bolt holes which provided points of location when the gravity tables were fitted with steel dowel pins (*American Machinist* 1913:841). Ford also developed 'multiple tooling' so that several cuts could be made at each clamping; in engine block production, the millers were ingeniously set up for 'double milling' of top and sides with the vertical cutters followed up by horizontal cutters (*American Machinist* 1913:843). The company also developed fixtures which allowed each cutting machine to finish many workpieces in a single pass; the millers and grinders skimmed 15–30 engine blocks at a single pass and, with small workpieces like the valve lifter, no less than 104 pieces were ground at each pass (*American Machinist* 1913:227).

At the same time changeover was never a major problem at Highland Park. The manufacturing task was such that the machine shop seldom lost time because the machine tools were being adjusted to process a workpiece of different type and dimensions. This kind of changeover was only necessary on a small number of machines like the five Bliss 900 ton presses which were worked in tandem; on this line, dies were regularly changed so that the presses could produce batches of different panels like crank cases or running boards (*Ford Times* June 1913). But, in the context of the Highland Park machine shop, these machines were very unusual because they were so fast acting; the heavy press line work rate of five strokes per minute (*American Machinist* 1914:49) was not much slower than the work rate of six to eight strokes per minute on comparable equipment in a modern car press shop (Williams *et al* 1991a). On the other hand, the machine shop was generally fitted out with belt driven 'single purpose' cutting machines whose cycle times were substantially longer than those of modern NC or CNC equipment; cycle times were further extended by Ford's general practice of running the machines slow so as to reduce the wear on cutters and drills. As Ford had an increasingly large volume throughput in the machine shop, the requirements of his manufacturing task and the characteristics of the available technology were together accommodated by doubling up identical machines which continuously processed one part.

It was (and is) possible to make two distinctive readings of Ford's machine shop practice. The imaginative reading would characterise it as an exemplary effort to reduce unproductive time through techniques of rapid set up which could be adapted elsewhere to reduce changeover time. The literal reading would emphasise that Ford's manufacturing task was unusual and his practice therefore contained few lessons for machine shops whose problem was changeover. This second literal reading appears to have triumphed in American manufacturing in the 1920s. In companies like GM which produced a changing family of models, high stocks were used to disguise unresolved problems about slow changeover; through the 1920s, General Motors ran with stocks equal to fifteen weeks sales cover whereas Ford in the early 1920s ran with stocks equal to seven weeks sales cover. This difference in performance is particularly interesting because companies like General Motors tried to buy in Ford's production expertise by recruiting senior Ford production men; in the 1920s, GM's expanding Chevrolet volume car division was headed by Knudsen who had supervised Ford's assembly branches before World War One. If energetic and competent executives like Knudsen were unable to reproduce the Highland Park achievements outside the company, the implication is that the application of Ford methods was difficult. This difficulty was hardly surprising when the discoveries were based on a break with existing management knowledges and were never consolidated into a new, transferable body of management techniques.

The break with the existing body of management knowledge was quite explicit because Henry Ford disliked and discouraged all fixed forms of corporate organisation and rejected all the approved forms of management calculation. Henry cultivated the persona of the 'Yankee mechanic' and employed hard-driving production engineers in a project-centred, flat and informal organisation. At Highland Park there was no organisation chart and few job titles, Ford did not share the modern American preoccupation with defining who is responsible for what and who reports to whom. Employees who had worked for the company for several years were often confused about the status and responsibility of senior managers. Thus a Ford tool designer struggled to define the roles of Martin and Sorensen who are labelled as 'factory superintendent' and 'assistant superintendent' in Ford's office photographs, 'I believe that Sorensen and Martin were considered on a par, I never could figure out exactly' (Pioch 1955). A Ford metallurgist, who joined the company in 1915, did figure out that Martin was, 'the man who was really in charge of manufacturing' but remained confused about the responsibilities of his two assistants, Sorensen and Avery, 'as far as I know there were no particular areas that each one was assigned to' (McCloud 1955:8).

Equally striking was Ford's rejection of all existing forms of management calculation, especially cost accounting. This again reflected Henry's personal view that, 'a lot of cost accounting people was unnecessary overhead' (Hutchins 1957:23). In 1919 the Chicago accounting firm of Thompson and Black recommended a larger accounting department and, shortly afterwards, work began on a new building to house the enlarged department. When Ford found out he personally ordered the contractor to fill in the ground work saying, 'we don't need any cost department' (Hutchins 1957:24). The company only acquired a cost centre system at the insistence of the American government in World War I when Ford became a defense contractor (Husen 1955:7). Before that, there was only a rudimentary product costing system which mainly inputted to financial reporting. In 1914, when the company had 13,000 employees, its cost department included just eight employees of whom three were stenographer operators (Husen 1955:8–9). The cost department could be so small because, just as in a modern Japanese company, Ford's production control was based on simple, direct physical measures. In effect, the factory was scheduled on the basis of a daily production stint; as Colvin observed, in the machine shop, 'every machine has an allotted daily output – and each foreman receives a daily schedule of work to be turned out by his department' (*American Machinist* 5 June 1913:761). And, as Husen recalled (1955:8–9), cost comparisons were never used as a basis for day by day production control, 'the foreman compared the actual production of each employee with the number of pieces established as a target'. The *kaizen* benefits of higher targets were measured very simply in physical and financial terms by calculating the savings in labour time and then, after applying the hourly rate, expressing the saving in terms of labour cost. In words which could be taken from our 1988 interviews with Toyota shop managers (Williams *et al* 1991a), Klann (1955:145) explained how these measures operated when he supervised motor assembly at Highland Park,

> we would go to work and mark the number of men we had in the shop. We would mark how many hours they worked. We divided the men by the number of motors we built, so many hours per motor. We watched minute cost.

Ford's informal organisation and direct forms of calculation were highly effective but they had their limits. If Ford engineers broke with existing management techniques, unlike a later generation of Japanese, they never institutionalised their discoveries into a set of techniques which would regulate Ford operations and could be applied elsewhere in the factories of suppliers and competitors. The failure to institutionalise is most obvious in the case of stock control where Ford never developed a regulatory system. Like everything else at Highland Park, de-stocking was an improvised response. Levels of factory work in progress were reduced by short travel layouts in a factory which was struggling to meet demand and to clear factory space for more machines. As Arnold and Faurote (1915:63) explain, the Highland Park factory was originally planned to run with twenty five days of stock cover, but by 1915 the company was actually running on three to five days stock. The company had learnt to live with permanent stock shortage by creating the role of 'stock chaser' who was responsible for getting part batch deliveries rushed forward to down stream processes which were running out of parts. This non-system worked to maintain production with low stocks as long as insatiable demand pulled through ever more output but it concealed a basic weakness. Ford had no regulatory techniques or system for enforcing low stocks in an enterprise which faced weaker demand; if the pressure of demand slackened, then stocks would rise to whatever level could be accommodated between processes and in off-line storage.

Ford's incentive to promote transference was reduced because, unlike modern Japanese mass manufacturers, the company was not interested in improving the productive competence of its suppliers. As we have argued Ford's policy on components was to reduce costs by cutting out inefficient suppliers and bringing the work in-house. If Ford found an efficient supplier then the company tried to internalise the benefits, as in the case of Keim Mills, a New Buffalo company which supplied Ford with steel pressings. In 1911 Ford purchased Keim Mills and shortly afterwards moved the press machines and key workers by rail to Highland Park where they became the nucleus of Ford's press department. After World War I, the policy of bringing work in-house was pressed harder with the construction of an all-new factory at Rouge which quickly supplanted Highland Park. As we will now argue, the development of Rouge sealed the fate of the precarious, unconsolidated experiment at Highland Park because Rouge represented a different principle of mass manufacture.

Rouge was the realisation of Ford's long standing policy of building more of the car in-house. After World War I, the 150 acre Rouge site was originally developed as a conversion plant which supplemented Highland Park manufacturing and branch assembly; Rouge turned iron ore, coal and lumber into castings and floor boards. The transfer of many Highland Park departments to Rouge in 1925 and the beginning of steel production in 1926 marked the point when Rouge became the world's first fully integrated car plant. Here Ford made everything from steel and glass to leatherette in a factory where raw materials were delivered to the dock and finished cars rolled out at the other end of the factory. Fairly obviously, this exhausted the possibility of further cost reduction through displacement of inefficient suppliers; it was no longer possible to cut the price of the car by doing more of the work in-house. Even materials cost was now directly controlled by Ford because the company's ships and railroads carried materials from Ford owned mines, factories and plantations to the docks and sidings of Rouge.

Ford could only continue along a trajectory of cost reduction, if the company refined and developed the cost-reducing techniques of mass manufacturing. At Rouge, production engineers rose to the challenge by constructing the world's first interconnected car factory; all work transfer problems were to be solved by using moving conveyors and every manufacturing operation (including steel and glass making) would be performed continuously on an interconnected set of conveyors. In terms of manufacturing technique, Rouge represented the triumph of the conveyor over all other forms of material handling; by the mid-1920s, there were no less than twenty seven miles of conveyor at Rouge (Nevins 1957). By way of contrast, Highland Park used a varied range of handling devices and at the same time tried to eliminate handling so that the total mileage of conveyor was very much smaller. In Highland Park's final stage of development after 1922, the longest conveyor system was the 'christmas tree' conveyor which ran for three and a half miles from the foundry through the machine shop to final assembly (Klann 1955:157–61).

It is wrong to suppose, as Hounshell (1984: chap. 7) does, that Ford's manufacturing system at Rouge had a limited capacity for product changeover. The six month halt in production in 1927 when the Rouge was changed over from the Model T to the Model A was caused more by the Ford organisation's increasingly arbitrary decision making than by the Ford factory's productive inflexibility. When the Model T was discontinued in May 1927, the company had still not ordered new body tooling, constructed running prototypes or even finished design work on the replacement. In these circumstances, the changeover from T to A was managed remarkably smoothly and quickly, partly because the re-usability of Ford's single purpose machine tools was once again triumphantly vindicated; no less than three-quarters of the machine tools used in T production were re-used in A production (Hounshell 1984:293). The real problem was that Rouge constrained process improvement. The system of interconnected conveyors was designed to deliver whole factory *kaizen* automatically whenever the belts were speeded up. But this gain was always more hypothetical than real because it depended on virtuoso production engineering to keep the whole system up and running and on the external condition of a market which would absorb increased output. If Ford could solve the internal problem, the company could not lift the external constraint which pressed as soon as Chevrolet became a major force in the cheap car market. Meanwhile, the Rouge system of interconnected manufacture made dialectical single process *kaizen* obsolete. The objective of single process improvement was lost as soon as all processes moved at one uniform pace, and the instrument of layout change was ruled out by a conveyor system where individual conveyors were all dedicated to the transfer of particular materials and components. In each case, the design of the belt and its fences presupposed a material or component with specific characteristics and dimensions. Paradoxically, the layout of Rouge was fixed by its endlessly moving lines.

Conveyorisation of everything undoubtedly generated substantial once and for all cost savings; it eliminated the labour expense of truckers at a very reasonable operating cost because the amount of power required to move the belts was modest. But the conveyorisation of everything closed off the old style of Highland Park continuous improvement and inaugurated the degeneration of Western mass manufacturing which continued as other car companies imitated the integrated form but not the interconnected substance of Rouge in factories like Volkswagen Wolfsburg, which can be plausibly represented as Rouge with buffers. Significantly, the major Japanese car manufacturers after World War II tried to realise the ideal of manufacturing flow in a different way through a system which was not vertically integrated.

## Conclusion

The concepts of mass production and Fordism (like their opposites flexible specialisation and post-Fordism) were designed to meet a meta-theoretical imperative rather than an empirical need. In our view, they represent the post-Marxist continuation of a Marxisant tradition of interpreting the genesis and resolution of economic and political crisis in the advanced capitalist countries; these concepts allow a section of the intelligentsia to conserve the form of a Marxist discourse whose substance has become unacceptable. Nevertheless, the discrepancy between theory and evidence has become an increasing embarrassment to the intellectual protagonists of concepts like mass production. Recently, they have tried to rationalise the discrepancy by arguing that their organising concepts, 'are ideal typical models rather than empirical generalisations or descriptive hypotheses about individual firms, sectors or national economies'. Hirst and Zeitlin (1991:5) defend flexible specialisation, in these terms, while Teague (1990: 42) uses more or less exactly the same formulation in defence of regulation theory, whose, 'purpose is . . . to build ideal types or frameworks which can be used to assess or investigate economic development'. With the theory thus immunised against disproof, it is possible to make two comfortable assertions: first, it is asserted that concepts like mass production and Fordism provide a positive basis for research into manufacturing (Teague 1990:50); second, it is asserted that the range of empirical variation in case material represents a hybrid 'mix' of elements of mass production and its opposite (Hirst and Zeitlin 1991:5).

Our research on Ford raises questions about the ideal type alibi and destroys the comfortable assumption about the fruitfulness of research based on concepts like mass production and Fordism. The massive disjuncture between Ford's achievements and the concepts of mass production and Fordism is an obvious problem. After all, Ford and the T is the classic case of mass production and Ford's practice is generally supposed to represent the apotheosis of its identifying characteristics. We may recall an analogy from the experience of an earlier generation of social scientists. For non-Weberians, the discrepancies between the 'protestant ethic' and Benjamin Franklin's life did constitute evidence which damaged the ideal type construct. If the apotheosis is not so, and the classic case does not fit, those who have not invested their intellectual capital in the construct will quite reasonably ask what is the point of the construct?

At the same time it is hard to see how the concept of mass production can provide a fruitful basis for further research into the theory or history of manufacturing. In theoretical terms, this is because mass production distracts us with inessentials and irrelevancies and, at the same time, crucially fails to identify the springs of dynamic cost reduction in features like layout change and low stock production. Researchers who are pre-occupied with mass production and its hybrids will literally not know what to look for when they study firms and industries. In historical research, the concept of mass production is equally misleading because it promotes a bogus periodisation and suppresses important issues about continuity and discontinuity. Researchers who are pre-occupied with mass production will not ask the serious questions about the conditions under which diffusion of Ford's flow practice failed, or about how the Americans apparently lost the art of productive intervention between 1920 and 1950 just before it was rediscovered in Japan.

For all these reasons, mass production is an idea whose time has gone.

Our own recent work on Japan, as well as on Ford, demonstrates that a series of important problems, about the logic and development of manufacturing, are open to investigation; but not by those who cling to the simplicities of organising concepts like mass production.

## Acknowledgements

We would like to thank Cynthia Read-Miller and the ever-helpful staff of the archive at Henry Ford Museum, Detroit who assisted our archive research. The Woodward Manchester Corporation gave us permission to view the semi-derelict Highland Park factory buildings. We are also grateful to Professor David L. Lewis and Ford Bryan who advised on and encouraged our research project.

## References

Abell, O. (1915) 'The making of men, motor cars and profits', *The Iron Age*, 7 January 1915, 33–41.

*American Machinist* (1913–5) various issues.

Arnold, H. and Faurote, F. (1915) *Ford Methods and the Ford Shops*, New York: Engineering Magazine Co.

*Autocar*, 18 August 1988.

Bornholt, O. (1913) 'Placing machines for sequence of use: the trucking involved when machines are grouped by classes', *The Iron Age*, 4 December 1913, 1276–1277.

Braverman, H. (1974) *Labour and Monopoly Capital*, New York: Monthly Review Press.

Bryan, F. (1990) *Beyond the Model T*, Detroit: Wayne State University Press.

Chandler, A. (1962) *Strategy and Structure*, Cambridge (Mass): MIT Press.

Fahnestock, M. (1968) *The Model T Ford Owner*, Arcadia (California): Post Motor Books. Reprinted from *Ford Owner*, 1922–26.

*Ford Times* (1908–16) various issues.

Hirst, P. and Zeitlin, J. (1991) 'Flexible specialisation versus post-Fordism', *Economy and Society*, 20, 1, 1–30.

Hounshell, D. (1984) *From the American System to Mass Production*, Baltimore: Johns Hopkins University Press.

Husen, O. (1955) *Oral Reminiscences*, oral history section, Ford Archive, Detroit.

Hutchins, D. (1957) *Oral Reminiscences*, oral history section, Ford Archive, Detroit.

Jones, E. (1916) *The Administration of Industrial Enterprises, with Special Reference to Factory Practice*, London: Longmans.

Klann, W. (1955) *Oral Reminiscences*, oral history section, Ford Archive, Detroit.

Kuhn, A. (1986) *GM Passes Ford*, University Park (Pennsylvania): Pennsylvania State University Press.

Lee, E. (1957) *Population Redistribution and Economic Growth in the United States, 1870–1950*, Philadelphia: American Philosophical Society.

Lewchuk, W. (1987) *American Technology and the British Vehicle Industry*, Cambridge: Cambridge University Press.

McCloud, J. (1955) *Oral Reminiscences*, oral history section, Ford Archive, Detroit.
Meyer, S. (1981) *The Five Dollar Day*, Albany: State University of New York Press.
Murray, R. (1988) 'Life after Henry (Ford)', *Marxism Today*, October 1988, 1–13.
Nevins, A. and Hill, F. (1954) *Ford: The Times, the Man, the Company*, New York: Scribner and Sons.
Nevins, A. and Hill, F. (1957) *Ford: Expansion and Challenge, 1915–33*, New York: Scribner and Sons.
Pioch, W. (1955) *Oral Reminiscences*, oral history section, Ford Archive, Detroit.
Piore, M. and Sabel, C. (1984) *The Second Industrial Divide: Possibilities for Prosperity*, New York: Basic Books.
Schonberger, R. (1980) *Japanese Manufacturing Techniques*, New York: Free Press.
Seltzer, L. (1928) *A Financial History of the American Automobile Industry*, New York: Houghton Mifflin.
Sorensen, C. (1956) *My Forty Years with Ford*, New York: W. W. Norton.
Teague, P. (1990) 'The political economy of the regulation school and the flexible specialisation scenario', *Journal of Economic Studies*, 17, 5, 32–54.
Van Deventer, J. (1923) 'Ford principles and practice at River Rouge', *Industrial Management: The Engineering Magazine*, January 1923, LXV, 1, New York.
Van Doren Stern, P. (1955) *Tin Lizzie*, New York: Simon and Schuster.
Williams, K. *et al* (1987) 'The end of mass production?', *Economy and Society*, 16, 3, 405–439.
Williams, K. *et al* (1989) 'Do labour costs really matter?', *Work Employment and Society*, 3, 3, 281–305.
Williams, K. *et al* (1991a) 'How far from Japan? A case study of Japanese press shop practice and management calculation', *Critical Perspectives on Accounting*, 2, 145–69.
Williams, K. *et al* (1991b) 'Management accounting: the Western problematic against the Japanese application', mimeo, University of Wales, Aberystwyth.
Williams, K. *et al* (1992) 'The myth of the line', mimeo, University of Wales, Aberystwyth.
Womack, J., Jones, D. and Roos, D. (1990) *The Machine that Changed the World*, New York: Rawson Associates.

Department of History
University College of Wales
ABERYSTWYTH
SY23 3DY

# STATISTICAL APPENDIX ON THE FORD MOTOR COMPANY, 1909–19

**Table 1:** Ford revenue costs and value added, 1909–19

**Table 2:** Ford cars per man and labour hours per car, 1909–16

**Table 3:** Ford cars per man, variant series, 1909–13

**Table 4:** Ford stock turn, fixed capital and payroll per employee

**Table 5:** Ford Model T profitability

**Table 6:** Ford turnover by activity, 1909–19.

***Table 1*** Ford Revenue, Costs and Value Added, 1909–19

| | Model T output (i) | Material cost per T $ (ii) | Labour share of value added % (iii) | Purchases as % of gross output (iv) |
|---|---|---|---|---|
| 1909 | 13,941 | 590 | 27.7 | 68.2 |
| 1910 | 20,738 | 544 | 32.7 | 62.2 |
| 1911 | 53,800 | 348 | 27.3 | 61.1 |
| 1912 | 82,500 | 407 | 29.1 | 61.1 |
| 1913 | 199,100 | 319 | 31.9 | 60.0 |
| 1914 | 249,700 | 371 | 30.3 | 58.0 |
| 1915 | 368,599 | 255 | 30.5 | 52.4 |
| 1916 | 585,400 | 262 | 39.1 | 55.5 |
| 1917 | | | 61.1 | 72.3 |
| 1918 | | | 89.1 | 76.6 |

*Sources:*

col. (i) Output of the model T is obtained from the company's shipping ledgers in Ford Archive: accession 463, box 1.

col. (ii) Material cost per T is calculated by dividing Model T output (Ford Archive: accession 463, box 1) into total material purchases (Ford Archive: accession 157).

col. (iii) Value added is calculated by subtracting purchases (Ford Archive: accession 157) from gross output (Ford Archive: accession 96, box 8 and 10). Total labour costs are given in Ford Archive: accession 843.

col. (iv) Material purchases are taken from Ford Archive: accession 157 and gross output from Ford Archive: accession 96, box 8 and 10.

***Table 2*** Ford Cars per Man and Labour Hours per Car, 1909–16

| | Cars shipped (i) | Number of employees (ii) | Cars per man year (iii) | Labour hours per car (iv) |
|---|---|---|---|---|
| 1909 | 13,941 | 1,655 | 8.4 | 357 |
| 1910 | 20,738 | 2,773 | 7.5 | 400 |
| 1911 | 53,800 | 3,976 | 13.5 | 222 |
| 1912 | 82,500 | 6,867 | 12.0 | 250 |
| 1913 | 199,100 | 14,366 | 13.9 | 216 |
| 1914 | 249,700 | 12,880 | 18.8 | 127 |
| 1915 | 368,599 | 18,892 | 19.5 | 123 |
| 1916 | 585,400 | 32,702 | 17.9 | 134 |

*Sources:*

col. (i) Numbers of cars shipped from Highland Park in built up or kit form for assembly in branches. The total number of cars shipped is taken from the company's shipping ledgers in Ford Archive: accession 463, box 1.

col. (ii) Number of employees at Highland Park and in American branch factories which assembled kits from Highland Park. The total number of American employees is taken from Nevins (1954:648).

col. (iii) Cars per man year is calculated by dividing total number of cars shipped by number of employees.

col. (iv) Labour hours per car are calculated on the assumption of 3,000 man hours up to and including 1913, when factory workers worked six ten-hour days in each of 50 weeks. From early 1914, after the introduction of the eight hour day, we assume 2400 hours when factory workers worked six eight-hour days for each of 50 weeks. This last assumption improves the 1914–16 performance slightly because branch assembly operations worked a nine hour day after 1914.

***Table 3*** Ford Cars per Man, Variant Series 1909–13

| | Average number of employees (i) | Cars per man (ii) |
|---|---|---|
| 1909 | 2,190 | 4.9 |
| 1910 | 3,672 | 5.8 |
| 1911 | 4,100 | 8.4 |
| 1912 | 7,042 | 10.8 |
| 1913 | 16,000 | 12.5 |

*Note:*
This table is based on the totals of 'average number of employees' given in the June 1913 issue of *Ford Times*; for the early years, the *Ford Times* totals are higher than those given in Nevins (1954) and used in table 2. Calculations based on the *Ford Times* employment series therefore show the company starting from a much lower level of physical productivity in 1909 and in 1910. This reinforces our argument that much of the gain in cars per man year came before 1913–14. It is also worth emphasising that the series does not show the stagnation of productivity in 1911–2 which Lewchuk (1987) emphasises.

**Table 4** Ford Stock Turn, Fixed Capital and Payroll per Employee

| | Stock turn in weeks (i) | Value of plant per employee $ (ii) | Depreciation as % of payroll (iii) | Payroll cost per car $ (iv) |
|---|---|---|---|---|
| 1909 | 10.7 | 875 | 1.1 | 84 |
| 1910 | 9.4 | 1055 | 5.5 | 100 |
| 1911 | 8.8 | 999 | 6.3 | 55 |
| 1912 | 8.9 | 1140 | 5.9 | 65 |
| 1913 | 7.4 | 915 | 4.9 | 61 |
| 1914 | 6.5 | 1695 | 5.2 | 65 |
| 1915 | 7.9 | 1646 | 5.4 | 64 |
| 1916 | 8.7 | 1353 | 4.1 | 70 |
| 1917 | 9.3 | 1606 | 5.1 | — |
| 1918 | 8.1 | 2126 | 6.4 | — |

*Sources:*

col. (i) Stock turn expresses average value of stocks in terms of the number of weeks sales cover. Average stock figures are taken from Ford Archives: accession 96, box 8. Sales revenue is taken from Ford Archive: accession 96, box 10.

col. (ii) The value of plant per employee is calculated by dividing total value of plant (excluding buildings) in the Ford Motor Company by number of employees. The value of plant is taken from Ford Archive: accession 96, box 8. Employee figures are taken from Nevins (1954:648).

col. (iii) Depreciation figures are taken from the company accounts in Ford Archive: accession 96, box 8; and payroll from Ford Archive: accession 843.

col. (iv) Wage costs per car are calculated by dividing cars shipped into total payroll costs. Cars shipped are taken from Ford Archive: accession 463, box 1 and payroll from Ford Archive: accession 843.

***Table 5*** Ford Model T Profitability

| | Profit 000's \$ (i) | Cars shipped 000's (ii) | Profit per car \$ (iii) | Price of Touring T \$ (iv) | Profit as % of selling price (v) |
|---|---|---|---|---|---|
| 1909 | 3062 | 14 | 218 | 850 | 26 |
| 1910 | 4163 | 21 | 198 | 950 | 21 |
| 1911 | 7339 | 54 | 136 | 780 | 17 |
| 1912 | 13543 | 83 | 163 | 690 | 24 |
| 1913 | 27089 | 199 | 136 | 600 | 23 |
| 1914 | 24698 | 188 | 131 | 550 | 24 |
| 1915 | 25532 | 247 | 103 | 440 | 23 |
| 1916 | 57157 | 506 | 113 | 360 | 31 |

*Sources:*

col. (i) Consolidated profit of the Ford Motor Company is taken from Ford's company accounts in various years. In the 1909–16 period this is effectively the profit on production and assembly for the American market. Only after 1920 do overseas activities become important.

col. (ii) The years used in calculating cars shipped and profit from 1909 to 1913 are calendar years. The company then changes its accounting period so that we calculate cars shipped and profit over nine months in 1914, 10 months in 1915 and finally for fiscal 1915/16 which ends on July 31. In each case the total number of cars shipped in the relevant accounting period has been abstracted from the company's Highland Park shipping ledgers in Ford Archive: accession 463, box 1.

col. (iii) Profit per car is calculated by dividing cars shipped from column (ii) into total profit from column (i).

col. (iv) Prices of the Touring T in dollars. This is the price of the Touring (soft top, four seat) Model T which was by far and away the most popular model variant in the early years. In most years there were autumn price cuts and we have taken the usually higher price which prevailed for most of the year. Car prices are taken from Ford Archive: accession 157.

col. (v) The margin on selling price is calculated very approximately by dividing column (iii) into column (iv).

***Table 6*** Ford Turnover by Activity, 1909–1919

| | Car manufacture as % of sales (i) | Parts and repairs as % of sales (ii) | Sundries as % of sales (iii) |
|---|---|---|---|
| 1909 | 85.5 | 9.0 | 6.4 |
| 1910 | 91.6 | 7.4 | 1.3 |
| 1911 | 91.4 | 6.5 | 1.9 |
| 1912 | 93.8 | 5.3 | 2.0 |
| 1913 | 92.9 | 5.4 | 1.8 |
| 1914 | 95.5 | 6.0 | 0.6 |
| 1915 | 91.0 | 8.1 | 0.9 |
| 1916 | 87.2 | 11.3 | 1.5 |
| 1917 | 87.4 | 12.6 | — |
| 1918 | 82.1 | 17.9 | — |
| 1919 | 76.8 | 23.2 | — |

*Sources:*
This breakdown of sales turnover by activity is taken from Ford Archive: accession 96, box 10. The percentages in the two main categories add up to 100 from 1909 to 1916 only with the inclusion of a 'sundries' balancing item. After 1916 'sundries' are presumably allocated to new car manufacture or parts.

# Part III
# Diffusion and Variations

# [9]

*Steven Tolliday / Jonathan Zeitlin*

## Between Fordism and Flexibility
## *The Automobile Industry and its Workers – Past, Present and Future**

The automobile industry has long occupied a central place in debates about management, technology and work in advanced industrial societies. Since the days of Henry Ford the industry has served as a model of economic expansion and technological progress based on mass production: the manufacture of standardised products in huge volumes using special-purpose machinery and unskilled labour, a system which has come to be known as »Fordism«. But from the mid-1970s, sweeping changes in markets and technology have transformed international competitive conditions and spurred automobile manufacturers in every country to experiment with new strategies based on greater product diversity and more flexible methods of production. These shifts in competitive strategy and manufacturing methods, whose outcome is still far from certain, pose far-reaching challenges not only for the automobile companies and their workers but also for trade unions and national governments.

This paper considers the development of the automobile industry from its origins to the present in a perspective informed by current upheavals in markets, technology and industrial relations. Section I looks at the rise of mass production and the modification of the Fordist model to suit national circumstances by managers, workers and trade unions. Section II examines industrial relations in the age of Fordism, focusing on international variations in trade union structure, bargaining strategy and job control practices. Finally, Section III traces the recent transformations in the international automobile industry, considers how far new production and marketing strategies mark a break with Fordism, and draws out the implications for industrial relations and trade union strategy.

### I. The Rise of Fordism

Mass production of automobiles, as developed by Henry Ford in the half-dozen years before the First World War, depended on three basic principles: the standardisation of the product, the use of special-purpose equipment, and the elimination of skilled labour in direct production. More than any of his contemporaries, Henry Ford grasped the vast latent demand for cheap, reliable basic transport in the American Midwest with its prosperous but isolated farms and small towns. His »Model T«, ruggedly designed, easy to repair and priced well below its competitors sold in unprecedented numbers, jumping from 6000 in 1908 to 189,000 in 1913 and 802,000 in 1917 to reach a total of 15 million by 1926.[1]

---

* This paper is based on the introduction to *Steven Tolliday/Jonathan Zeitlin* (eds.), The Automobile Industry and Its Workers: Between Fordism and Flexibility, Cambridge/New York 1987.

1 This account of Ford's innovations and the development of his competitive strategy draws on the following sources: *Alfred D. Chandler, Jr.* (ed.), Giant Enterprise: Ford, General Motors and the Automobile Industry, New York 1964, pp. 23–95; *David Hounshell*, From the American System to Mass-

In its heyday, the »T« came in a single standard model and was available in »any color as long as it was black«. The standardisation of the car and its components was intended to facilitate repair by distant customers, but more importantly to permit a dramatic reduction in costs through the use of special-purpose machinery and the division of labour. Ford drew upon and elaborated the heritage of the »American system of manufacturers« as it had developed from firearms through sewing machines, agricultural equipment and bicycles towards the high-volume production of interchangeable parts. He and his engineers simplified each component as far as possible, using new materials such as pressed steel; arranged the equipment in their new Highland Park factory so that the flow of materials followed the sequence of operations; and designed first jigs, fixtures and gauges and then new special-purpose machines to ensure full interchangeability. These »farmer tools« could be operated by unskilled workers without a background in the industry, and Ford likewise pursued the reduction of skill requirements in vehicle assembly through the subdivision of tasks and the elimination of fitting, hallmarks of »mass production« as he was later to define it. With the astronomical growth of sales, each of these developments reinforced the others in a constant struggle for faster throughput which gave rise to Ford's most famous innovation: the introduction of the moving assembly line in 1913.

Ford's rapid growth and dramatic innovations in production brought with them unprecedented problems of labour management. Skilled workers progressively moved into new »indirect« services such as toolmaking, toolsetting and maintenance, joining the growing army of supervisors, inspectors, progress chasers, clerks and engineers needed to administer mass production. The workforce shot up from 1548 in 1909 to 13,667 in 1913, an increasing proportion of which were unskilled immigrants from southern and eastern Europe, many of whom spoke no English. Turnover levels reached 370 % in 1913, with absenteeism rates of 10 % per day, creating enormous difficulties in production planning. It was in this context that Ford inaugurated his famous »Five Dollar Day«, a profit-sharing plan designed to stabilize and integrate the workforce alongside other measures such as a Sociological Department and intensive, driving supervision.

By the First World War, Ford had synthesised his innovations in product design, production and labour management into a coherent competitive strategy whose application appeared universal. Product standardisation led to economies of scale in an endless virtuous circle. The huge River Rouge complex, built at the end of the First World War, integrated from steelmaking to final assembly, with its own railways, docks and power plant was the tangible embodiment of this strategy. As sales of the Model T soared, capturing 55 % of the American market in 1921, Ford's prospects seemed limitless and »Fordism« to define an international standard of modern manufacturing practice.

But the tremendous success of Ford's competitive strategy depended in practice on specific features of the American market in this period: its enormous size, vast distances, egalitarian income distribution and homogeneous tastes. The British market was fundamentally different in each of these respects, remaining much smaller and dominated by quasi-luxury demand into the 1920s, and a strategy based on price competition and economies of scale ac-

---

Production, 1800–1932, Baltimore 1984, pp. 217–63; *Stephen Meyer III*, The Five Dollar Day: Labor Management and Social Control in the Ford Motor Company, Albany, NY 1981; *idem*, Mechanical Engineers and Automobile Workers: American Machine-Tool Technology and the »Transfer of Skill« from the 1900s through the 1930s, in: *Nelson Lichtenstein/Stephen Meyer III* (eds.), On the Line: A Social History of the American Automobile Industry (Champaign-Urbana, IL, forthcoming 1988); *Alan Nevins with Frank E. Hill*, Ford: The Times, the Man, the Company, New York 1954; and *Charles F. Sabel*, Work and Politics: The Division of Labor in Industry, Cambridge 1982, pp. 32–33 (German translation: Arbeit und Politik, Wien 1986).

cordingly proved far less effective than across the Atlantic.[2] While the underdevelopment of the British motor industry allowed Ford to sell significant numbers of Model Ts before 1914, the company's hold on the British market quickly evaporated with the emergence of large-scale domestic production in the 1920s. Ford officials in Detroit refused to allow local managers to develop a model aimed specifically at the British market until 1928, and sought to impose American practices of exclusive dealerships, high day wages and low manning levels quite unsuited to British conditions. Even once local management achieved autonomy and began to pull the company back from its nadir of the late 20s, the influence of Detroit practice remained strong: Ford's Dagenham factory, opened in 1932 as a scaled-down version of River Rouge, proved too large for the British market until the 1950s.

British manufacturers such as Morris and Austin pursued a more pragmatic strategy, competing on the basis of new models and designs as well as price and moving gradually towards higher volumes of production. Production methods accordingly remained less rigid and capital-intensive than at Ford's, with greater customisation in bodywork, more adaptable equipment in machining and greater use of hand labour in assembly. In contrast to Ford, British manufacturers also relied on piecework incentive schemes rather than tight supervision to motivate the labour force; and this distinctive complex of product, production and labour strategies remained characteristic of the British motor industry until its crisis of the 1960s and 70s.

Elsewhere in Europe, the market for automobiles also differed sharply from the American pattern, but the attitude of local manufacturers was different.[3] Leading French and Italian entrepreneurs such as André Citroën, Louis Renault, Marius Berliet and Giovanni Agnelli were fascinated by Fordist methods on technological grounds and sought to impose them on the market. Engineers and managers from each of these companies visited the United States on several occasions, and each time came away determined to emulate the Fordist model on a more ambitious scale. In the 1920s and 30s, vast new factories were constructed, modelled first on Highland Park and then on River Rouge; machine tools were imported from the United States in large numbers; product lines were drastically simplified; and valiant efforts were made to rationalise production along American lines through the use of work measurement and the introduction of the assembly line.

In practice, however, European motor manufacturers' infatuation with Fordism created major difficulties both for production management and marketing. The great leap forward to mass production ran ahead of management's ability to elaborate suitable systems of cost accounting, work scheduling, supervision and quality control, so that shop-floor realities often lagged well behind the glossy visions of the engineers. The smaller size and greater differentiation of the French and Italian markets also created obstacles for Fordist marketing strategies, and overambitious expansion projects such as those of Berliet and Citroën crashed when demand turned downwards in 1921 and again in the early 30s. All the main manufacturers were forced to multiply models and expand their product ranges in order to maintain their sales, and as in Britain the organisation of production accordingly remained less rigid and

2 For a fuller discussion of the British case, see *Steven Tolliday*, Management and Labour in Britain, 1896–1939, in: *Tolliday/Zeitlin*, op. cit.

3 For an overview of the development of automobile production in Europe, see *Jean-Pierre Bardou/Jean-Jacques Chanaron/Patrick Fridenson/James M. Laux*, The Automobile Revolution, Chapel Hill, NC, 1982, pts. I–II. For a more detailed discussion of France and Italy, see *Sylvie Van de Casteele-Schweitzer*, Management and Labour in France, 1914–39, and *Duccio Bigazzi*, Management and Labour in Italy, 1906–45, both in *Tolliday/Zeitlin*, op. cit. For an analysis of the German market, in which demand for automobiles was particularly constrained by social and institutional factors before the 1930s, see *Fritz Blaich*, The Development of the Distribution Sector in the German Car Industry, in: *Akio Okochi/Koichi Shimokawa* (eds.), The Development of Mass Marketing, Tokyo 1982.

less mechanised than in the United States. None of the European manufacturers, finally, dared to emulate Ford's labour strategies in full, eschewing high day wages in favour of piece rates, speed up and worker flexibility as means of boosting output and adjusting to fluctuations in demand.

Despite these practical limits to the development of Fordism in Europe between the wars, the leading French and Italian automobile manufacturers continued to regard it as an ideal production strategy, to be pursued as far as the national market allowed. Certain smaller companies, such as Alfa Romeo in Italy and Mercedes-Benz in Germany followed with varying degrees of commercial success an alternative strategy of specialisation on high-performance luxury cars produced with modified craft methods. But on the eve of the Second World War Fiat, Renault, Peugeot and Citroën were all experimenting with small »people's cars« intended to create a new mass market, as was the Volkswagen Werke in Germany under the personal sponsorship of Adolf Hitler, another of Ford's European admirers.[4]

But even as European automobile manufacturers were straining against the limits of their national markets in their efforts to emulate Fordist principles, the American market itself was changing in ways that ultimately necessitated their partial modification.[5] In the United States, the 1920s saw a growing saturation of the market for utility cars; a fall-off in the proportion of first-time buyers and the spread of used-car sales; and a slower overall growth of demand. With the emergence of what Alfred P. Sloan of General Motors called the »mass-class« market, cars could no longer be sold effectively on the basis of price alone, and GM was able to invade the mass market from above by pricing its more comfortable and up-to-date Chevrolet only slightly above the Model T. To ensure that no one else could follow his lead, Sloan set out to produce »a car for every purse and purpose« in four autonomous divisions, each aimed at a distinct price band; and the company also used advertising, installment sales and annual model changes to stimulate consumer demand and outdistance its rivals. To prevent model changes disrupting production, to curb fluctuations in sales and to increase economies of scale, GM executives developed new methods of forecasting demand, raised the proportion of components purchased from outside subcontractors, and began to interchange parts across divisions. On the shop floor itself, the company moved away from Fordist practice by introducing »semi-special« machine tools which could be adapted to new models by changes to cams and gears, increasing the proportion of toolmakers and toolsetters in the workforce; and it sought added flexibility through seasonal layoffs for production workers and the use of piece rates and group bonuses rather than a day wage system.

GM's innovations in product and production strategy soon began to undermine Ford's grip on the American market. The Model T had become increasingly outdated despite incremental changes introduced over the years, and sales dropped off sharply after 1924 despite re-

4 On plans for a »people's car« during the late 1930s and their postwar realisation, see *Bardou et al.*, op. cit., pp. 142–50, 172–88, 212–15. For German motorisation under the Nazis and the Volkswagen project, see: *Blaich*, loc. cit.; *Richard J. Overy*, Transportation and Rearmament in the Third Reich, in: Historical Journal 16, 1973; and *idem*, Cars, Roads and Economic Recovery in Germany, 1932–8, in: Economic History Review, 2nd ser., 28, 1975.

5 This account of the transformation of the American market and the competing responses of Ford and General Motors draws on: *Chandler*, op. cit., pp. 95–178; *Hounshell*, op. cit., pp. 263–302; *Meyer*, loc. cit.; *Alfred P. Sloan*, My Years with General Motors, New York 1964, esp. pp. 59–70, 148–68; *Arthur J. Kuhn*, GM Passes Ford, 1918–38: Designing the General Motors Performance Control System, University Park, PA 1986; *Harry F. Katz*, The Decline of Competition in the U.S. Automobile Industry, 1920–40, New York 1977; and *Michael J. Piore/Charles F. Sabel*, The Second Industrial Divide: Possibilities for Prosperity, New York 1984, pp. 60–63. (German translation: Das Ende der Massenproduktion. Studie über die Requalifizierung der Arbeit und die Rückkehr der Ökonomie in die Gesellschaft, Berlin 1985).

peated price cuts. By 1927 Ford was forced to admit defeat and begin work on a wholly new model to restore the company's competitive position. The extreme specialisation of production for the Model T made the changeover enormously difficult and expensive: all the existing machinery had to be scrapped and the factory shut down entirely for nearly a year, opening the door for the emergence of new competitors such as Chrysler. Despite an initial sales boom, the new Model A never dominated the market as the T had and periodic modifications soon proved unavoidable; by 1932 Ford had been forced to introduce a successor, the V-8. These failures in product policy in turn necessitated a retreat from some aspects of Fordism in production as well: assembly lines became shorter and machinery less specialised; vertical integration was reduced to gain flexibility; and wage rates were cut to compensate for the collapse of profits.

Thus Fordism was widely modified to permit greater flexibility in response to shifts in the market and the innovations of GM. But the triumph of what came to be called »Sloanism« proved ambiguous in practice. During the depression of the 1930s, GM reduced the number of its divisions and increased the interchangeability of components across models to secure greater economies of scale in a contracting market, and in a postwar competitive environment of steady growth in demand and oligopolistic »room for all« pricing policies, the Sloanist strategy of »a model for every market« degenerated into a strategy based almost exclusively on styling and marketing competition. Once they had perfected the big V-8 engine and 3-speed automatic transmissions in the early 1950s, the big US companies turned their backs on innovation and abandoned investment in experiments that did not yield quick results. For twenty years, the only major technical innovations in the U.S. came in the field of air-conditioning. The styling studios were in command and led the industry through successive milestones of kitsch – the tailfins of the late forties, the port-holes of the early fifties and the panoramic windshields of the mid-fifties, all adorned with flashing chrome and lurid paintwork.

Underlying this market pattern was a commercial and competitive logic. Continuous restyling became one of GM's big selling features to defeat smaller companies who could not generate new styles and who lacked marketing power. The Big Three used their powerful advertising and dealership networks to force up the costs of competition by forcing *dis*economies of scale in body tooling on their rivals. They were ready to carry much higher fixed costs in design and tool and die equipment per vehicle in order to carry through the annual model changes that the smaller independent companies could not afford to emulate; and the latter's inability to participate in this model competition was one of the crucial factors leading to their eventual demise.

But much of this apparent product diversity remained superficial. Following GM's lead, the Big Three centralised their styling variants around a limited number of basic interior platforms and body-shells, with distinctive »skins« placed on each, while the »family resemblance« enabled many major stampings of body parts to be used in common over many models. Many key components, such as engines, transmissions, or chassis were shared by several cars and often retained for twenty years or more. This interchangeability of components enabled American manufacturers to pursue ever greater economies of scale through vertical integration and the introduction of dedicated automation equipment such as transfer machines. In assembly on the other hand, the proliferation of options involved enormous complexity and inhibited rationalisation of the production process as a whole.[6]

---

6 For these developments in the American industry, see: *Sloan*, op. cit., esp. pp. 169–88, 238–47, 264–78; *Kuhn*, op. cit., esp. pp. 7–9, 87–95, 318–48; *Lawrence J. White*, The Automobile Industry since 1945, Cambridge, MA, 1971; *Charles E. Edwards*, The Dynamics of the United States Automobile Industry, Columbia, SC, 1965; *Brock Yates*, The Decline and Fall of the American Automobile

In the immediate postwar period, European automobile producers followed a more classically Fordist strategy, centred on the mass production of basic transport for the newly »democratised« markets at home and abroad. Until the late 1960s, each national market in Europe remained quite distinct and dominated by one or two domestically produced cheap small cars such as the Renault 4 and Citroën 2CV in France, the VW Beetle in Germany, the Fiat 500 and 600 in Italy and the Morris Minor and BMC Mini in the U.K. But with widely varying taxation structures, income distributions and road conditions, it was impossible for any single producer to dominate the European market or for any single design philosophy to prevail. This fragmentation of the European market left ample space for more specialised manufacturers of luxury cars alongside the mass producers, such as BMW, Daimler-Benz, Volvo, Jaguar and Alfa Romeo. And as incomes rose in the 1960s, European mass producers themselves began to offer a wider range of models in order to maintain their coverage of increasingly diverse national markets. But the European manufacturers, unlike their American counterparts, did not integrate their differentiated product ranges into coherent families of models based on common components. Instead, as Volpato's essay demonstrates, the Europeans maintained their practice of developing each model separately, which allowed greater scope than in the US for continuous improvement and technical innovation, but made it difficult to match the economies of scale of their mass market best sellers across the whole of a scattered and unbalanced product line. Thus in Europe as in the United States, by the late 1960s Sloanist marketing strategies based on product differentiation coexisted uneasily with automobile manufacturers' continuing commitment to Fordist production strategies based on standardisation and economies of scale.[7]

## II. Industrial Relations in the Age of Fordism

As we have seen, by the 1930s the mass production of automobiles on Fordist lines had established itself to varying degrees in most of the major industrial nations, and this process of international diffusion and adaptation gathered renewed force during the long postwar boom. In every country, the advent of large-scale car production created new problems for workers and trade unions. The size and power of the emergent companies; the rapid expansion of the labour force and the new distribution of skills within it; acute fluctuations in employment and the introduction of new payment systems all posed a major challenge to the established organisational practices of trade unions.

These distinctive features of the automobile industry came together with ideological debates within the international labour movement to suggest that the establishment of industrial unions was the key to the organisation of mass production workers, an analysis which seemed largely borne out by the experience of the 1930s. The breakthrough to mass-production unionism in the United States followed directly on the split between the American Federation of Labor (AFL) and the Committee for Industrial Organization (CIO), which paved the way

Industry, New York, 1983, esp. pp. 186–200; *Eric J. Coder*, Trade Policy and the US Automobile Industry, New York, 1978; *William J. Abernathy*, The Productivity Dilemma: Roadblock to Innovation in the Automobile Industry, Baltimore 1978; *Emma Rothschild*, Paradise Lost: The Decline of the Auto-Industrial Age, New York 1973.

7 For this picture of postwar European developments, see: *Giuseppe Volpato*, The Automobile Industry in Transition: Product Market Changes and Firm Strategies in the 1970s and 80s, in: *Tolliday/Zeitlin*, op. cit.; *Alan Altshuler/Martin Anderson/Daniel Jones/Daniel Roos/James Womack*, The Future of the Automobile, London 1985, esp. pp. 121–45; *Bardou et al.*, op. cit., pp. 172–209; *Daniel T. Jones*, Maturity and Crisis in the European Car Industry: Structural Change and Public Policy, Brighton 1981; *George Maxcy*, The Multinational Motor Industry, London 1981, pp. 94–97.

for the establishment of the United Auto Workers' Union (UAW) against the violent objections of the craft unions. Similarly in France, the strike wave of 1936 which organised the Renault factories was run by the Metalworkers' Federation of the Confédération Générale du Travail (CGT) which had long been committed to industrial unionism. And the situation in Britain where a variety of craft and general unions catered for the automobile industry with limited success thus appeared to be the exception that proved the rule.[8]

With greater distance from the debates of the 1930s, however, the idea of industrial unionism as the uniquely appropriate form of organisation for the automobile industry – a trade union counterpart to Fordism – seems less compelling. Two main objections stand out. Closer historical investigation of the 1930s suggests that industrial unionism was a far from sufficient condition for the breakthroughs to mass organisation in France and the United States. And a broader temporal and spatial perspective suggests that under favourable conditions other forms of union organisation could prove equally effective.

The first objection stresses the role of politics and the state in the breakthrough to mass organisation of automobile workers. Revisionist historians of American labour have shown that without the political context of the New Deal and the support of the Federal government first under the Wagner Act and then under the War Labor Board, it is unlikely that the newly-formed industrial unions could have overcome the resistance of the giant corporations such as GM and Ford.[9] In France, it was the advent to power of the Popular Front in 1936 which triggered off the factory occupations and enabled the French Communist Party and the CGT to realise their longstanding aspirations to organise these symbols of the modern proletariat. In Italy, too, successful organisation in the car factories depended heavily on support from the government and the law in the 1940s and again in the late 1960s and early 70s, while industrial unions proved unable to resist determined management offensives in periods of political reaction such as the 1920s and 50s.[10]

In Britain, the political context of the 1930s – dominated by the National Government after the collapse of the Labour Party in 1931 – was unfavourable towards a union breakthrough in the motor industry. Some of the difficulties in organising the industry did stem from the structure and policies of the unions concerned, notably the craft unions' attachment to the standard district rate and friendly benefits, issues of limited relevance to semi-skilled pieceworkers, and the general unions' preoccupations with sectors outside automobiles. But in periods when government policies inhibited managerial resistance – the two world wars and the late 1950s and early 60s – both craft and general unions displayed considerable capacity for the creative adaptation in organising motor workers.[11]

8 For the debates over industrial unionism within the AFL, see: *Walter Galenson*, The CIO Challenge to the AFL, Cambridge, MA 1960; and *James O. Morris*, Conflict Within the AFL: A Study of Craft Versus Industrial Unionism, 1901–1938, Ithaca, NY 1958. For similar but more muted debates in Britain, see *James Hinton*, The First Shop Stewards' Movement, London 1973, pp. 275–337; *Hugh A. Clegg*, A History of British Trade Unions since 1889, vol. II: 1911–33, Oxford 1985, pp. 455–59; and *Tolliday*, loc. cit.

9 See especially *David Brody*, The Emergence of Mass-Production Unionism, and: Reinterpreting the Labor History of the 1930s, both in: *idem*, Workers in Industrial America, New York 1980; and *Howell Harris*, The Snares of Liberalism? Politicians, Bureaucrats, and the Shaping of Federal Labour Relations Policy in the United States, ca. 1915–47, in: *Steven Tolliday/Jonathan Zeitlin* (eds.), Shop Floor Bargaining and the State: Historical and Comparative Perspectives, Cambridge 1985.

10 On the role of the state in the unionisation of French and Italian car factories, see *Van de Casteele-Schweitzer*, loc. cit.; *Bigazzi*, loc. cit.; and *Giovanni Contini*, The Rise and Fall of Shop-Floor Bargaining at Fiat, 1945–80, in: *Tolliday/Zeitlin*, op. cit.

11 Even in the United States, historians have recently argued that the AFL unions showed greater organisational dynamism and flexibility than the polemics of the 1930s suggested: see *Christopher J. Tomlins*, AFL Unions in the 1930s: Their Performance in Historical Perspective, in: Journal of American History 64, 1979.

A further demonstration of the centrality of the political context and the variety of possible union structures compatible with mass production can be seen in the cases of West Germany and Japan in the postwar period. The German trade unions emerged from the experience of Nazism and war with a politically heightened commitment to broad, encompassing industrial organisation. This went even further than in the US, since a single union was established for the whole of the metalworking sector, and the strategy of centralised bargaining was reinforced by a legal framework which limits the right to strike to recognised unions and extends collective agreements to cover even non-unionised firms. But the political upheavals of the reconstruction period also resulted in a statutory system of works councils outside the unions' direct control. The codetermination powers and »peace obligations« of the works councils created the conditions for informal plant bargaining which has acted as a counterweight to the centripetal pull of industry-wide negotiation.[12] In Japan, industrial unionism on the American model also seemed destined to become the dominant pattern in the early postwar years. But during the early 1950s, as national politics swung against the left and the automobile companies struggled for economic survival, the centre of gravity shifted to enterprise unions, a development involving both white and blue collar workers which relegated the industry-wide federation to a residual role in collective bargaining.[13]

Mass production of automobiles posed new problems of job control for trade unions alongside the establishment of collective bargaining itself. In many countries, grievances about the impact of Fordism on working conditions and job security were as important as wages in the early unionisation drives. The 1936 sitdown strikes at General Motors, for example, were sparked off by demands for an end to speed-up on the assembly lines. But by the late 1960s and early 70s, observers in countries such as the US and West Germany believed that the institutionalisation of collective bargaining there had resulted in the sacrifice of these job control aspirations to the pursuit of economic gains which did not threaten managerial prerogatives on the shop floor. In this light, radical critics of trade unions drew invidious contrasts with the situation in Britain and Italy, where powerful and autonomous workplace organisations appeared to exercise the direct controls over the production process so conspicuously absent at home.[14]

Recent research on the automobile industry has tended to modify such stark contrasts between national systems of industrial relations in terms of the presence or absence of job control. In Britain, for example, shop stewards in many car factories during the 1950s and 60s achieved a measure of control over manning, workloads and job definition that made them

12 On the impact of the statutory framework on the relationship between industrial unions and works councils in the West German automobile industry, see *Wolfgang Streeck*, Industrial Relations in West Germany: A Case Study of the Car Industry, London 1984, especially pp. 9–39; and *Eva Brumlop/Ulrich Jürgens*, Rationalisation and Industrial Relations in the West German Auto Industry: A Case Study of Volkswagen, in: *Otto Jacobi et al.* (eds.), Technological Change, Rationalisation and Industrial Relations, London 1985; for the political conflicts surrounding its emergence in the late 1940s and early 50s, see *Mary Nolan/Charles F. Sabel*, Class Conflict and the Social Democratic Reform Cycle in Germany, in: Political Power and Social Theory III, 1982.

13 On Toyota in the immediate postwar years, see *Reiko Okayama*, Industrial Relations in the Japanese Automobile Industry, 1945–70: The Case of Toyota, in: *Tolliday/Zeitlin*, op. cit.; for parallel developments at Nissan, see *Michael Cusumano*, The Japanese Automobile Industry, Cambridge, MA 1985, pp. 137–185.

14 For American views of Britain, see *David Brody*, The Uses of Power I: Industrial Battleground, in: *idem*, op. cit.; and *Nelson Lichtenstein*, Auto Worker Militancy and the Structure of Factory Life, in: Journal of American History 67,2, 1980. For a West German view of Italy, see *Rainer Zoll*, Centralisation and Decentralisation as Tendencies of Union Organisational and Bargaining Policy, in: *Colin Crouch/Alessandro Pizzorno* (eds.), The Resurgence of Class Conflicts in Western Europe Since 1968, 1978, vol. II.

the envy of local militants in countries like the United States. But job control under shop stewards was more limited and precarious than is commonly supposed. Controls over the production process did not form part of a wider union strategy but were tied instead to sectional bargaining over piecework prices, and there was little coordination even at the level of the individual factory, perpetuating wide differentials between shops and insecurity of earnings for the workforce as a whole. Such job controls were, moreover, rooted in British management's own past modifications of Fordism – its reliance on incentive systems rather than mechanisation and tight supervision – and their exercise involved shop stewards' assuming considerable responsibilites for coordinating production to achieve high levels of output. When management strategies shifted away from piecework towards more direct administration of production during the 1970s, the weaknesses of union organisation at the company level made it increasingly difficult for the stewards to defend the pattern of job control which had developed in the preceding period.[15]

In Italy, by contrast, direct controls over the production process grew out of the success of a Fordist strategy rather than its limitations. During the 1950s and 60s, Fiat management reorganised car production on Fordist lines and recruited vast numbers of southern migrants to fill the resulting unskilled jobs. It was these production line workers who struck *en masse* in 1969–70, establishing a new system of factory councils based on elected shop stewards. Once in place, the factory councils used their new-found power not to push up the earnings of individual work groups as in Britain, but instead to impose strict controls over workloads, compress wage differentials, simplify job classifications and make layoffs close to impossible. In certain shops, stewards began to experiment with a more cooperative style of shop floor bargaining, but political rivalries and the shifting strategies of the national unions pushed workplace organisation at Fiat away from any durable accommodation with management. As the economic and political climate began to change in the late 1970s, this confrontational line isolated the factory council from key sections of the Fiat workforce and left it vulnerable to a managerial offensive aimed at restoring productive flexibility. The failure of a company-wide strike against redundancies in 1980 because of widespread worker defections resulted in a collapse of shop floor organisation as sudden as its emergence a decade earlier.[16]

In the United States, the main constraints on managerial prerogatives came not from direct controls over manning and work organisation as in Italy and Britain, but rather from contractually negotiated seniority rules and grievance procedures which governed the allocation of workers between jobs. While management was free to determine the initial job structure according to Fordist principles, the association of wage rates and job security rights with these job definitions then imposed substantial rigidities on the deployment of labour in the plant. In recent years, these rigidities have become a major problem for management in its attempts at work reorganisation and their relaxation has assumed a key significance in plant negotiations. In its heyday, the US industrial relations system did not in itself preclude the decentralised bargaining over issues such as workloads which was central to job control in Britain, since production standard grievances remained strikeable during the life of the contract. But UAW leaders' preoccupation with company-wide bargaining objectives – together with the centralisation of corporate industrial relations policies and the growing political

15 For a fuller development of this argument, see *Steven Tolliday/Jonathan Zeitlin*, Shop-Floor Bargaining, Contract Unionism and Job Control: An Anglo-American Comparison, in: *idem*, op. cit.

16 For this interpretation of the Italian case, see *Contini*, loc. cit.

isolation of organised labour – left little room in practice for dynamic local bargaining over shop-floor issues.[17]

Two more varieties of job control, rooted in varying management and union strategies, can be seen in the cases of West Germany and Japan. Detailed studies of workplace industrial relations in West German car factories show that while there are few constraints on management's ability to allocate labour between jobs, they are much less free than their American or British counterparts to lay off workers in response to fluctuations in demand. As we have seen, West German works councils are legally prohibited from direct negotiations over wages or calling strikes; and sectional bargaining on British lines is further discouraged by the union's preoccupation with industry-wide objectives. But the works councils also have far-reaching statutory powers of codetermination in hiring, dismissals and overtime which they use systematically to protect the employment prospects of their core constituency in the factory. The result has been the development of long-term manpower policies in the enterprise – themselves consonant with the orientation of some German car firms towards specialised production and with public apprenticeship programmes – which provide maximum security of employment in exchange for great flexibility in the deployment of labour within the enterprise.[18]

A similar trade-off between relative employment security and the flexible deployment of labour within the firm can also be seen in the case of Japan. While there is considerable controversy about the independence of Japanese enterprise unions – which often contain representatives of lower management in leading positions – recent scholarship suggests that these organisations can exercise considerable influence on company policy through an extensive system of consultation and consensus-building and through occasional alliances with factions of management. Thus at Nissan, where the enterprise union has been closely linked with key managers, union leaders have used their influence to impose greater restrictions on work intensity and automation than at Toyota as well as to obstruct proposals for overseas production. Like German works councils, Japanese enterprise unions press corporate management to maintain maximum stability of employment for their core workforce; and they are also concerned to regulate the resulting inter-plant transfers so as to cushion their impact on individual workers. As the pace of industry expansion has slowed since the mid-1970s, union cooperation and consensus has been sustained by increasing stability of employment within the factory, secured through mobility of the core workforce between jobs rather than through the use of temporary and seasonal workers, who had previously played a larger part in responses to cyclical fluctuations. As we shall see in the next section, such mobility forms part of a flexible system of car production based on fluid job roles and broad worker training which marks a clear break with Fordist principles.[19]

---

17 For discussion of job control in US industrial relations, see *Tolliday/Zeitlin*, loc. cit.; *Nelson Lichtenstein*, Reutherism on the Shop Floor: Union Strategy and Shop-Floor Conflict in the US, 1945–70, and *Harry C. Katz*, Recent Developments in US Auto Industrial Relations, both in: *Tolliday/Zeitlin*, op. cit.; and *Steve Jefferys*, Management and Managed: Fifty Years of Crisis at Chrysler, Cambridge 1986.

18 For an interpretation along these lines, see *Streeck*, op. cit.; and for a detailed comparison with the United States, see *Christoph Kohler/Werner Sengenberger*, Policies of Work Force Reduction and Labor Market Structures in the American and German Automobile Industry, paper presented to the conference of the International Working Party on Labour Market Segmentation, Modena, Italy, 7–11 September 1981; and *idem*, Konjunktur und Personalanpassung: Betriebliche Beschäftigungspolitik in der deutschen und amerikanischen Automobilindustrie, Frankfurt 1983.

19 For discussion of enterprise unionism and shop floor practice in Japan, see: *Okayama*, loc. cit.; *Koichi Shimokawa*, Product and Labour Strategies in Japan, in: *Tolliday/Zeitlin*, op. cit.; *Taishiro Shirai* (ed.), Contemporary Industrial Relations in Japan, Madison, WI 1983, especially *Shirai*, A

National systems of industrial relations thus differ in the focus as much as the level of job control: direct controls over the production process in Britain and Italy; contractual seniority rules and job classifications in the US; and broad employment guarantees in West Germany and Japan. These international variations in the focus of job control stem as much from differences in strategy as from differences in power: from differences in management's interpretation of Fordism and in the responses of unions once established. Each national pattern involves a complex balance of advantage and constraint for management and unions alike. But as the recent decline of shop-floor organisation in Britain and Italy demonstrates, some forms of job control may prove less adaptable than others to the demands of new product and labour strategies.

## III. Beyond Fordism

For nearly three decades after the Second World War, both the American and European automobile industries remained faithful to the underlying principles of the Fordist model, with varying degrees of modification to accommodate more differentiated marketing strategies. But by the early 1970s a series of forces were at work within the world economy which would call this model into question more fundamentally than ever before. Throughout the Western economies, the postwar boom was losing momentum, with a small but visible slowdown in overall growth rates. Major national car markets were becoming saturated at existing income levels, and the proportion of replacement demand relative to sales of new models was rising. By the late 1960s, for example, replacement demand had reached 75–80 % of the US market with a motorisation level of more than one car for every two people.[20]

These trends made it difficult to achieve continued economies of scale within national boundaries. Previously, trade between manufacturing countries had concentrated on the market segments furthest from the specialisation of domestic producers, presenting little challenge to their product ranges, as in the case of small car imports to the US and exports of large luxury cars from Germany and Sweden to other European countries. But now slowly growing domestic demand and underutilisation of capacity triggered off a scramble for exports resulting in direct head-to-head competition between mass producers in the core segments of each other's markets. In 1960, for example, cross-national trade between West European producers accounted for 20 % of total sales; by 1980 this figure had risen to 36 %. But while each of the major producers moved down this road, only the Japanese really succeeded in making a major impact across a wide range of foreign markets. By 1980 the Japanese had seized 23 % of the US market, 10 % of the European market and a 28 % share of world imports. Since no major manufacturer has been eliminated and all companies have invested

Theory of Enterprise Unionism, and *Kazuo Koike*, Internal Labor Markets: Workers in Large Firms; and: Industrial Relations in Japan and the United States: How Different?, a symposium in Industrial Relations Research Association, Papers and Proceedings of the 35th Annual Meeting, Madison, WI 1982, especially *Koji Taira*, Labor Productivity and Industrial Relations in the United States and Japan, and *Haruo Shimada*, Japan's Postwar Industrial Growth and Labor-Management Relations; and *Robert E. Cole*, Work, Mobility and Participation, Berkeley 1979. For a comparison of enterprise unionism at Nissan and Toyota, see *Cusumano*, op. cit., pp. 174–79.

20 For the following account of changes in world demand for cars and the emergence of new forms of international competition during the 1970s, see: *Volpato*, loc. cit.; *Altshuler et al.*, op. cit., chs. 2 and 6; *Piore/Sabel*, op. cit., pp. 165–93; *Jones*, op. cit., passim; *Yves L. Doz*, Automobiles: Shifts in International Competitiveness, in: *Milton Hochmuth with William Davidson* (eds.), Revitalizing American Industry, Cambridge, MA 1985; *Douglas H. Ginsburg/William J. Abernathy* (eds.), Government, Technology and the Future of the Automobile, New York 1980.

massively in new production facilities, the resulting boost to overall capacity has exacerbated the struggle for market share. At the same time, however, the growing saturation of demand for basic transport has forced mass producers in each country to explore new forms of product differentiation in order to appeal to market segments increasingly sensitive to quality, technical innovation and the special attributes of products.
Superimposed on these structural trends were a number of external shocks which exacerbated the growing crisis of the Fordist model and brought it to a head in sudden and unexpected ways. Most dramatic were the oil shocks of 1974 and 79 which sent petrol prices soaring and created widespread consumer alarm about the security of sources of supply. For the first time since the war, absolute levels of world demand fell off, declining by 16 % between 1973–5. While car sales quickly recovered as oil prices fell back in real terms, they never returned to previous growth rates, and demand fell again by 12 % between 1979 and 1981. The oil shocks also had a major impact on the composition of demand within each market. Growing consumer anxiety about petrol consumption set off a headlong rush among producers of shift car designs in the direction of smaller, more fuel-efficient cars. This move towards »downsizing« was most pronounced in the United States where imports of small cars rose sharply at the expense of domestically-produced »gas guzzlers«.

These broader trends in consumer awareness were given added bite by dramatic shifts in public policies. Governments everywhere set out to contain their dependence on imported energy sources by regulating the use and production of automobiles. Particularly important were the US Corporate Average Fuel Economy (CAFE) regulations which set mandatory fuel consumption targets for domestic producers based on a weighted average of their product ranges and forced them to move towards a larger proportion of small cars. Alongside these interventions in many countries came the proliferation of regulations concerned with emissions controls and vehicle safety which gave a further impetus to product redesign.

To many observers, the events of the 1970s signalled the clear transformation of automobiles into a mature industry characterised by slow growth of demand for a well-defined and technologically stable product. The two oil shocks appeared to have accelerated this process by eroding the distinctiveness of national markets and promoting the convergence of international demand on a limited range of fuel-efficient cars. But viewed in the longer-term, these assumptions about the development of international automobile markets proved ill-founded. Firstly, the oil shocks of the 70s did not signal a once-and-for-all shift to higher oil prices. By the early 1980s OPEC was struggling to maintain its cohesion and oil prices were falling again in real terms, as they had between the two shocks themselves. Secondly, the fears about the exhaustion of oil resources commonly voiced in the mid-70s came to appear exaggerated as recession and energy conservation drastically reduced the fuel consumption of the developed countries. The real long-term consequence of the oil shocks proved to be a growing trend towards instability of raw material prices more generally, which, reinforced by the erratic fluctuations of exchange rates throughout the international economy, created major difficulties in forecasting demand and made large-scale investments and locational decisions increasingly uncertain.

Despite the worldwide pressures for greater fuel efficiency, moreover, the impetus to convergence of national markets proved less intense than had been expected. Economies in petrol consumption were secured across the whole spectrum of car sizes, and markets such as that of the United States remained distinctive in the share of larger cars in total demand. Within each national market, finally, competition came to centre increasingly on the multiplication of an ever-wider range of models and vehicle types and the introduction of new product features such as aerodynamic bodies, electronics and fuel injection, fragmenting demand into a larger number of less clearly defined segments. Observers have accordingly come to characterise the international automobile industry in terms of tendencies towards

»de-maturity« or indeed as a »neo-infant industry« in which the definition of the product and the conditions of competition have once again become open questions.[21]

No one has done as much as the Japanese to push the world automobile industry towards »de-maturity« and no one has benefited as much from the market trends associated with it. By 1982, Japan had become the world's biggest automobile producer, with an annual output of 6.9 million cars, 60 % exported. The Japanese had won a major stake in many Western car markets – 23 % in the US and 8.6 % in Europe – and their share would have been considerably higher had they not been forced to accept voluntary import quotas in many cases. To take the most extreme examples, Italy limits Japanese sales to 2,200 per year or less than 0.1 % of the domestic market and France to 2.7 %.[22]

The Japanese industry revived after the war as a low volume, high cost producer catering for a protected home market characterised by low incomes, poor roads and restricted demand. Japanese producers initially set out to emulate the Fordist model as far as possible. Some companies such as Nissan sought access to foreign technology and vehicle designs through joint ventures with Western companies such as Austin, employment of consultant engineers and imports of machinery. Others such as Toyota preferred to learn by doing rather than buy in ready-made equipment and designs, copying Western machines in their own workshops and adapting them to the smaller volumes demanded by the Japanese market. But in both cases, the Japanese struggle to catch up with their Western competitors through continuous rationalisation and model upgrading gave rise to distinctive modifications of Fordist methods which would ultimately permit greater product diversity and productive flexibility.

The Japanese pursued their dash for growth through the constant redesign of their manufacturing system and the establishment of new relationships between assemblers and suppliers and between managers and workers substantially different from those prevailing in Western car firms operating on Fordist principles. To offset the cost penalties of low volume production, the Japanese reduced their dependence on special-purpose machinery by lowering minimum efficient scales of operation in each phase of the manufacturing process, by replacing large machines with several small ones, attaching special jigs and fixtures to general-purpose equipment, perfecting quick die changes and machine set-ups, and assembling several different models on the same line. Similarly, to reduce the capital costs of rapid expansion, the Japanese cut down inventory and work-in-progress as far as possible. Rather than depending on large buffer stocks to ensure continuity of production, they developed »kanban«

---

21 For the debate on maturity and »de-maturity« in the international automobile industry, in addition to the works already cited see: *Gerald Bloomfield*, The World Automotive Industry, London 1978; *Krish Bhaskar*, The Future of the World Automobile Industry, London 1980; *Peter Waymark*, The Car Industry: A Study in Economics and Geography, London 1983; *Louis T. Wells*, The International Product Life Cycle and United States Regulation of the Automobile Industry, in: *Ginsburg/Abernathy*, op. cit.; *Maxcy*, op. cit., pp. 159–78; *Yves L. Doz*, The Internationalisation of Manufacturing in the Automobile Industry: Some Recent Trends, in: Social Science Information 20, 6, 1981; and especially *William J. Abernathy/Kim B. Clark/Alan M. Kantrow*, Industrial Renaissance, New York 1983.

22 On the development and characteristics of the Japanese automobile industry, in addition to the works cited in note 19, see: *Richard J. Schonberger*, Japanese Manufacturing Techniques, New York 1982; *Koichi Shimokawa*, The Structure of the Japanese Auto Parts Industry and Its Contribution to Automative Process Innovation (unpublished paper presented to the MIT Auto Program International Policy Forum, Hakone, Japan, 1982); *David Friedman*, Beyond the Age of Ford: The Strategic Basis of the Japanese Success in Automobiles, in: *John Zysman/Laura Tyson* (eds.), American Industry in International Competition: Government Policies and Corporate Strategies, Ithaca, NY 1983; *John B. Rae*, Nissan/Datsun: The History of Nissan Motor Corporation in the USA, 1960–80, New York 1982; *Abernathy et al.*, op. cit., chs. 4–7; *Piore/Sabel*, op. cit., pp. 156–62 and 223–26; and *Altshuler et al.*, op. cit., pp. 146–164.

systems whereby components were produced to order and delivered »just-in-time« for the assembly process. »Just-in-Time« systems highlighted bottlenecks in production and helped to overcome the poor reputation of Japanese products in export markets by facilitating continuous quality control and defect prevention in place of the systems of inspection and rectification used by Western car firms.

To lessen their dependence on imported components and facilitate continuous upgrading of models and rationalisation of production, the Japanese assemblers deliberately stimulated the creation of networks of specialist suppliers clustered around the main factory, frequently spun off from their own organisation. In contrast to prevailing Western practice of vertical integration, multiple sourcing and competitive contract tendering, the Japanese assemblers decentralised production as far as possible to component suppliers with whom they cultivated long-term relationships based on single sourcing and detailed collaboration on product engineering and technology transfer. After the defeat of the industrial unions in the early 1950s, the Japanese car manufacturers were also able to undertake a major restructuring of workplace relations as part of the broader changes in manufacturing practice. In return for considerable security of employment for their core labour force, the Japanese, as we have seen, created a highly flexible system of working practices based on widespread job rotation linked to broad worker training, a wage system that facilitated transfers between posts without constant renegotiation, and the devolution of responsibility for task assignment and quality control to supervisors and work groups. In the 1950s the Japanese enjoyed the advantage of wage levels well below those of Western producers, but by the 1970s earnings in the car assembly plants had begun to catch up with European standards and by the early 1980s had overtaken countries like Britain and Italy. As in Western countries, however, compensation levels in supplier firms remain substantially lower, and the Japanese continue to enjoy a relative cost advantage in this area because of their lower degree of vertical integration.

The Japanese innovations in car production began as a series of *ad hoc* modifications to Fordist practice designed to enable them to match Western standards of efficiency and quality in mass production as quickly as possible. But over time these innovations came to form the basis of a new system of production which could outpace Western car manufacturers in labour productivity, product quality and cost competitiveness while also permitting the production of a wider range of models and more flexible responses to market trends. With the main features of this system already in place, the Japanese automobile industry took off in the 1960s, pushed forward by high levels of investment and capacity utilisation within a rapidly expanding domestic market and by the first wave of export sales. The oil shocks of the 1970s gave an added boost to the Japanese industry by cracking open the North American market and realigning demand in precisely those segments where they were strongest. As sales volumes soared, the Japanese were able to achieve dramatic advantages in cost competitiveness – as great as $ 2000 per model over the Americans – while at the same time achieving higher standards of quality and reliability. This cost advantage, as a number of detailed studies show, is not the result of greater mechanisation nor can it be explained in any large measure by lower wages. Japanese companies' pioneering use of CAD/CAM systems has allowed them to capitalise on their productive flexibility to speed up the introduction of new models and widen their product ranges, opening up a variety of specialist market niches. And by the mid-1980s, their growing financial and productive strength had enabled the Japanese to close the design gap with the Europeans and force the pace in technical innovation, leaving them poised to achieve competitive superiority across the entire product spectrum.

The crisis of the 1970s presented Western automobile manufacturers with three main strategic options which they pursued singly and in various combinations. Faced with changes in demand and new forms of international competition, Western car firms could seek to revivify the Fordist model through increased automation, decentralisation of production to low wage areas abroad and reduction of manning and compensation levels at home; they could see to emulate the Japanese by revamping their own manufacturing systems in the direction of increased flexibility and product diversity; or finally they could seek to insulate themselves from the effects of international trends through the pursuit of protection and joint ventures in their domestic markets.

In the late 1970s, the American producers in particular based their strategy on the diagnosis discussed earlier that international demand for cars was converging around a narrow range of models. They believed that the rise in oil prices necessitated a once-and-for-all design shift towards smaller cars which would be even more standardised than their predecessors. This shift would revive the possibility of further economies of scale and cost reductions through the introduction of new forms of dedicated automation and the relocation of labour-intensive phases of production in low-wage areas abroad, a trend encouraged by the efforts of less developed countries to attract multinational direct investment. Costs in the remaining domestic production facilities could then be dramatically reduced by massive cutbacks in employment and pressure on the surviving workforce to accept lower wages and revised working practices. The most ambitious attempt to put this »World Car« strategy into practice was GM's »J-Car«, a small fuel-efficient model aimed to be built and sold worldwide, but Ford, too, sought economies of scale at a global level with its »Escort-Lynx« project, and European manufacturers such as Volkswagen began to step up their investments in Third World countries such as Mexico and Brazil. While most automobile companies trimmed their workforce, these cutbacks reached dramatic levels in the US and the UK, where employment fell by more than 25 % between 1978 and 1981. The resulting unemployment, together with the threat of increased overseas production, gave the companies the opportunity to extract major concessions from the unions.

The »World Car« strategy quickly ran aground on a series of unforeseen problems. Rather than the expected convergence of national markets, as we have seen, world demand for cars entered a phase of progressive »de-maturity« and diversification. Relocation in Third World countries, moreover, proved no guarantee of low production costs: long supply lines, large inventories, poor quality control and unstable exchange rates more than offset the advantages of low wages. By the late 1970s, finally, political upheavals and labour militancy in countries like Brazil had begun to close the wage gap with domestic manufacturing facilities. With the increasing importance of tailoring production to precise segments of demand in individual national markets, even General Motors had been forced to retreat from its full-blooded »World Car« strategy by the early 1980s, though it continues to source certain components such as engines and minor mechanicals on a global basis.[23]

With the eclipse of the world car strategy, automobile manufacturers in the US as in Europe have turned increasingly toward efforts to emulate the achievements of the Japanese in product diversity and productive flexibility. As in the case of responses to Fordism in a previous era, however, reactions to the Japanese model have taken a variety of forms depending on the configurations of national markets, the political and institutional context and the strate-

23 For the rise and fall of the »World Car« strategy, in addition to the works cited in note 21 above, see: *Rainer Dombois*, The New International Division of Labour, Labour Markets and Automobile Production: The Case of Mexico, in: *Tolliday/Zeitlin*, op. cit.; *Robert B. Cohen*, International Market Positions, International Investment Strategies and the Domestic Reorganization Plans of US Automakers (unpublished paper, New York Graduate School of Business, 1982); *Rhys Jenkins*, Internationalisation of Capital and the Semi-Industrialised Countries: The Case of the Motor Industry (unpublished paper presented to the Conference on Internationalisation and Industrialisation of the Periphery, Oaxtepec, Mexico, 1983); *Rich Kronish/Kenneth S. Mericle* (eds.), The Political Economy of the Latin American Motor Vehicle Industry, Cambridge, MA 1984; *John Humphrey*, Capitalist Control and Workers' Struggle in the Brazilian Auto Industry, Princeton, NJ 1982.

gic choices of the various actors concerned. And just as the lessons of Fordism proved ambiguous when the Europeans tried to appropriate them between the wars, so too the sources of Japanese success have been subject to countervailing and often contradictory interpretations by academics and Western car makers alike.[24]

Both in the US and in Europe the 1980s have seen far-reaching renewal and restructuring of model ranges. As Volpato argues, in the US manufacturers' efforts have centred on large-scale entry into hitherto neglected market segments involving smaller vehicles and front-wheel drive. In Europe, on the other hand, where vehicle designs were more advanced and model ranges wider, efforts centred initially on the rationalisation and integration of unbalanced product lines through increased use of common components. But in both regions car manufacturers have become increasingly concerned to offer a wider variety of distinctive packages targetted at more narrowly defined market segments. The model replacement cycle has been accelerated, particularly through the use of CAD/CAM, and there has been a multiplication of engine types as well as an increasing emphasis on specialised products such as people carriers, off-road vehicles and high-performance saloons.

But both European and American automobile companies have concentrated their greatest attention on production in order to narrow the Japanese cost advantage and introduce a larger measure of flexibility into the manufacturing process. Much of the effort and investment in this area has focused on the introduction of new forms of automation aimed at improved product quality and process efficiency rather than the elimination of direct labour per se. Whereas in the past automation had been associated with dedicated machines and extreme productive rigidity, technological developments such as reprogrammable robots have now made it possible to combine high levels of automation with increased flexibility. Such developments have accordingly reduced the importance of economies of scale in many parts of the manufacturing process, bringing the benefits of automation within the reach of specialist as well as mass producers. But the new technology also leaves considerable scope for managerial choice and some companies have gone further down this road than others. Fiat set a pioneering example with its Robogate system which allowed several different body types to be assembled on the same line, and German producers such as VW have largely followed this lead. Giant multinationals such as Ford of Europe have felt less need to move away from dedicated equipment because of their ability to achieve considerable flexibility through switching production between plants in different countries; while BL has turned increasingly towards flexible automation after its negative experience with rigid multiwelders installed on the Metro lines at end of the 1970s.

These changes in product and production strategies have driven car manufacturers to seek new relationships with their component suppliers. Pressures towards improved quality and lower costs along with the growing trend towards the use of common components across different models have led to the rationalisation of component supply networks and the elimination of weaker companies. But Western automobile producers have also set out to imitate the Japanese Kanban model by developing closer and more collaborative relationships with

24 For competing interpretations of the impact of Japanese competition on Western automobile manufacturers' product and production strategies, see: *Volpato*, loc. cit.; *Altshuler et al.*, op. cit., pp. 77–105 and 164–80; *Piore/Sabel*, op. cit., chs. 8–9 and 11, passim; *Harry C. Katz/Charles F. Sabel*, Industrial Relations and Industrial Adjustment in the Car Industry, in: Industrial Relations 24, 3, 1985; *Wolfgang Streeck*, Industrial Relations and Industrial Change in the Motor Industry: An International View (1985 Leverhulme Lecture, Industrial Relations Research Unit, University of Warwick); *Karel Williams/John Williams/Colin Haslam*, The Breakdown of Austin-Rover, Leamington Spa 1987, chs. 3–4; *Davis Dyer/Malcolm Salter/Alan Weber*, Changing Alliances: The Harvard Business School Project on the Auto Industry and the American Economy, Cambridge, MA 1987.

the suppliers who remain. Many companies have turned towards single sourcing and long-term contracts with a smaller number of suppliers whom they expect to take increasing responsibility for quality control, technical innovation and the reduction of inventory through frequent deliveries; and American automobile manufacturers in particular have pressed component firms to relocate around their main assembly plants. Even in the most advanced cases, however, these changes have tended to fall short of the Japanese example, since supply lines remain relatively long and manufacturers have found it difficult to resist the temptation to offload the burdens of frequent deliveries onto their suppliers rather than helping them to develop genuine forms of »just-in-time« production.

Efforts to imitate the Japanese example have also led European and American car manufacturers to seek far-reaching changes in shop-floor relationships and working practices. Growing product diversity and the shift away from rigid technologies creates pressures for more fluid job roles and more flexible deployment of labour within the factory. Management on both sides of the Atlantic has pursued the goals of broader job classifications and the abolition of barriers to labour mobility: widespread use of team working, the devolution of certain skilled tasks to production workers and the relaxation of demarcation lines on maintenance crews are all examples of this trend. But as in the case of automation, management retains considerable strategic choice in the development of working practices, and there remains great ambiguity about the implications of the Japanese example. In some cases, as in Britain, management has been most concerned with the reassertion of its authority on the shop floor against the unions and the improvement of productivity through tighter discipline and more intensive working, with limited aims in the area of flexibility. In other cases, management has made increased flexibility of labour deployment a central priority, whether imposed unilaterally as in Italy or bargained with trade unions as in West Germany and the US.

Where management has pursued increased flexibility of working practices as a central objective, the institutional context of industrial relations and trade union strategies have strongly conditioned the outcome. In Italy, as we saw in the previous section, the trade unions' political commitments and tactical rigidity made any durable accommodation with Fiat management impossible, and after its victory in the strike of 1980 the company swept away shop floor controls and moved rapidly to introduce team working and job mobility. In West Germany, industrial unions and works councils concerned with job security have been quite prepared to enhance established patterns of flexible labour deployment, as in the case of the LODI agreement at Volkswagen which links wages to skill within broad job bands. Finally, in the US, the UAW and its locals under intense pressure from management have moved slowly towards the relaxation of seniority rules and the widening of job classifications in exchange for greater job security, protection against outsourcing and the prospect of greater influence over business decisions. But in every case, the growing demands from management for increased productive flexibility are forcing the unions to rethink and revise job control practices and bargaining strategies established in the heyday of Fordism.[25]

---

25 For recent developments in shop floor practice and industrial relations, see: *Altshuler et al.*, op. cit., ch. 9; *Katz/Sabel*, loc. cit.; *Wolfgang Streeck/Andreas Hoff* (eds.), Workforce Restructuring, Manpower Management and Industrial Relations in the World Automobile Industry, International Institute of Management (IIM), Berlin 1983; *Wolfgang Streeck* (ed.), Industrial Relations and Technical Change in the British, Italian and German Automobile Industry, IIM, Berlin 1985; *idem*; *Katz*, loc. cit.; *Ulrich Jürgens/Knuth Dohse/Thomas Malsch*, New Production Concepts in West German Car Plants, in: *Tolliday/Zeitlin*, op. cit.; *Paul Willman*, Labour Relations at BL Cars, ibid.; *David Marsden/Timothy Morris/Paul Willman/Stephen Wood*, The Car Industry: Labour Relations and Industrial Adjustment, London 1985; *Stephen Wood*, Technological Change and the Cooperative Labour Strategy in the U.S. Auto Industry, in: *Richard Hyman/Wolfgang Streeck* (eds.), New Tech-

While Western car manufacturers have sought to emulate Japanese product, production and labour strategies, they have also sought to insulate themselves to some degree from the harsh winds of international competition through government protection and joint ventures with the Japanese themselves. Nearly every major car producing nation maintains formal or informal restrictions on Japanese imports of varying intensity. In some markets such as France, Italy and Spain, Japanese cars are almost entirely excluded and in others such as Britain and the US, domestic pressures have been mounting for some kind of local content legislation. At the same time, however, Western manufacturers have entered commercial tie-ups with Japanese automobile companies in hopes of learning to emulate their production methods or failing that to gain a financial stake in Japanese success and keep intact their dealer networks by maintaining a full range of products. The Japanese, for their part, have responded to these barriers against their products by moving into direct foreign investment, either independently as in the case of Honda and Nissan or in collaboration with local manufacturers. The success or failure of these ventures which are still in their early stages will prove a major test of the transportability of the Japanese model outside its national base.[26]

By the 1970s the modified Fordist systems practiced by American and European automobile manufacturers had become sluggish and internally contradictory. Their factories were bulging with inventory and work-in-progress; maintenance and work scheduling were a constant problem, resulting in frequent interruptions of production; and high volume output was pursued at the expense of product quality. The diversity of models and options required for marketing stood in tension with the uniformity and standardisation required for the efficient operation of a Fordist production system, and Western car manufacturers had become extremely vulnerable to any major disruption in the postwar growth pattern.

The Japanese set out initially to adapt Fordist systems to the requirements of an accelerated catching-up process, and some observers see their achievement as a quantum leap forward in the same direction. The Japanese, in this view, have brought out and eliminated certain key imperfections in the Fordist model, making it possible to combine product diversity with mass production on an unprecedented scale. In consequence, the gap they have opened up can now be closed by Western manufacturers through a straightforward process of organisational imitation and management reform without a fundamental reconsideration of established strategy and practice. Some automobile manufacturers appear to share this view, notably Ford, whose »After Japan« programme concentrates on automation, inventory reduction and quality control with more limited shifts in personnel policies and relations with suppliers.[27]

---

nology and Industrial Relations: International Experience, Oxford, forthcoming 1988; *Michael Indergaard/Michael Cushion*, Conflict, Cooperation, and the Global Auto Factory, in: *Daniel B. Cornfield* (ed.), Workers, Managers and Technological Change, New York/London 1987.

26 On protectionism, joint ventures and the internationalisation of the Japanese industry, see: *Altshuler et al.*, op. cit., pp. 223–47 and 291–99; *Yves L. Doz*, International Industries, Multinational companies and Government Policies: The European Experience (unpublished paper presented to the IIM Conference on MNC-Host Government Relationships in Developed Countries, Berlin, October 1980); *idem*, L'internationalisation de l'industrie automobile japonaise: potentiel et limites, in: Cahiers IREP/Développement, no. 6, 1984; *Koichi Shimokawa*, loc. cit., 1987; *idem*, New Developments in International Cooperation Within the Motor Industry, in: The Wheel Extended 13, 1, Tokyo 1982; *idem*, Automobiles – Groping for Coexistence Rather Than International Rivalry (unpublished paper, Hosei University, Japan, 1984); *James A. Dunn, Jr.*, Automobiles in International Trade: Regime Change or Persistence?, in: International Organization, 41, 2, 1987.

27 For interpretations in this direction, see: *Williams et al.*, op. cit., ch. 3; and *Wood*, Technological Change and the Cooperative Labour Strategy. For a senior management view from the Ford Motor Company, see the interview with James Bakken, Vice-President Operations Support Staff, Ford U.S., in: Automotive News, 28 February 1983.

Other observers, by contrast, see the Japanese innovations as a deeper challenge to the Fordist model which reverses its central principles and points towards the emergence of a qualitatively new system of production. This system, which can be termed flexible specialization and has been observed across a wide range of industrial sectors, depends on the combination of increasingly flexible, general-purpose equipment and a skilled, adaptable labour force to manufacture an ever more diversified range of products for which economies of scale are becoming decreasingly important. Not only the Japanese but also some Western automobile manufacturers seem to be moving in this direction, notably the Germans and the Swedes with their growing emphasis on multi-skilled work teams to exploit the full potential of flexible automation systems for the manufacture of a changing array of high quality products.[28]

Recent developments in the international automobile industry have undermined the foundations of the Fordist model as it developed in the postwar period. But many of the established features of automobile production can still be discerned, while the more recent innovations are susceptible to elaboration in different directions. CAD/CAM systems have reduced the lead time for the introduction of new models, facilitating the development of wider product ranges; but the increasing complexity of the product and the difficulties of integrating automobile subsystems experiencing different rates of innovation keep research and development costs high and create financial problems for smaller manufacturers. While there has been a qualitative expansion in the available range of vehicle types and models, the use of common components and subsystems has also become increasingly important in the generation of new variants. Flexible automation equipment and changes in job design have lowered minimum efficient scales of operation in many stages of component production and vehicle assembly, but substantial economies of scale still remain in the manufacture of major mechanicals such as engines and gearboxes, leading to growing reliance on joint ventures on the part of smaller producers. The new automation equipment can be used to decrease batch sizes and eliminate machine pacing for more broadly skilled workers, or can serve instead to move towards the ideal of a »worker-less factory« through the integration of complex sequences of transfer machines. While overall skill levels have clearly been rising, it remains uncertain, finally, whether this tendency will be confined to the development of a wider range of interchangeability between semi-skilled jobs or will lead instead to the emergence of new job categories which more fundamentally erode the boundary between skilled and semi-skilled tasks. Current practice in automobile companies accordingly ranges from the reduction of rigidities in the manufacture of a broadly standardized product line to conscious efforts at building maximum flexibility into all aspects of an increasingly diversified production process.

The changed market conditions and new technologies of the 1970s and 80s have brought with them shifts in competitive strategy and manufacturing practice which pose a fundamental challenge to the Fordist model. But as in the case of the emergence and diffusion of Fordism itself, the outcome of its crisis for the future of the automobile industry and its workers will be determined not by some intrinsic imperatives of markets and technology but rather by the strategic choices of corporate managements, trade unions and national governments.

---

28 For fuller discussions of »flexible specialization«, see *Piore/Sabel*, op. cit.; *Charles F. Sabel/Jonathan Zeitlin*, Historical Alternatives to Mass Production: Politics, Markets and Technology in Nineteenth-Century Industrialization, in: Past and Present, no. 108, 1985; and *Paul Hirst/Jonathan Zeitlin* (eds.), Reversing Industrial Decline? Industrial Structure and Industrial Policy in Britain and her Competitors, Oxford, forthcoming, 1988. Analyses of the automobile industry in these terms include *Katz/Sabel*, loc. cit.; and *Streeck*, op. cit. (1986), who also uses the term »differentiated quality production«.

# [10]

# Between Taylorism and Technocracy: European ideologies and the vision of industrial productivity in the 1920s

Charles S. Maier

As Antonio Gramsci recognized in his prison reflections from the end of the 1920s, the impact of United States technology offered a valuable key for understanding recent European development: 'The European reaction to Americanism . . . must be examined attentively. Analysis of it will provide more than one element necessary for understanding the present situation of a series of states of the old continent and the political events of the post-war period.'[1]

By Americanism Gramsci meant a whole complex of approaches to industrial production and labour relationships. 'Fordism' embodied one aspect, 'Taylorism' another; yet as a German commentator pointed out in 1924, these appeared merely as the most typical contribution to America's prodigious economic achievement as a whole.[2] By the 1920s, scientific management – which extended the original approaches of Taylorism into all areas of labour productivity, technological efficiency, and even corporate organization – evoked enthusiasm among European emulators as 'a characteristic feature of American civilization'.[3]

[1] Antonio Gramsci, *Note sul Machiavelli, sulla politica, e sullo stato moderno* (Turin, 1949), 312.

[2] F. von Gottl-Ottlilienfeld, *Fordismus? Paraphrasen über das Verhältnis von Wirtschaft und technischer Vernunft bei Henry Ford und Frederick W. Taylor* (Jena, 1924), 6.

[3] League of Nations, International Labour Office, *International Economic Conference Geneva, May 4, 1927, Documentation: Scientific Management in*

As Gramsci sensed, this vogue of so-called Americanism testified to important transformations within Europe; it reflected most directly the powerful demand for technocratic expertise that had been especially encouraged by the first world war. But the war aside, European society could easily press into service doctrines of technological efficiency: structural changes in the twentieth-century economy awoke a concern for 'rationalization'; artistic and architectural innovation revealed a fascination with the social possibilities of mechanization. Taylorism and Fordism provide a good starting point for analysing what was at stake. They evoked a European resonance less for their strictly technical features than for their social and political implications. The engineer, who was central to the new industrial gospel, appeared not so much a master of machines as a potential manipulator of all industrial relationships. The cultural and political appeal, rather than actual factory applications, forms in fact the focus of this essay. Because of the ideological implications, a survey of scientific management, and the related concern for economic and social planning, open new perspectives on the period between the first world war and the Great Depression.

Whereas in America the commitment to technological efficiency and productivity pervaded almost the entire culture, in Europe it appeared more selectively. The central question is what determined that pattern of receptivity – at least that receptivity as measured by public discussion and government sponsorship. It is noteworthy that the ideological breakdown between the enthusiasts and the indifferent or hostile, did not follow any simple left-to-right alignment. Generally during the early post-war years technocratic or engineering models of social management appealed to the newer, more syncretic, and sometimes more extreme currents of European politics. Italian national syndicalists and fascists, German 'revolutionary conservatives' and 'conservative socialists', as well as the so-called left liberals who sought to mediate between bourgeois and social democracy, and finally the Soviet leaders, proved most receptive. Later in the decade, as the American vision of productivity was divested of its more utopian implications, it came to serve a useful function for business conservatives. Be-

*Europe* (Geneva, 1926), 7–8. This report is an abbreviated version of Paul Devinat, *Scientific Management in Europe*, ILO, Studies and Reports, Series B, No. 17 (Geneva, 1927).

BETWEEN TAYLORISM AND TECHNOCRACY

tween the original enthusiasm for Taylorite teachings and the later *éclat* of Fordism lay an important evolution in the ideological thrust of Americanist doctrines. In general, however, all the variants enjoyed most appeal where representative government was deemed to be working badly. Ironically enough, American productivity contributed to the critical attitude towards parliamentary liberalism.

What the Americanist vision seemed to promise through its brash teachings of productivity, expertise, and optimalization was an escape from having to accept class confrontation and social division. Albeit for very different reasons, all the enthusiasts of scientific management and technological overhaul were seeking to deny the necessary existence of the pre-war model of ideological conflict and to validate a new image of class relationships.

*The promise of engineering in America*

Before 1914 Taylorism had already been picked up in Europe as one of the most provocative aspects of America's formidable economic expansion, although even in the United States it was rarely applied in full. Still, its career and intellectual elaboration reveal the dynamic inherent in the idea of technology as social arbiter. Following its influence from this point of view makes clear the stakes that any recourse to the technician or 'producer' would entail, in Europe as well as America.

Throughout the first decade of the new century Frederick W. Taylor (1856–1915) popularized a process of labour discipline and workshop organization based upon supposedly scientific studies of human efficiency and incentive systems. Preoccupied with the problem of 'soldiering' or labour slowdowns, Taylor timed basic work actions, developed programmed task instruction cards for employees, recommended factory planning departments, and devised wage scales based on piece work, such that the productive worker shared in the expansion of output, but would fall below a subsistence wage and be forced to quit were he to prove inefficient. Taylor's system was propagated by his zealous disciples and similar versions were advanced by eager competitors, while it became fixed in the public eye through a series of controversies concerning its benefits and its alleged inhumanity.[4]

[4] See Samuel Haber, *Efficiency and Uplift, Scientific Management in the Progressive Era, 1890–1920* (Chicago, 1964); M.J. Nadworny, *Scientific Manage-*

Certainly there had been notions of rationalized management practice before. What was novel about Taylorism was the application of the supposedly machine-oriented discipline of engineering to labour relations. How, in fact, had engineering intruded into this sphere? The American Society of Mechanical Engineers, founded in 1880, represented a profession significantly different in origin from European counterparts such as the *polytechniciens* of France. In France and Germany engineering schools had originally been sponsored by royal, revolutionary, or Bonapartist regimes concerned with national wealth and power. In England and America the mechanical engineering profession came of age with the surge of industrialization, and its early practitioners emerged not from the technical institute but from the factory itself. By the late nineteenth century a heightened professionalism was drawing many of America's engineers from the old, ethnically and socially established middle classes – men who perhaps did not wish to give themselves up entirely to business pursuits, who insisted on the credentials of expertise as well as the sanctification of money, and who retained a marked distrust of labour's collective ambitions. For the professionally committed, engineering suggested a self-image of impartial technical arbitration, a dedication to scientific standards and objectivity above the clash of interests in the factory.[5]

As the application of science to the world of economic constraints, engineering logically had to work with the concept of efficiency: the ratio of output to input and benefits to cost. Optimality – although the term was not used in early Taylorism itself – became the implicit key notion behind the application of

*ment and the Unions, 1900–1932* (Cambridge, Mass., 1955), esp. 1–42; F.W. Taylor, *Scientific Management, Comprising Shop Management, The Principles of Scientific Management, Testimony before the Special House Committee* (New York, 1911); C.B. Thompson, ed., *Scientific Management. A Collection of the More Significant Articles Describing the Taylor System of Management* (Cambridge, Mass., 1914). On Taylor himself: F.B. Copley, *Frederick Winslow Taylor* (New York, 1923). For the controversies over application: H.G.J. Aitken, *Taylorism at Watertown Arsenal* (Cambridge, Mass., 1960).

5 Haber, *Efficiency and Uplift*, 9–17; Monte Calvert, *The Mechanical Engineer in America, 1830–1910* (Baltimore, 1967); on English and European training: W.H.G. Armytage, *A Social History of Engineering* (London, 1961), esp. 108 ff., 149–52, 185 ff.; J.P. Callot, *Histoire de l'Ecole Polytechnique* (Paris, 1958); for a note on French origins, Georges Sorel, *Les illusions du progrès* (3rd ed., Paris, 1921), 357–8. Cf. also H. Klages and G. Hartleder, 'Gesellschaft und soziales Selbstverständnis des Ingenieurs', *Schmollers Jahrbuch*, 1965.

BETWEEN TAYLORISM AND TECHNOCRACY

engineering to industrial relations. Worker and employer had no scope for quarrelling about wages or hours or conditions of labour when both parties were yoked to the arbitration of science. 'What we need', wrote Henry L. Gantt, one of Taylor's most engaging followers, 'is not more laws, but more facts, and the whole question will solve itself'.[6]

In practice, not surprisingly, the supposedly impartial findings of science tended to confirm the approach of management, not labour. Collective bargaining had little place in a world of technological imperatives and piece-work wages. Management alone, Taylor insisted, could call upon the directing intelligence and alone set the norms of efficient production.[7] Nonetheless, in theory there could be no arbitrary decisions. And if Taylor himself usually emphasized the need to eliminate worker 'soldiering', reformist Taylorites were later to stress how conservative entrepreneurial practice must change. The important thing was that both findings still carried a commitment to transcend conflicts of interest. To borrow the language developed for game theory, Taylorism promised an escape from zero-sum conflict, in which the gain of one party could be extracted only from the equal sacrifice of the other.

In addition to the optimal allocation of given production and income, the expansion of output through improved workshop organization was also to benefit both sides. Increased production would be shared with labour as well as with investors, so that there need be no bitter scrapping over any given level of return. Efficiency, optimality, enhanced productivity and expanded output thus formed a coherent system. It both demanded and promised much. As Taylor told the House of Representatives, the essence of scientific management was not merely piece work, task cards, or time studies, but 'a complete mental revolution on both sides', such that old contentions were eradicated:

[6] Cited in L.P. Alford, *Henry Laurence Gantt: Leader in Industry* (New York, 1934), 262.

[7] Taylor, *Testimony before the Special House Committee*, 235; cf. also Nadworny, *Scientific Management and the Unions*, 9. The implicit assumptions in favour of the employer are also discussed in Reinhard Bendix, *Work and Authority in Industry* (New York, 1963), 276–87. Taylor, moreover, did not believe pay should rise in exactly the same proportion as output; smaller increments would force the worker to remain ambitious. See *Shop Management*, 29.

The great revolution that takes place in the mental attitude of the two parties under scientific management is that both sides take their eyes off of the division of the surplus as the all-important matter, and together turn their attention toward increasing the size of the surplus until this surplus becomes so large . . . that there is ample room for a large increase in wages for the workman and an equally large increase in profits for the manufacturer.[8]

In short, what Taylorism offered – certainly within the plant, and ultimately, according to its author, in all spheres of government and social life[9] – was the elimination of scarcity and constraint. It therefore implied a revolution in the nature of authority: the heralded utopian change from power over men to the administration of things. Such an evolution logically removed the basis for class formation as conceived by sociology.[10] Ostensibly Taylor's factory could become the nucleic building block of a post-bourgeois world, or at least a secure managerial one.

For Americans of the Progressive era this sort of doctrine had great appeal. Social efficiency in the years before the first world war became a shibboleth for reform as well as for productivity. It showed the interests of employers and employees to coincide, in the words of Taylor and then of the Gilbreths – Taylor's rivals who relied on motion-picture analysis of basic work-movements anagramatically christened Therbligs – with 'the one best way to do work'.[11] A publicist of related views, Harrington Emerson, organized the Efficiency Society, while in December 1916 Henry Gantt helped to found the short-lived 'New Machine', an association seeking to acquire political power and exercise it according to the criteria of industrial efficiency. Inspiration for this effort was found in the works of Thorstein Veblen and the now obscure

8 Taylor, *Testimony*, 27–30.

9 Taylor, *Scientific Management*, 8.

10 See Ralf Dahrendorf, *Class and Class Conflict in Industrial Society* (Stanford, 1959) 157–205. A theoretical framework for a view more in line with Taylorite implications is provided by Talcott Parsons, 'Social Classes and Class Conflict in the Light of Recent Sociological Theory', in *Essays in Sociological Theory* (New York, 1964).

11 Haber, *Efficiency and Uplift*, 41. Cf. F.B. Gilbreth, *Motion Study: A Method for Increasing the Efficiency of the Workman* (New York, 1911). For a general discussion of efficiency as a national theme see Daniel Bell, 'Work and its Discontents: the Cult of Efficiency in America', in *The End of Ideology* (Glencoe, Ill., 1960).

BETWEEN TAYLORISM AND TECHNOCRACY

Charles A. Ferguson, who combined evangelism with an elitist proto-syndicalism.[12]

Given an overall national commitment to democracy, its redefinition to square with criteria of optimality and efficiency seemed imperative, and this the Progressive era writers undertook. Democracy, wrote Ferguson, 'is not the rule of the majority but of the wilful servants of all', and he advocated the devolution of power to self-administering economic associations.[13] 'Democracy is a method, a scientific technique of evolving the will of the people', claimed Mary P. Follett, a future theorist of scientific management influenced by English neo-Hegelianism and Guild Socialism.[14] Minimal, but scientific government by experts, complex schemes for the self-regulation of industry, with production, not profit as the criterion, reflected a similar quest for a new concept of authority that would transform the economic interests now smothering the public welfare into the very bearers of the community's advance into abundance. 'There is no legitimate power but the power to deliver goods', stated Ferguson,[15] while Gantt wrote that 'The era of force must give way to the era of knowledge'. The engineers would be summoned to impose optimality upon society as they did in the factory: 'The new democracy does not consist in the privilege of doing as one pleases, whether it is right or wrong, but in each man's doing his part in the best way that can be devised from scientific knowledge and experience.'[16]

This functional model naturally became more topical when a War Industries Board offered a prototype of the new industrial co-ordination. Unprecedented material requirements also pressed

12 Haber, *Efficiency and Uplift*, 44–9; Alford, *Gantt*, 264 ff.; for Emerson's rotarian prose see *The Twelve Principles of Efficiency* (New York, 1913); for Ferguson, *The Great News* (New York, 1915).

13 Ferguson, *The Great News*, 59, 73–5; for the Progressives and scientific management: Haber, *Efficiency and Uplift*, 75 ff.

14 Mary P. Follett, *The New State: Group Organization the Solution of Popular Government* [1918] (3rd ed., London, 1934), 180. Miss Follett spotlighted the tension – to be found in later planning concepts – between the public interest and the policies of the quasi-syndical bodies to whom the English pluralists wished to give authority (258–319); for an illuminating discussion of ambiguities tending the other way, see Charles Forcey, *The Crossroads of Liberalism* (New York, 1967), 37 ff.

15 Ferguson, *The Great News*, 103.

16 Alford, *Gantt*, 253, 196.

home the need for efficient use of national resources. Hence the war made the problem of industrial relations both more urgent and more tractable, according to Morris Llewellyn Cooke, another reformist engineer, for it had convinced all parties of the necessity of increasing production.[17]

It was logical, too, that the war and its aftermath should help to crystallize two explicit alternatives for the role of the engineer: Veblen's isolated revolutionary prescription, and Herbert Hoover's ameliorist activism. By the early 1920s Veblen was addressing himself directly to the engineers in an effort to reshape modern capitalist society. Throughout his works he had envisaged an enduring social conflict between the industrious and the exploiters. The 'pecuniary' occupations justified their frankly parasitic role by the conventions of private ownership, while those conditioned by technological rationality – engineers and workers – were most liable to question the nexus of private property (or absentee ownership).[18] By the end of the first world war, Veblen was concentrating on the engineer as strategically pre-eminent, for through him the genuinely productive forces at last had a hand upon the nerve centres of modern society. The engineers could act, were they so disposed, to end the conventions of absentee ownership on behalf of all those engaged in non-exploitative labour. Veblen probably misunderstood the temper of the engineers; certainly he inverted the social role of engineering as its actual practitioners conceived it.[19] In their 1921 investigation of waste in industry, for example, the Hoover committee, sponsored by the Federated American Engineering Societies, pointed to the engineer not as a syndicalist revolutionary, but as a rationalizer of a basically successful system:

His lifelong training in quantitative thought, his intimate experience with industrial life, leading to an objective and detached point of view, his strategic position as a party of the third part with reference to many of the conflicting economic groups, and above all his practical emphasis

17 M.L. Cooke, 'Forward', *Modern Manufacturing, A Partnership of Idealism and Common Sense. Annals of the American Academy of Political and Social Science*, September 1919, vi.

18 Thorstein Veblen, *The Theory of Business Enterprise* [1904] (New York, New American Library, n.d.), esp. 144–76.

19 Edwin Layton, 'Veblen and the Engineers', *American Quarterly*, Spring 1962; Calvert, *The Mechanical Engineer in America*, 263–76; Veblen, *The Engineers and the Price System* [1914–21] (New York, 1963), 93–108.

BETWEEN TAYLORISM AND TECHNOCRACY

on construction and production, place upon him the duty to make his point of view effective.[20]

Certainly the attributes of the engineer were the potentially technocratic ones, but Hoover's view of his task was more circumscribed than Veblen's. For Veblen the business system had virtually to manufacture waste to preserve hierarchy, and commitment to optimality or abundance was impossible within American capitalism. For Hoover the engineer helped to eliminate the frictions of a basically superior economic order.[21] It was not surprising that Veblen's conception found little response until the depression, while Hoover's helped to set the tone of the 1920s. Together, they showed the malleability of the stress on productivity: as the next decade in Europe revealed, the technological vision could serve the ends of transformation or of the status quo.

*The ambivalence of the right-radical response*

The imagery of the technological vision was as potent as its utopian ideology; if the machine was to alter society it must transform the environment. Viewed retrospectively, the response of art and architecture revealed in what milieux the imaginative concepts of technology proved influential, as they did in Germany and Austria, Italy, and later in Russia and France. The formation of the German Werkbund in 1907, for example, brought together the left-liberal and national-social political leader Friedrich Naumann, representatives of forward-looking industries such as German General Electric (AEG), and architectural innovators, including Hermann Muthesius, Peter Behrens, and Walter Gropius. Its establishment, however, did not mean that the claims of modernity were carrying the day in central Europe. Instead it suggested that a self-conscious technological inspiration might very well arise where society revealed deep fissures and strong reactionary impulses. The vision behind the Werkbund represented in fact an effort to overcome the fracturing materialism of Wilhelmine

[20] Committee on Elimination of Waste in Industry of the Federated American Engineering Societies, *Waste in Industry* (New York, 1921), 33. Later popularization of the waste theme is found in Stuart Chase, *The Tragedy of Waste* (New York, 1925). For the transition from engineering to planning at the end of the decade, see Charles A. Beard, ed., *Toward Civilization* (New York, 1930).

[21] On general economic views, see Hoover to Woodrow Wilson, 28 March 1919, in A.J. Mayer, *Politics and the Diplomacy of Peacemaking* (New York, 1967), esp. 25; and Herbert Hoover, *American Individualism* (New York, 1922).

society. Gropius himself praised the stark American factories and grain elevators as models for the new style and claimed that a good factory aesthetic was important from a social point of view, for it permitted a more joyful cooperative effort.[22]

Contributing to a related tendency, but one with a different political outcome, were the Italian Futurists, whose work drew on the machine as a fount of eroticism, violence, and death. Severini's and Bolla's paintings as well as Marinetti's notorious *Manifesto* prefigured a crucial development in the engineering vision: the right-radical union of technology and irrationalism. Revealingly, liberal France and England seemed at the time to produce less work of specific technological inspiration but by the early 1920s, as the American industrial model attracted ever more attention, it stimulated an artistic response there too. Le Corbusier praised Ford and Taylor in his treatises and sought to bring France's sadly fallen architects up to the level of 'healthy and virile, active and useful, balanced and happy engineers'. The house, then the city, was to be transformed from monument to tool. The new aesthetic required a new technocracy: Le Corbusier's town-planning evangelism of the early twenties demanded a linear regularity imposed by a far-seeing authority – a 'technical work' on behalf neither of communism nor of capitalism.[23] Not that communism lacked similar inspiration: in Moscow an abstract formalism that celebrated the union of technological possibilities with social revolution flourished for a few exciting years, reflecting some of the same impulses that led to Lenin's flirtation with

22 For general discussion, Nikolaus Pevsner, *Pioneers of Modern Design* (Baltimore, 1965), 31–9, 179 ff.; Reyner Banham, *Theory and Design in the First Machine Age* (London, 1960), 68–87; Walter Gropius, 'Die Entwicklung moderner Industriebaukunst', *Jahrbuch des deutschen Werkbundes* (Jena, 1913), 17–22. Cf. also Hermann Muthesius, 'Das Formproblem im Ingenieurbau' in the same issue; W.H. Jordy, 'The Aftermath of the Bauhaus in America: Gropius, Mies, and Breuer', *Perspectives in American History*, II (1968), esp. 489–91.

23 For Futurism and Le Corbusier, Banham, *Theory and Design*, 99–137, 220–63; cf. James Joll, 'F.T. Marinetti: Futurism and Fascism', in *Intellectuals in Politics* (London, 1960), esp. 169–70 for Mussolini's delight in the mechanical, and T. Marinetti, *La democrazia futurista* (Milan, 1919), for a technocratic projection. See Le Corbusier, *Vers une architecture* (2nd ed. Paris, 1924), 6 ff. for the engineer, and 234 on the needs of the 'service class', and *The City of Tomorrow* (London, 1947), transl. of *L'urbanisme* 1929 ed.), 308–09.

BETWEEN TAYLORISM AND TECHNOCRACY

Taylorism and the Russian enthusiasm for scientific management and American engineering.[24]

Conversely, those places where the cultural avant-garde showed little response displayed less interest in the new doctrines in general. In England, before the war, schemes of scientific management awoke scant interest among engineers and managers. Not merely did this reflect an industrial leadership set in its ways; an underlying satisfaction with decentralized production, with the premises of a liberal regime in a country where the middle-classes felt little anxiety about the social order, postponed real interest until the economic difficulties of the 1920s and 1930s.[25] Initially France, too, seemed little moved by American technological messianism. There the response to scientific management also remained scattered until the later 1920s, when, serving more conservative ends, American-inspired visions of productivity and modernization were able to arouse businessmen and politicians.[26]

Certainly impulses towards 'Americanism' were present earlier; the necessities of war production encouraged interest in the innovations of France's ally; in early 1918 Clemenceau asked that attention be paid to Taylorism in war-plants and suggested the

24 Banham, *Theory and Design*, 193ff.; Camilla Gray, *The Great Experiment: Russian Art 1863–1922* (New York, 1962), 181–97, 215–27; cf. also K.G. Pontus-Hultén, *The Machine as seen at the End of the Mechanical Age* (New York, 1968), 107 ff., 128 ff. On Russian enthusiasm for scientific management, Devinat, *Scientific Management in Europe*, 86 ff.

25 A.L. Levine, *Industrial Retardation in Britain, 1880–1914* (London, 1967), 60–8; Cf. the judgment of a leading advocate that scientific management was long delayed in England as a general movement despite occasional Taylorite applications – a lag attributed to empirical modes of thought and a dislike of large-scale organization, which lasted beyond the first world war. L. Urwick, *The Development of Scientific Management in Great Britain* (London, 1938), 75–80.

26 For a brief account of the progress of scientific management in France and other European nations, see Devinat, *Scientific Management in Europe*, 233–45, and the preface by Albert Thomas, which describes initial French working-class resistance. Devinat includes extensive bibliographies of French and German works. See above all the works of Henri LeChatelier, metallurgist at the Sorbonne, editor of a Comité des Forges-sponsored journal, and translator of Taylor, including *Le Taylorisme* (2nd ed. Paris, 1934). For the related approach of Henry Fayol and his *doctrine administrative* see Fayol, *Industrial and General Administration* (Engl. transl. London, 1930), with an extensive bibliography, and J. Billard, *Organisation et direction dans les affaires privées et les services publics. Un essai de doctrine, le Fayolisme* (Paris, 1924); for a contrast between Taylorism and Fayolism: François Bourricaud, 'France', in A.M. Rose, ed., *The Institutions of Advanced Societies* (Minneapolis, 1958), 490–1. For the major French socialist critique of Taylorism as practised in America see André Philip, *Le problème ouvrier aux Etats-Unis* (Paris, 1927), 39–87.

establishment of Taylorite planning departments.[27] Even more promising was the ideological tendency in French politics that anticipated the quest for the engineer as social manager. Saint-Simonianism embodied a proto-technocratic ideology that rejected traditional class divisions in favour of the unity of all 'productive' and 'industrious' elements, bourgeois, peasant, and proletarian, against the useless aristocrats and rentiers.[28] Veblen's contemporary categories were strikingly reminiscent of the Saint-Simonian scheme; and, of course, American condemnations of idleness and waste might have been taken directly from the French utopian's writings. Saint-Simonianism had projected a disinterested social optimalization from above, a functional administrative structure, and a commitment to the aggregate wealth and welfare of society – all themes that appeared in American writings.

But in France after the first world war only a handful of confessed Saint-Simonians existed to publish the obscure *Le Producteur*.[29] To be sure, the idea of 'production' aroused many observers, including, for instance, the popular Mayor of Lyon, Edouard Herriot, who in 1919 called attention to Taylorism and appealed for bureaucratic, economic, and educational modernization in a technologically inspired 'fourth republic'. The new regime was to abandon the pre-war party cliques, local patronage, and *café-comptoir comités* that formed the warp and woof of French politics.[30] But Herriot's rhetorical ebullience did not imply practical commitment, nor were his own Radical Socialists likely to follow his advice on technological overhaul and abandon the small-town network of interests that was their own power base. Likewise, when Etienne Clémentel, Clemenceau's Minister of Commerce, sought to present an organization model for French industrial self-administration in a Fédération des Syndicats, he encountered suspicion and apathy from businessmen who desired primarily to shake off wartime supervision and return to their old and less daring habits.[31]

27 Cited in Copley, *Taylor*, I, xxi.

28 For summaries of the ideology see Frank Manuel, *The New World of Henri Saint-Simon* (Cambridge, Mass., 1956), and Manuel, *The Prophets of Paris* (Cambridge, Mass., 1962), 105–48.

29 For this group see Marc Bourbonnais, *Le néo saint-simonisme dans la vie sociale d'aujourd'hui* (Paris, 1923).

30 Edouard Herriot, *Créer* (2 vols., Paris, 1919), esp. I, 448–68, II, 335.

31 For Clémentel's efforts, *La Journée Industrielle*, 8–9 March, 16, 25–8 April 1919.

BETWEEN TAYLORISM AND TECHNOCRACY

That these strict Saint-Simonian themes found only a faint echo was understandable, for their origins linked them with the logic of strong executive authority, and they had enjoyed their greatest influence under the Second Empire. As long as the parliamentary regime functioned satisfactorily, the process of political selection kept the would-be technocrats from positions of influence. Nonetheless, the war could not leave the parliamentary status quo absolutely immune. The legacy of the 1917 crisis, the impatience with earlier Radical-Socialist domination, the pervasive feeling that total war must yield profound if vague transformations – all contributed to the anti-parliamentary overtones that emerged in the Bloc National elections of November 1919. For the first time since the 1890s, not merely a clerical or reactionary, but a genuine right-radical tone was evident[32]: here and there among Bloc National candidates themselves, occasionally in the scattered efforts of Action Française, finally in the ephemeral new movement led by Lysis (Ernest Letailleur). Lysis complained that France lacked '*l'idée d'une technique nationale*', and was stifled by backward oligarchies and parliamentary stockjobbers. He called for the representation of professional groups, distinguished 'productive' capitalism from its parasitic version, and advocated a new socialism that postulated class unity instead of class conflict.[33] His followers fared badly in the elections, because the Bloc National candidates responded well enough to the discontents he manipulated, not because his programme was rejected outright. For a significant political organization to exploit Lysis' ideological themes, France had to wait until the mid-1920s with its parliamentary paralysis and an angered bourgeoisie.

Elsewhere right-radical spokesmen for a 'productivist' ideology had more impact. Nevertheless, a central ambivalence towards technology itself marked their thinking, much as it did Futurist art. If right-radical spokesmen wished to assail the liberal capitalist order, or at least the liberal parliamentary order, their anti-intellectualism undermined a reliance upon the engineer, the manager, or other specialized expert as a potential leader. Before

32 André Siegfried, *Tableau des partis en France* (Paris, 1930), 131–2.

33 Lysis, *Vers la démocratie nouvelle* (Paris, 1919), 37 ff., 117 ff., 277; for the Bloc National programmes see *Programmes, professions de foi et engagements électoraux de 1919* (Paris, 1920).

the war this had been foreshadowed in the problematic writings of Georges Sorel. Like Veblen, with whom he forms an instructive contrast, Sorel retained the old dichotomy between useful production and financial exploitation. By training an engineer, Sorel saw the virtuous man as maker, but he treasured a pre-industrial *morale des producteurs* that only the small workshop could preserve. Whereas Veblen's glorification of workmanship was unambiguously pacific, Sorel's ideal included a vigorously militant component that in modern conditions could be restored only through commitment to a myth of imminent revolutionary struggle. The enlightenment-bred rationalism that helped to advance the machine age also denatured man, while for Veblen technology in no way diminished humanity.[34]

The Italian episode of early fascist technocracy reflects the stresses inherent in Sorel's theories within a real institutional context. In Italy the themes of new industrial leadership and anti-parliamentarism were tightly interwoven. Pre-war Italian nationalist writers had assailed liberal and social democracy on behalf of a right-radical syndicalism.[35] In 1917 the President of the Comitato Nazionale Scientifico Tecnico, G. Belluzzo, later fascist economics minister, called for an eventual transformation of the state, to be preceded by industrial rationalization and concentration.[36] By August 1918, Mussolini had changed the subtitle of his own newspaper from *Socialist Daily* to *Daily for Soldiers and Producers*. Indicatively, he condemned the socialist-party 'parasites of blood', and 'parasites of labour', adding that 'to defend the producers means to let the bourgeoisie complete its historical

34 Georges Sorel, *Réflexions sur la violence* (11th ed., Paris, 1950), 109–20, 377 ff.; *Les illusions du progrès* (Paris, 1947); cf. also I.L. Horowitz, *Radicalism and the Revolt against Reason; The Social Theories of Georges Sorel* (Carbondale, Ill., 1968), esp. 127–63; Michel Freund, *Georges Sorel, Der revolutionaere Konservatismus* (Frankfurt am Main, 1932).

35 See among other works, Enrico Corradini, *La marcia dei produttori* (Rome, 1916) and *Discorsi politici (1902–1923)* (Florence, 1923); P.M. Arcari, *L'elaborazione della dottrina politica nazionalista (1870–1914)* (3 vols., Florence, 1934–1939); and the very useful Paolo Ungari, *Alfredo Rocco e l'ideologia giuridica del fascismo* (Brescia, 1963).

36 G. Belluzzo, *La organizzazione scientifica delle industrie meccaniche in Italia* (Milan, 1917), 3–4; cited in Paola Fiorentina, 'Ristrutturazione capitalistica e sfruttamento operaio in Italia negli anni '20', *Rivista Storica del Socialismo*, January–April 1967, 135–6. For an Italian discussion of the progress of Taylorism and scientific management in this period, see Angelo Mariotti, 'L'organizzazione del lavoro', *Rivista Italiana di Sociologia*, 1918.

function'.[37] This theme was still pursued after the war: 'No political revolution, no extremism, no expropriation and not even a class struggle, if the chiefs of the enterprises are intelligent. Intensive, harmonious collaboration of industrialists and workers in production.'[38]

Mussolini's *produttovismo* depended less on the engineer, or technology, than on the expert in general. One of the young exponents of fascist technocracy was to capture the prevailing imagery later, when he claimed: 'The fascist state is more than a state, it is a dynamo' – a rhetorical flourish reflecting the Futurist influence in the early movement.[39] Once in power, the fascists sought to establish committees of experts from all fields. The party statutes of autumn 1921 required local *fasci* to prepare lists of cooperative specialists in the public services and economic life. In theory, these *gruppi di competenza* were to furnish the Fascist Party with a general staff ready to take over the state; more practically, to win potential sympathizers by making the movement seem less narrowly ideological. Nonetheless, the cadres were conceived almost entirely in terms of restoring state and bureaucratic authority, oriented towards ministries, not factories. Those instituted in 1923, moreover, were presided over by leading government figures, which meant they could scarcely escape political supervision.[40]

The tentative character of the effort and its lack of real anchorage as a technocracy were demonstrated by the fate of the groups during the party disputes of 1923–4. Massimo Rocca, former journalist and champion of the *gruppi di competenza*, was also the exponent of fascist 'revisionism'. This represented a policy of

[37] Benito Mussolini, 'Novita', *Il Popolo d'Italia,* 1 August 1918, included in *Opera Omnia di Benito Mussolini,* XI (Florence, 1953); cf. also Renzo De Felice, *Mussolini il rivoluzionario* (Turin, 1965), 405–6; Roberto Vivarelli, *Il dopoguerra in Italia e l'avvento del fascismo (1918–1922),* I, 234–5, 271–7.

[38] 'Il sindicalismo nazionale. Per rinascere!' *Popolo d'Italia,* 17 November 1918; cited De Felice, *Mussolini,* 493–4; *Opera Omnia,* XII, 11–14. For the influence of Lysis on Mussolini see De Felice, 410. While De Felice sees Mussolini's 'productivism' as a new reformism seeking to undercut the socialists, Vivarelli emphasizes the link with rightwing nationalism from the outset.

[39] Camillo Pelizzi, *Problemi e realtà del fascismo* (Florence, 1924), 165.

[40] Alberto Aquarone, 'Aspirazioni tecnocratiche del primo fascismo', *Nord e Sud,* April 1964; Camillo Pelizzi, *Una rivoluzione mancata* (Milan, 1949), esp. ch. I; Massimo Rocca, *Come il fascismo divenne una dittatura* (Milan, 1952), 132 ff.

normalization, a downgrading of the revolutionary claims of fascism, consequently of its local violence, its militia, *ras* and squadrist leadership such as that exercised by Roberto Farinacci of Cremona.[41] It was prepared to sacrifice ideological purity for the sake of collaboration with the liberal elites. This was a line Mussolini found useful to encourage during the initial year or so of his rule, especially as he looked forward to an electoral campaign that would consolidate his position in parliament. Enjoying an electoral law that promised him two thirds of the seats, Mussolini chose a tactic of collaboration to woo the parliamentary notables of the liberal groups for his own slate, while simultaneously working to shatter their old party structures. With the electoral victory of April 1924, however, collaboration with non-fascists was less necessary, and the impatient stalwarts of a radical fascist policy could be appeased. Rocca, whom the fascist intransigents had sought to expel from the party in the fall of 1923, now came under renewed fire and was dropped. Moreover, in the late 1924 crises following the assassination of Matteotti, the spokesmen for integral party dictatorship prevailed over the voices for moderation and normalization, and in the process non-party technocratic aspirations succumbed.[42]

In fact, from the outset of 1924, the *gruppi di competenza* were being reshaped into less independent *consigli tecnici*, intended explicitly to serve only as bodies that would support the new rulers. In similar manner the syndicalist organizations of Edmondo

41 Massimo Rocca, 'Il fascismo e l'Italia', *Critica Fascista*, 15 September 1924, reprinted in Rocca, *Idee sul fascismo* (Florence, 1924), esp. 64; also Rocca, 'Diciotto Brumaio', *Critica Fascista*, 24 September 1923, now in Rocca, *Il primo fascismo* (Rome, 1964), 99. For Rocca's memoirs: *Come il fascismo divenne una dittatura* (Milan, 1952), esp. 145 ff. For Farinacci's views see his article 'La seconda ondata', *Cremona Nuova*, 29 May 1923, cited in De Felice, *Mussolini il fascista, I. La conquista del potere (1921–1925)* (Turin, 1966), 413–15; also his letter to Mussolini of 4 August 1923, complaining about the preference given to non-fascist, even allegedly anti-fascist technical appointments. See *Segretaria particolare del Duce*, National Archives film, T 586, Roll 448, 062223–24. For a general discussion of the tendencies within the party, Giacomo Lumbroso, *La crisi del fascismo* (Florence, 1925).

42 For the political developments, see the works by Rocca cited in the preceding note; also *Mussolini il fascista*, 518–730; Luigi Salvatorelli and Giovanni Mira, *Storia d'Italia nel periodo fascista* (Turin, 1957), 269–332; Adrian Lyttelton, 'Fascism in Italy: The Second Wave', *Journal of Contemporary History*, 1966, republished as *International Fascism, 1920–1945* (New York, 1966), 75–100.

BETWEEN TAYLORISM AND TECHNOCRACY

Rossoni were being circumscribed in such a way that any independent labour-oriented objectives would be clearly subordinated to state and party requirements. The regime's growing commitment to state corporations instead of the former syndicates portended a general braking of any genuine radical experimentation. Fascist technocracy was to wane alongside this emasculation of any independent economic or administrative centres of expertise.[43]

In any case, the very concept of technocratic cadres suffered from a basic ideological equivocation. By appealing to production and technology, fascism, like Saint-Simonianism, wished to assert the role of a new ruling group originating outside the traditionally conceived classes. Fascist, or more precisely national-fascist ideology, resembled Taylorism in a key particular. It promised a 'non-zero-sum' world in which classes no longer prospered only at each other's expense, in contrast to the implications of the traditional spectrum of European ideologies, which were all zero-sum or redistributive. Their prescriptions entailed transferring portions of a given quantity of power, status, and wealth from one social group to another – or preventing such transfer. Marxism involved only the most radical redistributive objective. The appeal of Saint-Simonianism, or of the American engineering vision, consisted precisely in its claim to avoid such painful transfers. Expanding productivity meant that no repartition of a fixed quantum of national wealth was required. Postulating a new social category of producers, or more narrowly, an elite of scientific managers who arbitrated conflict, meant that the hostile confrontation between the traditional classes was superseded.

While making similar claims, fascist ideology differed in some key respects. The model of social engineering indicated that internal disputes about power could be sublimated into technical questions of optimalization. Fascism added the concept that class disputes must dissolve before overriding clashes among nation-states: Italy as a whole was a proletariat among European powers. Secondly, if Saint-Simonianism stressed the contribution of a vanguard of entrepreneurs, and Taylorism spotlighted the engineers, fascism drew upon other potential leaders. Arising out of a fervent interventionist commitment, it posited the *combattenti* as a

[43] Aquarone, 'Aspirazioni tecnocratiche del primo fascismo', *loc. cit.*, 125–28; *L'organizzazione dello stato totalitario* (Turin, 1965), 113–18.

directing elite by virtue of their trials at the front. It did not reject the claims of technology and productivity to coordinate hitherto opposed interests, but these could not be the only claims to leadership; 'blood', and exposure in the trenches also counted.

This effort to combine technology with vitalist sources of energy contributed powerfully to the appeal of the right-radical ideology. German right-radicalism or so-called revolutionary conservatism often revealed the same problematic synthesis – a dual hostility to liberalism and to a materialism that the Left also condemned. The Werkbund, for example, attracted future right-radical spokesmen as well as democratic ones; and its early architects looked back to the inspiration of the *Rembrandtdeutscher*, Langbehn, who demanded a break with the stuffy and syncretic styles of the 1880s.[44] Oswald Spengler, who was considered one of the elders of Weimar's revolutionary conservative movement, symptomatically fused machine imagery and hostility to liberalism. 'The centre of this artificial and complicated realm of the Machine', he wrote, 'is the organizer and manager'. But with the manager was 'the *engineer*, the priest of the machine, the man who knows it . . . the machine's master and destiny'.[45] Nonetheless, Spengler saw the powers of money – 'our inner England' he called them elsewhere: 'capitalism and parliamentary liberalism'[46] – enslaving the forces of technology to be defeated, in their turn, only by 'blood' and a new Caesarist collectivism, or perhaps a Prussian socialism of labour and subordination to the state.[47] In Spengler's conception, therefore, technology could never replace power. Technocracy, strictly speaking, was impossible, even though the engineer was the indispensable auxiliary of rule in the machine age.

Thus from Sorel to Mussolini and the German conservative revolutionaries, the technological vision was incorporated in an uneasy relationship with a commitment to nonrational values. For its adherent there was a compelling psychological validity in the image of the engineer at the service of an aggressive national allegiance: the hard master of machine civilization sweeping away nineteenth-century sentimentality and petit-bourgeois democracy.

44 Banham, *Theory and Design*, 72.

45 Oswald Spengler, *The Decline of the West* (Engl. transl. New York, 1932), II, 504–5.

46 Oswald Spengler, 'Prussianism and Socialism', *Selected Essays* (Engl. transl., Chicago, 1964), 87.

47 Spengler, *Decline of the West*, II, 506; *Selected Essays*, 129–31.

BETWEEN TAYLORISM AND TECHNOCRACY

The ideological effort to banish social conflict could invoke national power and a new authoritarianism as well as national welfare and slide-rule optimality. Hence it was consistent that America should be seen in a Janus-like perspective: the empire of technical rationality on the one hand, whose new cities so impressed and horrified men like Spengler (and Le Corbusier)[48]; the embodiment, on the other hand, of a hypocritical Wilsonian democratic pathos – detested by nationalists in Italy and Germany as masking Anglo-Saxon financial imperialism. In that contradictory estimate was reflected the radical Right's own inner division between technological reason and the utilitarian rationality of liberalism.

*The ambiguities of planning*

Approaching the national syndicalism of the Right was a growing interest in planning among men of the Left. Walther Rathenau of AEG and his collaborator, the Prussian aristocrat and engineer Wichard von Moellendorf, elaborated their experience of organizing wartime production and raw-material allocation into a conception of *Planwirtschaft* to be preserved after hostilities ended. Socialist Party members such as Rudolf Wissell, who served as Economics Minister in 1919, and Max Cohen, who advocated a corporatist upper house, as well as Georg Bernhard, editor of the liberal *Vossische Zeitung*, close to the new Democratic Party and a proponent of a National Economic Council, likewise sought to structure the economy by combining elements of the leftist *Räte* (councils) idea with organic concepts of the state and community. What they all envisaged, in general, was a pyramid of industrial planning organs that would include representatives of the entrepreneurs, labour, and the state. With the power to set prices, allocate raw material and market shares, and generally determine economic policy, the new institutions were to embody the vision of class collaboration in the public interest. Even when the advocates of these schemes belonged, like Wissell, to the Social Democratic Party (SPD), they emphasized not proletarian hegemony, but maximum production for the *Gesamtheit* of German society.

But there were as many ambiguities in the Left's conception of planning as in the Right's attitude towards technocracy. It was

[48] Spengler, *Decline of the West*, II, 100–1, on the soullessness of checkerboard planning; Le Corbusier, *The City of Tomorrow*, 63, 76.

assumed that by seating together the delegates of industry, labour, and consumers or the state, all decisions reached would be bound to ensure the public interest at large; furthermore, that political constraint could be banished from the economic sphere. Here, of course, was the same industrial utopianism that scientific management and wartime organization had suggested in America. In Germany, however, the attempt actually to institutionalize the vision in early Weimar led to difficulties that were not fully evident in the United States until the experiment with the National Recovery Administration.[49]

The institutional models for *Planwirtschaft* were borrowed primarily from Germany's wartime organization, including the war corporations of mixed state and private ownership that Rathenau had seen as a stage between capitalism and state socialism. The war also created the material preconditions for the collaboration of management and labour that was so central to the planning schemes. The ravenous appetite of the war effort for production at any price – more precisely, the demands of generals and industrialists for often irrational production at inordinate cost – facilitated the bargaining: the entrepreneurs could enjoy extraordinary profits, while trade union leaders won new influence over conditions of labour. Inflationary war finance obviated older conflicts over wages, as industry and labour together appropriated resources from the relatively fixed-income sectors of the economy.[50] Given wartime demand, a commitment to production could indeed foster cooperation – but the community paid as well as benefited.

Rathenau and Moellendorff did not approve of the economic megalomania of the Hindenburg Plan, but they did wish to consolidate the new collaboration. Besides tackling the same administrative tasks together, they also shared similar spiritual predispositions. The great dynamos of German General Electric, and the austere image of Prussian discipline and tradition, cap-

49 Cf. Ellis W. Hawley, *The New Deal and the Problem of Monopoly* (Princeton, 1966), 35–46, for the tensions within planning conceptions.

50 For Rathenau's views on war companies see the memo cited in Gerald Feldman, *Army, Industry, and Labor in Germany* (Princeton, 1966), 49. I have drawn upon this book in general for the description of the war's effects. Interestingly enough, the historian of the German wartime organization of raw material production, superintended by Rathenau and Moellendorf, also went on to discuss Taylorism: Otto Goebel, *Taylorismus in der Verwaltung* (Hanover, 1925).

BETWEEN TAYLORISM AND TECHNOCRACY

tured their imagination equally. For Rathenau, who could never refrain from philosophizing, the modern era was characterized by the new machine order: 'It is a consolidation of the world into an unconscious association of constraint, into an uninterrupted community of production and economy'. But the way to master this technological destiny was not through any rancorous Marxism, but through a new moral consciousness and, as he suggested first, the 'depersonalization' of property by transforming private enterprises into foundations and giving title to employees, universities, or administrative authorities. In his subsequent discussion of the new economy Rathenau envisaged cartels, with state participation, to co-ordinate planning and eliminate the destructive aspects of competition: a public syndicalism of the producers analogous to the ideas of association contemplated in the United States as well.[51]

Moellendorff entertained similar ideals, although he occasionally dissented on details. The coming era, he believed, must be either socialist or anarchic; the engineer, moreover, would be central in making the choice. In his view of the engineer Moellendorf had before the war borrowed heavily from Taylorism, which he found Germanic in its intuitive daring. It imposed the criterion of competence as the ordering principle of the economic world, and it demonstrated that the resources of human labour were not a fixed limit upon production. Taylorism was the paradigm of what made America vital; it infused the economic system with the collective élan of those model, primitive German communities described by Tacitus.[52] 'If we really come to grasp Taylor fully we will choke off the evils of our economy from above and below: the confusion of the incompetent, the constraints of interest on the industrious, the arbitrariness of the shortsighted, the supremacy of the success-

[51] Walther Rathenau, *Von kommenden Dingen*, [1916] *Gesammelte Schriften*, III (Berlin, 1918), 35, 64 ff., 139–40, 158–9; cf. *Die neue Wirtschaft* [1917], *Gesammelte Schriften*, V (Berlin, 1918), esp. 203 ff., 231 ff. for the state cartel concept: 'These structures are differentiated from the old guild system . . . no sanction for association of individual interests, no interest group of sovereign individual and small firms, but a community of production in which all members are organically interwoven' (235). On the constructive role of the state, 249–50; for the new order's role in advancing welfare, but not imposing a forced equality, 255. For Rathenau's critique of orthodox Social-Democratic solutions, see *Der neue Staat* (Berlin, 1922), 38, 61 ff.

[52] Wichard von Moellendorff, *Konservativer Sozialismus* (Hamburg, 1932), 34–46.

ful, the pity of the timid'. Moellendorff's vision was not without its authoritarian side: Taylorism functioned in his eyes as a 'militarism of production', training workers to drop their complaints about the inevitable division between management and labour.[53] Like Taylor's hierarchical but conflict-free economy, Moellendorff's embodied a stern collectivism. Although he remained to assist Wissell in the capacity of under-secretary at the Economics Ministry he always disliked the debates over the meaning of 'socialization', to which he preferred the concept of *Gemeinwirtschaft*, roughly an economic commonwealth.[54]

Was it surprising that the Social-Democratic leaders, once they had time to reflect on the memoranda being prepared in Wissell's ministry, were far from happy? They confronted a central dilemma: the responsibility of parliamentary leadership without the power to reshape the economy after they had renounced a *Räte* regime and quick expropriation measures. Once they had opted for a parliamentary democracy, would not the results of a self-administered *Gemeinwirtschaft* really depend upon the power that each side could bring to bear within the organs of political and economic administration? The SPD was understandably confused as to whether planning ideas would advance or hinder socialism. The debates over Taylorism within the party were indicative in this respect. In March 1919, Otto Bauer argued that 'in a democratic and rationally socialized state', Taylorism would serve to increase productivity and thus help the country to acquit the reparation debt more quickly. Two years later Kurt Lewin argued that Taylorism – by which he really meant industrial psychology as a whole – could serve a socialist regime by allocating people to professions not on the basis of a class-biased training, but according to competence.[55] Taylorism, in short, could legitimately assist socialism in power. But was socialism in power?

53 Ibid., 49–51, 56.

54 Ibid., 118–24. From *Der Aufbau der Gemeinwirtschaft; Denkschrift des Reichswirtschaftsministeriums vom 7. Mai 1919* (Jena, 1919).

55 Bauer decree of 19 March 1919 and comment in Gustav Pietsch, 'Das Taylorsystem', *Neue Zeit*, 19 September 1919; Kurt Lewin, *Die Sozialisierung des Taylorsystems* (Berlin, 1921). Naturally, when labour spokesmen looked at the implications for conditions of work, and not the increment to production, they were less happy; Pietsch's complaint – that man is made into a mere cog in the machine – was a typical criticism. By the mid 1920s, however, German and French labour were more willing to accept Taylorite proposals, so long as they did not imply mere speed-up. See Devinat, *Scientific Management*, *passim*.

BETWEEN TAYLORISM AND TECHNOCRACY

The occasional discussion of Taylorism in its narrow sense raised the very issues which marked the more momentous controversies over the economic agencies of the new regime. The more radical Independent Socialists, often the labour rank and file, desired as great a scope as possible for the factory councils that were composed of worker representatives. Trade union leaders looked with distrust upon the efforts to anchor these *Betriebsräte* in the new constitution, but in the wake of the huge strikes of spring 1919 had to acquiesce in this demand. Union leaders, and the quasi-corporatist socialists of the *Sozialistische Monatshefte*, including Max Cohen, preferred so-called parity-based economic governing bodies – i.e. composed equally of workers and employers – along the lines of the Arbeitsgemeinschaft that union leaders and industrialists had worked out of their own accord. At the second congress of *Räte* delegates in April 1919, and then at the SPD congress at Weimar in June, Cohen advocated a Labour Chamber as upper house of the new parliament. The Labour Chamber would emanate from production councils, in which professional men, entrepreneurs, and labour representatives would co-operate to safeguard production and ward off rash nationalization projects. Here was a scheme close in intent to the Rathenau-Moellendorff-Wissell ideas, and indeed Cohen shared their assumptions. Socialism in his eyes meant little more than an easier path to enhanced productivity, and it required a continuing partnership with the entrepreneurs, if not as capitalists, then as industrial experts.[56] In short, his plans really abandoned any significant redistribution of power to the working class, concentrating instead on seeking a harmony that would eliminate the need for socialism or workers control.

Actually, the SPD was so used to arguing in terms of a general democratic commitment to the community as a whole that the

[56] For the Cohen arguments: *Protokoll über die Verhandlungen des Parteitages der SPD abgehalten in Weimar am 15/16. Juni 1919* (Berlin, 1919), 422–8. Cohen was debating against Hugo Sinzheimer, who brilliantly defended the *Räte*, concurred in the idea of a planned economy with priority for the 'needs of the whole community', but wanted no incorporation of *Räte*-based delegates in parliament, lest a chamber of councils degenerate into a mere representation of interest groups. Ibid., 413–16. On the general problem of the *Räte* and the conflicting pressures in early 1919 see Peter von Oertzen, *Betriebsräte in der Novemberrevolution* (Düsseldorf, 1963). For the link between *Räte* and planned-economy notions, cf. Rudolf Wissell, 'Zur Räte Idee', *Neue Zeit*, 30 May 1919, 195 ff.

socialism it did press for had little institutional bite – witness the ineffective Reich Coal Council established in the spring of 1919. Moderate socialists did not really desire a victory of the proletariat at the cost of production, a course that seemed suicidal given the position of the Allies. To consolidate parliamentary democracy, furthermore, was thought the surest way to uplift the working class. Nevertheless, did not Cohen's plans mean the emasculation of any gains in the economic sphere? Enough pitfalls seemed to loom ahead for the SPD to reject the general conceptions of *Planwirtschaft*, first within the cabinet in July 1919, then finally at the Kassel congress in late 1920. As Karl Landauer admitted, according to organizational criteria, industrial rationalization and planning might appear a step towards socialism. Misleadingly so, however; for without working-class power such planning institutions would only rationalize capitalism.[57]

Here in fact lay the seductiveness of planning for many of the socialists to begin with. From Hilferding's 1910 analysis of finance-capital on, the era of capitalist cartellization and concentration was interpreted as a transformation of the bourgeois economy which might temporarily postpone its collapse but would ultimately render at least the economic transition to socialism all the easier. Even Lenin had accepted this view, and thereafter was also able to endorse a stage of state capitalism that amounted to trustification under proletarian auspices. To Russian observers, in fact, the German war economy, with its nascent planning, actually embodied the economic aspect of the transformation. In his first months of power, Lenin openly endorsed Taylorism as a means to reinforce Soviet power. His economic advisers, Milyutin and Larin, drew explicitly on the notions of Rathenau and Moellendorff; and continuing into the twenties, industrial trusts under Bolshevik command served as a flexible framework within which

[57] For rejection of Wissell's proposals on *Planwirtschaft* see the cabinet meeting of 8 July, 'Alte Reichskanzlei, Kabinett-Protokolle', National Archives German Foreign Ministry Films 1349/742683–731; also the National Assembly session of 28 July: *Verhandlungen der verfassungsgebenden deutschen Nationalversammlung*, Bd. 328, 1848 ff. For Karl Landauer's comment see 'Planwirtschaft. Ein Nachwort zum Parteitage', *Neue Zeit*, 10 December 1920, 249–56. For a recent socialist view making the same point see Wolfgang Abendroth, 'Die Alternative der Planung: Planung zur Erhaltung des Spätkapitalismus oder Planung in Richtung auf eine klassenlose Gesellschaft', in *Antagonistische Gesellschaft und politische Demokratie* (Neuwied, 1967).

BETWEEN TAYLORISM AND TECHNOCRACY

to reorganize a shattered economy.[58] What was crucial, however, was that Lenin had seized effective power before instituting steps towards planning; he had settled the central *kto-kogo* (who-whom) question, and this the German Social Democrats had not done. Taylorism and planning could indeed serve Soviet rule, but were no substitute for it; communism, as Lenin said, might be Soviet power plus electrification – but not electrification alone.[59] Finally, if in the West Taylorism was incorporated into ideologies that denied the necessary existence of class conflict, in Russia it could be accepted precisely because that conflict had been decided and a new era of relationships had opened.

What happened when the Left embraced the utopias of productivity before securing power was demonstrated by the sequel to planning in Germany. Cohen's idea for a Chamber of Labour and Production Council finally was incorporated in the Constitution in a compromise form. A Reich Economic Council, *Reichswirtschaftsrat*, which would group employee, employer, and public representatives and would advise the parliament on legislative proposals, was to crown a pyramid of economic advisory organs. It found its strongest champions among those bourgeois democrats, such as Georg Bernhard, who wished to keep labour in a partnership of moderation. Nevertheless, Bernhard was not entirely happy. He had endorsed Cohen's original plans for the same reason the Left had rejected them – as a step beyond the idea

[58] V.I. Lenin, 'The Taylor system, the last word of capitalism in this respect, like all capitalist progress, is a combination of the subtle brutality of bourgeois exploitation and a number of its greatest scientific achievements in the field of analysing mechanical motions during work, the elimination of superfluous and awkward motions, the working out of correct methods of work, the introduction of the best system of accounting and control, etc. The Soviet Republic must at all costs adopt all that is valuable in the achievements of science and technology in this field. The possibility of building Socialism will be determined precisely by our success in combining the Soviet government and the Soviet organization of administration with the modern achievements of capitalism. We must organize in Russia the study and teaching of the Taylor system and systematically try it out and adapt it to our purposes.' 'The Immediate Tasks of the Soviet Government', *Izvestia*, 28 April 1918; translated in V.I. Lenin, *Selected Works* (2 vols., Moscow, 1947), II, 327. For the economic policies of Lenin and his advisers, and the question of trusts and planning, see E.H. Carr, *The Bolshevik Revolution* (3 vols. London, 1950–3), II, 86–95 on state-capitalism, 109–15 on productivity and Taylorism.

[59] Ibid., II, 360–75, on the origins of planning and its relationship to the dominant class.

of nationalization towards achieving a 'real equalization of the producers – entrepreneurial producers and labour producers'.[60] What had emerged, however, was a compromise that the SPD had permitted in order to appease the discontents of those working-class elements who wanted an autonomous role for the *Räte*. Rather than submerge all the workers councils into parity committees, the plans for a hierarchy of economic organs amalgamated *Räte* delegates and industry's representatives only at the institutional summit of the projected system:

Nothing more of any Building of Production is to be discovered here. All ideas seeking to create a new professional ethos in the worker and to summon him to co-operation in productive labour to construct our national economic life have been rejected. In place of a productive socialism the fulfilment of an old union propaganda demand has intervened. In place of a reconciliation of the antagonisms between employer and employee on behalf of a common fruitful cooperation in the service of the enterprise, the old wall between workers and entrepreneurs has been thickened.[61]

In fact, Bernhard had no need to worry. The subordinate economic councils never came into existence; hence the *Räte* influence never reached the Reichswirtschaftsrat directly, and – as its records in Potsdam reveal – the latter rarely progressed beyond stalemate and paralysis.[62]

The final pathetic testimony to the frustrations of the German Left in their invocation of productivity was provided by the renewal of the coal nationalization controversy in mid-1920. Labour reiterated its demand for public control, but was outmanoeuvred largely by the very institutional vagueness their acceptance of production as a goal entailed. Rathenau's scheme for self-administration became the main proposal for discussion, but

60 Georg Bernhard, *Wirtschaftsparlament* (Vienna, 1923), 42. For an English discussion of the Economic Council: Herman Finer, *Representative Government and a Parliament of Industry. A Study of the German Federal Economic Council* (London, 1923).

61 Bernhard, *Wirtschaftsparlament*, 46.

62 The Reichswirtschaftsrat functioned actively from 1921 through 1923 and was thereafter restricted in role. Though not a policy-making body, it provided a forum for argument and testimony, usually splitting over reports favourable to industry and those welcome to labour; and could therefore usefully delay proposals that internally-divided ministries wished to cool off. For its records: Deutsches Zentralarchiv, Potsdam, Aktenbestand 04.01; cf. C.D.H. Hauschild, *Der vorläufige Reichswirtschaftsrat 1920–1926* (Berlin, 1926).

BETWEEN TAYLORISM AND TECHNOCRACY

where in the suggested structure of self-administration would authority rest? And who would guard the guardians? Even Hugo Stinnes and his collaborator Paul Silverberg managed to exploit the flexibility of self-administration schemes by presenting a grandiose programme that would have given public authority to their own enterprises – all in the name of productivity and the common good![63]

The Stinnes plans of 1920 revealed the dangerous institutional ambiguity that all the productivist ideas for transcending class conflict incorporated. Stinnes' insistence upon the necessity of increasing output, which actually meshed with his own interests, represented the logical end-point of the moderates' ideal of productivity. What had in fact occurred was that the original stress on engineering by a Moellendorf had subtly evolved into an emphasis primarily upon corporate reorganization. Like the scientific-management enthusiasts in America, Stinnes, and even Rathenau – for despite their celebrated clashes they shared many key attitudes – wanted private networks of producers to form the nuclei of public authority. To be sure, Rathenau intended a more truly public commitment than Stinnes did; but institutionally his schemes did not guarantee this to any greater degree, for they depended upon little more than a moral commitment to community. Moreover, what was now the substance of these organs for planning and production was no longer technical expertise, but financial manipulation. The two aspects of enterprise that Veblen, for one, had always separated, now became fused in business conceptions in Weimar Germany – and with the exploitative results Veblen had feared. In the last analysis, however, this development was not confined to Germany alone: it was to remain a central ambiguity in all conceptions of technocracy or planning that devolved authority upon private interests. Whether in fascist Italy or liberal Weimar, those who by invoking industrial utopias sought to deny the relevance of power, subordinated themselves to those who

[63] I have drawn here on my unpublished Ph.D. dissertation, 'The Strategies of Bourgeois Defense, 1918–1924: A Study of Conservative Politics and Economics in France, Germany, and Italy' (Harvard University, 1966). Major sources include *Verhandlungen der Sozialisierungs-Kommission für den Bergbau im Jahre 1920* (2 vols., Berlin, 1920); Bundesarchiv Koblenz, Paul Silverberg Nachlass; Bundesarchiv Koblenz, Reichskanzlei papers, 'Verhandlungen des Unterausschusses der Sozialisierungsfrage', R 43 I/ 2114.

really had power, political or economic. But perhaps that was what they actually desired.

*Fordism and the rationalization of capitalism*

With the period of 'stabilization' in Europe came significant changes in the ideological implications of industrial productivity. Here only the salient transformations can be indicated. From mid-decade it was the German-elaborated concept of 'rationalization' that dominated discussions of scientific management. Rationalization focused upon enhancing productivity and technical efficiency, but above all it was associated in Germany with extensive corporation activity – the formation of new cartel-like arrangements upon the ruins of such fragile vertical configurations of the inflation period as the Stinnes empire. German spokesmen, however, still credited the United States with originating the underlying ideas; and American businessmen, such as Edward Filene of Boston and his Twentieth Century Fund, continued to sponsor international studies and congresses to advance scientific management.[64] But the favoured images of advanced techniques that America presented to the world were changing. The teachings of Taylorism, in its strict sense, were viewed more critically, while Fordism became the vogue. A German commentator explained the change as a widening of scope: while Taylorism concerned only the management of labour, Ford's doctrines stressed reorganization of the entire productive process.[65] In part this was a rationaliza-

[64] See Devinat, *Scientific Management in Europe*, esp. preface and 63 ff. National organizations of importance included the Reichskuratorium für Wirtschaftlichkeit, the Masaryk Labour Academy in Prague, the Institut Solvay in Brussels, dedicated to the 'productivist', Saint-Simonian views of its founder, the Russian 'Time Leagues' and the All-Russian Scientific Management Conference, and the Italian Ente Nazionale per l'Organizzazione Scientifica (ENIOS). The best overall survey for rationalization in its European home is Robert Brady, *The Rationalization Movement in German Industry* (Berkeley, 1933). Congresses were held at Prague in 1924, Brussels 1925, and Rome 1927; an International Management Institute was formed in Geneva in January 1927.

[65] G. Briefs, 'Rationalisierung der Arbeit', in: Industrie- und Handelskammer zu Berlin, *Die Bedeutung der Rationalisierung für das deutsche Wirtschaftsleben* (Berlin, 1928), 41: 'Only one name need be cited here, that of Ford, who developed American rationalization of labour beyond its Taylorite excess and built it into the larger rhythm of the flowing production process. . . . If in Taylor the unveiled profit idea is dominant, with Ford it reigns only within the limits of the idea of social service.' Cf. Ernest Mercier: 'what the average European understands by "Taylorism" tends towards a rigid doctrine that

BETWEEN TAYLORISM AND TECHNOCRACY

tion, for Taylorism, too, had earlier been interpreted in the widest sense. But now for practical reasons European apostles of scientific management and rationalization chose to confine Taylorism to its original concern with labour efficiency, thus limiting its utopian implications at the same time. Conversely, the contributions of Ford – the moving assembly line, standardization, and the enlargement of a mass market by low prices and high wages – were seized upon to prove the social potential open to capitalism and large-scale industry, as they existed. Paradoxically, Ford's images of abundance best served the bourgeois-conservative, often Malthusian ends of European business and industry in the later 1920s. Fordism, in sum, offered a technological élan for the beneficiaries of the economic system that Taylorism could not safely provide.

The change arose in large part from the general economic situation of the later 1920s. Currency stabilization and revaluation, accompanied by sharp, if brief, deflationary pressures, characterized much of the period. Although national incomes rose, the increases were associated with severe sectoral or class dislocation: American agriculture and the European coal and steel industry were burdened with excess capacity. More ominously, the fear that the market would be saturated acted as a major spur to rationalization in Europe, with its emphasis upon cutting factor costs, including that of labour. A French spokesman for rationalization pointed to the 'notable diminution of the internal market', and the threat of rigorous foreign competition for customers at home and abroad. A German advocate wrote that the benefit of scientific management to labour must always be smaller in Germany than in the United States, for if America could produce primarily for her own domestic market, German goods had to be competitive abroad and consequently wages had to remain low.[66] The coal and steel men

industrial practice has abandoned in many cases to adopt a more supple solution'; he added that the last word in scientific organization was symbolized by *travail à la chaine*, i.e. Ford's assembly line. See 'Les conséquences sociales de la rationalisation en France', in *L'aspect social de la rationalisation*, Redressement Français (Paris, 1927). On the other hand, one commentator was willing to put both men in perspective: 'No matter how important the influence of a Ford or a Taylor, what is it compared to that of a Luther or Rousseau?' A. Verdurand, 'L'homme d'affaires et la France', *Revue de France*, 31 December 1927, 618. For the doctrine that inspired all this, as touched up by subordinates, see Henry Ford, *My Philosophy of Industry* (1921). For a survey of German responses to Fordism see Peter Berg, *Deutschland und Amerika 1918–29* (1963), 96–132.

66 Auguste Detoeuf, *La réorganisation industrielle*, Redressement Français

of the continent were engaged in long and wearying negotiations to stabilize market quotas without price competition.[67] All this tended to reorient the thrust of scientific-management ideas. No longer an economic policy that promised a radical reorganization of society with gains for all, the emphasis on engineering and scientific management took on what we have termed a redistributionist or zero-sum role. Despite protestations to the contrary, rationalization entailed an effort, first to subordinate small producers to large-scale industry, second to reduce the percentage claims of labour upon output as a whole. Fordism justified these policies as a commitment to abundance. In American practice often conservative, Taylorism's stress on the technocrat had a disquieting potential for subversion; Fordism refurbished the entrepreneur directly.

The new conservative role for scientific management doctrines was certainly evident in Germany and Italy. The enthusiasm for rationalization in the Weimar Republic accompanied the four-year political domination of the conservative and bourgeois parties, not the governments with SPD participation. It was prefaced by a stabilization crisis which brought a contraction of credit that not only spurred industrial concentration but encouraged harsh measures against the trade union gains of 1918.[68] In Italy, during an era of strait-jacketing of labour, manufacturers organized one of the most active cadres for scientific management in Europe – ENIOS – and played host to a congress of like-minded associations in Rome in September 1927. In practical terms, it aimed at the subordination of the many small producers to the large firms and the ascendancy of *Confindustria* policies within the councils of state.

Italian rationalization took place within a context analogous to the German; it accompanied a government shift to protectionism and a deflationary reconversion to the gold standard. In such a transition, with its own liquidity crisis, concentration of industries

(Paris, 1927); Bruno Birnbaum, *Organisation der Rationalisierung Amerika-Deutschland* (Berlin, 1927), 70–1.

67 Valuable documentation on these parleys is provided by the National Archives, German Foreign Ministry films: L 177, Handakten Min. Dir. Ritter.

68 See Ludwig Preller, *Sozialpolitik in der Weimarer Republik* (Stuttgart, 1949), 294–316; Hans-Hermann Hartwich, *Arbeitsmarkt, Verbände und Staat, 1918–1933* (Berlin, 1967), *passim*; for the employers' viewpoint: Hermann Buecher, *Finanz-und Wirtschaftsentwicklung Deutschlands in den Jahren 1921 bis 1925* (Berlin, 1925), 41–53.

and pressure on wages was a logical response; nor were the fascist labour organizations likely to resist. *L'organizzazione del Lavoro*, therefore, tended to add up to a search for methods of cost-cutting by which the major metallurgical and electrical industries might survive advantageously under changed conditions.[69]

This general redirection of emphasis characterized France and Britain too in the later 1920s. With the contradictory policies of a divided Left between 1924 and 1926, discontent with the French parliamentary regime became more pronounced. The American image of technological expertise contrasted with the spectacle of floundering policies at home; furthermore, it reinforced the drive for consolidation and overhaul on the part of spokesmen for the new large-scale mechanized industries. The most typical product of the growing vogue for Americanism was probably Ernest Mercier's *Redressement Français*. This association, founded during the last protracted agonies of the Cartel des Gauches, represented an effort to form a directing elite of economic experts supposedly above party politics, a cadre for institutional and industrial modernization.[70]

The Redressement recapitulated the themes inherent in the quest for productivity. America again loomed as a model of class collaboration and, thanks to Secretary of Commerce Hoover, of efforts at standardization and elimination of waste.[71] Even if there were no 'definitive solution to the social question', Mercier noted, the United States had achieved far more than 'a simple truce'; and another writer stated his conviction 'that there are formulas of economic and social union which enrich the whole of a country without impoverishing its poorest elements'.[72] Rationalization, its

69 Rosario Romeo, *Breve storia della grande industria in Italia* (Bologna, 1963), 153–6; Felice Guarneri, *Battaglie economiche tra le due grandi guerre* (Milan, 1953), 113–39, 146–59; Fiorentina, loc. cit., 137–45.

70 R.F. Kuisel, *Ernest Mercier, French Technocrat* (Berkeley, 1967), 45–88, provides the basic discussion of the Redressement that I have drawn upon.

71 Ernest Mercier, *La production et le travail*, Redressement Français (Paris, 1927), 10–16. Mercier had visited the US in 1925 and met Filene, Dennison, and other enthusiasts of scientific management. Cf. Detoeuf, *La réorganisation industrielle*, 67–80, on American rationalization. For the link with Saint-Simonian themes, cf. E.S. Mason, 'Saint-Simonism and the Rationalization of Industry', *Quarterly Journal of Economics*, August 1931.

72 Mercier, *La production et le travail*, 25; Detoeuf, *La réorganisation industrielle*, 1. Among social consequences, of course, was possible unemployment, but both authors thought it would be only transitory: Mercier, 'Les conséquences sociales', loc. cit., 16, 41–2; Detoeuf, 41–2.

advocates believed, or at least claimed, 'promised a real social revolution'.[73] Throughout French industry enthusiasm for American neo- or super-capitalism marked the late 1920s; the years that saw Lindbergh vault the Atlantic witnessed a growing adulation for Fordism as well. André Tardieu, who filled the technical ministries in the Poincaré cabinet of 1926–9, pledged the government to ship and road construction. When he succeeded to the premiership himself, he declared a five-year programme for 'national retooling'.[74]

Despite the rhetorical promises of social revolution and the sweeping vistas of modernization, the ideological implications of the new fashion were actually far less radical than those of the movements earlier in the decade. The political and economic analysis of a Mercier was obviously unlikely to indulge in any anti-capitalism. Spokesmen of the Redressement condemned not a parasitic financial network but the inefficiency of the traditional small producer. Their calls for concentration implied primarily an effort to take over middle-level manufacturers. An artisanate whose semi-luxury trades did not encroach upon industrial production could be praised as a valuable sheet-anchor of French social stability, but the small factory that resisted centralization and standardization was allegedly a threat to progress.[75] Furthermore, even the technological imagery of the rationalization movement in France was not without its pastoralism. If in Soviet Russia vast hydroelectric projects stood for revolutionary transformation, industrial and political leaders in France praised grid development for saving the small domestic producers of the countryside and slowing down rural depopulation.[76] Indeed, Mercier's own social analysis reflected a bourgeois conservative

[73] Mercier, 'Les conséquences sociales de la rationalisation', 19.

[74] Rudolph Binion, *Defeated Leaders* (New York, 1960), 289–92. For the general enthusiasm: P. Bourgoin, 'La rationalisation', *Revue de France*, 15 November 1929; Pierre-Etienne Flandin, 'Le problème social, *Revue de Paris*, 1 February 1928; Edmond Giscard d'Estaing, 'Le Néocapitalisme', *Revue des Deux Mondes*, 1 August 1928.

[75] On the artisanate, Mercier, 'Les conséquences sociales de la rationalisation', 32. Lucien Romier, editor of the *Journée Industrielle*, then *Le Figaro*, stressed the need to overcome excessive individualism, and was second in eminence within the Redressement. See Kuisel, *Mercier*, 64–5.

[76] On this theme: Maurice-Charles Bellet, *La politique générale de la Fédération Républicaine de France* (Paris, 1924), which saw electricity saving the French rural family, and Detoeuf, *La réorganisation industrielle*, 33.

traditionalism, far from any right-radical *ressentiment*. The masses needed and deserved welfare benefits, but had no capacity to assume direction of the country or of its industrial plant: 'What one expects from the workers appears at first glance most simple. It is a question merely of their understanding and accepting the necessities.'[77] Labour unions had to switch from sterile political agitation to collaboration on restricted professional issues. In general, Mercier's managerial elitism emerged as a new defence of that very traditional French bulwark, the bourgeoisie; and his own parliamentary dream was the venerable Union Nationale of the centre groupings. The social policy of the Redressement faded off into the traditional justifications of capitalism by many prominent business politicians of the late twenties, including André François Poncet and Pierre-Étienne Flandin, some of whom ended up with Vichy.[78]

Rationalization in Europe, therefore, was only a stunted offspring of the American productive vision as originally conceived. It served a conservative business community seeking to exploit, first the transition to overall non-inflationary monetary conditions, then the prosperous but increasingly saturated market of the later 1920s. Gramsci's insight was thus partially correct, when at the end of the decade he wrote, 'What is today called Americanism is in large part the pre-emptive critique of the old strata, who are precisely the ones who will be shattered by the new order, and

[77] Ernest Mercier, *La production et le travail*, 59–60. Cf. Mercier, 'Réflexions sur l'élite', *Revue des Deux Mondes*, 15 February 1928.

[78] On Mercier's politics, Kuisel, *Mercier*, 81. For A. F. Poncet: *Réflexions d'un républicain moderne* (Paris, 1925); for the success of American-style capitalism in solving the social question, Flandin, loc. cit., and Giscard d'Estaing, loc. cit. The business 'moderates' who embraced Fordism can be usefully set off against the more radical syndicalists; for their ideas: René Pinon, 'Les nouvelles conceptions de l'état', *Revue Economique Internationale*, October 1929; Sammy Beracha, *Rationalisation et révolution* (Paris, 1930), esp. 38. Georges Valois was a leading exponent of a syndical system that would replace the parliamentary regime. He had started as a disciple of Georges Sorel, participated uneasily in the Action Française, and organized the Faisceau in the mid-1920s, then broke with the Redressement after 1930. For Mercier's denunciation of syndicalism as 'the Soviet method', Kuisel, 72. For the concept of *synarchie* – a conspiratorial view of an elitist and semi-fascist technocracy – which has marked some French discussions of Vichy, Mercier, the 1930s 'X-crise', as well as Clémentel, see André Ullmann and Henri Ayzeau, *Synarchie et pouvoir* (Paris, 1968). Mercier denounced syndical representation as 'the Soviet method' (Kuisel, 72). For a sample of syndical writings see Sammy Beracha, *Rationalisation et révolution* (Paris, 1930), esp. 38.

are already prey to a wave of social panic, dissolution, and desperation.'[79] In fact, the strata involved were not only the old ones, for they included the most dynamic of the entrepreneurs. Nevertheless Gramsci's conception of social defence was justified. A radical Americanism had arisen as a concomitant of war production – production, after all, without effective price and demand constraints – but in the changed conditions a decade later, its social function altered. In different ways, Tardieu and Hoover (and, it might be argued, even Stalin)[80] took over the most easily manipulated aspects of Americanism, but each came to subordinate its claims as an autonomous social vision to his own ideology. The trajectory of the technological vision ended with the Great Engineer an impotent Depression President and a querulous defender of the propertied classes.

Even at the bottom of the Depression, however, the chiliastic idea of productivity and social engineering could flare again briefly. Howard Scott's Technocracy captured the American imagination in late 1932 and struck a chord in Europe.[81] Scott, an eccentric heir of Gantt and The New Machine, had organized a so-called Technical Alliance in 1921, and sold Technocracy as a messianic prediction of energy utilization. Recalling Veblen's contrast of pecuniary manipulation and industrial production, Scott detached real energy resources from the conventions of the price system. Even as he wrote, he said, a corps of engineers was preparing a huge energy inventory that would prepare the way for an age of fabulous leisure. Because of its messianic promise and its recognition of the disequilibrium between industrial potential and real income distribution, Technocracy did respond to felt needs, but it grew not out of the flush of American economic success,

[79] Gramsci, *Note sul Machiavelli*, 343–4.

[80] For an introduction to the story of American engineers in Russia during the plan, see Peter Filene, *Americans and the Soviet Experiment, 1917–1933* (Cambridge, Mass., 1966); W.H.G. Armytage, *The Rise of the Technocrats* (London, 1965), 219 ff. But cf. the speech of Bukharin to Soviet engineers in February 1932, reminding them that engineers must still subordinate themselves to the dictatorship of the proletariat, not aspire to a technocracy. Cited Dorfman, *Thorstein Veblen and his America*, 514.

[81] On technocracy, Dorfman, 510 ff.; Howard Scott et. al., *Introduction to Technocracy* (New York, 1933); Allen Raymond, *What is Technocracy?* (New York, 1933); A.M. Schlesinger, jr., *The Crisis of the Old Order, 1919–1933* (Boston, 1957). European comment: Erich Kraemer, *Was ist Technokratie?* (Berlin, 1933); Karl Resar, *Technokratie, Weltwirtschaftskrise und ihre endgueltige Beseitigung* (Vienna, 1935).

but out of the crisis of capitalism; it exemplified the quackery of despair, not the vision of triumph. In fact, the conditions of the Depression necessarily undermined all Americanist industrial utopias. Economic contraction destroyed the postulates for class collaboration and discredited the managers of the system. At least until the second world war and its aftermath, America's model of industrial productivity lost its catalytic inspiration. Not that Roosevelt's social experimentation would not attract followers, but the supreme confidence in technology and production, in engineering as social redemption, perished with the other dreams of the twenties.

*This essay was awarded the Klaus Epstein Memorial Prize, given jointly by the Institute of Contemporary History, London, and Brown University, Providence, Rhode Island.*

# [11]



Thomas P. Hughes

# Gropius, Machine Design, and Mass Production*

Die Architekturhistoriker nach Nikolaus Pevsner und Sigfried Giedion unterstreichen immer wieder die linare Entwicklung, die die architektonische Form von den Konstruktionen der Ingenieure des 19. Jahrhunderts bis zu den Gebäuden der modernen Architekten des 20. Jahrhunderts genommen habe. Bei dieser Interpretation kommt jedoch der Einfluß, den die Produktionstechnologien des frühen 20. Jahrhunderts auf die architektonische Theorie und Praxis von modernen Architekten wie Walter Gropius, Martin Wagner, Bruno Taut und Ernst May ausgeübt haben, zu kurz. Die Planung und Errichtung von Neubauten in Deutschland war geprägt von den Massenproduktionstechniken eines Henry Ford und den Methoden der wissenschaftlichen Betriebsführung eines Frederick W. Taylor. Gropius' serienmäßige Fertigung von Häusern in Dessau-Törten in den 20er Jahren ist dafür ein typisches Beispiel. Auch das äußere Erscheinungsbild der Bauten spiegelt sehr stark die Produktionstechniken wider, die die frühen modernen Architekten von Ford, Taylor und anderen modernen Ingenieuren übernommen haben. Ihre Gebäude sahen aus wie maschinell hergestellte Produkte, weil schon ihre Planung mit Maschinen rechnete.

Historians of technology may see facets of Walter Gropius's architecture more clearly than historians of architecture. This is especially true of a new generation of historians of technology who choose to study technological change in a social or institutional setting, an approach to be contrasted with the more traditional emphasis on straightforward description of machines, devices, processes, and structures. Many architectural historians prefer to concentrate on the analysis of architectural forms and to place these in a chronological narrative. These historians of architecture tend to isolate architecture from the social context within which it has evolved.

Historians of architecture, especially those influenced by Nikolaus Pevsner and Sigfried Giedion, take into account technological influences on architecture, but they tend to see technology as materials and construction techniques associated with civil engineering; they do not suitably stress technology as a highly organized means of production grounded in economic principles.[1] Giedion, for instance, in *Mechanization Takes Command,* describes the introduction of the machine and the assembly line into production of goods, but he does not sufficiently stress the economic principles

* Paper presented at the Symposium on the *Social Origin of Modern and Post Modern Architecture* (July 25–27, 1984) held at the Institute for Advanced Study Berlin (Wissenschaftskolleg zu Berlin).

and organizational structures that allowed the assembly line to function. Pevsner and Giedion have the point of view of engineers who concentrate primarily on the technical component of complex technological problems which, in fact, additionally involve economics, politics, and other institutionalized social factors.

In contrast, Walter Gropius and some of his contemporaries of the modern movement were aware, in the early twentieth century, of the nature of technological change. By then technology had taken on the highly complex, institutionalized character it has today. By 1900 business corporations, government bureaus, and military departments presided over technological change. Financiers, managers, and bureaucrats were as deeply involved in technology as engineers, craftsmen, and workers. Technology had become a cultural artifact reflecting the values and institutional structures of the era of high capitalism in the West. Gropius could not have managed great corporations like the Ford Motor Company and Allgemeine Elektrizitäts-Gesellschaft (AEG), but he did have a good understanding of the financial, managerial, and technological principles that shaped the institutional structure and the production strategy of large-scale technological and business enterprises. This knowledge he wanted to apply in the practice of architecture. Gropius had the breadth of vision of an architect-entrepreneur.

To understand the architecture of Gropius, historians cannot, as already noted, simply and exclusively place him in an architectural tradition, even if the tradition is the so-called modern that Pevsner and others traced back into the 19th century. Gropius's references were to the highly organized world of technology and business as well as to the architectural. He was marching to a different drummer than his tradition- and discipline-bound architectural contemporaries. Like so many truly innovative persons, he distanced himself from the on-going institutions and ideas of his field and sought applicable analogies in other realms of human activity. With the notable exception of a few years immediately before and after World War II, Gropius behaved like a technological enthusiast. He believed that technology shaped the modern world more fundamentally than any other historical force, and he was determined to express the spirit of this age in his architectural practice. He declared that architects have always had to come to terms with the technology of their era.[2]

Descriptions of modern technological materials and forms in Gropius's buildings are common; rarely, however, do historians stress Gropius's concern with the means of modern production. Some fail to do this because they are blinkered by the programmatic pronouncements that Gropius and his associates published when they founded the Bauhaus at Weimar. Before the admission of Laslo Moholy-Nagy to the Bauhaus faculty and the influential visits of Theo Van Doesburg in 1923, Gropius and the Bauhaus went through an expressionistic phase that celebrated the presumed values of the medieval, holistic handicrafts.[3] Even before the move in 1925 to Dessau, however, Gropius showed increased interest in modern industrial culture and a growing determination to relate the philosophy and practice of the Bauhaus to it. This shift was not entirely unrelated to the attenuation of the spirit of withdrawal, and of the longing for utopianism stimulated in sensitive souls by the horrors of the war. Doesburg's reminding Gropius of his prewar commitment to industrial architecture especially factory design; the move to industrialized Dessau; the end of inflation; and the growing expenditure of government funds for housing development all reinforced Gropius's renewed interest in the interactions of architecture, technology, and industry. Considering Gropius's subsequent commitment to applying modern technology to architecture and his prewar dedication to industrial architecture, the expressionist and handicraft phase of his career appears the exception, not the rule.

Historians of architecture also fail to appreciate fully Gropius's commitment to modern production technology because they see him through the filter of the memorable image and phrase-maker, Le Corbusier. He so vividly defined the shapes of modern architecture that those who have read him tend to organize their knowledge of the modern movement and of modern architects, such as Gropius, in terms of Corbusian metaphors. First it should be noted that Gropius elucidated concepts and built buildings of a modern character a decade before Le Corbusier published *Vers Une Architecture.*[4] Then it should be stressed that the grain elevators, airplanes, ships, and automobiles portrayed by Le Corbusier were the products of modern technology, not the methods of modern technology. Le Corbusier's juxtaposition of the Parthenon and the Delage »Grand-Sport« car – a memorable image of forms – has wrongly led many architects and historians to define modern technology entirely in terms of form rather than of form and methodology. Le Corbusier did in fact have a substantial interest in mass production techniques,[5] but this can be overlooked in a superficial reading – or seeing – of his published work. If Le Corbusier had used illustrations of a Ford production line and a Taylor plant layout, he would have come closer to the heart of modern matters, at least as understood by Gropius.

Letters, memoranda, and publications testify to Gropius's technological enthusiasm and to his familiarity with the means of modern production. During the mid-twenties he explicated the technology of mass production in his advocacy of the standardization of precision component parts; the variability of the assembly of these; the use of capital-intensive, labor-saving, special-purpose machinery; and the division of labor. He showed his familiarity with the economics of production through his application of cost accounting techniques, his emphasis on unit cost, and his concern about dead capital.

He revealed his awareness of modern managerial practices in the layout of production facilities for housing in accord with scientific management philosophy. He also borrowed such general concepts from modern technology as the notion of invention and patents, research and development in a laboratory setting, and factory production. His complex and coherent philosophy of production was not borrowed from 19th century civil engineers; it was modelled on the strategies and structures of twentieth century, institutionalized, science related, economics permeated, production technology.

Before providing evidence of Gropius's entrepreneurial style as an organizer of the means of mass production, one should acknowledge that the Bauhaus faculty and Gropius's architectural associates influenced him – his was a collective enterprise, not simply a heroic *Siegeslauf.* Other modern architects in Germany including Peter Behrens, Ernst May, Bruno Taut, Martin Wagner, also used modern production methods in their housing construction.[6] These architects undoubtedly influenced one another, but the prime sources of their and Gropius's production techniques were engineers, managers, and financiers who created and presided over great mass production enterprises. The most famous of these, even in Europe, was Henry Ford whose autobiography, *My Life and Work,* was translated and widely read in Germany.[7] »Fordismus« became the popular term for modern mass production. Frederick W. Taylor, the American pioneer in scientific management, enjoyed even greater respect among professionals, including the European architects of the modern school. Germans named his philosophy »Taylorismus.«

In Germany, the theory and practice of engineers and managers at AEG (German General Electric) also significantly influenced the modern architects. But AEG leaders acknowledged their debt to American practice as well. Peter Behrens provided a linkage between industrial practices at AEG and architecture. He was AEG's consultant from 1907 to 1914 on product design, trademarks, worker housing, and physical reorganization, or space utilization. During this period, Walter Gropius, Le Corbusier, and Ludwig Mies van der Rohe worked in Behrens's architectural bureau. Not only did Behrens influence other architects through direct contact, but also through the prominent part he played in Werkbund debates about architecture and technology.

Ford production methods are too well known to require detailed explication here. Massive functional coordination of raw material suppliers, manufacturing processes, and dealer sales to insure a smooth and massive flow of products were hallmarks. Less well remembered today is Frederick W. Taylor's insistence that workers be relegated to the role of programmed component parts in production systems and that these systems be methodically planned and organized to minimize the expenditure of animate and inanimate resources. German commentators on the modern production methods often combined Fordism and Taylorism under the embracing concept of »Rationalisierung.«[8]

In the case of AEG influence, Walter Rathenau, the head of the firm, presented a grand vision of »Rationalisierung« and the political economy of production through a series of books popular among intellectuals. A leading engineer at AEG, Michael von Dolivo-Dobrowolsky, explained mass production at AEG to a select audience.[9] The ideas of Rathenau, Dolivo-Dobrowolsky, and those of other engineering and managerial personnel at AEG probably filtered into the architectural world through Behrens or directly to interested architects such as Gropius. Ideas expressed by the highly influential architect and publicist, Hermann Muthesius, resembled in striking ways the philosophy of manufactures prevailing at AEG and other giant German manufacturing enterprises.[10] As is often the case, general notions or ideas that influence a movement or an age cannot be attributed to a single source because they are widely shared and broadly publicized. The point to be stressed here, however, is that the breeding ground of ideas and practices applied by Gropius and other architects of similar mind was the world of contemporary technological affairs.

Gropius elucidated his views on modern production technology in several publications. Among the most explicit was an essay first published in 1924 in the *Berliner Tageblatt.* Entitled simply »Wohnhaus-Industrie« the article predicted a revolution in production techniques and related architectural style. The latter depended on the former. Unlike Le Corbusier, Gropius did not say that a house was a machine to live in; he believed that a house was a machine-made product. At hand, Gropius insisted, were the technological means to fulfill the age-old dream of an inexpensive, attractive, and healthy house for everyone. Inertia and sentimental attachments, however, had prevented the production of houses in a way similar to that in which Henry Ford had produced an inexpensive, attractive automobile. Not far off, Gropius predicted, was the day when houses like shoes could be ordered from a catalog. Also forthcoming should be mobile homes for modern transient people. Gropius wanted factories to produce standardized, interchangeable house components that could be assembled rationally into various combinations or forms. In automobile parlance, there would be different models of houses, but variation from standardized components was limited – the consumer could have any color house he wanted as long as it was white.

In a striking contrast to handicraft techniques, Gropius advocated using large, capital-intensive, labor-saving machines in housing factories. He wanted materials to be used that especially suited machine manufacture. Gropius understood that both material and form had to be adapted to machine production, not on aesthetic grounds, but because machine characteristics determined to a large degree product characteristics. Plane surfaces and right angles, for instance, suited many production machines better than undulating and convoluted shapes. This was commonplace knowledge among design engineers; some architects had a difficult time grasping the concept. Gropius, more aware, pointed out that the art of architecture ought to be practiced within the range of variation that machine production allowed.

Historians of architecture have often confused the forms shaped by machine technology with forms borrowed from the shapes of machines. Gropius did not. An architectural form modelled on machine forms can resemble an architectural form designed for machine manufacture, but deriving forms from the method of manufacture is more original and fundamental than simply imitating machine design. Machines look like the products they make because machines are also designed for machine manufacture. Architects familiar with industrial practices, like Peter Behrens and Walter Gropius, were inspired not only by bridges, railway stations, and other engineering works or products; they were also moved by the methods that the engineer used to construct these.

If his dependence on the concepts of the technological world is not sufficiently evident from the preceding remarks, one should consider Gropius's call for a research and development laboratory. He recommended the establishment of experimental laboratories staffed by architects and engineers and influenced by businessmen; laboratories that would invent and develop the building materials, standardized building components, and production machines required by the new housing industry. On occasion, Gropius and his associates spoke of patents for invention being taken out by architects. After the move of the Bauhaus to Dessau, Gropius described it as an experimental laboratory for the housing industry.[11]

In 1924 Gropius, Martin Wagner, Bruno Taut, and Ernst May borrowed from industry the organizational concept of a *Kopfgemeinschaft* (oversight consortium) and founded DEWOG (Deutsche Wohnungsfürsorge A.G. für Beamte, Angestellte und Arbeiter), an enterprise dedicated initially to the invention, design, and development of experimental houses that could become models, or exemplars, for mass production. Obviously this was analogous to the laboratory model or semiworks developed by industrial scientists and engineers for the same mass production goal. Each of the participating architects agreed to design and construct an experimental, or test, house and to share with the others the experience gained and the income derived from patents. The members of the consortium explicitly defined economic and technical parameters. The house designs should suit a mode of production that would be large scale, rationalized, and systematized. DEWOG circulated a memorandum stating economic principles of production that would be observed. The goals were intensive labor saving, full plant utilization, economic material handling, and vertical integration of production facilities.[12]

After the move of the Bauhaus, Gropius discussed with Hugo Junkers, the Dessau airplane manufacturer, the possibility of establishing a house building factory. In a proposal, Gropius estimated the capital requirements of the initial outlay as 220,000 marks, and apportioned it among administrative costs, factory buildings, and machines. (Gropius seems to have been especially interested in acquiring cranes and a motor-driven truck.) He specified a factory employing twenty men with a capacity to produce 1000–1200 houses over a five-year period. In this modest proposal, Gropius was limiting production to the Dessau market.[13]

In 1926 Gropius secured the opportunity to fulfill his long-standing desire to become a production entrepreneur when the city of Dessau commissioned him to plan and supervise the construction of a housing project in Törten, a suburb of Dessau. The plan culminated in over 300 houses between 1926 and 1928. Gropius designed several basic house types, all of which were suited to modern production methods and intended for worker families. The plan and project gave Gropius and his associates the occasion to apply economic, technical, and managerial concepts expressed in his programmatic essays, in DEWOG memoranda, and in housing factory plans. The National Society for Research into Economic Building and Housing encouraged him and provided financial support, as it did other architects who applied modern production techniques.[14]

The most effective way of showing Gropius's theories as practice is through diagrams and photographs of the Dessau project. These reveal the use of low-cost materials, special purpose machinery, and flow diagrams and time charts to coordinate machinery and labor in a dynamic process. The construction site layout also brings to mind Taylorismus and Fordismus.

Gropius and his associates took inordinate pride in the machines used in component manufacture and construction on the site. They were especially pleased by the installation of a large crane and of rails for local transportation of materials. A machine for on-site manufacture of iron-reinforced beams *(Betonrapidbalken)* also delighted the machine enthusiasts. Along with these, there were concrete mixers, stone crushers, and building-block machines. As Gropius had explained on numerous occasions, the size of the project made possible the use of machines, for the investment could be spread over many housing units.

Existing manufacturers supplied the machines, but probably according to specifications of the Dessau-Törten planners. On the other hand, the layout of the on-site construction and manufacturing facilities were innovations. The layout resembled the ground plans of an assembly-line production facility. Gropius and the contractors organized the various machines, material-storage areas, and railways systematically around, and between, a parallel line of row houses. There were eight building-block machines for making the hollow blocks from various materials, including cinders, and these machines were moved from one group of eight houses to another. Appropriately located storage bins fed raw materials such as rock, sand, cement, and cinders to the machines. The most complex of the machines, the roof-beam maker, had the primary position at the head of the building site, and rails spread out from it to the various houses under construction. Gropius said that the scheduling of production, delivery, and work resembled that used by large railway management. There were numerous charts scheduling the movement of machines, materials, and work crews from one set of houses to another.

The fact that Gropius designed for machine production and mass production is of considerable consequence for the interpretation of the social origins of modern architecture. It also throws postmodernism into perspective. Gropius and other moderns designed under the influence of the modern means of production because their goals were similar to those of the engineers, managers, and financiers who invented, developed, and organized modern mass-production technology – the production of low-cost products for a mass market. The market for mass housing in Weimar Germany was a major source of commissions for the modern architects. The list of housing developments in Berlin, Frankfurt, and elsewhere in Weimar Germany is a long one. The analogy between a mass-produced Ford automobile and mass-produced products in which to live was too obvious to be missed by architects whose objectives were similar to those of Ford. Like Ford, they were deeply influenced not only by the technical aspects of modern production but by the social and economic principles shaping it as well.

## Conclusion

The point of view taken in this interpretation of the origins of modern architecture has stressed factors – especially modern technology – in the historical context in which the movement emerged. Relatively little has been said about the architectural tradition against which the architects of the modern movement reacted. If an analogous point of view were taken with regard to the postmodern movement, then emphasis would also be on the historical context. Essays on postmodernism usually stress, in contrast, the reaction of the movement against the modern movement, hence the name »post-modernism.« It would be worthwhile to search for influential contextual factors shaping postmodernism that have had an effect comparable to that of the modern means of production, or technology, on modern architecture.

# Footnotes

1 Nikolaus Pevsner, *Pioneers of Modern Design* (New York: Museum of Modern Art, 1963). Sigfried Giedion, *Space, Time and Architecture* (Cambridge, Mass.: 1941).
For an exceptional essay that does stress the means of production, see Karin Wilhelm, »Von der Phantastik zur Phantasie: Ketzerische Gedanken zur »Funktionalistischen Architektur,« in Neue Gesellschaft für Bildende Kunst, *Wem gehört die Welt: Kunst und Gesellschaft in der Weimarer Republik* (Berlin: NGBK, 1977), pp. 72–86.

2 Walter Gropius, »Monumentale Kunst und Industriebau,« typescript (37 pages) of an address given in the Folkwang Museum, Hagen, Westfalen, 29 January 1911, pp. 9–11. File so titled in Bauhaus Archiv, Berlin. Hereafter cited as B.A. Ideas expressed by Gropius in this address reflect in many instances those of Peter Behrens in »Art and Technology,« a lecture delivered at the 18. Jahresversammlung des Verbandes Deutscher Elektrotechniker, Braunschweig, 26 May, 1910 and reprinted in Tilman Buddensieg, *Industriekultur: Peter Behrens and the AEG, 1907–1914,* trans. Iain Boyd Whyte, (Cambridge, Mass.: MIT Press, 1984), pp. 212-19.

3 Reyner Banham, *Theory and Design in the First Machine Age* (Cambridge, Mass.: MIT Press, 1980), pp. 276–85.

4 On Gropius's early views on industrial architecture and his Fagus Werk, see Karin Wilhelm, *Walter Gropius Industrie Architekt* (Braunschweig / Wiesbaden: Friedr. Vieweg & Sohn, 1983).

5 Stanislaus von Moos, »Le Corbusier und Gabriel Voisin,« in *Avant Garde und Industrie,* hrsg. v. Stanislaus von Moos and Chris Smeenk (Delft: Delft University Press, 1983), pp. 77–103.

6 Justus Buekschmitt, *Ernst May* (Stuttgart: Verlagsanstalt Alexander Koch, 1963), pp. 33–57; Kurt Junghanns, *Bruno Taut, 1880–1938* (Berlin: Henschelverlag Kunst und Gesellschaft, 1970) pp. 61–86. Rosemarie Bletter of Columbia University, N.Y. is preparing a book on Taut.

7 *Mein Leben und Werk* (1923) sold more than 200.000 copies. Jost Herman und Frank Trommler, *Die Kultur der Weimarer Republik* (München: Nymphenburger Verlagshandlung, 1978), pp. 52–54.

8 Otto Bauer, »Rationalisierung und Gesellschaftsordnung,« in *Kapitalismus und Sozialismus nach dem Weltkrieg* (Berlin, 1931) reprinted in *Rationalisierung 1984,* Ausstellung Katalog Staatliche Kunsthalle Berlin und Neue Gesellschaft für Bildende Kunst (Berlin, 1984), pp. 117–126.

9 Michael von Dolivo-Dobrowolsky, »Modern Mass Production in the Electrical Appliance Factory of the AEG,« a lecture delivered in AEG Conference Hall, 17 January 1912 and reprinted in Buddensieg, *Industriekultur,* pp. 256–61.

10 Hermann Muthesius, »Handarbeit und Massenerzeugnis,« *Technischer Abend im Zentralinstitut für Erziehung und Unterricht,* IV (1917), a 30 page essay.

11 Walter Gropius, »Wohnhaus-Industrie,« *Berliner Tageblatt,* 24 September 1924. See also Gropius, *Scope of Total Architecture* (New York: Harper, 1955), pp. 128–35. Gropius to Schumann Haus- und Küchengeräte, Berlin, 11 July 1925. Korrespondenz W. Gropius über Rationalisierung im Bauwesen. B.A. »Arbeitsprogrammskizze für einige Versuchswohnhäuser,« a 6 page typescript dated 29 December 1924. Korrespondenz 1924/26. Dewog. B.A.

12 »Arbeitsprogrammskizze für einige Versuchswohnhäuser.«

13 Gropius, »Plan zur Errichtung einer Wohnhausfabrik.« Undated manuscript (5 pages) with a pencil notation »to Junkers, Dessau.« Manuskripte verschiedener Bauwirtschaft, 1922–1934. B.A.

14 Reichsforschungs Gesellschaft für Wirtschaftlichkeit im Bau- und Wohnungswesen e.V., *Bericht über die Versuchssiedlung in Dessau.* Sonderheft 7, Gruppe IV, Nr. 4, 2. Jahrg. April 1929. Walter Gropius, *Bauhausbauten Dessau,* Bauhausbücher 12 (München: Albert Langen, 1930), pp. 152–200. See also, Reginald Isaacs, *Walter Gropius: Der Mensch und sein Werk* (Berlin: Gebr. Mann, 1983), I, 383–5.

# [12]

# "Reign of the Robots": The Homer Laughlin China Company and Flexible Mass Production

REGINA LEE BLASZCZYK

In 1936, a trade journalist writing for *Ceramic Industry* described recent technological innovations at Homer Laughlin China Company of Newell, West Virginia. The writer asserted that the craft potter "of poem, drama, and tale" was disappearing during "a new regime of push-button, sprocket, and bearing"; the artisan's realm had been invaded by "automatons" that were "streaked with flashes of electricity and backed with subtle overtones of science and research." Among the many awe-inspiring devices was a new glazing machine, which was designed to eliminate the hazardous, time-consuming manual task of dipping ware into vats of lead glaze. Instead, a ribbon belt conveyor fed plates through a spray booth, flipped them over, and sprayed them on the reverse at a "breathless" rate of 150 pieces per minute. Another appliance, the Iron Horse, moved people, supplies, and products in the name of efficient kiln management. At Homer Laughlin, the mesmerized journalist contended, the "reign of the robots" had begun.[1]

*Ceramic Industry*'s hyperbolic story announced major technical changes along Fordist lines at the world's largest pottery. Long estab-

DR. BLASZCZYK is assistant professor in the Department of History and the Program in American Studies at Boston University. A version of this article presented at the 1993 meeting of the Society for the History of Technology in Washington, D.C., won SHOT's Robinson Prize. The author thanks David A. Hounshell, Anne M. Boylan, Glenn Porter, Philip Scranton, Susan H. Myers, Steven Lubar, Bonnie Lilienfeld, and Lindy Biggs for assisting with this article in innumerable ways. Special thanks go to executives at the Homer Laughlin China Company, including Marcus Aaron II, Joseph M. Wells III, Joseph M. Wells, Jr., Jonathan O. Parry, and Ken McElhaney, who generously provided access to their factories, staff, and archives, to Carl C. Mooney and Willis G. Gaston, who shared their knowledge of mechanical engineering in the pottery industry, and to archivists at Kent State University.

[1]"Reign of the Robots," *Ceramic Industry* 27 (December 1936): 408–13; Walter Howard Emerson, "We Move the Work and the Worker," *Ceramic Industry* 23 (September 1934): 138–42, on the Iron Horse.


0040-165X/95/3604-0006$01.00

lished as the American ceramics industry's low-cost, high-volume producer, Homer Laughlin's reputation for innovation was virtually unrivaled among tableware manufacturers at home or abroad. This West Virginia firm was among a handful of American potteries specializing in quantity production and experimenting with mass-production methods at a moment when many contemporaries still exclusively used conventional batch-production techniques. The company survived by catering to working-class and middle-class consumers who acquired their household ceramics as premiums with purchases of packaged cereal, from mail-order catalogs, or at five-and-dime stores. Serving mass retailers and their customers was the name of the game at Homer Laughlin. From the 1920s through the 1940s, the firm's managers undertook an expansion and renovation program designed to strengthen their alliances with mass merchandisers and to safeguard their reputation as potters to "Her Majesty—The American Housewife."[2] With the objective of meeting the demand for more goods and varied goods, Homer Laughlin's managers selectively introduced Fordist methods to their Ohio River valley industrial site, deliberately shielding certain craft operations against mechanization.

For decades, historians of manufacturing focused their studies on companies and industries that fit Alfred D. Chandler, Jr.'s paradigm of big business wherein mass- and other quantity-production practices represent a pinnacle of technological achievement.[3] But in the 1980s, with the automobile industry in financial straits, scholars from various disciplines—economists, business theorists, sociologists, and historians—began challenging the dominant model of big business and mass production. Leading the quest for new approaches to business organization and manufacturing technologies, labor economist

[2]Joe Dickey, "Industrial Welfare Department, Homer Laughlin China Company," *Potters' Herald* (East Liverpool, Ohio) (January 22, 1925), p. 4. In claiming the patronage of "The American Housewife," Homer Laughlin's managers were distinguishing their company from Josiah Wedgwood & Sons, Ltd. (and predecessor firms), which appealed to status-conscious middle-class and upper-middle-class consumers by advertising itself as "Potters to Her Majesty, the Queen," a title granted to founder Josiah Wedgwood by Queen Charlotte in the 18th century. In contrast, Homer Laughlin's managers acknowledged their debt to working-class and middle-class consumers, calling their firm "Makers of American Pottery for the American People." See Homer Laughlin China Company (hereafter HLCC), *Catalogue and Price List, April 1, 1901*, p. 8, Newell, West Virginia, HLCC, Executive Files (hereafter HLCC EF).

[3]Alfred D. Chandler, Jr., *The Visible Hand: The Managerial Revolution in American Business* (Cambridge, Mass., 1977); David A. Hounshell, "Hughesian History of Technology and Chandlerian Business History: Parallels, Departures, and Critics," *History and Technology* 12 (1995): 205–24.

Michael J. Piore and sociologist Charles F. Sabel, in a sweeping analysis of the nature of industrial change, theorized how mass production triumphed over alternative manufacturing modes and came to dominate the American economy by the late 19th century.[4] Sabel's later collaboration with historian Jonathan Zeitlin clarified the origins and functions of mass production and its broadly defined alternative, "flexible specialization," shifting the focus of studies on manufacturing styles away from the Chandler model.[5]

Concurrently, other historians began focusing their research on industries that generally fit the paradigm of flexible specialization—for example, textiles, paper, iron, specialty steel, and toys—to develop a more holistic understanding of American and British industrialization. Philip Scranton, Judith McGaw, John Ingham, Maxine Berg, and others have explored how companies in tightly knit manufacturing districts used general-purpose tools and skilled labor forces to make goods for highly differentiated markets, goods that complemented the standardized products of bulk manufacturers.[6] In Scranton's definition of flexible specialization among consumer-goods firms, ambitious "batch" manufacturers sometimes reached out of their "core format" to borrow aspects of "custom" and "bulk" production methods as part of their efforts to constantly adjust products to meet vigorous and mutable demand. In Scranton's view, tidy categories such as "batch" and "mass" production neatly point to obvious differences between production styles, but rigid technological taxonomies cloud the uneven, messy nature of industrial change, leaving little room for "hybrid," or crossover, modes of operation.[7]

[4] Michael J. Piore and Charles F. Sabel, *The Second Industrial Divide: Possibilities for Prosperity* (New York, 1984).

[5] Charles F. Sabel and Jonathan Zeitlin, "Historical Alternatives to Mass Production: Politics, Markets, and Technology in Nineteenth-Century Industrialization," *Past and Present*, no. 108 (August 1985), pp. 133–76.

[6] Philip Scranton, *Proprietary Capitalism: The Textile Manufacture at Philadelphia, 1800–1885* (New York, 1983), and *Figured Tapestry: Production, Markets and Power in Philadelphia Textiles, 1885–1941* (New York, 1989); Judith A. McGaw, *Most Wonderful Machine: Mechanization and Social Change in Berkshire Paper Making, 1801–1885* (Princeton, N.J., 1987); John N. Ingham, *Making Iron and Steel: Independent Mills in Pittsburgh, 1820–1920* (Columbus, Ohio, 1991); Maxine Berg, *The Age of Manufactures, 1700–1820: Industry, Innovation, and Work in Great Britain* (New York, 1986).

[7] This analysis of manufacturing technology at HLCC owes much to Philip Scranton's insights and theories on flexible specialization, which are firmly rooted in his research on the textile, furniture, and jewelry industries. See, e.g., Scranton, *Proprietary Capitalism, Figured Tapestry*, "Diversity in Diversity: Flexible Production and American Industrialization, 1880–1930," *Business History Review* 65 (Spring 1991): 27–90, "Determinism and Indeterminacy in the History of Technology," in *Does Technology Drive History? The Dilemma of Technological Determinism*, ed. Merritt Roe Smith and Leo Marx (Cam-

The American pottery industry provides an ideal window for examining the intersection of production modes in the especially style-conscious home-furnishings business. The history of the ceramic tableware industry's largest firm—the Homer Laughlin China Company—buttresses Scranton's assertions that analytical categories are best understood as elastic prototypes.[8] This case study of Homer Laughlin's selective experiments with Fordism beginning in the 1920s—gleaned from corporate archives, union records, manufacturing equipment, and product lines—explores the mechanisms that enabled technological diffusion between disparate industries, illuminating how managers in an industry dominated by 19th-century traditions perceived and exercised the dominant technological model of their time, mass production. To Homer Laughlin's managers, mass production was never the exclusive purview of heavily capitalized metalworking firms making automobiles, electrical goods, or small appliances. Like other modes, mass production was a grab bag, offering ideas and opportunities that all managers might contemplate, modify, and use as much or as little as they saw fit.[9]

In the expanding consumer economy of the 1920s, Homer Laughlin's managers realized that their firm's competitiveness, if not its survival, necessitated the formulation of a new strategy for lowering costs, boosting volume, and enlarging the productivity of individual workers and machines—that is, for increasing what Chandler called "the velocity of throughput" in their factories. To realize "economies of speed," firms that aspired to quantity production generally used more reliable raw materials, intensified their use of energy, and installed output-enhancing machinery.[10] In the hope of achieving

---

bridge, Mass., 1994), "The Legacy of Specialization: American Industrial Districts and the Decline of Paterson, 1910–1940" (paper presented at the Conference on Paterson in the State and the Nation, Paterson, N.J., December 4–5, 1992), and "Manufacturing Diversity: Production Systems, Markets, and an American Consumer Society, 1870–1930," *Technology and Culture* 35 (July 1994): 476–505. For more on "hybrid" modes of production, see Piore and Sabel; Sabel and Zeitlin; and Jonathan Zeitlin, "Flexibility and Mass Production at War: Aircraft Manufacture in Britain, the United States, and Germany, 1939–1945," *Technology and Culture* 36 (January 1995): 46–79.

[8] Scranton, "Diversity in Diversity" and "Manufacturing Diversity."

[9] This analysis also owes much to David A. Hounshell, *From the American System to Mass Production, 1800–1932: The Development of Manufacturing Technology in the United States* (Baltimore, 1984). As Hounshell has explained, the "ethos" of mass production was a powerful device shaping American culture, including management thought, beginning in the 1920s, but few historians have examined precisely how managers in diverse—and seemingly unlikely—industries embraced the philosophy and adapted the techniques of mass production. See Hounshell, chap. 8.

[10] Chandler, *The Visible Hand*, chap. 8.

these goals, Homer Laughlin's managers hired a group of "scientific men"—experts in ceramics science, product design, and mechanical engineering—to revamp manufacturing practices and product lines at their West Virginia potteries.[11] As they designed and built conveyance units and special-purpose equipment to augment output and to improve worker performance, the firm's mechanical engineers unquestionably borrowed materials-handling procedures and moving-assembly-line principles refined by the nation's automobile manufacturers. The resultant mass-production machines amazed visitors, including the writer from *Ceramic Industry* who equated the devices with science-fiction automatons.

Metaphors about robots notwithstanding, complete mechanization was never a viable proposition at Homer Laughlin. In the demand-driven tableware industry, a firm's success hinged on its ability to satisfy a spectrum of customers with diverse tastes. At Homer Laughlin, the consumer was queen, and the objective of meeting her expectations for novel household accessories ultimately shaped the contours of the firm's technological system. The allure of ceramic goods, like the attraction of fabrics, furniture, or cutlery, fundamentally rested in the products' physical attributes, including color, shape, and decoration. The know-how of craft workers, the judgment of those artisans venerated in "poem, drama, and tale," undergirded a pottery's design vitality—that is, its ability to make appealing goods quickly and repeatedly. Contrary to journalistic yarns, craft potters persevered in Homer Laughlin's robotic kingdom. In fact, they were indispensable to the firm's strategy of flexible mass production. Their skills as fabricators of unusual shapes and colorful decorations were essential if a pottery were to realize consumer delight—and dollars. Combined with the expertise of the firm's ceramic engineers and product designers, craftsmanship provided Homer Laughlin's managers with the means for endowing products with the tactile and visual qualities coveted by a fastidious purchasing public.

In the burgeoning home-furnishings market of the 1920s, the real challenge before Homer Laughlin's managers was the task of satisfying their customers' demands for both volume and distinctiveness. Mass production, carefully and selectively grafted on to batch produc-

[11] "Meeting of the Joint Conference Committee of the United States Potters' Association (USPA) and the National Brotherhood of Operative Potters (NBOP)," transcript of proceedings (hereafter "USPA/NBOP Proceedings"), August 11–21, 1936, pp. 238–44, Kent, Ohio, Kent State University Archives, National Brotherhood of Operative Potters Papers (hereafter KSU NBOP Papers), box 49, folder 1.

tion, offered an opportunity particularly suited to the moment. Multiple interests and factors converged to shape the contours of technological change on the shop floor and the visible form, or look, of Homer Laughlin's products. Big machines created to lower costs and to enhance productivity circumscribed design possibilities, and vice versa. Technology and design were partners in a dance marathon, sustaining each other to the finish line, to win the favor of the purchasing public. In the end, Homer Laughlin's mechanical engineering department necessarily created a mass-production system that not only increased throughput but also remained flexible enough to accommodate the seemingly ceaseless design changes requisite to meeting the myriad aesthetic expectations of American consumers.[12]

### *Looking Backward: Homer Laughlin and the Retailing Revolution*

Comprehending the problems facing Homer Laughlin's managers and engineers during the 20th century requires an understanding of the company's history, including its earlier manufacturing practices, marketing procedures, and product lines. From its establishment in the 1870s to the early 20th century, this firm—like others in the upper Ohio River valley's ceramics center, the East Liverpool district—relied primarily on labor-intensive manufacturing methods perfected by potters in Staffordshire, England, during the late 18th and 19th centuries. A fitting testimony to this ancestry, the tristate East Liverpool district—which consisted of the central "Crockery City" of East Liverpool and nearby towns flanking the Ohio River in Ohio, Pennsylvania, and West Virginia—was nicknamed the "Staffordshire of America."[13]

[12] The term "flexible mass production" was coined in the late 1920s by Samuel Crowther, Henry Ford's ghostwriter, to describe modifications to the Ford Motor Company's mass-production system in response to competition from General Motors. According to Hounshell, *From the American System to Mass Production,* Ford focused on altering its machine-tool system, so as to "accommodate changes in design of the Ford automobile without totally tearing out machinery used to produce the old model" (p. 294). By the 1930s, "flexible mass production," a production system geared toward fast annual model changes, had triumphed over Fordism, or orthodox mass production, in the automobile industry. At HLCC, "mass-production" methods were grafted on to a "flexible" batch-production system, rather than vice versa. As such, this case study provides a counterpoint to the dominant model of flexible mass production.

[13] Useful major technical and historical studies of the American and English pottery industries include Aaron Hamilton Chute, *Marketing Burned-Clay Products, Including an Analysis of Location, Importance and Development of the Industry* (Columbus, Ohio, 1939); William C. Gates, Jr., *The City of Hills and Kilns: Life and Work in East Liverpool, Ohio* (East Liverpool, Ohio, 1984); Andrew McGeorge Lamb, "A History of the American

From its beginnings, Homer Laughlin was shaped by English traditions. Established in 1874 in the city of East Liverpool by two American-born brothers, Homer and Shakespeare M. Laughlin, the firm won awards for its white earthenware products at the United States Centennial International Exhibition of 1876 in Philadelphia. Between 1877 and 1896, Homer Laughlin ran his pottery as a sole proprietorship, using Staffordshire methods in all departments, from clay processing to forming, firing, decorating, packing, and shipping. Laughlin's facilities were limited, but so were his markets. Like other American potters, Laughlin sold his wares directly to shopkeepers or to jobbers who, in turn, distributed goods to country stores and urban shops.[14]

By the turn of the century, major changes were under way for the Laughlin firm—and for the American pottery industry. Laughlin and his bookkeeper, W. Edwin Wells (1863–1931), incorporated the Homer Laughlin China Company in 1896. A year later, Pittsburgh investors Louis I. Aaron (1840–1919) and his sons, Marcus (1869–1954) and Charles I. (1872–1947), bought a large interest in the firm, while the founder gradually withdrew.[15] The Aarons oversaw the company's long-term prospects, leaving daily operations to Wells, the new general

Pottery Industry: Industrial Growth, Technical and Technological Change and Diffusion in the Generalware Branch, 1872–1914" (Ph.D diss., London School of Economics 1984); Rexford Newcomb, Jr., *Ceramic Whiteware: History, Technology and Applications* (New York, 1947); Ernest Albert Sandeman, *Notes on the Manufacture of Earthenware* (London, 1901); Marc Jeffrey Stern, "The Potters of Trenton, New Jersey, 1850–1902: A Study in the Industrialization of Skilled Trades" (Ph.D. diss., State University of New York at Stony Brook, 1986), and *The Pottery Industry of Trenton: A Skilled Trade in Transition, 1850–1929* (New Brunswick, N.J., 1994); Harold Edison Stewart, "The Development of the American White Ware Industry: A Geographical Interpretation" (Ph.D. diss., Ohio State University, 1937); Herman John Stratton, "Factors in the Development of the American Pottery Industry, 1860–1929" (Ph.D. diss., University of Chicago, 1929); Department of Commerce, Bureau of Foreign and Domestic Commerce, *The Pottery Industry: Report on the Cost of Production in the Earthenware and China Industries of the United States, England, Germany, and Austria,* Miscellaneous Series 21 (Washington, D.C., 1915); Richard Whipp, *Patterns of Labor: Work and Social Change in the Pottery Industry* (London and New York, 1990).

[14] Gates, pp. 76–78; Lucille T. Cox, "How the World's Largest Pottery Had Its Beginnings 65 Years Ago," *Potters' Herald* (January 28, 1937), pp. 1–2; HLCC, *Catalogue and Price List of Semi-Vitreous China, August 1, 1896,* HLCC EF.

[15] "Necrology: W. E. Wells," *Bulletin of the American Ceramic Society* (hereafter *BACS*) 10 (October 1931): 326–28; "Marcus Aaron," *The National Cyclopedia of American Biography* (New York, 1963), 46:203–4. For this article, I have used the collective name, the Homer Laughlin China Company, to refer to several companies owned and operated primarily by the Aaron and Wells families: the Homer Laughlin China Company (Ohio) (1896–1934), the Homer Laughlin China Company (West Virginia) (1906–36),

manager. Whereas many pottery owners never looked beyond the boundaries of East Liverpool for solutions to managerial and technical problems, the Aarons, with their ties to a wider midwestern manufacturing district, abandoned this parochialism. Instead, they searched broadly through American business practice, selectively adapting modern manufacturing methods—including those pioneered by big businesses—as befitted the needs of their firm.[16]

Homer Laughlin entered the 20th century outfitted with a production system geared toward constantly making changing assortments of goods in response to variable consumption habits. When the new management came on board in the late 1890s, English craft practices dominated factory departments devoted to clay preparation, ware forming, glazing, decorating, firing, selecting, and shipping. A detailed inspection of three processes—forming, firing, and decorating—reveals the powerful and sustained influence of Staffordshire traditions.[17]

Craftsmanship ruled in the pottery's two ware-forming departments, the *clay shop* and the *casting shop*. In the clay shop, ware-forming crews, consisting of a *journeyman* and several subcontractors—a *batterout,* a *moldrunner,* and perhaps a *finisher*—made shallow flat circular forms (plates, saucers) or hollow forms (bowls, cups) on mechanically modified potter's wheels known as *jiggers* or *pull-downs* (fig. 1). Journeymen *dishmakers* made large flat forms (bakers, platters) by pressing clay pancakes into shallow molds. In the casting shop, workers made complex products (creamers, sugar bowls, pitchers, chamber pots, and shakers) by pouring *slip,* or liquid clay, into plaster-of-paris molds. These artisans made specialized forms (teapots, coffeepots, and tureens) by separately casting components (bodies, handles, spouts, and lids) and assembling those parts into a whole. Thus, the creation of distinctive shapes, variable forms, and multiple sizes depended on artisans versed in English craft techniques.

Homer Laughlin was equipped with Staffordshire-style *bottle kilns,*

---

Potteries Company of America (West Virginia) (1906–13), Homer Laughlin China Company, Inc. (Delaware) (1927–present). For my understanding of the administrative and financial history of these firms, I am indebted to HLCC's current president, Marcus Aaron II, who provided invaluable narrative and quantitative data from these firms' confidential Board of Directors' Minutes (hereafter HLCC Minutes).

[16] On industrial districts, see Michael Storper and Richard Walker, *The Capitalist Imperative: Territory, Technology and Industrial Growth* (Oxford, 1989); Scranton, "Manufacturing Diversity."

[17] Unless otherwise noted, the brief descriptions of pottery making in this article are based on the detailed historical and technical discussions in HLCC, *Catalogue and Price List of Semi-Vitreous China, August 1, 1896,* HLCC EF; W. G. Lawrence, *Ceramic Science for the Potter* (Radnor, Pa., 1972); Stern, *The Pottery Industry of Trenton,* chap. 2.

FIG. 1.—In late-19th- and early-20th-century American potteries, clay-shop crews, each consisting of a journeyman and his helpers, fabricated shallow flat circular forms, such as plates and saucers, on mechanically modified potter's wheels known as jiggers or pull-downs. This stereocard of the clay shop in an unidentified Trenton pottery shows a journeyman (foreground) and a batterout (background) making dinner plates. Absent is the moldrunner, who fetched and carried clay, molds, and finished ware for the other two workers. This scene could have taken place in the East Liverpool district. (Courtesy Smithsonian Institution, 88-14826.)

or ovens, for *firing*, or baking products. A well-managed kiln shed was the sign of a prosperous firm; a kiln catastrophe could portend disaster for a pottery. Bottle kilns were grouped into two functional categories: *bisque* (for hardening ware) and *glost* (for melting glazes on to ware). Another type of kiln, a *muffle*, was used for fusing decorations on to glazed ware. Hierarchical crews of skilled *kilnmen*, claiming exclusive knowledge of pottery pyrotechnics, filled, tended, and unloaded these high-temperature reaction chambers. In a bisque kiln

FIG. 2.—At the Homer Laughlin China Company, Newell, West Virginia, ca. 1906, a group of skilled kilnmen bring ware into the kiln sheds (background), fill saggers with ware (left), and load saggers into bottle ovens (right). Strength and stamina were essential attributes for these workers, who carried 75-pound saggers on their heads. One retired pottery executive described bottle kilnmen as muscular giants whose "necks, shoulders, and biceps" rivaled those of "any modern-day professional football lineman." (Postcard, author's collection; quotes from Charles E. Doll, Jr., "History of the Mount Clemens Pottery Company, 1913–1965," typescript, 1987, p. 5.)

shed, kilnmen known as *placers* put stacks of ware in refractory boxes called *saggers,* which they carried on their heads into the kiln. In a glost shed, placers stacked ware, coated with powdery glaze, with pins and stilts in the saggers.[18] (See fig. 2.) Fires were lit at the base of the filled oven, and the heat penetrated the kiln through flues, then exited through the chimney atop the bottle. The *fireman,* a senior worker responsible for the firing, gauged the kiln's progress by the color and clarity of the heat as seen through peepholes. A bisque oven was warmed for a day to an internal temperature of 2,200–3,000 degrees Fahrenheit and held there for two more days before it was cooled and unloaded. In bisque firing, the ceramic body was converted to a hard, porous, stable state in preparation for the glazing and decorating stages. A glost kiln was heated to 2,000 degrees Fahrenheit over two days, cooled, and unloaded; in this firing, the powdery glaze fused into a thin glassy coating that adhered to the clay body, making products waterproof. When a bottle kiln had

[18]Charles E. Doll, Jr., "History of the Mount Clemens Pottery Company, 1913–1965," St. Clair Shores, Michigan, typescript, 1987.

FIG. 3.—Following Staffordshire practices, American pottery manufacturers created highly differentiated products for diverse markets by covering standardized shapes with unique decorations, or patterns. The number of variations was limited only by the extent of consumer demand. (Catalog sheet, ca. 1905, Homer Laughlin China Company, Executive Files.)

cooled to about 100 degrees Fahrenheit, kilnmen opened it to *draw*, or unload, ware. Potteries fired kilns with any number of fuels—coal, natural gas, producer gas, wood, or oil—depending on local circumstances. No matter how much the clay shop produced, ware could be fired only as fast as the plant's kilns could be filled, stoked, cooled, and emptied.[19] Because of their intermittent production cycles, Staffordshire-style bottle ovens were also known as *periodic* kilns.

Homer Laughlin's managers and workers were experts at satisfying the inevitable vicissitudes of consumer taste, as was essential in the highly mutable home-furnishings trade. The firm's managers achieved the design vitality prerequisite to serving an expanding and increasingly segmented market by following English methods for creating new products. They combined standardized *shapes* with variable decorations, or *patterns* (fig. 3). In the factory's decorating rooms, female workers applied patterns, which were interchangeable parts,

[19] Bottle kilns thus could function as "reverse salients" in a pottery's production system. See Thomas P. Hughes, *American Genesis: A Century of Invention and Technological Enthusiasm, 1870–1970* (New York, 1989), chap. 2.

to wares before the final firing, using numerous techniques, including painting, banding, air brushing, printing and filling-in, gold stenciling, and stamping. By the late 1890s, the application of *decalcomania* transfers—fine imported chromolithographic prints—was becoming the preferred decorating method among East Liverpool potters. Smart managers like Wells, who understood that low-cost design innovation was the key to profits, used custom-designed decals to distinguish their lines from those of competitors.[20]

The legacy of the English pottery industry was both a blessing and a burden to entrepreneurs like Homer Laughlin's owner-managers who hoped to build up the American ceramics industry. On the one hand, American companies benefited from the ceramics know-how of British immigrants and their offspring. The expertise of these artisans and their progeny—their eclectic talents and skills—could provide a pottery with a tremendous degree of design flexibility, as reflected in the appearance and the salability of its wares. Product differentiation was an essential ingredient for an aspiring manufacturer of household furnishings, particularly in light of consumers' penchant for cozy, comfortable homes.[21] But the same craft traditions that made possible the manufacture of novel, beautiful household accessories empowered skilled male workers, creating obstacles to growth for ambitious pottery managers. Subcontracting—a system wherein skilled male workers hired, managed, and paid their own helpers—dominated work organization into the 20th century.[22] Any

[20] Foreign decal manufacturers, who were specialists in ceramics printing, sold decalcomania in stock designs or made patterns to a manufacturer's specifications. For more on decorating processes, see Regina Lee Blaszczyk, "The Aesthetic Moment: China Decorators, Consumer Demand, and Technological Change in the American Pottery Industry, 1865–1900," *Winterthur Portfolio* 29 (Summer/Autumn 1994): 121–53, and "Style and Strategy," *American Ceramic Circle Journal* 10 (in press).

[21] The literature on the home, its furnishings, and identity is vast. Useful studies include Mihaly Csikszentmihalyi and Eugene Rochberg-Halton, *The Meaning of Things: Domestic Symbols and the Self* (New York, 1981); Pierre Bourdieu, *Distinction: A Social Critique of the Judgement of Taste* (Cambridge, 1984); Witold Rybczynski, *Home: A Short History of an Idea* (New York, 1986); Nancy Dunlap Bercaw, "Solid Objects/Mutable Meanings: Fancywork and the Construction of Bourgeois Culture, 1840–1880," *Winterthur Portfolio* 26 (Winter 1991): 231–47; Kenneth L. Ames, *Death in the Dining Room and Other Tales of Victorian Culture* (Philadelphia, 1992).

[22] For other industries that used an inside contract system, see John Buttrick, "The Inside Contract System," *Journal of Economic History* 12 (Summer 1952): 205–21, on firearms; and Hounshell, *From the American System to Mass Production* (n. 9 above), on firearms, sewing-machine components, and sewing-machine cabinets. For an overview of subcontracting, see Ernest J. Englander, "The Inside Contract System of Production and Organization: A Neglected Aspect of the History of the Firm," *Labor History* 28 (Fall 1987): 429–46.

resistance on the part of journeymen and their helpers, who since 1890 had been organized into the National Brotherhood of Operative Potters (NBOP), could circumscribe a firm's ability to increase productivity or otherwise meet consumer demand.[23] To gain competitive advantage, general manager Wells strained to reconcile the pros and cons of his firm's Staffordshire potting heritage with the demands of the American market.

By 1900, a "retailing revolution" was under way, for mass merchandisers rapidly were displacing jobbers and general merchants as the primary organizations for coordinating the flow of consumer goods through a marketplace becoming more urban and suburban.[24] Homer Laughlin, like other batch manufacturers in the home-furnishings business, increasingly relied on the country's major chain stores, premium distributors, mail-order houses, and department stores for its lifeblood. These colossal retail enterprises placed unprecedented demands on the pottery. Representatives from the American Cereal Company of Chicago pressured Homer Laughlin to ship them 4,000 dozen, or 48,000, decorated bowls per week for use in Quaker Oats cereal promotions.[25] The buyer for one of the country's tea stores, the Union Pacific Tea Company of New York, urged the pottery to "fill . . . orders immediately" to meet the demand for ceramic premiums, or gifts, that his company had created through "extensive advertising."[26] Clinton Pierce Case, the domestic crockery

[23] For union activity, see David A. McCabe, *National Collective Bargaining in the Pottery Industry* (Baltimore and London, 1932); and Don Anthony Shotliff, "The History of the Labor Movement in the American Pottery Industry: The National Brotherhood of Operative Potters–International Brotherhood of Operative Potters, 1890–1970" (Ph.D diss., Kent State University, 1977).

[24] Chandler, *The Visible Hand* (n. 3 above), chap. 7, on the "retailing revolution"; Susan Strasser, *Satisfaction Guaranteed: The Making of the Mass Market* (New York, 1989); Richard S. Tedlow, *New and Improved: The Story of Mass Marketing in America* (New York, 1990); Alfred D. Chandler, Jr., *Strategy and Structure: Chapters in the History of the American Industrial Enterprise* (Cambridge, 1962), chap. 5, for Sears; John K. Winkler, *Five and Ten: The Fabulous Life of F. W. Woolworth* (New York, 1940); William I. Walsh, *The Rise and Decline of the Great Atlantic and Pacific Tea Company* (Secaucus, N.J., 1986); William Leach, *Land of Desire: Merchants, Power, and the Rise of a New American Culture* (New York, 1993), for department stores; Susan Porter Benson, *Counter Cultures: Saleswomen, Managers and Customers in American Department Stores, 1890–1940* (Urbana, Ill., and Chicago, 1986).

[25] Wells to American Cereal Company, November 21, 1900, Newell, West Virginia, Homer Laughlin China Company, Executive Correspondence (hereafter HLCC EC), box 301. Cereal and soap manufacturers, which offered dinnerware at reduced prices or as premiums with purchases of their merchandise, had been big customers of American potteries since the 1890s. See Chute (n. 13 above), pp. 137–38.

[26] George Everett Clapp, Union Pacific Tea Company, to Wells, March 2, 1902, HLCC EC, box 370. Tea stores pioneered the premium game in the 1880s. See "Regu-

buyer from F. W. Woolworth and Company, was blunt; he demanded "good goods and plenty of them at fair prices."[27]

American pottery executives tackled changing market conditions in several ways, all characteristic of 19th-century batch manufacturers.[28] Collectively, tableware manufacturers embraced associationism as a control mechanism. Through the United States Potters' Association (USPA), proprietors attempted to regulate markets and prices by negotiating for favorable railroad shipping rates, bargaining with the NBOP over wage increases, creating standard sizes for pottery products, and lobbying Congress for tariff protection against European and Asian imports. At the level of the firm, some managers coped with the retailing revolution by accelerating production in their clay shops. Others expanded old factories and built new ones.

The Homer Laughlin China Company aggressively pursued many of these tactics after the infusion of new blood into top management in 1897. Executives began refurbishing old facilities and erecting new plants, so that by 1902 the firm was operating three factories in the Crockery City's growth area, the East End. That year, Wells and the Aarons pooled their capital, incorporating the Homer Laughlin subsidiary, the North American Manufacturing Company (NAMC), whose wholly owned companies built the town of Newell, West Virginia, across the Ohio River from East Liverpool. At Newell, the NAMC constructed a village for workers and factories for Homer Laughlin and the Edwin M. Knowles China Company, another firm partially owned by the Aarons. Newell was connected to East Liverpool by a suspension bridge across the Ohio River and streetcar lines, all operated by NAMC firms. Homer Laughlin's Newell facility, Plant No. 4, completed in 1906, was reputed to be the most up-to-date pottery in the world. Whereas one Laughlin plant had employed 124 workers in 1892, the company operated four factories in two states—Ohio and West Virginia—staffed by 1,800 workers by 1910. Gross sales increased more than twelvefold during the first decade of the 20th century; in 1898, sales amounted to less than $200,000, but by 1910 they totaled around $2.5 million.[29] By 1912, Homer Laughlin was the largest pottery manu-

---

lar Crockery vs. Gift Tea Dealers," *American Pottery and Glassware Reporter* 9 (October 4, 1883): 29.

[27] Clinton Pierce Case to HLCC, December 1899, HLCC EC, box 247. For more on HLCC and Woolworth's, see Regina Lee Blaszczyk, "Imagining Consumers: Manufacturers and Markets in Ceramics and Glass, 1865–1965" (Ph.D. diss., University of Delaware, 1995), chap. 3.

[28] Scranton, "Diversity in Diversity" (n. 7 above).

[29] HLCC Minutes; "Questionnaire filed with the State of Ohio's Bureau of Labor

facturer in the United States, producing approximately one-tenth of the nation's ceramics.[30] Executives had added yet another factory, Plant No. 5, to their Newell site in 1914.[31] The Great War greatly reduced foreign firms' market share, and Homer Laughlin's sales soared during and immediately afterward. By 1920, the company had a solid grip on markets for inexpensive and moderately priced pottery used by working-class and middle-class consumers.[32]

In the 1920s, Woolworth's crockery buyer enjoyed more leverage with Homer Laughlin's general manager than ever before, and he never hesitated to wield it. The new buyer, William F. Newberry, pressured Wells for more goods, prodding him to tackle underproduction. "The only thing that will help us out," Newberry apprised Wells, "is more merchandise." He implored the general manager to "find a way to give us an increased output which will put our stores in shape to meet the full demand."[33] The adjurations of the Woolworth's buyer were not new; the difference was the timing. The expansion of chain stores dealing in groceries, drugs, hardware, and general merchandise was so immense and so advanced in the 1920s that the decade was nicknamed the chain-store age. Woolworth's buyer got what he wanted.

### *Putting a Pottery on a Conveyor*

To accommodate the rising expectations of mass retailers like Woolworth's, Homer Laughlin's directors began to invest wartime

Statistics," August 11, 1892, HLCC EC, box 111; HLCC, *Catalogue and Price List of Semi-Vitreous China, August 1, 1896;* HLCC, *Catalogue and Price List, April 1, 1901;* HLCC, *Price List and Illustrations of Plain White Wares* (East Liverpool, Ohio, 1905); *Price List and Illustrations, Plain White Wares Made by the Homer Laughlin China Company, Newell, West Virginia, January 1, 1907;* HLCC, *Daintily Decorated Table and Toilet Wares from the World's Largest Potteries* (Newell, W. Va., and East Liverpool, Ohio, 1908), all in HLCC EF.

[30] *Proceedings of the Thirty-third Annual Convention of the United States Potters' Association, Held at Hotel Raleigh, Washington, D.C., December 3 and 6, 1911* (East Liverpool, Ohio, 1911), p. 30. By 1912, HLCC had a production capacity of 300,000 pieces per day, an estimated one-tenth of the nation's consumption. See HLCC, *Daintily Decorated Table and Toilet Wares from the World's Largest Potteries* (Newell, W.Va., and East Liverpool, Ohio, 1912), p. 3, HLCC EF.

[31] *Price List and Illustrations, Plain White Wares Made by the Homer Laughlin China Company, Newell, West Virginia, and East Liverpool, Ohio, January 1, 1914,* HLCC EF.

[32] Stratton (n. 13 above), p. 301. In 1914, gross sales by HLCC totaled $2.9 million. In 1918, they were $4.7 million; in 1919, $5.3 million; in 1920, $5 million. See HLCC Minutes.

[33] William F. Newberry to Wells, May 25, 1923, HLCC EC, box S-769.

profits in scientific, aesthetic, and technological innovation.[34] Specialized departments headed by new groups of professionals—ceramic engineers, product designers, and mechanical engineers—figured in the firm's corporate strategy. Motivated by competition from foreign potteries and open shops, general manager W. Edwin Wells and his successor, son Joseph M. Wells, embraced the tripartite objectives of increasing output, controlling labor costs, and lowering prices. They inaugurated a building and renovation program that extended into the post–World War II period, a program that made earlier enlargements pale in comparison. These managers adapted mass-production methods to pottery manufacturing, but they did so selectively and incrementally. They gradually installed tunnel kilns, conveyance units, and mechanical devices to facilitate throughput in their factories. These devices were the "robots" that impressed visitors. Nonetheless, Fordism never dominated every nook and cranny at Homer Laughlin, as full mechanization was not a viable proposition in a fashion-conscious pottery.

In 1920, production problems focused the attention of W. Edwin Wells on the vital need for tighter control over manufacturing processes.[35] This soon led to the hiring of Albert V. Bleininger, department chief at the ceramics laboratory of the National Bureau of Standards in Pittsburgh, to head a new factory division devoted to quality control. The establishment of Bleininger's ceramics laboratory signaled the beginning of the replacement of Staffordshire craft secrecy with standardized methods at Homer Laughlin.[36] When Homer Laughlin built three new potteries in Newell between 1923 and 1929, Bleininger figured into the decision-making process, counseling executives on matters from plant layout to kiln design.[37]

Erected and equipped in 1923 at the expense of more than $1 million, the first new factory, known as Plant No. 6, was heralded as the world's most modern pottery. Constructed by a Cleveland-based industrial engineering firm, H. K. Ferguson Company, the new concrete, brick, steel, and glass structure reminded one observer of an

[34] Author's conversations with Joseph M. Wells II, July 1993, Newell, West Virginia.

[35] Frederick Hurten Rhead, "How Can American Potters Maintain Their Gains in the Domestic Market?" *Ceramic Industry* 37 (November 1941): 60.

[36] On Albert V. Bleininger (1873–1946), see J. McKeen Cattell, *American Men of Science*, 2d ed. (New York, 1910), p. 46; J. McKeen Cattell and Jacques Cattell, *American Men of Science*, 6th ed. (New York, 1938), pp. 128–29; "Two Great Names Pass into Ceramic History," *Ceramic Industry* 46 (June 1946): 49. On Bleininger's daily work, see Ceramics Laboratory Notebooks, 1920–46, HLCC.

[37] On Bleininger's role in kiln installation, see Fred B. Lawrence to Carl B. Harrop, May 27, 1923, HLCC EC, drawer 131, folder: Construction, Plant No. 6.

automobile factory.[38] Plant No. 6's rational facade, layout, and equipment testified to a new managerial affinity for continuous-flow production. The factory's *tunnel kilns,* or continuous ovens, became the linchpins of Homer Laughlin's flexible mass-production system. Without this technology, Homer Laughlin could not expect to accommodate the volume requirements of clients like Woolworth's.

Homer Laughlin's managers witnessed the practicality of tunnel kilns by visiting the Mount Clemens Pottery Company. The tunnel kiln connection was the first of several technological exchanges between the two firms in the interwar era. Since January 1920, this pottery in Mount Clemens, Michigan (established 1913), had been a wholly owned subsidiary of the Midwest's fastest-growing variety chain store, S. S. Kresge and Company. After the Great War, Mount Clemens general manager Charles E. Doll convinced Kresge's executives, his most important customers, to buy the factory and invest in a $1 million modernization program. Kresge's managers wanted volume, and Doll promised to give it to them—if they provided the money for mechanization. Doll realized that quantity production would be the name of the consumer products game during the 1920s, and he was anxious to adopt Fordist techniques that would facilitate the manufacture of inexpensive, standardized pottery. Kresge's managers swallowed the bait, integrating backward into pottery supply by purchasing Mount Clemens. Doll subsequently devised the most radical production program heretofore undertaken by any American pottery, a plan that began with the building of tunnel kilns in 1920 and culminated with full-scale mechanization in the late 1930s.[39]

[38] HLCC Minutes; "The Manufacture of Dinner Ware," *Clay-Worker* 81 (April 1924): 451–66; "Bird's Eye View of Plant No. 6 of Homer Laughlin China Company, the Largest and Most Modern Plant under One Roof in the World," *Potters' Herald* (January 31, 1924), pp. 1, 5; H. K. Ferguson Company, "Specifications for Factory Building for HLCC," and "Commission Contract with HLCC," both March 26, 1923, HLCC EC, drawer 131, folder: Construction, Plant No. 6; "Modernize or —?" *Ferguson Cross Section* (July 1929), pp. 4–9, clipping, HLCC EF.

[39] The corporate records of the Mount Clemens Pottery Company have not survived, and Kresge's annual reports reveal little about this wholly owned subsidiary. See S. S. Kresge & Co., Annual Reports, 1922–59, Corporate Reports Collection, Cole Room, Baker Library, Harvard University Graduate School of Business Administration, Boston. For my understanding of technological innovation at Mount Clemens, I am indebted to Charles E. Doll, Jr., who assisted his father in the management of this pottery starting in 1925. After his father's retirement, Doll was general manager at Mount Clemens until 1965 when Kresge sold it. Unless otherwise noted, my discussion of Mount Clemens is based on Doll (n. 18 above); interview with Charles E. Doll, Jr., May 10, 1988, St. Clair Shores, Michigan; "Comments by Charles E. Doll, Jr., for Regina Lee Blaszczyk," typescript, August 1992.

Backed by Kresge capital, Mount Clemens became one of the first U.S. tableware factories to install tunnel kilns.[40] In the 1920s, potteries that pioneered tunnel-kiln use realized significant gains in terms of increased capacity, improved product quality, reduced losses, and lower fuel bills.[41] Homer Laughlin's owner-managers keenly watched Mount Clemens assume technological leadership in the pottery industry and also experience organized labor's negative reaction. Union men had negotiated with Doll over piecework rates for tunnel kilns, only to find that Mount Clemens's manager violated the compact when it suited him. Doll, unswerving in his quest to cut costs, next employed apprentices, rather than journeymen, to operate his jiggers, again breaching a union agreement. After the NBOP called a strike to protest Doll's actions, a frustrated Kresge vice president retaliated by withdrawing from the USPA and establishing Mount Clemens as an open shop.[42]

Despite the labor unrest at Mount Clemens, Homer Laughlin's executives remained curious about tunnel kilns. Mount Clemens's goods were so cheap that the Woolworth's buyer complained that his stores were losing business to Kresge's crockery department. Answerable to a single vendor, Doll could keep his costs down by limiting shapes

[40] Mount Clemens's bisque and glost tunnel kilns, designed by ceramic engineering professor Carl B. Harrop of Ohio State University, and the firm's decorating tunnel kilns, by Holcroft & Co. of Detroit, were widely discussed in the trade literature. See *Journal of the American Ceramic Society* (hereafter *JACS*) 3–7 (1920–24): passim; *Ceramic Industry* 1 (1923): passim; "Pioneer Dinnerware at Mount Clemens Pottery," *BACS* 17 (March 1938): 156–57.

[41] Pioneered by the Germans in the late 19th century, tunnel kilns had been used by American brickmakers and European ceramists for decades, but the technology did not gain acceptance among U.S. whiteware manufacturers until after the Great War. For early debates on tunnel kilns, see *Transactions of the American Ceramic Society* (hereafter *TACS*) 1–3 (1899–1901): passim. During the early to late 1910s, the ACS published papers on tunnel kilns, stimulating and reflecting interest in the technology. See *TACS* 16–19 (1913–17): passim; *JACS* 1 (1918): passim. On the adoption of tunnel kilns by U.S. firms, see "The Tunnel Kiln," *Brick and Clay Record* 66 (January 6, 1923): 22–36; Philip Dressler, "Tunnel Kiln Development in the Ceramic Industries," pts. 1 and 2, *Ceramic Age* 10 (September 1927): 81–91; (October 1927): 122–30.

[42] On labor's response to Mount Clemens's continuous kilns, see "Minutes of Meeting of the Standing Committee," February 7, 1921, and April 4, 1921, Columbus, Ohio Historical Society, United States Potters' Association Papers, coll. 855, box 1, folder 9: Standing Committee Minutes, 1919–1924. For the NBOP's retrospective account of the apprentice jiggermen issue, see "USPA/NBOP Proceedings," August 11–21, 1936, pp. 697–707, KSU NBOP Papers, box 49, folder 1. On resistance to the NBOP's efforts to reorganize Mount Clemens, see "Court Refuses to Dissolve Injunction Granted Mount Clemens against L.U. of Brotherhood," *Potters' Herald* (May 15, 1924), pp. 1–2; "Injunction of Mount Clemens Pottery against NBOP Dismissed," *Potters' Herald* (April 4, 1927), p. 1. For NBOP strikes in the 1920s, see Shotliff (n. 23 above), pp. 143–214.

and decorations. A piece of Mount Clemens dinnerware was functional, cheap, and widely available—not so much as hinting at fashion. Homer Laughlin still catered to multiple clients with diverse expectations, even though Woolworth's was its largest customer. The company's products, thanks in part to its decorating departments, had been variable for decades. In the 1920s, Homer Laughlin's managers looked to Mount Clemens with an eye to selective borrowing, seeking volume-enhancing tools that would quicken production without abandoning their firm's manufacturing flexibility.

With Mount Clemens on to something good, Homer Laughlin's managers sent a group of production experts, including Bleininger, to Michigan to study Doll's setup. Doll welcomed them for reasons that reveal much about the technical information corridors in the pottery industry.[43] Before moving to Michigan in 1914 to help establish Mount Clemens, Doll had been assistant manager at the W. S. George Pottery Company in East Palestine, Ohio, then the nation's second-largest producer of chain-store dishes.[44] In East Liverpool, old-time managers jealously guarded formulas, production statistics, and sales figures, but progressive managers like Wells and Doll had begun to share information, casting off the shackles of Staffordshire craft secrecy. Formally, forward-looking executives used meetings of the USPA and the American Ceramic Society as data dissemination channels, but casual exchanges over meals and golf also yielded valuable information. By the time he left for Michigan, Doll had forged an informal alliance with the like-minded Wells. Thus, a larger vision for growth and modernization unified a generation of managers, including Doll and Wells, who otherwise were rivals in the marketplace.

For more than a decade, the bond between Doll and Wells remained strong, facilitating the flow of technical information between their firms. Common frustrations energized the long-distance relationship. As the pottery industry's most ambitious managers, both strove to build a wider clientele among mass merchandisers, while struggling to accumulate reasonable earnings for their firms. Before Doll sold Mount Clemens to Kresge, he and Wells secretly collaborated to erect a bulwark against chain-store buyers' entreaties for

[43] W. Edwin Wells, Jr., et al., "Memorandum on Visit to Mount Clemens for W. Edwin Wells," February 5, 1923, HLCC EC, drawer 131, folder: Construction, Plant No. 6; Doll to Edward L. Carson, March 31, 1923, and Carson to Doll, April 4, 1923, both in HLCC EC, box L-901.

[44] In 1916, HLCC was the largest U.S. producer of chain-store ceramics, and W. S. George was the second largest. See Wells to Doll, August 7, 1916, HLCC EC, box L-528.

cheaper goods. The managers fixed their firms' wholesale prices to major distributors, including F. W. Woolworth and Company, S. S. Kresge and Company, and S. H. Kress.[45] These covenants further strengthened the allegiance between the two general managers.

Another critical factor—a mutual frustration with pottery workers—fused the bond between Doll and Wells. As founder and chair of the USPA's labor committee, Wells had been in charge of union negotiations since 1906, and he routinely communicated with managers, including Doll, about labor issues. In one such exchange, Doll confided to Wells that he had deserted Ohio to escape the "annoyances and bad practices that have crept into the [pottery] business through the attitude of the workmen."[46] After the Great War, Wells and other East Liverpool potters were stymied by the NBOP's insistence on shorter hours and higher wages.[47] Pressures from the union, combined with mass retailers' pleas for greater volume and lower prices, threatened to erode profits and stability in the tristate pottery district.[48] In the shadow of NBOP strikes partially precipitated by his actions, Doll willingly shared his technological knowledge with a confrere in the managerial defense against organized labor.[49]

Wells spent much of the 1920s battling NBOP officials over the caliber of union workmanship, and many of Homer Laughlin's Fordist innovations were installed as buffers against the negative effects of a batch-production system gone partially awry. In 1921, Wells returned from Staffordshire, determined to institute radical changes in his factories, if not the entire American pottery industry. He was convinced that something had gone wrong with English manufactur-

[45] Correspondence between Doll and Wells, August 5, 1916–December 26, 1916, HLCC EC, boxes L-528 and L-629; "Memorandum concerning Prices Made to F. W. Woolworth and Company by the Mount Clemens Pottery Company," August 17, 1916, HLCC EC, box L-628.

[46] "I am convinced that the pottery business of America can never rise to the point of being a really great and efficient industry," Doll continued, "until the time comes when we get completely [out] from under the domination of the [labor] forces that have caused so much reaction during recent years." See Doll to Wells, August 16, 1919, HLCC EC, box L-727. For more on labor issues, see correspondence between Doll and Wells, October 14, 1916–August 16, 1919, HLCC EC, boxes L-528, L-629, and L-727.

[47] "USPA/NBOP Proceedings," vol. 2, August 2, 1921, pp. 583–92 (on Saturday work), KSU NBOP Papers, box 46, folder 13; "USPA/NBOP Proceedings," August 23–26, 1922, pp. 1–107 (on wages), KSU NBOP Papers, box 46, folder 16.

[48] Wells complained that HLCC was often "on thin ice." See "USPA/NBOP Proceedings," August 23–26, 1922, p. 102, KSU NBOP Papers, box 46, folder 16.

[49] On various "governance mechanisms," such as trade associations and personal trust, which provided cohesion and strength to batch producers, see Scranton, "Manufacturing Diversity" (n. 7 above).

ing practices as transplanted to American soil. Wells learned that an average English glost bottle-oven firing yielded a mere 5 percent loss, a startling discovery given that an American glost firing generated defective ware at the rate of 28–35 percent. But annoying *thirds* were only one fly in the American pottery manager's ointment. Wells had toured English factories where teams of skilled and semiskilled artisans—workers often related by blood—earned a collective living. With entire households dependent on a pottery for their daily bread, workers had a vested interest in a smoothly operating system. If Wells understood that the family wage system greased the cogs of flexible production in Staffordshire, he never betrayed it to the NBOP. Instead, he held conscientious British workers in esteem, imploring NBOP leaders to encourage a similar attitude among union members. American pottery workers, with their Staffordshire ancestry, Wells argued, were capable of rising to the challenges of quantity and quality; they simply refused to do so.[50]

Wells believed that Staffordshire batch-production methods, as they had been transformed in East Liverpool over several generations, must be modified—but not abandoned—if Homer Laughlin were to persist in the chain-store age. He formulated an umbrella scheme for lowering his company's overall manufacturing expenditures, and tunnel kilns were the cornerstones of his strategy.[51] (See fig. 4.) Continuous kilns, however, were more than a quick managerial solution to an intractable labor force. They also were effective tools for controlling the pace of work—and, concomitantly, the quality of products. Wells claimed that the union-backed piecework system encouraged fast and sloppy work among bottle-oven placers, resulting in heavy losses. In contrast, tunnel kilns guaranteed laborers high earnings and steady jobs while curtailing the impulse, inherent in piecework, to speed up production, thus promising to eliminate the troublesome and expensive problem of thirds.[52]

[50] "USPA/NBOP Proceedings," vol. 2, August 2, 1921, pp. 99–105, 577–78, KSU NBOP Papers, box 46, folder 13. "Thirds" referred to second-class ware. Obviously, English potteries were burdened with their own set of problems, including disputes over wages and factory conditions. See Whipp (n. 13 above).

[51] Wells fully understood the cumulative and generative effects of tunnel kilns, calling them the "cornerstones" of Homer Laughlin's "mass-production methods." See W. Edwin Wells, "Acres of Dishes," *Factory and Industrial Management* 77 (April 1929): 697–700. On troublesome labor as a force pushing managers toward technical innovation, see Nathan Rosenberg, "The Direction of Technological Change: Inducement Mechanisms and Focusing Devices," in *Perspectives on Technology* (New York, 1976), pp. 108–38.

[52] Wells cited production figures to back his argument. On average, a glost kilnman at one of Homer Laughlin's bottle ovens filled and placed a sagger every 160 seconds,

FIG. 4.—The glost tunnel kiln at Homer Laughlin China Company's Plant No. 6. Tunnel kilns were firebrick tunnels, approximately 300 feet long, equipped with a furnace section at midpoint. A train of cars made of iron frames, iron wheels, and fireclay decks traveled through the tunnel on a narrow-gauge track, propelled by a hydraulic ram. Cars holding saggers entered the oven at a cold receiving end, grew hotter as they approached the furnace area, and gradually cooled before exiting. Cars left tunnel kilns at short intervals, allowing for an endless flow of wares to subsequent operations. (From *Clay-Worker* 81 [April 1924]; courtesy Smithsonian Institution, 93-13027.)

But most important, fast and dependable tunnel kilns could provide a pottery with the reliable throughput essential to achieving economies of speed. Homer Laughlin's production experts returned from Mount Clemens enthusiastic about the high-performance potential of the new technology. A glost tunnel kiln could yield a carload of ware every forty to forty-five minutes, sending an uninterrupted

or 2⅔ minutes, which included the climb into the kiln with a sagger on his head. In contrast, a tunnel-kiln laborer in new Plant No. 6 filled and placed a sagger on a kiln truck every 200 seconds, or 3⅓ minutes. While a bottle-oven placer, who was paid a piecework rate, handled 166 saggers, a tunnel-kiln placer, who was paid a daily rate, handled only 100 saggers. The slower speed resulted in better ware. See "USPA/NBOP Proceedings," August 12–14, 1924, pp. 175–210, KSU NBOP Papers, box 4, folder 22. Wells reiterated his arguments to the USPA. See "Wages, Employment, and Production Problems Reviewed by W. E. Wells in Annual Report to Members of the United States Potters' Association," *Potters' Herald* (January 1, 1925), pp. 1–2. Tunnel kilns enhanced product quality in two other ways. First, continuous ovens could be fired with any number of fuels, but natural gas, used by HLCC, insured a clean atmosphere and further reduced inventories of dirty ware. Second, saggers were stacked in 9-foot pillars on tunnel-kiln cars, rather than in 20-foot tiers as in bottle ovens, greatly lessening the risk of a calamitous toppling over.

flow of pottery to the factory's warerooms. Attaining this kind of regular pace was requisite to lowering wholesale prices so that Homer Laughlin might compete during the chain-store age. Thus, the promise of increased dependable throughput was the tunnel kiln's greatest draw.

Total payroll expenditures for operating a tunnel kiln with hourly laborers under union rates exceeded those associated with bottle-kiln firing, as a continuous oven simply required more manpower at higher wages.[53] Continuous ovens were managed by Bleininger and his staff of ceramic engineers—white-collar professionals operating under the aegis of science.[54] These engineers controlled tunnel-kiln firings from their laboratory—planning, plotting, calculating, watching, and recording all details. Kilnmen were operatives who performed assigned duties for hourly wages. Tunnel kilns proved to be an excellent investment for Homer Laughlin. The new technology yielded startling results. Less than one-fifth of the products fired in a glost tunnel kiln were off-quality.

Some NBOP members resisted the deskilling technology, but their protests fell on deaf ears.[55] One local complained when Wells hired

[53] "The Dramatic Story of an American Pottery," *Pottery, Glass, and Brass Salesmen* (hereafer *PGBS*) 40 (August 1, 1929): 11–12. On the kilns at Plant No. 6, see Victor J. Roehm, "Notes on the Operation of Direct-Fire Tunnel Kilns," *JACS* 8 (August 1925): 514–24. Wells summarized the comparative labor costs of tunnel- and bottle-oven operations for the USPA and NBOP. In its tunnel-kiln plants, HLCC paid 13.5 percent more for glost placing, 16 percent more for dipping (or glazing), and 14 percent more for bisque placing than it paid for those tasks under the piecework system in its bottle-oven plants. See "USPA/NBOP Proceedings," August 12–14, 1924, pp. 175–210, KSU NBOP, box 4, folder 22.

[54] Bleininger's laboratory staff was small, but his chemists were knowledgeable in ceramics and competent to manage the new kilns. Victor J. Roehm, e.g., had a B.S. in chemical engineering from Ohio State University and experience as a ceramics researcher for the National Bureau of Standards. See Joe Dickey, "Industrial Welfare Department, Homer Laughlin China Company," *Potters' Herald* (July 16, 1925), p. 4, and (April 1, 1926), p. 4; "Contributions during 1925 to the Work of the Society," *BACS* 4 (January 1926): 79.

[55] On deskilling, see Harry Braverman, *Labor and Monopoly Capital: The Degradation of Work in the Twentieth Century* (New York, 1974); Richard Edwards, *Contested Terrain: The Transformation of the Workplace in the Twentieth Century* (New York, 1979); Stephen Meyer III, *The Five Dollar Day: Labor Management and Social Control in the Ford Motor Company, 1908–1921* (Albany, N.Y., 1981), and "Technology and the Workplace: Skilled and Production Workers at Allis-Chalmers, 1900–1941," *Technology and Culture* 29 (October 1988): 839–64. Recently, historians of technology have tested the deskilling paradigm on myriad occupations. See Deborah Fitzgerald, "Farmers Deskilled: Hybrid Corn and Farmers' Work," *Technology and Culture* 34 (April 1993): 324–43; James P. Kraft, "Musicians in Hollywood: Work and Technological Change in Entertainment Industries, 1926–1940," *Technology and Culture* 35 (April 1994): 289–314.

common laborers rather than journeymen to load decorating kiln cars, claiming a violation of the apprentice clause in the national wage agreement. Some union workers objected into the late 1920s, but Wells added more tunnel kilns and more day laborers.[56] To many East Liverpool workers, the new technology was less a powerful deskilling agent than an economic savior. By middecade, the faltering tristate district had a surplus of unemployed potters, including old kilnmen grown too weak for bottle-oven work and journeymen jettisoned from longtime positions by plant closings, who were more than willing to buck their outspoken NBOP brothers for an easy job and a steady paycheck.[57]

Tunnel kilns appeared to be a miraculous corrective to the persistent problems of speed, economy, and quality. Wells estimated that a tunnel-kiln plant with three continuous ovens could produce as much as a factory outfitted with 20 bottle kilns, while realizing fuel savings of more than 65 percent.[58] Building on the success of Plant No. 6, Homer Laughlin's managers constructed additional tunnel-kiln facilities in Newell in the late 1920s.[59] In a powerfully symbolic

[56] Carl Vagel to John T. Wood, March 10, 1924; Harry Waggoner et al. to Wood, March 30, 1924; "Decorating Kiln, Laughlin No. 6, Some of Wells's Statements," April 2, 1924; L.U. No. 95 to USPA/NBOP Arbitration Board, April 7, 1924; memorandum on decorating tunnel kiln, May 1924; "W. E. Wells's Statements," January 16, 1925; L.U. No. 124 to John McGillivray, December 6, 1927; Charles F. Goodwin, USPA, to Wood, May 9, 1927, all in KSU NBOP Papers, box 19, folders 1, 6, 8, 9, 26, and 33; "USPA/NBOP Proceedings," August 12–15, 1930, pp. 163–210, KSU NBOP Papers, box 47, folder 1.

[57] "USPA/NBOP Proceedings," August 12–14, 1924, p. 183, KSU NBOP Papers, box 4, folder 22; "USPA/NBOP Proceedings," 1928, p. 50, KSU NBOP Papers, box 46, folder 25.

[58] Wells reported that the energy costs of the East Liverpool potteries had more than tripled since the prewar era. See "NBOP/USPA Proceedings," vol. 2, August 2, 1921, pp. 103, 270, KSU NBOP Papers, box 46, folder 13; Wells, "Acres of Dishes" (n. 51 above). Like Mount Clemens, Plant No. 6 was equipped with bisque and glost tunnel kilns designed and built by consulting engineer Carl B. Harrop, Columbus, Ohio, and with a decorating kiln by Holcroft & Co., Detroit. See HLCC EC, drawer 131, folder: Correspondence, Plant No. 6 Construction; "Bird's Eye View of Plant No. 6" (n. 38 above); "The Manufacture of Dinner Ware," *Clay-Worker* 81 (April 1924): 451–66; J. B. Nealey, "Gas-Fired Continuous Kilns in the Chinaware Industry," *Ceramic Age* 14 (September 1929): 83–88.

[59] On Plant No. 7, see HLCC EC, drawer 133, folder: Construction, Plant No. 7; "The Contract Having Been Awarded, the Homer Laughlin China Company Will Break Ground for New Million Dollar Plant," *Potters' Herald* (February 24, 1927), p. 1; "Interior View of Plant No. 7," *Potters' Herald* (October 20, 1927), p. 3; "New Tunnel Kiln Plant," *Potters' Herald* (November 10, 1927), pp. 1–2. On Plant No. 8, see HLCC EC, drawer 134, folder: Construction, Plant No. 8; "Ground Breaking for Homer Laughlin Pottery No. 8," *Potters' Herald* (April 11, 1929), p. 1; "New Unit of

gesture, managers also tore down bottle ovens in their early-20th-century factories.[60] By 1934, all of Homer Laughlin's plants had continuous ovens, the supposed cure-all for throughput and productivity problems.

During the 1920s, several factors—the revival of European industry, economic downturns, reduced tariffs, cheap Japanese imports, high labor costs, shipping boycotts, and rising energy costs—eroded the prosperity that the Great War had brought to many American potteries. Production declined, and some firms shut their doors. Homer Laughlin was a juggernaut in an otherwise declining industry. By 1928, the firm controlled one-third of USPA production.[61] As one observer drolly noted, Homer Laughlin was the "giant among dishes."[62] But this development did not mean that the firm was freed from its commitment to flexible production and to fashion.

During the 1920s, powerful cultural agents—the movies, radio, mass magazines, and national advertising—began to reshape consumers' desires, affecting the chain stores and other mass distributors that fueled Homer Laughlin's growth.[63] Before the Great War, the quintessential dime-store shopper had been a "woman with a shawl," who browsed through Woolworth's for inexpensive chinaware to dec-

---

Homer Laughlin China Company Will Begin Operation," *Potters' Herald* (October 17, 1929), p. 1; "A Giant among Dishes," *Ceramic Industry* 14 (April 1930): 408–17.

[60] "Plant No. 4 of Homer Laughlin China Company, in Which New Tunnel Kiln Is to Be Erected," *Potters' Herald* (January 21, 1926), p. 1; "Plant No. 4 to Be Equipped with New Bisque and Decorating Tunnel Kilns," *Potters' Herald* (July 19, 1928), p. 1; "Laughlin China Company to Build Another Decorating Tunnel Kiln in Newell Plant," *Potters' Herald* (December 13, 1928), p. 1; "Homer Laughlin Awards a $60,000 Kiln Contract for No. 4 Plant," *Potters' Herald* (March 20, 1930), p. 1; "Laughlin Plant No. 5 to Be Remodeled in Preparation for Anticipated New Business," *Potters' Herald* (April 12, 1934), p. 1.

[61] Joseph M. Wells, "Memorandum on the Pottery Industry of the United States," October 6, 1929, pp. 2–6, 10, HLCC EF, box: Joseph M. Wells correspondence. The USPA adjusted its statistics to reflect some members' adoption of tunnel kilns. Three tunnel kilns (for bisque, glost, and decorating) could produce as much ware as 20 bottle ovens. In 1921, gross sales of the Homer Laughlin China Company were $5.1 million, and in 1928, they were $6.6 million. See HLCC Minutes.

[62] "A Giant among Dishes" (n. 59 above).

[63] By 1921, sales to Woolworth's accounted for 36 percent of the pottery's business. See Wells to Newberry, May 17, 1923, HLCC EC, box S-769. By 1927, a year for which comprehensive sales statistics have survived, more than half of Homer Laughlin's output, measured in dollars, went to Woolworth's approximately 1,600 stores. Homer Laughlin's second-largest customer, premium vendor Quaker Oats, accounted for 8 percent of 1927 sales. Two other vendors, S. H. Kress and Butler Brothers, respectively bought 7 percent and 6 percent. The next four largest customers each absorbed 3 percent; one of these was Sears, Roebuck & Co. Notes on sales figures, 1927, HLCC EF, Louis K. Friedman files.

orate her tenement or farmhouse.[64] But things were changing fast. The customer with a "Woolworth's pocketbook" had evolved into an exacting patron with "Tiffany tastes," who expressed her personality and displayed her class status through the purchase, display, and use of consumer goods.[65] Savvy home-furnishings manufacturers scurried not only to increase output and lower prices but also to quicken the pace of design innovation to compete for these fastidious shoppers' dollars.

In 1927, Homer Laughlin's managers thus undertook the second phase of their modernization program, establishing a division devoted to product design and development. The new art department was headed by Staffordshire-born Frederick Hurten Rhead (1880–1942), a self-proclaimed "ceramic art engineer" who had worked in American potteries since 1902.[66] Rhead was charged with the job of studying consumers' variable tastes and purchasing habits and subsequently collaborating with managers, retail buyers, and technical staff to develop salable mass-market products. Rhead flourished at the helm of Homer Laughlin's new department by accepting two cardinal principles: first, design was a collaborative effort; second, the consumer was sovereign. During his early years at the pottery, Rhead, working within the constraints of the firm's technological systems, modified standard Woolworth's shapes and designed new patterns to reflect the upbeat tastes of dime-store shoppers. More and more, product design and process design were entwined, as Rhead labored to reconcile the dictates of the market with technical change on the shop floor.

The quest to increase the velocity of throughput in their potteries brought Homer Laughlin's executives to the third phase of their modernization project. Tunnel-kiln technology, the miraculous cure-all, proved to be a gifted problem child. Continuous ovens insured round-the-clock output and provided Homer Laughlin's managers with increased throughput, but they also created what Nathan Rosenberg has termed a "technical imbalance" among a factory's interde-

[64] Frederick Hurten Rhead, "Chats on Pottery," *Potters' Herald* (August 11, 1931), p. 6.

[65] Ibid. (March 31, 1933), p. 6.

[66] Frederick Hurten Rhead, "Organization of a Decorative Ceramic Research Department," *PGBS* 26 (December 14, 1922): 97; Sharon Dale, *Frederick Hurten Rhead: An English Potter in America* (Erie, Pa., 1986). For more on Rhead's role in product innovation, see Regina Lee Blaszczyk, "Art Engineering: Frederick Hurten Rhead, the Homer Laughlin China Company, and the American Consumer," Hagley Research Seminar paper 20, November 10, 1994, and "Imagining Consumers" (n. 27 above), chap. 5.

pendent technological systems.[67] Continuous ovens were hungry monsters that gobbled up tremendous quantities of ware faster than Homer Laughlin's plants could make it with existing batch-production methods.[68] The dilemma of reconciling the output and the pace of Homer Laughlin's clay-shop and kiln-placing crews with voracious tunnel kilns linked the West Virginia pottery once again to Doll at Mount Clemens, who was revamping his manufacturing system following automotive production models.

For almost a decade, Doll had wrestled with the advantages and disadvantages of tunnel kilns. Although his oil-fired continuous ovens saved an estimated 75 percent in fuel bills, Mount Clemens's tunnel kilns also created the same type of production disequilibrium that was plaguing Homer Laughlin.[69] Doll stood at a crossroads. He could lure unemployed NBOP potters away from the depressed East Liverpool district, increasing output by adding more skilled workers. But Doll's obsession for cutting costs and his stalwart anti-unionism led him down another path. Rather than wrestle with the NBOP, Doll chose to mechanize pottery production, drawing on Detroit's female labor pool to operate his new equipment. The remarkable productivity of Doll's tunnel kilns thus stimulated a sequence of technological innovations that brought a gendered division of labor to Mount Clemens.

To determine how to increase and speed up production to feed his tunnel kilns, Doll looked outside the pottery industry for models and expertise. An admirer of Henry Ford, Doll turned to the nearby auto industry for ideas. He visited Detroit's automobile factories, including the Ford Motor Company's Highland Park plant, where he studied the moving assembly line. Backed by Kresge capital, Doll gradually adapted materials-handling procedures and assembly-line principles refined by automakers to pottery manufacturing at Mount Clemens.[70]

Doll reorganized manufacturing with the help of a professional

[67] Rosenberg (n. 51 above), p. 113.

[68] Wells reminded the NBOP that "a tunnel kiln must, if successfully operated, run steadily." See "USPA/NBOP Proceedings," August 12–14, 1924, p. 186, KSU NBOP Papers, box 46, folder 22. For other managers trying to reconcile the demands of tunnel kilns with batch-production systems, see "USPA/NBOP Proceedings," 1928, pp. 293–315, 327–32, KSU NBOP Papers, box 46, folder 25.

[69] Wells et al., "Memorandum on Visit to Mount Clemens" (n. 43 above); "Progressiveness Shown in Equipment and Management of Mount Clemens Pottery," *Ceramic Industry* 1 (July 1923): 68–72.

[70] For Doll's admiration for Henry Ford, his interest in Highland Park, and his visits to automobile plants, see Charles E. Doll, Jr., to author, October 9, 1992.

engineer, Walter Howard Emerson (1902–47). After graduating from the Massachusetts Institute of Technology with a B.S. in chemical engineering in 1926, Emerson had gone to work for the Lamson Company, a Syracuse-based industrial engineering firm with offices throughout the United States. Assigned to Lamson's Detroit offices, Emerson designed conveyance systems for shoe, stove, and paint factories, using automatic and manual handling techniques.[71] By the late 1920s, Emerson was collaborating with Doll on installations for Mount Clemens. Massachusetts Institute of Technology's School of Chemical Engineering Practice taught students how to control the flow of energy and materials to and from chemical "reacting zones," know-how that Emerson modified and applied to ceramics production.[72] Emerson's emphasis on systematic decision making and the selective limitation of worker control over production—his quest for efficiency and economy along Fordist lines—was novel for Mount Clemens and the American pottery industry in general.

At Mount Clemens, Emerson experimented with what Lindy Biggs has dubbed a "rational factory" without worrying about granting concessions to the NBOP.[73] Emerson applied scientific management to his renovations, conducting extensive time-and-motion studies on various processes before designing continuous-flow installations that promised to yield "greater production per operator" than existing manual systems. Most notably, the engineer strove to design conveyance mechanisms that eliminated workers who did "not add value to

[71] For Emerson's background, I am grateful to Kara Schneiderman, MIT Museum (telephone conversation, September 9, 1993). My preliminary research on Walter Emerson has not revealed any blood relationship to famous efficiency engineer Harrington Emerson. For Lamson's corporate philosophy and conveyance installations, see Lamson Company, *Industrial Conveying in Factories, Warehouses, and Industrial Plants* (Syracuse, N.Y., ca. 1920s); Lamson Company, *Relieving the Shortage of Common Labor: Bulletin No. 1 of a Series Dealing with Significant Problems in Industry* (Syracuse, N.Y., ca. 1923); Lamson Company, *Knowledge That Pays Dividends* (Syracuse, N.Y., ca. 1920s). For Emerson's machine designs for Lamson, see "Warehouse Operations: St. Jerome Plant," typescript (late 1920s), HLCC Engineering Department (hereafter HLCC ED), Emerson correspondence, folder L-1937; photographs of conveyor installations at Acme White Lead and Color Works, HLCC ED, folder: Materials Handling.

[72] R. T. Haslam, "The School of Chemical Engineering Practice of the Massachusetts Institute of Technology," *Journal of Industrial and Engineering Chemistry* 13 (May 1921): 465–68. I am grateful to Kathryn Steen for this reference. See also Terry S. Reynolds, "Defining Professional Boundaries: Chemical Engineering in the Early 20th Century," *Technology and Culture* 27 (1986): 694–716.

[73] Lindy Biggs, "Industry's Master Machine: Factory Planning and Design in the Age of Mass Production, 1900 to 1930" (Ph.D. diss., Massachusetts Institute of Technology, 1987), p. 15.

the product."[74] By early 1930, he had built networks of conveyors that moved heavy lumps of clay, delicate newly formed ware, and fragile bisque-fired ware around the Mount Clemens factory.[75]

Among Emerson's innovations at Mount Clemens were the so-called placing conveyors he designed for tunnel-kiln departments. At each bisque and glost tunnel kiln, Emerson installed a self-contained elliptical conveyance unit that used moving flat belts to transport heavy saggers between work stations. Doll claimed that women were especially suited to filling glost saggers because of their neatness, attentiveness, and dexterity—not to mention their lower wages. Emerson, in turn, designed glost placing stations that minimized stretching and lifting, transforming the job of filling glost saggers from highly skilled men's work to light unskilled work, or women's work.[76] (See fig. 5.) A USPA executive who saw Emerson's placing conveyor at Mount Clemens told NBOP officials that the women's speed in "pinning saggers would have just simply driven some of you fellows dizzy."[77] In 1933, NBOP leaders who tried to reorganize Mount Clemens noted that most employees were "girls"—young women who apparently had little time, interest, or cash to spare for the union.[78] With the help of Emerson's machine designs and the accompanying gendered division of labor, Doll had not only increased output while cutting labor costs, but he had also reduced the potential for a dangerous resurgence of union activism. Emerson's engineering of the

[74] Walter Howard Emerson, "A Study of Plant Operations to Determine Conveyor Layout," service report for Mount Clemens Pottery Company by the Lamson Company, June 1, 1930, p. 5, HLCC ED, Emerson files, notebook 44: Mount Clemens.

[75] These installations cost Mount Clemens $70,000. See Emerson, "Mount Clemens Conveyor Specifications," March 20, 1930, HLCC ED, Emerson correspondence, folder L-1937.

[76] Photographs and blueprints of Mount Clemens installations designed by Emerson on behalf of Lamson, HLCC ED, folder: Materials Handling, and HLCC ED blueprint files; Charles E. Doll, Jr., to author, October 9, 1992, and July 29, 1993. Doll developed his perceptions of women's suitability for pottery production during the Great War, when he employed female jiggering crews. See "USPA/NBOP Proceedings," August 1928, p. 129, KSU NBOP Papers, box 46, folder 25.

[77] "USPA/NBOP Proceedings," August 12–15, 1930, pp. 301–7, KSU NBOP Papers, box 47, folder 1. Doll reported that female glost placers working on Emerson's belt conveyor loaded an average of 202 saggers, or 406 dozen pieces of ware, each day. See Doll to Joseph M. Wells, January 3, 1930, HLCC EC, box L-1336. The gendered division of labor allowed Doll to cut his payroll costs significantly. By 1936, he was paying female glost placers $.40 per hour, while HLCC's managers were paying male glost placers $.89 per hour. See "USPA/NBOP Proceedings," August 11–21, 1936, pp. 348–49, KSU NBOP Papers, box 49, folder 1.

[78] John McGillivray, Minutes of the NBOP Executive Board, July 27–29, 1933, p. 3, KSU NBOP Papers, box 2, folder 25.

FIG. 5.—Emerson's so-called glost placing unit eliminated strenuous fetch-and-carry tasks and the backbreaking bending and lifting that had characterized glost kiln loading for generations. The job of filling heavy saggers became light work, or women's work. Mount Clemens Pottery Company, ca. 1930. (Homer Laughlin China Company, Engineering Department, Emerson files.)

pottery-making process along Fordist lines set a precedent for the use of continuous-flow techniques by other firms, including Homer Laughlin.

Impressed with reports about Mount Clemens's placing conveyors, Wells contracted with Lamson's Pittsburgh office to install a glost placing device in Homer Laughlin's new Plant No. 8 during the late 1920s.[79] Undoubtedly, Homer Laughlin's managers hoped that a Lamson placing unit would partially solve the persistent, expensive problem of technological imbalance generated by tunnel kilns. In the

[79]"NBOP/USPA Proceedings," August 1928, pp. 172–74, KSU NBOP Papers, box 46, folder 25; Lamson Company, "Service Report for HLCC," May 6, 1929, and "Proposal from the Lamson Company for HLCC, for a System of Conveyors for Plant No. 8," May 21, 1929, in HLCC EC, drawer 134, folder: Construction, Plant No. 8. For an illustration of the conveyor, see L. K. Urquhart, "Automatic Handling: Another Mass-Production Tool for the Dishmaker," *Factory and Industrial Management* 79 (January 1930): 80–81. Lamson, along with other industrial engineering firms, previously had installed modest conveyors at HLCC. See HLCC EC, drawer 131, folder: Construction, Plant No. 7, and "Proposal by the Lamson Company," September 22, 1928, HLCC EC, box L-1255.

short run, the Lamson conveyor was a flop, but in the long run the faulty unit was the critical device that linked Homer Laughlin to Emerson, scientific management, and automotive mass-production methods. Some kilnmen went on strike when Homer Laughlin, like Mount Clemens, hired women to work on the placing conveyor for lower wages.[80] Managers eventually concluded that the appliance was so poorly engineered it could not be operated without severe losses, and ordered its removal. After unsuccessfully trying to redesign the system, Lamson's disgruntled sales staff dismantled the unit, dismayed at the forfeiture of Homer Laughlin's business.[81]

The Lamson incident coincided with shifts in Homer Laughlin's management and with major changes in the economy. In 1930, sixty-eight-year-old W. Edwin Wells stepped down from his position as general manager and was succeeded by his son, Joseph M., who also assumed an active role in the USPA labor committee. Two decades earlier, Louis I. Aaron had retired, and his son Marcus had assumed the company's presidency. In 1923, the new president's offspring, Marcus Lester Aaron (1900–94), joined Homer Laughlin's executive staff; he would become president in 1940.[82] The new generation faced the trial of coping with the consequences of tunnel-kiln technology while steering their family firm through the Great Depression and World War II.

[80] "Conveyor System in Laughlin No. 8 Plant Causes Dispute," *Potters' Herald* (November 14, 1929), pp. 1, 3, 6; "Settlement Made in Kiln Dispute at Homer Laughlin Company's No. 8 Plant," *Potters' Herald* (November 28, 1929), p. 1. For Wells's response to the union, see "USPA/NBOP Proceedings," August 12–15, 1930, pp. 312–13, KSU NBOP Papers, box 47, folder 1.

[81] Homer Laughlin's building committee criticized the design of the Lamson conveyor on several accounts. First, vibrations from the unit caused ware to fall off the belts and break. Second, the conveyor's capacity was insufficient for Homer Laughlin's needs; the system failed to deliver ware to the ovens at the requisite forty-minute intervals required for tunnel-kiln firing. Third, the conveyor compounded, rather than simplified, the tasks of dipping, placing, and drawing, so that kiln-loading became a labor-intensive, expensive procedure. See "Memorandum on Glost Conveyor System for Plant No. 8," April 7, 1930; "Memorandum on the Offer of the Lamson Company to Install New Conveyor," January 1930; "Added Labor Costs in the Present Conveying System as Compared to Regular Dipping, Placing and Drawing," January 1930, all in HLCC EC, box L-1400; "Comparison of Drawing and Placing Costs at No. 8 Plant between No. 1 Glost Kiln Where Conveyor Is Used and No. 2 Glost Kiln Where Bench Placing Method is Used," January 20, 1930, HLCC, Subsidiary Correspondence (hereafter HLCC SC), box 573. On the attempts to redesign the conveyor, see correspondence between Marcus Aaron and Lamson Company, December 30, 1929–April 3, 1930, in HLCC EC, boxes L-1326 and L-1400.

[82] "Announce Changes in Personnel of Homer Laughlin China Co.," *Ceramic Industry* 14 (January 1930): 187; HLCC Minutes.

Frustrated with the Lamson kiln conveyor experiment, alarmed by the state of the economy, and dismayed by the repercussions of cheap competition, Homer Laughlin's managers wrestled with their position on labor and technology. The depression was taking its toll, and even the dishmaking colossus had closed two of its plants. The company could not afford any more losses. Just as the economic crisis eroded consumer purchasing power, competition from cheap imported ceramics—from "millions of dozens of cups and saucers" from Japan—threatened American tableware factories serving the mass market.[83] Potteries were selling their goods at rock-bottom prices simply to keep their plants running. At the end of his long career, W. Edwin Wells had solemnly announced to the NBOP that only "well-managed and modernly equipped factories" would pull through.[84] Wells's farewell statement to the union presaged major technical changes that would take place at Homer Laughlin during the ensuing years under the direction of Walter Howard Emerson.

In 1932, Emerson, riding on his success at Mount Clemens, approached his superiors at Lamson with a plan for mending relations with Homer Laughlin. He suggested that Lamson provide Homer Laughlin with a free factory efficiency survey. During the depths of the depression, neither firm had much to lose, and Emerson soon moved to the East Liverpool district. The finished study impressed the pottery's executives, including new general manager Joseph M. Wells, who invested in another Lamson conveyor and subsequently hired Emerson to establish a mechanical engineering department at Homer Laughlin in May 1934.[85]

Leading Homer Laughlin's mechanical engineering department tested Emerson's mettle.[86] Times were tough, but the Great Depres-

[83] "USPA/NBOP Proceedings," August 12–15, 1930, pp. 11, 13, 15, 221, 231–32 (for quote), KSU NBOP Papers, box 47, folder 1.

[84] "USPA/NBOP Proceedings," August 12–15, 1930, p. 11, KSU NBOP Papers, box 47, folder 1.

[85] Correspondence between Marcus L. Aaron and Lamson Company, 1932–34, HLCC EC, box L-1647 and box L-1753; Paul N. Critchlow, "Memorandum of Mr. Emerson's Early Activities in the Pottery Industry," May 21, 1942, HLCC ED, legal file: Emerson vs. Miller Patent Interference, Docket 103, Book VIII, Memos.

[86] Prior to establishing their engineering department in 1934, Homer Laughlin's managers approached machine design in several ways. In some instances, the company purchased standard equipment from firms that specialized in clayworking machinery. In other cases, members of the firm's building committee pooled their expertise to design machines that were fabricated by regional equipment firms such as the Patterson Foundry and Machine Company in East Liverpool. Finally, Homer Laughlin hired consultants, such as Lamson or the Dravo-Doyle Company of Pittsburgh, to design and install conveyors for ware drying, refuse removal, and kiln drawing. For my understanding of mechanical engineering in the pottery industry, I am indebted to Carl C.

sion was only one of the chief engineer's problems. Homer Laughlin was a very different shop from Mount Clemens. Emerson faced a taxing trinity in the form of Homer Laughlin's batch-production setup, its vocal union workers, and its shifting client portfolio. At Homer Laughlin just as at Mount Clemens, Emerson had to construct a technological system that would increase output and lower unit costs. But this time he had to remodel potteries that were sometimes several decades old. Without question, his Lamson career had prepared him for this task. Yet any technology Emerson designed for Homer Laughlin had to accommodate uniform standards for wages and working conditions established by joint conferences between the USPA and the NBOP. In an average union pottery, wages accounted for about 60 percent of production costs.[87] If Homer Laughlin were to compete, Emerson must cut these expenditures without totally alienating the NBOP—whose members possessed many skills essential to the firm's design vitality. Emerson's installations thus had to sustain Homer Laughlin's proficiency in making distinctive products for a range of mass retailers and consumers. Emerson soon learned that Homer Laughlin's design flexibility—its avenue to profits—was synonymous with the maintenance of peaceable labor-management relations.

Soon after his arrival at Homer Laughlin, Emerson applied scientific management to the analysis of various jobs in the company's West Virginia potteries.[88] After conducting time-and-motion studies on skilled, semiskilled, and unskilled tasks, Emerson, with the assistance of a staff of engineers and machinists, redesigned

---

Mooney (interview, August 24, 1993, Newell, West Virginia) and Willis G. Gaston (interviews, August 24 and 26, 1993, East Liverpool, Ohio). Mooney had worked at Homer Laughlin since the 1930s. From 1946 until 1960, he was a machine builder, helping to construct much of Emerson's mass-production equipment. From 1960 until his retirement in 1993, Mooney was chief engineer at Homer Laughlin. He now works as a consultant for the firm. Gaston was a mechanical engineer and later chief engineer at the Edwin M. Knowles China Company from the 1940s until the firm closed in 1962. Homer Laughlin's mechanical engineering department designed numerous devices and systems for Knowles, but none have survived.

[87] W. Edwin Wells estimated that labor costs amounted to 50 percent of the retail price of a pottery item, or 60 percent of a piece's manufacturing costs. Wells, "Acres of Dishes" (n. 51 above); "USPA/NBOP Proceedings," August 12–15, 1930, p. 10, KSU NBOP Papers, box 47, folder 3. Emerson later estimated that labor costs amounted to 55 percent of an item's retail price. See Walter Howard Emerson, "Distribution of Sales Dollar in Pottery Industry," September 7, 1944, HLCC ED, folder: Time-and-Motion Analysis.

[88] Walter Howard Emerson, notes on time studies, HLCC ED, folder: Time-and-Motion Analysis.

targeted operations in extant factory buildings to obtain a continuous flow of products and to reduce expenditures primarily on unskilled labor. Emerson envisioned certain factory departments—most notably the kiln department and the clay shop—as self-contained, highly mechanized units that were linked by interlocking mechanical and manual systems. In buildings dating from 1906 to 1929, he installed a system of overhead monorail trolleys, flat-belt conveyors, elevators, and trucks that moved raw materials, products, and even people in, to, and from firing and ware-forming areas. Efficient, rational coordination—the modification of factory layouts and the careful reconceptualization of selected operations to emulate the workings of a perfect machine—was Emerson's ne plus ultra.

One of Emerson's first innovations for Homer Laughlin, designed while he was still a Lamson consultant, was the Iron Horse, a refinement of his Mount Clemens placing conveyor. The Iron Horse was the most Fordist of Emerson's inventions. The patented unit moved unskilled laborers, saggers, and ware at variable speeds, subjecting workers to considerable stress. The company estimated that it would save $19,000 annually for each tunnel kiln equipped with one of these conveyors.[89] In 1935, kilnmen complained that standing on the Horse's moving platform while loading saggers put them under constant tension and hence was a detriment to their physical and mental health. Some men went home exhausted, collapsing at the end of a day's work. One worker experienced an extreme reaction; after three months on the Horse, lifting an average of four filled saggers each minute, a Mr. Ward reported that his arms thrashed uncontrollably while he slept. The union president called the Iron Horse an "inhuman monster."[90] NBOP representatives, concerned with worker safety and potential technological unemployment, met with Homer Laughlin's managers about a dozen times to negotiate the terms of operation for the new appliance. More than 300 frus-

[89] "Proposal from the Lamson Company to HLCC for Conveyor System," January 5, 1934 (rev. January 13, 1934), HLCC SC, box 595; Lamson Company, "Proposal for Glost Placing Conveyors," August 27, 1934, HLCC EC, box L-1858; "New Conveyor at Laughlin Shop Is Improved Style," *Potters' Herald* (May 3, 1934), pp. 1–2; "Homer Laughlin Spends $225,000 on Plant No. 5," *Ceramic Industry* 22 (May 1934): 238; Emerson, "We Move the Work and the Worker" (n. 1 above), pp. 138–42; U.S. Patent 2,033,695; memorandum comparing labor costs on tunnel kilns with and without conveyors, ca. 1935, KSU NBOP Papers, box 19, folder 34.

[90] For the "monster," see "Record of Protest on the Conveyors," September 14, 1935, KSU NBOP Papers, box 19, folder 35; NBOP/USPA, "Homer Laughlin Conveyor Study," 1935, KSU NBOP Papers, box 22, folder 34.

trated workers called wildcat strikes.[91] Despite their initial opposition, NBOP leaders eventually agreed to a settlement on the "wild steed" that included adjustments to wages, conveyor speed, and working hours because of acute unemployment conditions in the East Liverpool district during the Great Depression.[92]

While Emerson contemplated designs for the Iron Horse and other installations, the purchasing habits of American consumers and the choices of big-time buyers were undergoing yet another metamorphosis. Early in the depression, Homer Laughlin had been jolted by serious curtailments of Woolworth's orders.[93] By middecade, Woolworth's purchases constituted only a fragment of Homer Laughlin's sales, as the chain-store buyer was filling his crockery departments with inexpensive, brilliantly decorated Japanese, Czechoslovakian, and German porcelain. In particular, Japanese potters, whose expenditures on labor and raw materials were low, copied favorite American patterns, selling these imitations at half the price of comparable union-made goods.[94] Homer Laughlin's managers were forced to look elsewhere for volume sales, counting on premium vendors such as Quaker Oats to gobble up tremendous quantities of decal-decorated pottery. Much like depression-era automakers who employed styling to stimulate car sales, pottery managers exercised the triumvirate of ceramic science, product design, and manufacturing technology in an effort to revive their chain-store trade and to build up department-store accounts.[95]

[91] "Iron Horse Brings Cessation of Work on the Part of Kilnmen," *Potters' Herald* (August 29, 1935), p. 1.

[92] "USPA/NBOP Proceedings," August 11–21, 1936, pp. 350 (for the "wild steed"), 318–19, 341–42, 359–87, KSU NBOP Papers, box 49, folder 1; J. M. Wells, HLCC, to James M. Duffy, NBOP, October 2, 1935, copy in "USPA/NBOP Proceedings," August 25–30, 1942, pp. 383–85, KSU NBOP Papers, box 50, folder 2; Shotliff (n. 23 above), pp. 234–35.

[93] Between 1929 and 1936, Homer Laughlin's sales to Woolworth's had dwindled from $3.8 million to $1 million. Sales figures, HLCC EF; Charles Leigh Sebring, another manager from the East Liverpool district, reported that his company's chain-store business had plummeted from 24 percent of gross sales in 1935 to 10 percent in 1936. See "USPA/NBOP Proceedings," August 11–21, 1936, p. 693, KSU NBOP Papers, box 49, folder 1.

[94] On competition, see "USPA/NBOP Proceedings," August 12–15, 1930, p. 10, and "USPA/NBOP Proceedings," December 2–5, 1930, pp. 49, 271–73, both in KSU NBOP Papers, box 47, folder 3; "USPA/NBOP Proceedings," August 11–21, 1936, pp. 18, 690, KSU NBOP Papers, box 49, folder 1; Charles Leigh Sebring, "Unfair Foreign Competition Ruining Domestic Pottery Industry," *Ceramic Industry* 19 (November 1932): 244–46.

[95] "USPA/NBOP Proceedings," August 11–21, 1936, p. 380, KSU NBOP Papers, box 49, folder 1; "Premium Business Growing Heavier—Pottery Imports Disturbing

Homer Laughlin's executives engaged the expertise of Rhead and Bleininger to create products that would resuscitate business. In the art department and the ceramics laboratory, product innovation remained as it had for decades—a process of repeated assessment and negotiation, a procedure described by Eugene Ferguson in writing about engineering design as a "contingent process, subject to unforeseen complications and influences."[96] External fashion intermediaries, in the form of retail buyers, transmitted data about consumer preferences to Rhead, who in turn translated that information into product prototypes with the technical assistance of Bleininger. The art director and his staff not only listened to buyers, but they also constantly tapped into the knowledge banks of materials suppliers, salesmen, managers, and mechanical engineers as they improved and developed products.[97] Through his routine work, Rhead was learning that obstructions to satisfying consumer demand increasingly came, not from people, but from technology, from "tunnel kilns, conveyors, . . . and other machines and contraptions."[98] At Homer Laughlin, Rhead noted, "a cent a dozen extra for making and finishing" was "an effective curb against any innovation."[99] In product development "for mass production," he wrote, "there is no escape from the limitations imposed by the manufacturing conditions involved."[100] By the mid-1930s, technology thus constantly circumscribed product design at Homer Laughlin, where volume production went hand in hand with tight profit margins.

The greatest challenge to art engineer Rhead, who earned his livelihood by divining and interpreting taste, was the reconciliation of technological constraints and managerial profit imperatives with the dictates of the market. As always, the ultimate aim of product innovation in the home-furnishings business was consumer satisfaction. Rhead thus had to render solutions to product differentiation that would please prospective chain- and department-store shoppers while reducing his employer's manufacturing costs. One avenue to cutting

Market," *Ceramic Industry* 25 (December 1935): 319; David Gartman, *Auto Opium: A Social History of American Automobile Design* (London and New York, 1994), chap. 5.

[96] Eugene S. Ferguson, *Engineering and the Mind's Eye* (Cambridge, Mass., 1992), p. 37.

[97] Blaszczyk, "Art Engineering" (n. 66 above), and "Imagining Consumers" (n. 27 above), chap. 5.

[98] Frederick Hurten Rhead, lecture given in England, typescript, 1938, p. 4, HLCC, Art Department files.

[99] Frederick Hurten Rhead, "Chats on Pottery," *Potters' Herald* (April 20, 1933), p. 6, and (December 8, 1932), p. 6.

[100] Frederick Hurten Rhead, lecture.

those costs was to design products that owed their appeal to stylish shapes, versatile clay bodies, and brilliant chromatic glazes, rather than to applied decoration. "Harlequin" tableware, created for Woolworth's in response to a vogue for colorful kitchen accessories (fig. 6), was fired at very low temperatures and mechanically sprayed with bright chromatic glazes. Carefully engineered by Rhead, Bleininger, and the Woolworth's buyer, a piece of Harlequin cost less to make than a comparable item covered with decals, primarily because it never entered the firm's decorating departments. Harlequin's allure depended on its streamlined shapes and eye-catching colors. But modernistic lines like Harlequin constituted only a small percentage of Homer Laughlin's output. To most consumers of Homer Laughlin china, modernism took a back seat to traditional styling. Rhead never hesitated to create lines decorated with conventional applied decorations, including polychrome decalcomania. He knew that tableware embellished with cheerful or sentimental flowers would sell. The fuller the market basket with homey floral patterns, the happier were Homer Laughlin's customers, who were drawn primarily from the working and middle classes.[101]

As Homer Laughlin and other USPA firms limped along, the nation's few open shops prospered. Mount Clemens was reported to operate at full capacity. Kresge, in fact, was cornering the market for American-made dime-store dishes by stocking Doll's inexpensive products. Chain-store buyers held Kresge's retail prices in esteem and pledged to forsake importers if union shops could make goods as cheaply as Mount Clemens.[102] Under these pressures, Homer Laughlin's general manager contemplated his firm's policy on technology and labor. Union workers frustrated Wells, but several factors mitigated against severing relations with the NBOP. Homer Laughlin's reputation rested on both volume and variety. Skilled labor was a salient feature in this pottery dedicated to product diversity. Committed by family tradition to batch production and to the East Liverpool district, where most potters were union members, Wells hesitated to break his ties with the NBOP. Instead, he implored the union to cooperate with management to save the ailing pottery industry. The only way to beat the competition, Wells told NBOP leaders, was to use unskilled labor—when it was possible to do so.[103] At Homer Laughlin,

[101] Blaszczyk, "Art Engineering," and "Imagining Consumers," chap. 5.

[102] "USPA/NBOP Proceedings," August 11–21, 1936, pp. 9–27, 743–44, 902–4, KSU NBOP Papers, box 49, folder 1.

[103] Wells also implored the union to accept wage cuts. See "USPA/NBOP Proceedings," August 11–21, 1936, p. 24, KSU NBOP Papers, box 49, folder 1.

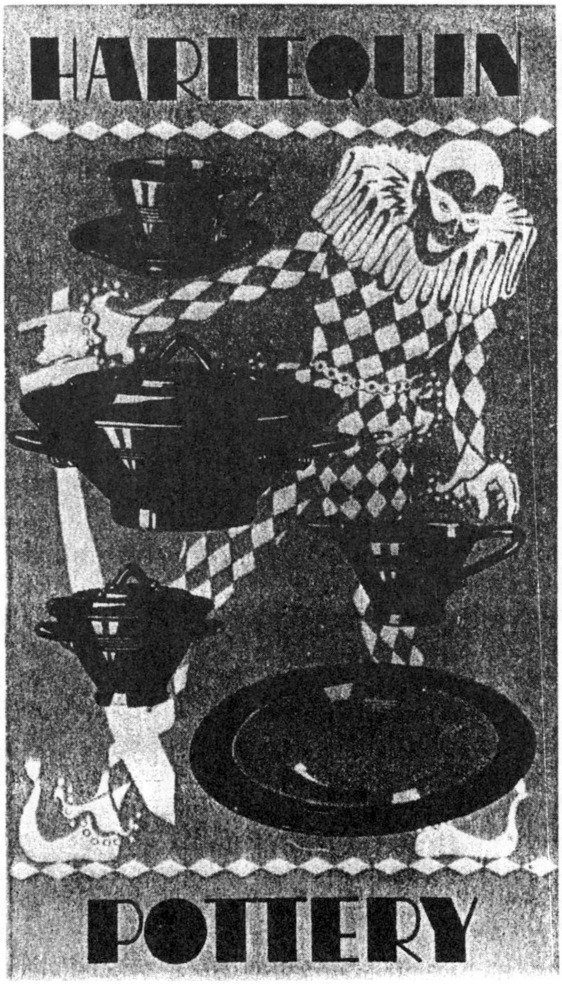


FIG. 6.—In the mid-1930s, Homer Laughlin's art director, Frederick Hurten Rhead, orchestrated the development of "Harlequin"-brand tableware for F. W. Woolworth and Company. Designed to appeal to consumers with modernistic tastes, Harlequin was sold alongside more conventional lines embellished with decal patterns. Woolworth's customers were a myriad lot. (Brochure, Homer Laughlin China Company, Executive Files.)

Emerson might build big machines to help reduce certain costs, but in a batch-production firm, technology could never be a cure-all for slumping sales. Sweeping mechanization simply was incompatible with Homer Laughlin's diverse product lines.

Emerson tried to balance Homer Laughlin's requirements for both quantity and quality as he designed the company's mass-production equipment. Among the many devices that he engineered for Homer Laughlin, those Emerson created for the clay shops testify to his Fordist desires "to bring the work to the worker whenever possible" and "to keep the product at all stages of the process either moving or mobile."[104] Pottery executives generally agreed that the clay shop was a tremendous financial burden; one observer noted that a pottery accrued 40 percent of its labor costs in this department. In the mid-1930s, Homer Laughlin's clay shops still resembled those in 19th-century Staffordshire, where jiggering teams fabricated small lots of ware. Craft traditions, including subcontracting and apprenticing, were entrenched.[105] The jiggerman hired, supervised, and paid his helpers, as journeymen had done for generations, and the NBOP protected these craft practices. Jiggermen preferred unemployment to compromise; they rejected USPA proposals for reduced piecework rates and an alternative hourly wage system. In one instance, they refused to make a product line designed to compete with Japanese imports, resulting in the loss of a significant sale for Homer Laughlin. The tenacity of jiggermen also created throughput problems, and kilnmen sometimes sat idle for lack of ware to fill saggers.[106]

Emerson recognized that the mechanization of Homer Laughlin's clay shops would not only weaken these strongholds of male craft autonomy but also would allow the firm to realize significant gains in throughput without jeopardizing its flexibility and vigor in design and decoration.[107] Simple flat and shallow forms—plates, bowls, and saucers in various sizes—comprised the bulk of any pottery's output. The majority of a pottery's production—that is, flat forms in any

[104] Walter Howard Emerson, "A Pottery Plant on a Conveyor," *Ceramic Industry* 24 (June 1935): 342–47 (p. 342 for quotes). Emerson echoed Henry Ford's sentiments. See Henry Ford, *My Life and My Work* (New York, 1926).

[105] William J. Miller to Edwin M. Knowles China Company, July 15, 1936, HLCC ED, Emerson files, notebook no. 40: X-101; Walter Howard Emerson, "Description of Some of the Apparatus and Methods Used in the Plants of the Homer Laughlin China Company and the Edwin M. Knowles China Company at Newell, West Virginia," December 12, 1942, p. 3, HLCC ED, Emerson files, notebook: Homer Laughlin and Edwin M. Knowles.

[106] "USPA/NBOP Proceedings," August 12–15, 1930, pp. 14–15, KSU NBOP Papers, box 47, folder 1; "USPA/NBOP Proceedings," December 2–5, 1930, pp. 169–90, KSU NBOP Papers, box 47, folder 3; "USPA/NBOP Proceedings," August 11–21, 1936, p. 356, KSU NBOP Papers, box 49, folder 1.

[107] William Howard Emerson, typescript of an article for the American Gas Association, June 6, 1935, HLCC ED, folder: No. 5: Article on New Plant; Emerson, "A Pottery Plant on a Conveyor," pp. 342–47.

shape—could be fabricated on mechanical jiggers, sprayed with clear or colored glazes, and perhaps decorated with applied patterns. The pottery still had to employ dishmakers, casters, and decorators to fashion large flat pieces, to make fancy hollowware, and to embellish goods with unique patterns. By mechanizing flatware jiggering, Emerson employed mass-production methods to make the base of a product, leaving the firm's managers with a spectrum of production opportunities, particularly in the realm of finishing or decorating. The mechanization of jiggering, therefore, did not compromise Homer Laughlin's commitment to flexible production.

Emerson accomplished his reorganization of Homer Laughlin's jiggering shops in several stages, all too complex to describe here in detail. Briefly, Emerson first devised a stopgap measure to reduce clay-shop costs in one of Homer Laughlin's oldest factories, Plant No. 5. He repositioned two members of each jiggering crew—a journeyman and a batterout—around an enormous monorail conveyor linked to a continuous dryer (fig. 7). With this modified hand-jiggering system, Emerson had created a self-contained clay shop and eliminated what he considered to be nonessential unskilled personnel, the moldrunners.[108] Emerson gradually refined his interest in the continuous flow of materials into a passion for mechanization that was given full expression in his designs for the "King Kong" and "Big Flat" jiggers during the late 1930s and the 1940s.

When the first King Kong machine was unveiled to the trade in 1940, one visitor claimed it would "do for pottery what the [automatic] loom has done for textiles."[109] Whereas a traditional three-person jiggering crew could make about 2,400 dinner plates in an eight-hour shift, one fully automatic King Kong machine produced 9,600 items—four times the conventional quantity—in the same period (fig. 8). Each of the six high-speed King Kong machines did the work of several jiggering teams while operatives monitored the process. Handwork was entirely eliminated from the forming stage in this flatware operation. Like Emerson's earlier installations, each King Kong was a self-contained unit that was linked to previous and subsequent operations through an extensive network of conveyors.[110]

Meanwhile, American pottery managers in non-union plants were

[108] Emerson, "A Pottery Plant on a Conveyor," pp. 342–47. The setup at Plant No. 5 was unique in the American pottery industry. See "USPA/NBOP Proceedings," August 11–21, 1936, p. 286, KSU NBOP Papers, box 49, folder 1.

[109] "Jigger Twenty Plates Every Minute!" *Ceramic Industry* 34 (June 1940): 30–34.

[110] Ibid.; "Jigger Tool Life Increased Fourfold by Heat Treating at 1,850 Degrees Fahrenheit," *Ceramic Industry* 37 (July 1941): 35, 38.

FIG. 7.—In 1935, Emerson first applied his knowledge of continuous-flow production to revamping the jiggering installation at Homer Laughlin's Plant No. 5. Here, a journeyman (foreground) and his batterout (rear) work on a moving assembly line. The wall at their left conceals the enormous drying conveyor, which delivered molds to the batterout and carried newly formed ware away from the journeyman. (From *Ceramic Industry* 24 [June 1935]; courtesy Smithsonian Institution, 93-13007.)

also turning to mass-production equipment in their steady quest to expand their foothold in the national market. The Scio-Ohio Pottery Company was experimenting with single-fire ware, casting machinery, and automatic jiggering equipment to make undecorated white pottery, which it sold at rock-bottom prices to mass merchandisers.[111] This threat to Homer Laughlin from nearby Ohio was exacerbated by competition from decorated wares made in distant Michigan. In

[111] "A Most Unusual Dinnerware Plant," *Ceramic Industry* 15 (July 1930): 22–23; "USPA/NBOP Proceedings," February 26–March 2, 1935, pp. 296–99, KSU NBOP Papers, box 48, folder 5. Joseph M. Wells reported that Scio sold plain white cups at $.31 per dozen, undercutting HLCC's lowest price of $.32 per dozen. See "USPA/NBOP Proceedings," August 11–21, 1936, pp. 14–15, KSU NBOP Papers, box 49, folder 1; "Mechanization Most Important Trend," *Ceramic Industry* 36 (January 1941): 72, 74; "How Scio Makes Pottery!" *Ceramic Industry* 38 (February 1942): 40–55. By 1946, Scio-Ohio was the world's largest producer of plain whiteware. See "Scio Wages Up 15 Percent," *Ceramic Industry* 46 (February 1946): 42; and Norman E. Nygaard, *Lew Reese and His Scio Pottery: The Story of a Business with a Heart* (New York, 1948).

Fig. 8.—Six of Emerson's King Kong machines ran 24 hours a day to make 173,000 pieces. Here an operative monitors some of the jiggers, ca. 1940. (Homer Laughlin China Company, Engineering Department, Emerson files.)

Mount Clemens, the enterprising Doll had grabbed the mechanization bull by the horns and was developing jiggering equipment. Doll's new engineering cohort was William J. Miller, a retired glass machinery builder from Swissvale, Pennsylvania. Miller's interest in automatic clayworking equipment dated to the late 1920s, but not until the early 1930s did he install one of his jiggering machines at Doll's former factory in Ohio, the W. S. George Pottery Company. Although unsuccessful, the experiments at W. S. George captured the attention of Doll, who brought Miller to Mount Clemens, providing the inventor with space and money for a machine development project. This time, Doll was not willing to share; during the depression, the risks of collegiality were too great. By the late 1930s, Mount Clemens and Homer Laughlin were in a race to perfect automatic jiggers. By 1942, Doll was operating three so-called Miller machines. The early 1940s witnessed a protracted legal battle between Miller and Emerson, as the Pittsburgh inventor sued Homer Laughlin and its chief engineer for patent infringement, and vice versa.[112]

[112] In 1934, Emerson and other Homer Laughlin representatives visited Miller's shop near Pittsburgh to determine if they wanted to rent his jiggering equipment. Nonplussed by the performance of Miller's machine, Emerson immediately made sketches

During the 1940s, Emerson created an improved automatic jigger, the Big Flat, designed to circumvent Miller's patent claims. Three Big Flat jiggers were installed in Homer Laughlin's Plant No. 4, dating from 1906. Whereas one King Kong produced twenty plates per minute, a Big Flat could make up to twenty-four shallow, or flat, pieces during the same period. More versatile than King Kong jiggers, the new machines could also make a greater range of flat shapes and sizes. Under an agreement with the NBOP, Homer Laughlin either used displaced workers to run its automatic jiggers or reassigned the men to different operations. Former jiggermen tended the monolithic machines, while ex-batterouts and ex-moldrunners removed finished products from passing conveyors. Whereas they previously earned piecework rates, machine tenders now received hourly wages plus bonuses based on their jigger's output. Many of these laborers earned more than batterouts, moldrunners, and finishers but less than journeymen on conventional jiggering teams.[113]

Whereas Emerson's Iron Horse moved people and products, his automatic jiggers brought products to people. The Big Flat jigger represented the culmination of Emerson's mass-production designs for Homer Laughlin. In the thirteen years since he left Lamson, Emerson's career had been shaped by the objective of realizing cost-effective quantity-production methods for a company devoted to flexible production. The path to selectively Fordizing the world's largest pottery had been a craggy one. Homer Laughlin's batch-production system—and its union workers—were formidable opponents for Emerson, who venerated mass-production techniques. Not surprisingly, union leaders had castigated Emerson and his staff, branding these college-educated mechanical engineers as inept "efficiency experts" who brought "the last word in chain systems" from

for his own automatic jigger. In late 1934 or early 1935, Homer Laughlin's executives authorized Emerson to build his first automatic jiggers. See "Miller System Machine Revolutionizing Method of Pottery Manufacture," *National Glass Budget* 53 (April 2, 1938): 3, 7; Critchlow, "Memorandum of Mr. Emerson's Early Activities" (n. 85 above).

[113] "'Walking' Conveyor Carries Molds to Homer Laughlin's New Jiggers," *Ceramic Industry* 51 (November 1948): 86–87; "A Pottery Conveyor," *Die Castings* 7 (January 1949): 38, 66. For the NBOP response, see Shotliff (n. 23 above), pp. 268–69. On pay rates, see Charles F. Goodwin to USPA, November 28, 1939, KSU NBOP Papers, box 19, folder 3. In 1942, workers with three years of experience on automatic jiggers earned $1.35 per hour. In contrast, conventional jiggermen at Homer Laughlin earned from $1.40 to $1.77 per hour, depending on their experience, the aptitude of their helpers, and the piecework rates for various products. See *White Granite and Semi-Porcelain Wage Scale and Size List Agreement between the NBOP and the USPA* (February 1942), pp. 16, 25, KSU NBOP Papers, box 49, folder 18; "Study of Journeyman Jiggermen at Homer Laughlin, 1942," KSU NBOP Papers, box 19, folder 32.

the automobile industry to the potteries only to create ineffectual "new contraptions" such as the Iron Horse.[114] Emerson never lived to see his Big Flat jigger "bring the work to the worker"—or to hear his critics in the NBOP condemn this device. Perhaps the protracted legal battles against Miller had been too stressful for the hard-driving chief engineer. As machinists were building his latest appliance in 1947, Emerson succumbed to a heart attack on the shop floor.[115]

### *Preserving Flexibility*

Thus far, this history of technology at Homer Laughlin during the interwar and immediate post–World War II period has focused on the firm's efforts to emulate mass-production models pioneered by automobile manufacturers. The array of new technologies installed at the West Virginia pottery from the 1920s through the 1940s certainly enhanced the firm's throughput and insured its supremacy as the tableware industry's high-volume, low-cost producer—at least among union plants. But whether or not the "new regime of push-button, sprocket, and bearing" testified to the widespread displacement of craft potters—to the marginalization of workers with handicraft skills—depended on where one looked in Homer Laughlin's factories.

At first glance, Homer Laughlin appears to have embraced Henry Ford's formula for mass production, a recipe that included "mechanization, high wages, low prices, and large-volume output."[116] A closer examination demonstrates that Homer Laughlin's brand of mass production succeeded precisely because it was not thoroughgoing Fordism. Unlike Doll at Mount Clemens, Homer Laughlin's managers achieved high-volume output without totally alienating its work force. Labor-management relations were tense at times, but they never fractured. Managers and engineers deliberately devised a mass-production system that was flexible by selectively blending continuous-flow techniques with Staffordshire methods. Preserving certain British pottery traditions went hand in hand with the decision to preserve Homer Laughlin's diversified customer base.

Without question, technological change in any mid-20th-century pottery was circumscribed partially by the nature of vitreous materials. Ceramic ware was extremely fragile, even after numerous firings. Products in various stages of manufacture all had to be handled with

[114] "USPA/NBOP Proceedings," August 11–21, 1936, pp. 230–36, 937–76 (quotes on pp. 230 and 974), KSU NBOP Papers, box 49, folder 1.

[115] Mooney interview (n. 86 above).

[116] Hounshell, *From the American System to Mass Production* (n. 9 above), p. 305.

the utmost care. In many operations, human hands had no substitutes. Ceramic raw materials were almost impossible to move around a plant with conventional quantity-production tools and methods. Some raw materials were used in powdered form, others in a coarsely ground state, and still others in rough lumps, so it was cheaper and more efficient to transport those ingredients around a factory with pushcarts than to design special handling equipment. Wet glazes were extremely corrosive and incompatible with most industrial spray guns that had been developed for ordinary paints. Conveyors that worked on magnetic principles, such as those designed to transport metals around canneries, were useless in potteries, since clay products were nonmagnetic.[117]

But Homer Laughlin's mechanical engineers, who designed and built most of their own equipment, consciously avoided wholesale mechanization. One of Emerson's objectives, after all, had been the elimination of nonproductive work, rather than the eradication of useful labor. While Emerson's machines created technological unemployment in some departments—most notably in clay shops devoted to manufacturing small flat items—wholesale deskilling was never part of the chief engineer's game plan. Consumer demand long had dictated that Homer Laughlin retain a high degree of resiliency in prices and style. Workers who added value to goods were among the pottery's greatest assets.

At Homer Laughlin, automatic ware-forming devices never fully replaced manual labor but complemented traditional techniques used in the firm's clay shops and casting shops. The company still depended on hand jiggering teams, dishmakers, and casters to make flat- and hollowware that not only filled its hungry tunnel kilns, but also satisfied customers' expectations for novel goods (fig. 9). Shapes such as large platters, covered casseroles, pitchers, and coffeepots simply could not be made by Emerson's big machines. In 1947, general manager Joseph M. Wells anticipated that 43 percent of Homer Laughlin's production in the subsequent year would be made by conventional methods.[118]

Emerson's mass-production equipment also coexisted alongside conventional decorating departments at Homer Laughlin. Although Bleininger, Rhead, and their successors in the ceramics laboratory and art department designed wares that depended on colored glazes

[117] Wells, "Acres of Dishes" (n. 51 above), pp. 697–700; Mooney and Gaston interviews (n. 86 above).

[118] Joseph M. Wells, "Memorandum on Production in 1946 and 1948," May 26, 1947, HLCC ED, Emerson files.

FIG. 9.—Skilled workers remained essential components of Homer Laughlin's production system under the "reign of the robots." Here, in the 1940s, a caster finishes gravy boats. (Homer Laughlin China Company, Engineering Department, Emerson files.)

and unique shapes for their eye appeal, Homer Laughlin continued to make and sell dozens of lines ornamented with decals and other applied decorations. In 1942, Wells admitted that there was "no more uniformity between decorated patterns" than "among men's faces," acknowledging that decorating was a calculated job that constantly "required watching."[119] Those charged with the task of deliberating over embellishments were the firm's skilled or semiskilled *decorators*, usually women who had served two-year apprenticeships under NBOP agreements. *Decal girls* labored at conveyors that brought them supplies and blanks, but, like their male counterparts in clay shops and casting rooms who worked on primitive assembly lines, these women possessed considerable practical knowledge. They had to recognize different patterns, know how to position them on ware, understand whether the ware was dry enough, and so on.[120] Still other decorators worked as liners, gilders, stampers, and silk screeners, using a combination of handwork and simple mechanical devices,

[119] "USPA/NBOP Proceedings," August 25–30, 1942, p. 278, KSU NBOP Papers, box 50, folder 2.

[120] "USPA/NBOP Proceedings," August 11–21, 1936, pp. 538–43, KSU NBOP Papers, box 49, folder 1.

such as hand-propelled banding wheels, to embellish objects. The more highly decorated the product, the greater the firm's gains. Even after Emerson's robots were in place, decorating remained one of the keys to design vitality—and to profits.

Nothing speaks to the flexibility of Homer Laughlin's mass-production system more than the company's products. During the early 20th century, Homer Laughlin's owner-managers approached design innovations much like 19th-century Staffordshire potters, making standard shapes covered with more individualized patterns. In 1915, the firm decorated wares from an assortment of five hundred decalcomania patterns.[121] In the late 1920s, decoration was still an important element in many potteries' competitive strategy. USPA officials reminded NBOP leaders of the "constant effort on the part of the manufacturers to vary their decorations and vary their shapes" to distinguish their products from those of their competitors.[122]

Once Emerson's mass-production equipment was installed, Homer Laughlin still provided numerous vendors with a variety of products. Whereas Mount Clemens and Scio-Ohio made cheap, nondifferentiated goods, Homer Laughlin manufactured enormous quantities of pottery in a range of forms, sizes, shapes, and patterns for a wide portfolio of customers.[123] Department stores added new patterns to their china and glass selections each year and sold replacements for broken dishes in inactive patterns. Chain stores expected the pottery to provide wares that were distinguishable from those of their competitors.[124] In the post–World War II era, Homer Laughlin's product line was more diverse than in previous decades (fig. 10). Whereas the firm had offered three dinnerware shapes in 1914, the company made twenty-seven shapes in three clay bodies and nineteen glazes by 1953. Decorating options included decalcomania, underglaze transfer printing, banding, hand lining, machine lining, stamping, spray tinting, silk screening, and stenciling. Homer Laughlin's managers complained about the dearth of skilled workers who could produce the range of styles and designs desired by American consum-

[121] W. Edwin Wells to R. F. Tattersall, March 15, 1915, HLCC EC, box L-461; *Price List and Illustrations, Plain White Wares Made by the Homer Laughlin China Company, Newell, West Virginia, and East Liverpool, Ohio, January 1, 1914,* HLCC EF.

[122] "USPA/NBOP Proceedings," 1928, p. 229, KSU NBOP Papers, box 46, folder 25.

[123] Joseph M. Wells noted that USPA products generally were superior to Mount Clemens wares because Doll used inferior clay bodies. See "USPA/NBOP Proceedings," August 11–21, 1936, p. 766, KSU NBOP Papers, box 49, folder 1.

[124] "Technology and Plant Data Feature Talks at Ohio Whiteware Meeting," *Ceramic Industry* 24 (June 1935): 359.

Fig. 10.—During the "reign of the robots," Homer Laughlin's product lines were more varied than ever before. This 1953 advertising leaflet shows dinnerware decorated with several decalcomania patterns, silk-screened motifs, and bright monochromatic glazes. (Homer Laughlin China Company, Art Department files.)

ers.[125] Mass-production technology had expedited the manufacture of some blank forms, but the making of unusual shapes and the application of many types of decorations still required a good deal of time and human judgment.

This story of technological innovation at the Homer Laughlin

[125] "Report of the Decorating Committee," HLCC ED, folder: Decorating Committee Report, 1950; "Notes on Ware Inventory," December 7, 1953, HLCC EF, box: Inventory Data.

China Company from the 1920s through the 1940s suggests that the models of batch production and mass production are not exclusive categories. At the world's largest pottery, managers and engineers grafted significant aspects of mass-production technology onto existing batch-production conditions. The new manufacturing system increased the velocity of throughput in departments dedicated to forming, glazing, and firing. The constraints on the full expansion of mass-production techniques in this industry were not principally technological. They lay primarily in the nature of the market, and partly in the nature of the labor system. Managers and engineers henceforth necessarily preserved many of the manual tasks that enhanced the company's ability to respond to consumers' demands for novel dinnerware. Homer Laughlin's version of flexible mass production was suited to a specific time and place—and most notably, to a particular firm and its markets in the home-furnishings trade. Walter Emerson's robots never ruled over a totalitarian technological regime. Instead, they coexisted with the potters of "poem, drama, and tale," with the working-class men and women who remained vital components of Homer Laughlin's manufacturing system well into the mid-20th century.

# [13]

# *Technology and the Workplace: Skilled and Production Workers at Allis-Chalmers, 1900–1941*

STEPHEN MEYER

> Every epoch has its technology, and this technology has the style of the epoch. A style that shows to what extent human acts link and interface with each other: How, if you like, technology undergoes the influence of what we call general history: and, at the same time, acts on history. Here again we have a double problem. It has been called, since Marx, by a contradictory word: the problem of historical materialism. That is, at bottom, nothing more than the problem of total history. [LUCIEN FEBVRE, "Reflections on the History of Technology," 1935][1]

The problem with the current practice of the history of technology is nothing less than what Lucien Febvre, a founder of the famous "Annales" school of social and economic history, labeled "the problem of total history." With a few notable exceptions, historians of technology have tended to avoid the relationship between innovations in production technology and their influence on the work processes and social relations of the workplace. They have written about the machines and the innovators but not about the people who worked at the machines or about the politics of production. Unless historians of

DR. MEYER, author of *The Five Dollar Day: Labor Management and Social Control in the Ford Motor Company, 1908–1921*, is currently completing a book manuscript, "Work, Class, and Power: The Making and Unmaking of a Militant UAW Local at Allis-Chalmers, 1900–1950," and is researching workers and technology in the United States automobile industry. He teaches history and labor studies at the University of Wisconsin—Parkside. He wishes to thank Phil Scranton, Margo Anderson, Stephen Victor, and Lindy Biggs for their comments on earlier versions of this essay. One version was presented at the 1984 Cambridge, Massachusetts, meeting of the Society for the History of Technology.

[1]Lucien Febvre, "Reflections on the History of Technology," *Annales d'histoire économique et sociale* 6 (1935): 532–35; reprinted in *Technology and History* 1 (October 1983): 15.


0040-165X/88/2904-0004$01.00

technology move beyond their partial explanations of the technical past, they will remain condemned to the periphery of the larger historical profession.[2]

A truly meaningful examination of technological innovation requires more than the view from the top—that is, from the perspective of the manufacturers, the managers, and the engineers. For workers, modern production machines, methods, and processes were much more than artfully crafted devices. The new "labor-saving machines" frequently brought anxious stirrings to factory life. Such devices altered the workers' world—their shop traditions, their level of skills, the patterns of control, and the social relations of the workplace.

In this article, the Allis-Chalmers Manufacturing Company serves as a means of entering "the hidden abode of production" and examining the technical dimensions of "the hidden history of the workplace." Most important, it offers an inclusive view of technical practice from both sides of the capital-labor divide, rather than simply prescriptive tales of engineering ideals.

### *Skilled Workers and the "Big Stuff"*

The Allis-Chalmers Manufacturing Co. originated as the E. P. Allis Co. in 1863. At the turn of the century, that firm specialized in the manufacture of a wide range of industrial equipment—flour and sawmill machinery, steam and pumping engines, and electrical generators. It manufactured "big stuff," equipment sometimes weighing over 100 tons that required the highly refined skills of molders, patternmakers, machinists, and electrical workers.

To be sure, some of the Allis products required many duplicate parts, such as turbine blades whose manufacture entailed repetitive mass-production techniques. Yet, the typical Allis product often required the design of something unique, the creation of patterns to

[2]John M. Staudenmaier, "What SHOT Hath Wrought and What Shot Hath Not: Reflections on Twenty-five Years of the History of Technology," *Technology and Culture* 25 (October 1984): 719–20, 725–28; David Hounshell, "Commentary: On the Discipline of the History of American Technology," *Journal of American History* 67 (March 1981): 854–65; and Darwin Stapleton and David Hounshell, "The Discipline of the History of American Technology: An Exchange," *Journal of American History* 68 (March 1982): 897–902. For the most recent analytical work on the politics of the work process, see Michael Burawoy, *The Politics of Production* (London, 1985), pp. 21–68, 122–32; Michel Aglietta, *A Theory of Capitalist Regulation: The U.S. Experience* (London, 1979); Donald MacKenzie, "Marx and the Machine," *Technology and Culture* 25 (July 1984): 473–502; and David Montgomery, *Workers' Control in America: Studies in the History of Work, Technology, and Labor Struggles* (New York, 1979), and *The Fall of the House of Labor: The Workplace, the State, and American Labor Activism* (New York, 1987).

make molds, the founding of a large casting along with smaller ones for components, the machining of castings and other parts, sometimes the addition of electrical wiring, and complicated assembly processes. The "big stuff" demanded the skills, the customs, and the culture of craft production.

Against this background, the new Allis-Chalmers Company originated in a technologically dynamic environment. At the turn of the century, the metal trades were undergoing extensive and rapid mechanization. Cities were rapidly expanding their electrical systems and were electrifying their transit. Factories and plants were shifting from lineshafting to individual electric motors to power the new production machines. Farmers were adopting mechanized forms of production. The American rise to world economic predominance involved increased military production. As Allis-Chalmers redirected its resources toward the manufacture of more and smaller electrical products, tractors and agricultural equipment, and war materials, its engineering and manufacturing operations moved from specialized batch production to modified mass production.

Even prior to the formation of the Allis-Chalmers Company, E. P. Allis officials had planned the expansion of their production facilities. In November 1900, they purchased a 100-acre site on the outskirts of Milwaukee with access to three major railway lines. Officials incorporated the newly merged Allis-Chalmers Company in early May 1901. Two weeks later, right in the midst of corporate expansion and reorganization, Allis-Chalmers machinists struck the new firm. At the time, the mechanization of the metal trades undermined the customary skills of craftworkers and offered management opportunities to recast the social structure of work and to reassert control over the workplace. Milwaukee machinists joined their union brothers in a national strike to maintain shop customs and higher wages.[3]

"The men," said Edwin Reynolds, Allis-Chalmers vice-president, chief engineer, and general plant superintendent, "have taken the bit into their mouths, and I guess we shall have to let them chew it for a while." A former employee and associate of George Corliss and a former president of the American Society of Mechanical Engineers, Reynolds was then president of the National Metal Trades Association, which had recently negotiated and signed the famous 1900 "Murray Hill Agreement" with the International Association of Machinists. This pact had established a short-lived truce between the metal trades employers' organization and the machinists' union. Reynolds's state-

[3]Walter F. Peterson, *An Industrial Heritage: Allis-Chalmers Corporation* (Milwaukee, 1978), pp. 102–5; Montgomery, *Workers' Control*, pp. 48–83, and *Fall*, pp. 80–213, 259–69.

ment was the management response to demands from an Allis-Chalmers shop committee of machinists. The Milwaukee machinists called for a 12½ percent wage increase and a reduction of the working day to nine hours. Other issues included the machinists' dissatisfaction with newly established piece-rate and premium-wage payment systems.[4]

On May 20, 1901, about 800 workers walked out of the Allis-Chalmers Reliance Works on Clinton Street in Milwaukee. A few days after the machinists walked out, about fifty Allis-Chalmers patternmakers also struck. In all, about 83 percent of the machinists and 63 percent of the patternmakers left the Allis-Chalmers plant in the strike's first days. The principal strike issues echoed worker concerns about technical change in their shops. The skilled Allis-Chalmers workers wanted their fair share of the fruits of new production methods and technologies.[5]

Allis-Chalmers management and other Milwaukee industrialists took a firm position. In June, company officials announced that workers who failed to appear for work would be "considered no longer in our employ." The Milwaukee employers held the line on the principles of management rights and the open shop. Gradually, the skilled machinists returned to work. For Allis-Chalmers workers, the strike ended in total defeat in late July 1901.[6]

During the machinists' strike, Allis-Chalmers officials had begun the construction of a new plant in West Allis, an area, the press predicted, "destined to become an important manufacturing village." The new West Allis plant was a part of the company's strategy for dealing with intractable skilled labor. A new and expanded physical plant meant the introduction of the latest technical improvements. Indeed, the machinists' reasons for the strike suggested their responses to existing changes in their work tasks and work routines in the old Allis-Chalmers plant. As machinists' work became more mechanized, more machine paced, and more routinized, the machine operators actually worked harder and felt entitled to greater compensation for shorter hours. Moreover, their protest against piece-rate and premium sys-

[4]"Employers Have Taken Stand of Defiance to Machinists," *Milwaukee Journal* (hereafter cited as *MJ*), May 18, 1901; "Machinists Walk Out of Milwaukee Shops," *MJ*, May 20, 1901.

[5]"Employers Have Taken Stand," *MJ*, May 18, 1901, and "Machinists Walk," May 20, 1901. On the number of workers in the different shops of the Allis-Chalmers plant, see "The General Scheme of the New Milwaukee Plant of the Allis-Chalmers Company," *American Machinist* 25 (June 12, 1902): 840. On skilled workers, see Montgomery, *Fall* (n. 2 above), pp. 144–213.

[6]"Breach Widening," *MJ*, May 30, 1901, and Peterson, *Industrial Heritage* (n. 3 above), pp. 117–19.

tems reflected changes in the management and organization of their work.[7]

First opened in September 1902, the new Allis-Chalmers West Allis Works was a thoroughly modern plant for the batch production of large machinery. Edwin Reynolds designed and constructed the plant on a "unit" plan. "One complete unit," *Machinery* reported, "has now been built and consists of a pattern shop and storage building, foundry, machine shops, power house, blacksmith shop, erecting shop, and yard storage, with cranes, tracks, etc." A single plant unit contained the necessary shops for the sequential production of Allis-Chalmers metal and electrical products. Reynolds calculated the size of the different units from the shop statistics of the old Allis-Chalmers Reliance Works to establish a formula for future expansion and "to preserve the balance between the pattern and foundry work and the machine shop work."[8]

A reflection of the traditional Allis-Chalmers products, the West Allis plant design followed the classic pattern of batch production of expanded manufacture through the extension of traditional production units rather than the redesign of the labor process. To use Alfred Chandler's term, the manufacturing principle was to achieve a more intensive "throughput," that is, faster movement through the sequence of productive operations. For the most part, the design of the plant retained customary worker skills and also pushed skilled craftsmen to work harder through the reduction of downtime between operations.[9]

The scale of the new buildings reflected the hugeness of the early Allis-Chalmers products. The pattern shop and the pattern storage area occupied almost 68,000 square feet, the foundry over 124,000 square feet. The foundry had three fully equipped modern cupolas and three separate bays with a higher center bay for the large work. A third section contained two machine shops, a blacksmith shop, and

[7]"West Allis Makes New Factory Hub," *MJ*, May 26, 1901. For changes in production technology, see Stephen Meyer, *The Five Dollar Day: Labor Management and Social Control in the Ford Motor Company, 1908–1921* (Albany, N.Y., 1981), pp. 9–65; Stephen Meyer, "Mechanical Engineers and Automobile Workers," paper presented at International Conference on the Automobile Industry and Its Workforce at Coventry, England in June 1984; and David A. Hounshell, *From the American System to Mass Production 1800–1932: The Development of Manufacturing Technology in the United States* (Baltimore, 1984), pp. 217–301.

[8]"General Scheme" (n. 5 above), pp. 840–44; "The Great West-Allis Plant of the Allis-Chalmers Company, Milwaukee, Wis.," *Machinery* 9 (February 1903): 285–86; and "Shop Extensions at the Plant of the Allis-Chalmers Co., Milwaukee, Wis.," *Engineering News* 56 (July 12, 1906): 828–29.

[9]Alfred D. Chandler, Jr., *The Visible Hand: The Managerial Revolution in American Business* (Cambridge, Mass., 1977), p. 241.

an erecting shop. This section was an E-shaped series of buildings with traveling craneways between the different shops, which occupied over 389,000 square feet. In all, the plant encompassed almost 585,000 square feet of factory floor. The different buildings were arranged with attention to the sequence of manufacturing operations and with a network of railroad tracks and craneways for the transportation of work in progress from shop to shop. In 1906, officials decided to expand the West Allis plant with the addition of more pattern-shop, foundry, and erecting-shop space to provide for three larger machine shops. They also added a shop for the manufacture of smaller electrical products. These seven new shops followed the dimensions of the original unit plan for expansion. After the new construction, the West Allis works totaled over 1.4 million square feet of floor space.[10]

Although the entire plant created sizable, specialized metal goods, corporate engineers equipped the no. 1 machine shop for work on the "big stuff." It possessed a central crane with a 40-ton capacity, and two smaller galleries in the shop had auxiliary cranes with 10-ton capacities. An extensive "floor plate" system, which measured 200 feet by 24 feet, covered a substantial portion of the main no. 1 machine shop. A recent industrial innovation, the grooved floor plate facilitated machine work on massive castings. Attached to a substantial adjustable foundation, it held the castings and also the machine tools for work on them. The recent application of electric-motor drives facilitated the use of floor plates, since they increased the portability of large machine tools. Also attached to the floor plate, electric motor drives ended reliance on complicated systems of overhead shafting and belting to provide power for the machine tools. Sometimes, several machines worked on a large casting simultaneously. Under this system, the machine tools and their motor drives were attached to the solid and stable floor-plate foundation and adjusted for their work. Instead of moving the heavy work to the machines, cranemen hauled a casting to the floor plate and skilled machinists attached it to the floor plate and leveled it. Cranemen then brought the successive machine tools for different operations to the large casting. Machinists adjusted and leveled the huge machines for their work. A work crew of skilled machinists and their helpers, apprentices, and laborers oversaw the full sequence of machine operations.[11]

[10]"General Scheme" (n. 5 above), p 841; "Great West-Allis Plant" (n. 8 above), pp. 285–88; and "Shop Dimensions at the Plant of the Allis-Chalmers Co., Milwaukee, Wis.," *Engineering News* 56 (July 12, 1906) 28–30. See also Peterson, *Industrial Heritage* (n 3 above), p. 111.

[11]"Great West-Allis Plant," pp. 287–88, and "Floorplate Work in the Allis-Chalmers Shops," *Machinery* 12 (August 1906): 627–28.

Despite its reliance on skilled craftsmen, the Allis-Chalmers Company also utilized the most advanced industrial equipment. This machinery catered to specialty production rather than mass production. In the no. 1 machine shop, massive machine tools also testified to the incredible dimensions of the machine work in the new West Allis plant. "The machine tool work of the floorplate," a technical journalist reported, "is a remarkable collection of tools and indicates the disposition on the part of the management to provide every facility for turning out the heaviest class of work economically and accurately." For example, the new Allis-Chalmers equipment included two Morton traveling shapers, one 4 × 4 × 6 feet and one 6 × 6 × 11 feet. The latter was "the largest shaper of this type ever built." The no. 1 machine shop also had a pair of 10-foot Bement-Niles rotary planers with 26-foot beds. The cutting heads were 120 inches in diameter.

These and other equally large machine tools finished huge castings from the plant's foundries. In the 1930s, Allis-Chalmers officials boasted: "Castings of any size up to 120 tons are made here [in the West Allis plant]. To date the foundry has produced 160 castings weighing over 100 tons each." These large and unique industrial and electrical products demanded complex and varied skills from workers in the metal and electrical trades.[12]

After the completion of the West Allis plant, a 1906 molders' strike illustrated more clearly the complex relationships among technical innovation, transformed work, and the open shop. At Allis-Chalmers, industrial expansion meant a differential pace of mechanization for the various metalworking operations. In addition to rapid technical change in machinists' work, this was also a period of rapid mechanization in the highly skilled molders' trade. The process began with the invention of hand, and later pneumatic, hydraulic, and steam ramming devices to shape the sand around patterns for the mold. Then, mechanical engineers devised pattern-drawing machines to remove the pattern from the mold. In the 1890s, the Tabor and other molding machines had fully automated the molders' craft. Mary Loomis Strecker, a labor economist who examined the trade, reported: "The brains which formerly had [been] vested in the molder were transferred entirely, through the efforts of engineer and pattern maker, to the machine." Although these machines had the greatest effect on employers who manufactured long runs of similar products, they also affected the large and unique work in the specialty plants, where huge

[12] "Great West-Allis Plant," pp. 288–89; "Floorplate Work," pp. 627–28; Allis-Chalmers Co., *Works and Products of the Allis-Chalmers Co.* (Milwaukee, 1911), passim; Allis-Chalmers Manufacturing Company (ACMC), *Eighty-Eight Years of Progress: Allis-Chalmers* (Milwaukee, 1935), p. 10.

machines such as turbines or engines often required numerous smaller castings of component parts.[13]

The mechanization of the molder's trade brought the specialized molder into the foundry. Trained in just a few weeks, an unskilled laborer could readily become a "handyman" molder who could easily produce large runs of similar products. Without identifiable craft skills, these specialized handymen were production workers in the foundry. The dilution of skill, Strecker noted, had advantages for the manufacturer: "The value of the machines to the manufacturer is largely determined by the amount of labor and skill they eliminated from the molding process. Again, skilled molders had not been trained to the heavier work made necessary by the machines; their apprenticeship had been served to brain and not to brawn."

Without the skilled worker's formal apprenticeship, which sustained a craft culture on the shop norms of technology, productivity, and work, the specialized molder did not bear the molders' shop tradition of the "set," an informal quota of output which the molders and later their union attached to particular patterns in the shop. In 1910, one foundryman claimed that the new molding machines raised productivity from 100 to 500 percent. The new production technologies raised productivity through the elimination of workers with established traditions in the foundries.[14]

As in the earlier management offensive against the machinists' union, the metal manufacturers used mechanization to undermine skilled craftsmen and to reassert management control over work processes. In 1904, the National Foundrymen's Association, an employers group, refused to continue its former recognition of the molders' union and began to install the new molding machines on a large scale in their foundries. The result was a series of clashes, which culminated in the national "great molders' strike" of 1906.[15]

Although Milwaukee foundrymen were offered a 10 percent wage increase and the nine-hour day for skilled molders, the union molders "enthusiastically" decided to strike on May 1, 1906, for a larger pay increase and for restoring employer recognition of union molders and

[13]F. T. Stockton, "The International Molders Union of North America," *Johns Hopkins University Studies* 39 (1921–22): 186–87; and Mary Loomis Strecker, "The Founders, the Molders, and the Molding Machine," *Quarterly Journal of Economics* 32 (February 1918): 284.

[14]Stockton, "International Molders Union," pp. 186–87; Strecker, "Founders," pp. 289, 297–98; and A. O. Backert, "The Introduction of Molding Machines in Foundries," *Transactions of the American Foundrymen's Association* 25 (1917): 713.

[15]Stockton, "International Molders Union," pp. 194, 196; and Strecker, "Founders," pp. 295, 301.

also of union coremakers. (Most Milwaukee foundries had ceased recognizing the molders' union in 1904; Allis-Chalmers officials did so upon contract expiration in 1906.) The founders' wage offer reflected the technical changes within the metal trades industry. Their proposal, which raised the rates only of skilled floor molders, distinguished between high wages for the floor molders and lower wages for the more specialized and mechanized bench molders. In all, from 900 to 1,200 molders, reports suggested, struck Milwaukee foundries. The largest contingent of strikers, about 400 molders, came from the Allis-Chalmers plants. In the first days, some machinists talked of a general citywide metal trades strike. But the machinists resented the molders' refusal to aid their 1901 strike and so crossed the molders' picket lines. After the first week, approximately 1,400 molders had walked out of Milwaukee's principal foundries. The new industrial technologies provided the means for the defeat of craft unionism in the foundries.[16]

Once again, Allis-Chalmers officials took the lead in defending management rights and the open shop and overcame the resistance of militant Milwaukee craft unionists. The new foundry technologies allowed inexperienced workers, less skilled handymen, to perform the formerly skilled work tasks of the striking molders. According to one report, the new machine technology allowed Allis-Chalmers managers to bring in "unskilled men from the streets and boys from high schools and with good foremen who thoroughly understood the business, teach the new men to make excellent castings of repetitive types in no more than four months." Officially, the strike lasted well over a year, but new mechanized foundry practices doomed the cause of the skilled molders in the strike's first months. Strecker's conclusion on the introduction of the new technology fitted well for Allis-Chalmers molders: "It is to a certain extent true that the molding machine and the handyman molder have made possible the open shop foundry."[17]

## *The Uneasy Transition to Mass Production*

During World War I, Allis-Chalmers workers approached still another conjuncture of technology, work, and skilled worker militancy.

[16]"No May Day Strikes Here," *MJ*, May 1, 1906; "Molders Will Strike Today," *Milwaukee Sentinel* (hereafter cited as *MS*, ), May 2, 1906; "Coremakers Cause Strike," *MJ*, May 2, 1906; "Labor Troubles Still Unsettled," *MS*, May 4, 1906; "Both Sides Remain Firm," *MJ*, May 4, 1906; "More Molders Walk Out," *MS*, May 8, 1906; "Declare for an Open Shop," *MS*, May 9, 1906; "Foundrymen Issue Statement," *MS*, May 16, 1906; "Now Want Open Shop," *MS*, May 24, 1906; and Strecker, "Founders," p. 302.

[17]"College Men Break Strike," *MJ*, May 26, 1906; Peterson, *Industrial Heritage* (n. 3 above), pp. 120–21; John P. Frey, *History of a Criminal Conspiracy to Defeat Striking Molders* (n.d, n.p.), p. 3; and Strecker, "Founders," p. 303.

Until this time, the company's mass-production techniques were an adjunct to batch production. In some instances, large numbers of specialized products required limited and modified mass-production techniques. In others, the numerous similar components of specialized goods needed more extensive mass-production techniques. In wartime, however, the high-volume manufacture of military goods forced specialized manufacturers to adopt mass-production techniques and reshaped work processes. At the same time, introduction of a new Allis-Chalmers product line, tractors and agricultural equipment, also pressured officials and engineers to shift from batch to mass-production methods. By the 1930s, a dual industrial structure—batch production and mass production—coexisted in the West Allis works.

For Allis-Chalmers, wartime production began in the summer of 1915. The initial order was a subcontract for machine work on 2 million shell casings. For an initial investment of $230,000 in machines and tools, the company received a $3.7 million subcontract for this military work and hired 500 specialized machinists on the piecework system. So, in addition to increased orders for its conventional product line of industrial equipment, World War I brought a full range of other work to Allis-Chalmers. Mass-production methods made 3.75 million shell casings for Great Britain and the United States. Specialized and limited mass-production methods produced gun turrets, slides, and mounts for naval vessels. And the specialized batch-production methods produced forgings, steel armor plate, marine turbines and engines, and propeller and line shaftings. Between August 1915 and February 1916, the West Allis work force expanded from 3,200 to 5,500.[18]

During the war, American machinists engaged in a second series of national strikes that ultimately touched Milwaukee and again reached down into the West Allis plant. For this period, David Montgomery has noted a series of "munitions strikes" in the metal trades that were associated with a "control strike wave" from 1916 to 1920 and that resulted in the growth of the International Association of Machinists (IAM) nationally from 54,000 to 331,000 members. In April and May of 1916, Jacob Friedrich, a Milwaukee IAM organizer, related that a "large number" of Allis-Chalmers workers enrolled in the machinists' union and "immediately started agitation for a strike, figuring that it was the prudent time because of the European war work coming in." On July 17, 1916, approximately 1,100 to 1,200 workers struck the West Allis plant, for the third time. The main strike demand was the

[18]Peterson, *Industrial Heritage*, pp. 167, 169–75.

eight-hour day with no reduction in pay, an inherently technological issue.[19]

In the new Allis-Chalmers shrapnel department, the large runs of this highly specialized work required mass-production machine tools that required fewer and less skilled operators. One skilled machinist lamented: "It is a continuous circus out there. Lathes are being run by tailors, carpenters, and shoemakers and all kinds of men and the stuff that is spoiled is a shame. The company gets big money, but it has to forfeit $2 for every shell scrapped. There are a lot of men quitting every night, but more are being taken on. Some that are mechanics will do all right, but others will not make their salt." Although machinists generally made good money for this work, the aggrieved machinist complained about the speedup of work in the West Allis plant.[20]

Clearly, when the machinists' strike broke out in 1916, Allis-Chalmers and other Milwaukee workers were in the process of changing production technologies and changing social contours in their shops. By July 21, the strike involved a total of about 3,000 metal-trades workers in nine plants and threatened to involve over 10,000 machinists in all Milwaukee plants. Milwaukee manufacturers followed the strategy of refusing to negotiate, of waiting out the strike. The other metal-trades craftworkers continued to work, and gradually the strike fever wore off. After about eight weeks, Milwaukee and Allis-Chalmers workers drifted back to their shops, having suffered another disastrous defeat.[21]

The decision to manufacture shell casings represented a step toward the manufacture of profitable, mass-produced goods. Increasingly, through the 1910s and 1920s, smaller standardized products—such as electrical switchgear, texrope drives (multi-belted pulleys) to power machine tools, smaller electric transformers and relays, and, most

[19]Montgomery, *Workers' Control* (n. 2 above), pp. 95, 98, 99; Jacob Friedrich, NLRB Testimony, July 23, 1937, p. 1606 in Box 9, Series 11, Record Group 25, National Labor Relations Board, National Archives, Suitland, Md.; "Machinists Plan to Resume Strike Tuesday Morning," *MS*, July 25, 1916.

[20]" 'Speed-Up' Plan Is in Operation at Allis-Chalmers for Shell Work," no newspaper title, May 15, 1915, in Allis-Chalmers, 1901–34, Industrial Clippings File, Milwaukee County Historical Society (hereafter cited as AC Clippings).

[21]"Shell Orders to Increase Here, Report," no newspaper title, May 17, 1915, and "Here's Milwaukee's Contribution to War Supplies," no newspaper title, August 31, 1915, AC Clippings; "2,000 Machinists Quit in 5 Plants," *MJ*, July 19, 1916; "Meeting May End Strikes of Machinists," *MJ*, July 21, 1916; "Fail to Agree to End Strike," *MJ*, July 22, 1916; "Prolong Strike If Necessary," *MJ*, July 24, 1916; "Strike Ties Up Another Plant," *MJ*, July 25, 1916; "More Shop Men Join Strike," *MJ*, July 26, 1916; "Bosses Stand Pat on Strike," *MJ*, July 27, 1916; "Strikers of One Shop Go Back," *MJ*, September 18, 1916.

important, tractors—accounted for an increasing proportion of the West Allis plant's output.

The Allis-Chalmers decision to manufacture tractors proved a successful market strategy in product diversification. In contrast to the boom and bust feature of specialty production, tractor manufacture offered the advantage of steady revenues in the era of the mechanization of American agriculture. It also illustrated how new production technologies delineated and altered the social contours of work and skill in the specialty plant and were in turn influenced by patterns already in place.

In 1914, Allis-Chalmers entered the tractor field with the introduction of the Model 6-12, a small one-plow tractor. The company intended that the 6-12 meet "the requirements of the small acreage farmer, at a price that such farmers could afford to pay." Apparently the plan gained some success, for in 1917 the company formally established its tractor division. Initially, the move into the tractor field barely affected work routines and work processes in the West Allis plant. The production of tractor parts was allocated to the existing foundries and machine shops, and early plant photographs revealed the most conventional assembly methods. Skilled workers actually built tractors from scratch at stationary locations. Detroit's innovative production methods had not yet reached the large Milwaukee industrial and electrical equipment plant.[22]

The success of Allis-Chalmers tractors initiated a process of plant expansion and productive innovation through the 1930s. In 1919, company officials remodeled a pattern storage building and equipped it for tractor production. This no. 1 tractor shop gradually achieved an output of about thirty tractors per day. In 1920, officials decided to add a new tractor foundry to the West Allis industrial site. Since production methods differed for tractors, the new foundry was devoted principally to the manufacture of castings for engines and transmissions. Like earlier new Allis-Chalmers buildings, the tractor foundry was large, occupying 70,000 square feet. It had four 500-foot craneways for the movement of materials through the foundry. Its scale alone revealed ambitious plans for Allis-Chalmers's tractor manufacture; the new foundry had a production capacity of about 100 engine and transmission castings per day. Through the 1920s, tractors constituted an ever-increasing share of the Allis-Chalmers product mix, and the expanding tractor division introduced several new models.

[22]R. A. Crosby, "The Allis-Chalmers Tractor Division," manuscript in F. Institutional Materials: Tractor Group—Historical, Box 3, Allis-Chalmers Papers, Milwaukee County Historical Society (hereafter cited as AC Papers); ACMC, *Eighty-Eight Years* (n. 12 above), p. 7.

However, the design trend they exemplified moved away from the original notion of a small one-plow tractor and toward larger models. Despite the original corporate intentions and strategies, existing production techniques and the complex of craft skills pushed the company in the direction of larger, more powerful, and unique tractors.[23]

In the late 1920s, Allis-Chalmers's growing tractor sales made the firm the United States' third largest producer of agricultural equipment. Hence, officials authorized the construction of the no. 2 tractor shop. Originally planned to produce 3,000 of its new smaller United tractors per year, the company later announced its intention to produce approximately 4,500. First opened in August 1929, the new tractor building occupied over 127,000 square feet. Corporate president Otto Falk emphasized the modern industrial equipment in the new tractor shop: "the desired combination of high quality and low price led to a bold decision. We decided that only the most modern production equipment would be used. Today every piece of equipment in our *new* tractor shop is in itself brand *new!*"

Many other Allis-Chalmers officials reiterated this modernist theme.[24] For example, William Watson, who directed and supervised the West Allis plant's productive operations, alluded to the problem of the divided shop structure of worker skills in the plant. The older products and methods of manufacture still necessitated skilled craftsmen, and the newer ones used more specialized and less skilled production workers. According to Watson, " . . . the [tractor] work was to be segregated from the rest of the works, in which was manufactured a diversified line of equipment ranging from small motors to huge turbines, largely in order to develop a personnel highly specialized in tractor production. Obviously, an entirely different kind of production thinking is needed, for example, for the work of fabricating a gyratory crusher, weighing hundreds of tons." For tractor production, Allis-Chalmers managers and engineers required more attention to plant layout, to the integration of productive operations, and to the use of the latest machine tools.[25]

In another instance, E. A. Harrison, general superintendent of the foundry and pattern shops, discussed the latest technical improve-

[23]ACMC, *Sales Bulletin* (November 1920), *Sales Bulletin* (November 1921), and *Sales Bulletin* (February 1924), in F. Tractor, Box 3, AC Papers; and Peterson, *Industrial Heritage* (n. 3 above), pp. 241–45.

[24]ACMC, *Sales Bulletin* (November 1928), and *Sales Bulletin* (November 1929), AC Papers; and Otto Falk, "An Act of Faith in Modern Equipment: The Allis-Chalmers Model U Tractor," *American Machinist* 72 (March 13, 1930): 435.

[25]William Watson, "All-Modern Equipment Produces Allis-Chalmers Tractor," *American Machinist* 72 (March 13, 1930): 436–37, and Frank J. Oliver, "Special-Purpose Equipment for Accurate Tractor Parts, *American Machinist* 72 (March 13, 1930): 458–62.

ments in foundry practice for production of the new tractor. He also addressed the crucial problem of the relationship between the new industrial technology and savings in the cost and the amount of labor. "Here again," Harrison related, "as in other parts of the plant, mechanization has been used to reduce the number of workers ordinarily employed." In the mechanized foundry, Harrison noted, "only ten men are required . . . to maintain the given production. . . ." The "entirely different kind of production thinking" included detailed attention to the relationships among technology, work processes, work force skills, and manning levels in the West Allis plant.[26]

Despite proclamations about the most modern equipment and the latest shop practices, the no. 2 tractor shop did not achieve state-of-the-art production engineering comparable to the much more technologically advanced automobile industry. An examination of the arrangement of the shop's floor plan reveals only a partial accommodation of the best automotive shop methods. For the most part, machine tools were placed in traditional and orthodox groupings. Instead of being arranged according to the sequence of machine operations, they were grouped according to the type or class of machine tools. The no. 2 tractor shop did, however, utilize line assembly methods, which represented a departure from former Allis-Chalmers practice. Yet these were primitive and slow-moving assembly lines.[27]

Because of the continued success of the tractor division, even through the economically difficult 1930s, Allis-Chalmers officials decided to build yet another tractor shop in 1937. A year later, Joseph Geschelin, an automotive journalist, reported on technical innovations of the no. 3 tractor shop and its new mechanized foundry. He reported of the two new tractor buildings: "They embody the very latest principles of modern plant layout; they feature one of the most interesting mechanized final assembly lines to be found in the tractor industry. Last but not least, the plant has been equipped with production machinery which compares both in modernity and practice with the best we have seen in the leading parts of the automobile industry." This 126,000-square-foot shop was for the production of the popular United tractor. At the time, Allis-Chalmers engineers also retooled and rearranged the no. 1 and no. 2 tractor shops, which produced parts for the other models. The no. 2 tractor shop, Geschelin reported, was now "laid out for economical straight-line production in keeping with

[26]E. A. Harrison, "Completely Mechanized Foundry for Tractor Castings," *American Machinist* 72 (March 13, 1930): 445.

[27]William Bryce, "Impressions of the Allis-Chalmers Model U Tractor Plant," *American Machinist* 73 (July 31, 1930): 202; Allis-Chalmers Co., *Works and Products* (n. 12 above), p. 60; and Harrison, "Mechanized Foundry" (n. 26 above), pp. 438–39.

the most advanced principles known to factory management." In 1937, after over twenty years of tractor production, the West Allis plant had almost caught up with modern production methods in the American automobile industry. Still, as worker grievances will reveal, it did not fully achieve Detroit-style production.[28]

As the expansion of tractor production facilities demonstrated, the Allis-Chalmers corporation moved into different areas and different forms of production through the 1920s and into the 1930s. If Allis-Chalmers required highly skilled workers when the West Allis plant was first constructed, more specialized semiskilled workers entered the works in the 1920s and 1930s. Harold Story, an Allis-Chalmers vice-president and attorney, described the West Allis machinery which reflected the company's product mix in the mid-1930s: "roughly 50 per cent are special machinery and 50 per cent are mass production machinery." In part, this shift revealed a continued corporate strategy of product diversification similar to the earlier one when Allis-Chalmers went into various types of electrical and industrial equipment. The new products, however, also meant a shift into new production methods and practices in the West Allis shops. No longer soley concentrating on the "big stuff," the Allis-Chalmers managers and engineers moved in the direction of mass production.

Consequently, the West Allis shops developed a bifurcated structure of production. The skilled craftworkers who had produced the large products saw less skilled production workers move into other sections of their plant. Moreover, during the social and economic catastrophe of the Great Depression, a corporate policy of retaining skilled men often meant transfers from skilled shops to the mass-production ones. In addition to resenting Allis-Chalmers's antilabor and open-shop policies, workers also experienced a deep sense of unease and disaffection because of transformed production methods and work processes. When the New Deal offered new opportunities for labor organization, a large majority of Allis-Chalmers skilled and production workers moved swiftly toward militant industrial unionism.[29]

### *Decoding Grievances: Worker Reactions to Technical Change*

For Allis-Chalmers workers, the process of union formation involved the history of the people, the work, and the labor-management relations in West Allis, the social and economic convulsion of the Great

[28]Joseph Geschelin, "After Twenty-Three Years," *Automotive Industries* 79 (September 3, 1938): 282, 284.

[29]Harold Story, NLRB Testimony, July 12, 1937, p. 11 in Box 10, Series 11, Record Group 25, National Labor Relations Board, National Archives, Suitland, Md.

Depression, and the favorable labor legislation of the New Deal. The Allis-Chalmers shop tradition and the Great Depression instilled a restiveness that induced worker militancy and insurgency in the mid-1930s. From 1933 through 1937, workers moved rapidly and inexorably from a company union to American Federation of Labor (AFL) conservative craft unions and finally to a militant Congress of Industrial Organizations (CIO) industrial union. The earlier AFL organizational failures had left them disenchanted with cautious and conservative craft unions. Allis-Chalmers's open-shop policies also soured them on a peaceful relationship with management. Always in the background, new production technologies challenged traditional patterns of work skills and social relations in the West Allis shops. Consequently, when veteran unionists joined with young Socialist and Communist activists to establish militant plantwide unionism, large numbers of Allis-Chalmers skilled and production workers followed their shop leaders into United Automobile Workers (UAW) Local 248.[30]

Nonetheless, the social structure of production partially mirrored the bifurcated technical structure of production in the West Allis works. In the late 1930s and early 1940s, first through the supporters of Homer Martin in the UAW and later through independent and AFL union movements, conservative craft unionism retained strong support among a small number of Allis-Chalmers metal tradesmen. For these anti-CIO holdouts, craft unionism was the unionism of the batch producers.

After the formation of their militant UAW-CIO local, Allis-Chalmers workers utilized their union-acquired grievance procedure to register discontent and to test the limits of worker authority in the West Allis shops. Most often, worker grievances turned on disputes over contract interpretation and touched on questions of wages, hours, and working conditions. On the surface, these grievances appear to echo the classic issues of "pure-and-simple unionism." Yet a careful

[30]On the history of UAW Local 248 at Allis-Chalmers, see ACMC, "Resume of Pertinent Phases of Allis-Chalmers History Depicting the Destructive Effect of Communist Control of a Labor Organization," Box 1, AC Papers; Harold Christoffel testimony, "Official Report of Proceedings before the National Labor Relations Board, Case No. XII-R-40 . . . , Milwaukee, Wisconsin, July 15, 1937," pp. 499–665, passim, Box 11, Series 11, Record Group 25, National Archives, Suitland, Md.; NLRB, "CIO Union Logical Result of Labor History of Plant," *Allis-Chalmers Workers' Union News*, December 10, 1937; Bert Cochran, *Labor and Communism: The Conflict That Shaped American Unions* (Princeton, N.J., 1977), pp. 163–73, 272–75; Roger Keeran, *The Communist Party and the Auto Workers Unions* (Bloomington, Ind., 1980), pp. 212–15, 266–76; Thomas W. Gavett, *Development of the Labor Movement in Milwaukee* (Madison, Wis., 1965), pp. 176–97; and Robert W. Ozanne, "The Effects of Communist Leadership on American Trade Unions" (Ph.D. diss., University of Wisconsin, 1954).

examination reveals that the disputes went far beyond notions of conservative trade unionism. At bottom, issues connected to new technologies and new production methods were fundamental, consistent, and pervasive inner themes of the numerous and complex shop-floor grievances. Over time, UAW Local 248 gradually developed practices which used the grievance procedure to rationalize shop-floor relations within the framework of a workplace rule of law and to defend workers against the inexorable penetration of new industrial and managerial methods into the worker's domain on the shop floor.

An Allis-Chalmers subject index indicates that some 2,500 grievances from 1937 to 1941 encompassed an extremely wide range of issues—procedures for grievances and collective bargaining, seniority, transfer, layoff, discrimination, anti-union statements, discourtesy or intimidation by management of union representatives, discipline, firing, hiring, working conditions, wage rates, paychecks, and numerous other issues. Yet, despite this very broad range of topics, many of the grievances clearly had a technological dimension.

Almost 1,200 grievances, or 48 percent of the total, involved wages. Many of these protested the unfair setting of wage and piece rates, requested the presence of union representatives for time study, or demanded access to management time study records. Frequently, technical changes in production methods were the reason behind changes in the wage rates. Thus questions of technical change often loomed behind the simple wage grievance. For example, grievances on time studies clearly reflected the problems of technical change on the shop floor. Since early in the century, Frederick Taylor's notions, and especially those of Frank Gilbreth, had been directly connected to the reorganization of work processes and improved forms of factory layout. When the company changed its production methods, it often attempted to change the day or piece rates for a job classification. Time-and-motion studies were a managerial means to recapture the increased gains which came from technical improvements in the shops.[31]

At Allis-Chalmers, numerous grievances involved the problems of rate cutting, set-up prices, engineering changes, requests for new time studies, and pressures for increased efficiency. Obviously, these disputes mirrored more than a simple concern for money. They often directly or indirectly touched on worker concerns, fears and anxieties about new technologies and new work routines. Wage grievances frequently constituted a day-to-day technique for resisting new industrial

[31]Allis-Chalmers Manufacturing Company, "Subject Index of Grievance Rulings, 1–2,500," Don D. Lescohier Papers, Wisconsin State Historical Society, Madison, Wis.

technologies and new production methods in the West Allis shops. For the UAW union activists, worker grievances channeled and mobilized the worker's customary sense of equity and fairness that had lain dormant through a long period of unchallengeable managerial prerogatives. In many ways, the UAW institutionalized informal forward-looking values about worker rights in the process of technical advance.

Two piece-rate grievances reveal the different work problems with batch and mass production. In the no. 1 machine shop, which manufactured water turbines for the Tennessee Valley Authority, one grievance concerned the piece rate for the operator of a large boring mill. The machine operator complained that he essentially received an hourly rate instead of his piece rate for the large job. Although the main issue was the setting of a piece-work price before completion of the job, the facts of the grievance provide insight into the character of the work on the "big stuff." The worker machined two "runner hubs" for the water turbine; each needed six holes. For this worker, the total machining time for one hub was 340 hours and for the other 360 hours. Assuming an eight-hour day, it took an amazing forty-two and a half days to machine the first hub and an even more amazing forty-five days for the second![32]

The other grievance concerned the time study and piece rates for milling machines on the no. 3 tractor shops rod line. The aggrieved workers were dissatisfied with forty-two seconds per piece. "At 42 seconds," the shop stewards noted, "the personal privileges plus all other efficiencies are taken away." The time-study man, they claimed, "has created ill feelings between the workers and management." In its response, management officials noted technical improvements in the drive system of the machines and several retimings of the milling job. Furthermore, they noted similar earnings under the old and the new piece rates.[33]

These grievances demonstrate worker and management reactions to different forms of production technology. Under the batch system, the skilled worker attempted to secure a piece rate on a large job that could be paced according to the rate. If the rate was too low, the job could be rushed, anticipating a move to a better one. If sufficiently high, it could be nursed for a few more days. Management dangled the prospect of a piece rate before the boring-machine operator, but on completion of the work decided to pay him a lower day rate. Under

[32]Labor Advisory Committee, Ruling Number (hereafter cited as LAC) 58, March 3, 1938, F. Co Ans—37–38, Box 1, UAW Local 248 Papers, Archives of Labor History and Urban Affairs, Wayne State University, Detroit, Mich. (hereafter cited as ALHUA).

[33]LAC 178, February 28, 1939.

the mass-production system, managers and workers bitterly fought over piece rates or the actual seconds to produce each single item.

The installation of new industrial systems in the West Allis shops raised the tensions between managers and workers and resulted in numerous grievances from union workers. Management concern for cost efficiencies pressured supervisors who, in turn, pushed their subordinates on the shop floor. Already anxious about their work routines and wages, workers used their union to resist the baneful influence of technical innovation. For example, along with the construction of the mechanized no. 3 tractor shop, Allis-Chalmers managers and engineers installed new mechanized molding units in the no. 2 foundry. Early in 1938, Julius Blunk, a UAW Local 248 union official, wrote to two neighboring UAW locals with a similar "conveyor system" in their plants. "We would like to have some man in your organization, who is thoroughly familiar with your foundry problems," Blunk requested, "come before our foundry people to enlighten them on the worker's problems." His letter concluded: "The situation is new to us and we need all possible advice." In other words, shortly after the introduction of the new conveyorized foundry unit, Allis-Chalmers workers responded to technical change through their new union.[34]

After this initial union inquiry to other UAW locals, the Allis-Chalmers foundry shop steward began to file a number of grievances that illustrated the no. 2 foundry workers' concerns and fears about the arrival of the new mechanized equipment in their shop. According to the steward's reports, wage rates were a key problem. Although workers in the older no. 1 foundry received piece rates for their work, the workers in the new "32 man unit" received hourly rates. Foundry workers generally preferred piece rates because they could control their output and their level of wages. Since a conveyor system regulated the rate of production, Allis-Chalmers managers no longer needed piece rates and their ambiguous incentives for greater output. Under the new mechanized system in the new tractor foundry, workers received less pay than their fellow workers in the older foundry.

Within a short time, all classes of foundry workers—iron molders, pourers, shake-out men, sand mixers, casting separators, and sand operators—filed grievances and requested substantial raises in their new hourly rates. They wanted the new unit to pay wages similar to those in the older unit. The molders felt that their wage rate "is not enough money for the type of work they are doing and being [*sic*] in a foundry." The pourers argued that "they have to pour all day long and keep the

[34]Julius Blunk to Mr. Sehafer, January 31, 1938, and Julius Blunk to George Nordstrom, February 2, 1938, Box 1, UAW Local 248 Papers, ALHUA.

same pace as the line." The sand separators believed it unfair that they receive the same rate as ordinary laborers. On their grievances, the steward noted: "These men must work harder than laborers and they feel they should receive more pay." Clearly, the foundry workers believed that the labor-saving machines did not save *their* labor. If they worked harder at the pace of the conveyor, they should receive greater compensation for their efforts. The management-controlled Labor Advisory Committee denied the requests of the foundry workers and refused to increase any wage rates. Although the focus was wages, the mechanization of the foundry clearly had altered work routines, increased the pace of work, damaged earnings expectations, and changed the structures of work in the shops.[35]

Grievances continued to flow from the no. 2 foundry. In addition to the thirty-two-man unit, the Allis-Chalmers engineers installed a newer sixteen-man unit. Once again the foundry workers protested the introduction of an improved technical system into their shop. The less skilled workers, pourers, casting sorters, and shake-out men all contested their low hourly rates. The more skilled molders, who at this time were on piece rates, complained about how technical difficulties contributed to lost time and reduced pay. "While waiting for iron and the line is full," the foundry steward wrote, "they cannot put molds on the line, also breakdowns due to the cause of the machinery." Sometimes the skilled molders out-paced and out-performed the mechanized conveyors in their quest for higher wages. At other times, the machinery broke down, increased their idle time, and reduced their pay. Apparently, the molders earned more pay under the older and more traditional production methods. Since they could control the pace of their production in the older shop, they could work at a more consistent rate than in the new shop and could increase their earnings. Their complaint amounted to an implicit critique of the technical assumptions behind the new conveyor system. The Labor Advisory Committee refused to increase the hourly rates of the less skilled workers, but it did offer the skilled molders regular day rates for their lost time. The management strategy was to satisfy the complaints of the more skilled, more valuable, and hence potentially most threatening workers and to ignore the needs of the less skilled workers.[36]

Nonetheless, the mechanized foundry remained a source of many grievances from Allis-Chalmers workers. In the summer of 1938, molders complained about the cutting of piece-rate prices for their

[35]LAC 70–75, March 14, 1938.

[36]LAC 87 and 89–95, May 25, 1938, and LAC 88, June 13, 1938.

work. "When you bring a job down from the No. 1 Foundry to the No. 2 Foundry," the two foundry stewards wrote, "you are cutting the price by more than half, and we have the same amount of work." The worker vision of the relation between effort and wages linked fair earnings to how much labor was performed in the course of the day. Allis-Chalmers managers labeled this view the "quantum theory of work." A change in production technology often meant a change in the amount of work required in order to generate increased output for the employer. The foundry workers clearly believed that they should share in the benefits of the new technology either through decreased effort or higher wages. In line with the union contract provision that permitted rate cuts for "new production methods," management ruled that the reason for the reductions was the "different working conditions" and the "different equipment" in the two foundries. However, these differences mattered little to workers who saw themselves produce more of the same product and exert more physical effort for lower wage rates.[37]

In the winter of 1938–39, Harold Christoffel, the Local 248 president, issued still another grievance on the mechanized no. 2 foundry units. He noted that "problems were discussed with the foundry management continuously since the units were started to be installed (except when they were shut down)." Expressing the sentiments of the men in the shop, he called for an increase in the base rates of pay, a retiming of the piece-rate jobs with allowance for "various operations and disturbances," and a more equitable resolution of earlier grievances on workers' rates. Finally, Christoffel observed: "When these units were installed, the men were repeatedly told that they would make at least as much as they made before. Figures show that these men do not earn as high an average as before, yet fatigue under the new conditions is much greater. It is not a question of making more pieces, but a greater fatigue for the same time worked for which less monies received." Contrary to management views, the mechanized foundry meant more arduous work for the men in the shop. Mechanization increased the pace and the intensity of their work. The persistence of complaints about fatigue indicated perceptions of harder toil. Countering the management position that if the new methods made production work "easier," piece rates should be cut, the union president argued that if the men worked harder in the mechanized foundry, they should receive more compensation. Repeatedly, this "quantum" view of work represented the union effort to address problems resulting from the inception of new methods or technologies

[37]LAC 138, August 18, 1938.

in the workplace. The workers, Christoffel believed, should benefit from improvements in industrial technology. Once again, the Labor Advisory Committee refused to adjust the workers' wage rates.[38]

Taken together, this group of foundry grievances demonstrated that mechanization was truly a complicated problem for the workers in the West Allis shops. Although grievances took the form of wage-rate disputes, their substance was far more complicated. They concerned the nature and character of work, the structure of relations between managers and workers and between skilled piece workers and less skilled hourly workers, mechanical problems with the new industrial systems, and issues of worker or union control of technical innovation in the shop.

A similar situation developed in the new no. 3 tractor shop when managers and engineers improved production methods and technologies for the machining and assembling of tractors that were so important to the Allis-Chalmers product line in the 1930s. During 1937, workers filed grievances from the new shop, which again reflected worker concerns about new methods and techniques of production. Despite the considerable technical sophistication of line production methods in this period, the worker grievances revealed that in practice the assembly lines continued to fall far short of the state-of-the-art of modern production technology.[39]

For example, in October 1937 a Local 248 letter to national UAW headquarters raised some questions about how assembly lines operated in Detroit: "In the case of a small part, such as a cotter pin or screw of a motor being assembled incorrectly on the assembly line, do they take the individual off the line to tear the motor apart, and then place him back on the line? During the time this individual is off the line, does the line continue to operate as if the employee was still there, someone else doing his work too? Then, in case of error because of a flaw in material etc., the motor is assembled and later must be taken apart and corrected? What is done?" In the Detroit automobile plants, the notion of taking a person from the assembly line to correct an error would have been absurd. As automotive mass-production techniques diffused to other industries in the 1920s and 1930s, managers, engineers, and workers obviously confronted new situations and new problems in their recently mechanized shops. The design ran counter to the actual shop-floor use of the new production

[38]LAC 177, February 2, 1939.

[39]Peterson, *Industrial Heritage* (n. 3 above), pp. 242, 244–45, 263, 272–73, and Allis-Chalmers Workers' Union to United Automobile Workers of America, October 29, 1937, Box 1, UAW Local 248 Papers, ALHUA.

system. Management theory found limitations in actual practice. Although Allis-Chalmers managers and engineers had begun to introduce line production methods in the tractor shops, both the union and management had meager knowledge about the ordinary, day-to-day procedures entailed by these novel methods.[40]

One group of tractor-shop grievances demonstrated both the novelty of modern production methods and the low level of technical sophistication in the West Allis shops. The question of lunches emerged as a contested issue for workers in tractor shop no. 3. According to the union committeeman, "the men were under the impression that they were allowed to eat a sandwich whenever they felt hungry." This seemed a traditional craft custom and privilege throughout the West Allis shops. Craftworkers, who controlled their rates of production, could readily stop work and eat at their own convenience. It was, however, a most unusual practice for workers under line methods of production. At any rate, a supervisor "reprimanded" one production worker "severely for eating . . . while he was waiting for a tractor to come up so he could begin his operation." The committeeman noted that the supervisor said: ". . . a man on a machine could sneak a sandwich[,] but these men on production (on [an] assembly line) and definite operations are not assigned[.] [S]o when they are through with their operations[,] they should help someone else." In the end, the company ruled that workers should eat their lunches during the regular noon lunch period. This grievance demonstrated the persistence of custom and tradition in the shops, the truly glacial pace of the mechanized tractor assembly line, and the influence of new production methods on worker shop traditions.[41]

The new tractor shop also experienced technical problems and difficulties with repeated breakdowns. In March 1938, shop manager H. E. Ladwig wrote to Christoffel about the company's new policy for the technical problems "with new systems of manufacture." Essentially, the difficulties involved the tune-up of new equipment and the development of new skills by the workers. Once again, a technical question manifested itself as a wage question. "It is estimated," William Watson, author of the new policy statement, noted, "that the tuning up of new systems of tractor production in the foundries, tractor and general machine shops will require from thirty to ninety days more." Clearly, it took considerable amounts of time, energy, and thought to coordinate the men, machines, and materials for new methods of

[40]Allis-Chalmers Workers' Union to United Automobile Workers' Union, October 29, 1937, Box 1, UAW Local 248 Papers, ALHUA.

[41]LAC 63, February 28, 1938, and LAC 59, February 24, 1938.

tractor production. Therefore, Allis-Chalmers officials established "temporary" hourly rates and piece-rate prices for the production workers. The mechanization of the shops for a new tractor required numerous major and minor adjustments for a whole range of men and machines throughout the entire West Allis plant.[42]

Other tractor shop grievances reflected the shop-floor tensions which arose with the mechanization of the no. 3 tractor shop. One grievance called for the reprimand of foremen who swore at the workers in his shop. Naturally, foremen were under considerable pressure and stress to "tune up" the men and machines for the new mechanized systems. Often, they took out their frustration on the men in their shops. Another grievance involved the superintendents and foremen who threatened tractor shop workers with layoff and dismissal. The shop steward listed the reasons for the threats: "because of lunching, not working up to speed, because of mechanical mistakes, because the fellow wouldn't do everything he was told to do, because parts didn't fit right and a man didn't notice the mistake, because a fellow refused to do something after the foreman swore at him and many more minor happenings." Indeed, improved production methods created a seething shop with abusive foremen and malcontented workers.[43]

These different sets of worker grievances detailed how mass-output mechanization affected the foundry and tractor shops and influenced different aspects of work on the shop floor. Wage questions surely were based on much more than the insatiable worker desire for more, more, more. Rather, they clearly expressed a sense of equity and a customary notion of a fair day's effort and earnings in a rapidly changing technical environment. Since the new production methods and practices did not operate at maximum efficiency, Allis-Chalmers workers often demanded a more equitable solution to their problem of a reduced income. They also resented the changes in their working conditions, particularly those that eliminated their craft customs and traditions and undermined their perceptions of their dignity and worth. In their pursuit of these early grievances, the workers and their union attempted to control the penetration of new industrial methods and techniques and to alleviate their effect and influence on the shop floor. Interestingly, they never seemed to challenge directly the necessity for improved technology. Instead, they attempted to contain its influence on their terrain and to realize a fair share of its benefits. Surely, management's negative rulings on these subjects and its insensitivity

[42]H. E. Ladwig to Harold Christoffel, March 1938, Box 1, UAW Local 248 Papers, ALHUA.

[43]LAC 62 and 79, May 27, 1938.

to workers' sense of equity and fairness contributed to the militant union attitude in the 1939, 1941, and 1946–47 strikes.

*Conclusion*

By the eve of World War II, Allis-Chalmers workers had earned a reputation as a militant and "Red" UAW local. Within UAW Local 248, a small number of skilled craftsmen struggled with a much larger group of militant skilled and production workers in UAW factional dispute between the conservative UAW president Homer Martin and the UAW Executive Board. Although the overwhelming majority of Allis-Chalmers workers accepted industrial unionism, the bifurcated work processes and social structure of work persisted and spilled over into UAW politics. This bitter struggle erupted in a twenty-six-day strike in 1939. Furthermore, similar issues returned in a renewed AFL organizational campaign in 1940 and in the controversial "defense" strike of 1941. Over 7,000 workers struck the huge West Allis industrial complex for seventy-six days. To be sure, the strike had many deep-rooted causes: the Communist orientation of some union leaders, the continuously hostile anti-union policies of Allis-Chalmers management, the contentious factionalism of the AFL and CIO, and the combative social mood of the times. Yet the question remained of technical innovation and its influence on the important issues of work, skill, control, shop traditions, and workplace social relations.

The Allis-Chalmers Company illuminates the complex transition from batch to mass production and the difficult coexistence of the two forms of manufacture within a single plant. One form required a skilled labor force to manufacture specialized products; the other a less skilled work force to produced standardized products. Through the transition, Allis-Chalmers officials and managers inherited and possessed a physical plant and a management shop culture designed for specialty products. At the same time, deep worker dissatisfaction arose from management's requirement for skilled workers and resistance to their union culture. Given the social dimension of technical change, management theory often fell far short of management practice in the adoption of new technologies.

Nowhere was the social dimension more evident than in the different management and labor views on the new technologies. Allis-Chalmers managers and Allis-Chalmers workers viewed technical innovation from quite different perspectives. For managers, the new production technologies meant a decrease in the number of skilled workers, an increase in management control of work processes, and a decrease in the costs of production. For workers, they meant a loss of skills, an increase in the intensity and pace of work, and greater insecurity because of altered shop practices and traditions. Quite clearly, the Allis-Chalmers case demonstrates that one's standing relative to the white-collar/blue-collar divide often revealed one's attitude toward the influence of industrial technologies. Ultimately, a deeper understanding of the process of technical innovation requires a more thorough investigation and analysis of the social totality of the production process.

# [14]

# *Mass Production and the U.S. Tire Industry*

DANIEL NELSON

The study of mass production has been strongly influenced in recent years by two dissimilar bodies of work—that of Alfred D. Chandler and that of Harry Braverman and others who argue that mass production helped managers "deskill" and "control" industrial workers. The history of the U.S. tire industry between 1910 and 1930 underlines the value of Chandler's analysis and the limited applicability of the deskilling and control hypotheses.

After Henry Ford introduced the term mass production in a 1925 encyclopedia article, it enjoyed a remarkable vogue as a synonym for large-scale manufacturing, machine operations that insured standardization and the interchangeability of parts and the rise of the semi-skilled worker.[1] In scholarly works mass production has usually been used as a metaphor for the mechanization of the economy; in more popular treatments, and in the minds of most people, including most students, it is another name for the assembly line. Alfred D. Chandler's *The Visible Hand* provides a welcome break. Chandler defines mass production as a series of unrelated technologies that produced increases in capital, materials, energy, and management relative to labor.[2] He portrays these developments as responses to the growth of national and international markets and associates them with the rise of the vertically integrated corporation during the 1880s and 1890s. But if he rescues mass production from the world of auto factories and assembly lines, he resolutely refused to address another persistent issue, the meaning of mass production for the factory worker. Though this issue is familiar to any student of modern industry, a group of scholars associated directly or indirectly with the Marxist journalist Harry Braverman gave it a new poignancy in the 1970s and early 1980s by portraying mass production as a feature of a larger capitalist conspiracy to "deskill" work and to permit industrial managers to exercise greater "control" over workers.[3]

*Journal of Economic History*, Vol. XLVII, No. 2 (June 1987).  ISSN 0022-0507.

The author is professor of history at The University of Akron, Akron, OH 44325.

[1] See David A. Hounshell, *From the American System to Mass Production, 1800–1932: The Development of Manufacturing Technology in the United States* (Baltimore, 1984), pp. 1–3.

[2] Alfred D. Chandler, Jr., *The Visible Hand: The Managerial Revolution in America* (Cambridge, Mass., 1977), pp. 240–44; and Chandler, "The Emergence of Managerial Capitalism," *Business History Review*, 58 (Winter 1984), pp. 474, 480–82.

[3] The best known works are Harry Braverman, *Labor and Monopoly Capital: The Degradation of Work in the Twentieth Century* (New York, 1974), esp. pp. 146–50; Richard Edwards, *Contested Terrain: The Transformation of The Workplace in the Twentieth Century* (New York, 1979), pp. 20, 111–25; and David M. Gordon, Richard Edwards, and Michael Reich, *Segmented Work, Divided*

In many respects the best example of the process that Chandler describes is the rise of the rubber tire industry in the United States. One of several second generation mass production industries that expanded in the wake of the automobile revolution, tire manufacture exemplifies the possibilities of the new high-speed capital and energy intensive production. It also employed tens of thousands of machine operators, whose duties changed as the industry realized the possibilities of the new technology. Though obscured in the furor over the automobile and the Ford innovations of 1913, the tire industry provides an excellent opportunity to extend Chandler's analysis from middle management to the shop floor.

The rubber industry in 1900 is a classic illustration of labor intensive industries that did not have mass production potential. Rubber shoes and boots were made for a mass market with the high-speed handicraft methods characteristic of the leather shoe and clothing industries. Belts, hoses, and other "mechanical goods" were made for specialized, fragmented markets by machine techniques. By most of Chandler's measures, the rubber industry was similar to the cotton textile, clothing, and leather boot and shoe industries.[4] Firm size was small; the exceptions, the U.S. Rubber Company, formed in 1892, and the Rubber Goods Manufacturing Company formed in 1898, are examples of turn-of-the-century trusts that were less efficient than their components and subject to attacks by outsiders.[5] Neither of them, nor the remaining independents, was vertically integrated. The industry's labor force consisted of low-paid men, in most cases immigrants with few labor market options, and comparatively well-paid women shoe assemblers.

The growth of auto production introduced a new diversity to the industry. Although many of the mechanical goods companies had made bicycle and carriage tires, they were no better equipped for the auto era than outsiders. Between 1905 and 1920 most of the larger boot and shoe and mechanical goods producers tested the new market and withdrew; foreign tire manufactures like Dunlop and Michelin were no more successful.[6] The exceptions were B. F. Goodrich, one of the largest and most diversified of the non-tire manufacturers, which entered the tire industry via a merger with fast-growing Diamond Tire Co. in 1912, and U.S. Rubber, which was able to use its dominant market position in

*Workers* (Cambridge, Mass., 1982), pp. 131–33. For a discussion of the deskilling debate see Craig R. Littler, *The Development of the Labour Process in Capitalist Societies* (London, 1982), pp. 6–35.

[4] See Chandler, *Visible Hand*, pp. 242–43.

[5] See Glenn Babcock, *History of the United States Rubber Company: A Case Study in Corporate Management* (Bloomington, 1966), pp. 29–47.

[6] See "The Rubber Trade in New Jersey," *India Rubber World*, 60 (May 1, 1919), p. 606. Michelin abandoned the U.S. market in the early 1920s. Dunlop remained but was an insignificant producer until the 1930s.

mechanical goods and its ties with wholesalers to develop a large tire business, at least until the mid-1910s. Thereafter it too began to succumb to the competition.[7] In contrast the most successful tire manufacturers came from other careers. The industry's best known entrepreneurs, Frank A. Sieberling and Harvey S. Firestone, had backgrounds in real estate speculation and patent medicine sales respectively.[8] Thus by 1910 the rubber industry consisted of two distinct sectors, separated by substantial differences in markets and business psychology. At the time there was little difference in technology. And while women were not employed in tire plants, the labor force of a mechanical goods factory and a tire plant differed little.

For the next two decades the spectacular growth of the tire market widened the gap and provided the precondition for a new technology and a new labor force. Between 1910 and 1930 the demand for tires increased more than a hundredfold, reflecting the dynamic growth of the automobile industry. In comparison, the demand for rubber boots and shoes remained steady. Tire manufacturers satisfied most of the demand with innovations in tire design and rubber chemistry. Tire costs per thousand miles of driving fell from $9.39 to $0.65, due mostly to qualitative changes in tires, but incentives to improve manufacturing techniques remained strong.[9] More important, the aura of growth, change, and opportunity that characterized the industry encouraged inventors and imaginative managers to adopt a skeptical, creative approach to factory operations, regardless of the actual potential of any given innovation. It was this setting—comparable to the environment of the 1880s and 1890s for steel, cigarette, and match manufacturers—that provided the background to the introduction of mass production in the tire industry.

The cluster of technical innovations that transformed the manufacture of tires reflects a process common to many of the industries that Chandler discusses. Each new machine or engineering change, introduced in expectation of reduced unit production costs, created "internal complusions and pressures" that demanded more mechanization, reorganization and investment. The "compulsive sequences" continued for more than a decade, eliminating in the process all but the richest and most durable participants.[10] Unlike many of the industries that Chandler cites there was no single technical breakthrough that unlocked the

[7] Babcock, *U.S. Rubber*, pp. 212.

[8] See Hugh Allen, *The House of Goodyear: Fifty Years of Men and Industry* (Akron, 1949), p. 12; and Alfred Lief, *Harvey S. Firestone, Free Man of Enterprise* (New York, 1951), pp. 40–52.

[9] Ralph William Frank, "The Rubber Industry in the Akron-Barberton Area: A Study in the Factors Related to Its Development, Distribution and Localization" (Ph.D. diss., Northwestern University, 1952), p. 144. For similar data see Babcock, *U.S. Rubber*, pp. 338–39; and Howard and Ralph Wolf, *Rubber: A Story of Glory and Greed* (New York, 1936), p. 466.

[10] The quoted phrases are from Nathan Rosenberg, *Perspectives on Technology* (Cambridge, Mass., 1976), pp. 111–12.

TABLE 1
COMPARISON OF TIRE AND RUBBER BOOT AND SHOE PRODUCTION, 1914 AND 1929

| | Percent of U.S. Rubber Industry Mean | | | |
|---|---|---|---|---|
| | 1914 | | 1929 | |
| | Tires[a] | Rubber Boots | Tires | Rubber Boots |
| Capital/worker | 135% | 67% | 151% | 90% |
| Materials/worker | 131 | 57 | 133 | 34 |
| Horsepower/worker | 92 | 48 | 102 | 44 |
| Managers/worker | 195 | 45 | 106 | 93 |
| Wages/worker | 118 | 88 | 110 | 83 |

[a] Based on statistics for Akron, Ohio, the principal tire center. The Census Bureau did not distinguish tires as a separate industry until 1921.
*Sources*: U.S. Department of Commerce, Bureau of the Census, *Census of Manufactures, 1914* (Washington, D.C., 1918), vol. 1, pp. 1208–9; U.S. Department of Commerce, *Fifteenth Census of the U.S., Manufactures: 1929* (Washington, D.C., 1933), vol. 2, pp. 777–78; Daniel Creamer, Sergei P. Dobrovolsky, and Israel Borenstein, *Capital in Manufacturing and Mining, Its Formation and Financing* (Princeton, 1960), p. 243.

industry's potential for high-speed production; nor was any individual or firm of overriding importance. The advent of mass production was a "cumulative process resulting from a vast number of successive small changes."[11]

Judging from the surviving anecdotal evidence, the principal contributors were individuals familiar with the production process in toto, often at several firms, and thus were in a position to gauge the importance of particular bottlenecks. William State, inventor of the first commercially successful tire-building machine, was master mechanic at Goodyear Tire & Rubber. Fernley H. Banbury, inventor of the Banbury mixer, often considered the single most important technical breakthrough of the period, worked for Birmingham Iron Foundry, a manufacturer of various machines for the industry, and Ernest Hopkinson, inventor of the flat band tire-building machine, probably the key development of the 1920s, worked in the patent department of one of U.S. Rubber's mechanical goods subsidiaries.[12] The research and development departments which the larger firms introduced in the 1910s and 1920s employed chemists almost exclusively and devoted their efforts to product improvements.

The results were impressive by any gauge. Table 1 summarizes the industry's transformation in terms of the variables Chandler emphasizes. The rate of productivity increase in tire manufacture was among the highest of any manufacturing industry.[13] Its capital to output ratio

[11] John Dean Gaffey, *The Productivity of Labor in the Rubber Tire Manufacturing Industry* (New York, 1940), p. 90. For descriptions of machinery see Henry C. Pearson, *Pneumatic Tires* (New York, 1922).

[12] See D. H. Killeffer, *Banbury the Master Mixer, A Biography of Fernley H. Banbury* (New York, 1962); Gaffey, *Productivity*, p. 90; Babcock, *U.S. Rubber*, pp. 230–31.

[13] John W. Kendrick, *Productivity Trends in the United States* (Princeton, 1961), p. 136.

was the second highest of any industry in 1910 and 1920.[14] Technological change also contributed to the consolidation of the industry, which had the manufacturing sector's second highest concentration ratio by 1929.[15] By that time the largest firms were all vertically integrated and multinational.

The impact of the new technology on the labor force was equally notable. The single most striking feature of mass production was its labor-saving character. Employment in the Akron plants, which were the technological leaders, peaked during the post-World War I boom and never exceeded three-fourths of the 1920 high.[16] Boris Stern, who studied tire manufacture in the early 1930s, reported substantial reductions in the labor force in each major step of the manufacturing process. In the preparatory departments, for example, the introduction of "plasticators" led to a "very large displacement of labor."[17] In one plant 80 percent of the employees became redundant. The old method required 27 workers: (1) trucker to deliver bales of crude rubber to cutter, (1) cutter, (1) assistant to push pieces of rubber down chute to floor where crackers and washers were located, (1) worker to remove pieces of rubber from chute to pile, (4) cracking-mill operators, (8) washing-mill operators, (3) workers to remove broken-down sheeted rubber to drying kiln, (2) workers to remove dried rubber to breaking-down mills, (4) breaking-down mill operators, and (2) workers to place broken-down rubber on trucks ready for delivery to compounding room. The new method required only six workers: (1) trucker to open bales and delivery rubber to cutting machine, (1) cutting-knife operator, (1) helper, (1) feeder, (1) worker to remove rubber from plasticator to hook to conveyor to cool off, and (1) worker to remove cooled-off rubber from conveyor to trucks.[18]

The pattern was similar in other departments. In the plants Stern studied, Banbury mixers had displaced 40 to 60 percent of mill room employees.[19] In tire building, the flat band machine did not lead inexorably to the assembly line; on the contrary, the assembly process, in most cases, remained an individual operation. Yet machinery greatly

[14] Daniel Creamer, Serge, P. Dobrovolsky, and Israel Borenstein, *Capital in Manufacturing and Mining, Its Formation and Financing* (Princeton, 1960), p. 266.

[15] See Alfred D. Chandler, "The Structure of American Industry in the Twentieth Century: A Historical Overview," *Business History Review*, 43 (Autumn 1969), pp. 258–59. For competition among tire manufacturers see Lloyd G. Reynolds, "Competition in the Rubber Tire Industry,' *American Economic Review*, 28 (Sept. 1938), pp. 460–66 and Michael French, "Structural Change and Competition in the United States Tire Industry, 1920–1937," *Business History Review*, 60 (Spring 1986), pp. 28–54.

[16] Federal Housing Administration, "Akron, Ohio Housing Market Analysis" (Washington, D.C., 1938), p. 218.

[17] Boris Stern, "Labor Productivity in the Automobile Tire Industry," U.S. Bureau of Labor Statistics, *Bulletin No. 585* (July 1933), p. 39.

[18] Ibid., pp. 40–41.

[19] Ibid., p. 49.

reduced the labor force in the tire room. Stern reported that workers "are rapidly being displaced by various labor saving devices. . . ." Indeed, the "largest displacement of labor" occurred there.[20]

In the curing room, manufacturers introduced conveyors to transport molds to and from vulcanizing machines and a new curing machine, the "watchcase" vulcanizer. The traditional method was to load 25 to 40 tires, each in a separate mold, into a "pot heater." With the new machinery, each tire went into a separate "watchcase."

> The work of the operator consists merely in locking and unlocking the mold (if these operations are not performed automatically), removing the cured tire to a traveling hook conveyor and inserting another green tire in the mold. The vulcanizers are conveniently arranged in rows . . . A single operator can tend as many as 100 vulcanizers, which are so timed that when 1 vulcanizer is closed its immediate neighbor automatically opens to release a cured tire.[21]

Stern contrasted the methods in 1931. The pot heater used four lines of vulcanizers, operating on four 6-hour shifts and curing 18,000 tires per day. The watchcase vulcanizer used a battery of vulcanizers operating on three 8-hour shifts and curing 7,500 tires per day.

| Method | Number of Workers Required |
|---|---|
| Pot heater | |
| Gang leaders | 12 |
| Pressmen | 20 |
| Loaders and unloaders | 44 |
| Tire removers | 16 |
| Tire placers (into molds) | 12 |
| Other workers and helpers | 104 |
| Total | 208 |
| Watchcase vulcanizer | |
| Pressmen | 12 |
| Helpers | 3 |
| Total | 15 |

Tire output per worker was 86.5 and 600 respectively.[22] Inner-tube manufacture also changed with the substitution of the "molded" tube process for the "mandrel" process. Molded tubes required less labor and were cured in circular molds that eliminated seams. Stern reported an increase in output per worker of 100 percent in one plant.[23]

Stern's data point to other effects of mass production which have a more direct bearing on the deskilling and control hypotheses. First, as

[20] Ibid., pp. 50, 52.

[21] Ibid.

[22] Ibid., p. 70.

[23] U.S. Department of Commerce, *Fourteenth Census of the U.S.* Vol. 10: *Manufactures 1919* (Washington, D.C., 1923), p. 1001; *Fifteenth Census, Manufactures 1929* (Washington, D.C., 1933), vol. 2, p. 781.

Chandler emphasized, the new technology fundamentally altered the production process, making it difficult if not impossible to make meaningful before and after comparisons. With exceptions noted below, machines did not eliminate workers directly; they redefined a particular feature of the manufacturing sequence, displacing older machines, tools and operators. The focus of inventors and managers was on the work, not the worker. Indeed, contrary to the implication of the deskilling and control hypotheses, executives seem to have paid little attention to the social implications of their activities. The older progressive and liberal critiques of mass production, which indicted managers for indifference to the workers' fate, appear to be more pertinent and consistent with the evidence.

Second, as Stern's figures show, the largest displacement occurred among the laborers, the low paid, often transient workers who pushed, pulled or carried materials and parts between machine operations. In small, primitive, turn-of-the-century plants, machine operators moved their own materials. As the plants became larger, managers hired large numbers of laborers to take over these subsidiary tasks. But they were at best a temporary answer. The laborers were poorly motivated and erratic; their low status is suggested by the fact that they almost never appeared in photographs of factory operations. Yet as Stern's examples suggest, they numbered from one-third to one-half the labor force as late as the 1920s. The development of labor-saving machines and especially of fast, versatile conveyors brought the era of hand transportation to an end. Stern wrote that the most notable feature of the plants he studied was "the effective utilization of all types of conveyors. . . ."[24] By the mid-1920s they linked every step from the crude rubber warehouse to the tire storeroom. The effect was a "great reduction in the labor force as well as a very large increase in the average man-hour output of the plant."[25] The elimination of superfluous laborers probably accounted for half or more of the job losses between the 1910s and 1930s. A guide to employment in the industry prepared in the late 1930s does not mention laborers.[26] A detailed accounting of the B.F. Goodrich labor force at the same time lists only a handful of individuals who performed what remained of the laborers' duties. They ranged from 4 to 7 percent of the labor force in three plants.[27]

Third, despite profound changes in the character of tire manufacture, mass production did not significantly change the identity of the major machine operations or the "skills" required of the machine operators. The tire plant of 1930 performed the same sequence of operations as the

[24] Stern, "Labor Productivity," p. 61.
[25] Ibid., p. 61.
[26] See Mary J. Drucker, *The Rubber Industry in Ohio* (Columbus, 1937).
[27] "Summary of Enrollment," National Labor Relations Board Papers, National Archives, RG 25, Box 2116, File 1832.

tire plant of 1910. With few exceptions the work had never demanded substantial skills in the usual sense of the term: prolonged training, creativity or initiative. Rather it required judgment and manual dexterity. Within this narrow range, mass production raised the skill level of the labor force by eliminating the need for exceptional strength and increasing the speed of production. Yet the difference was comparatively slight. The time required to master machine operations did not change between 1920 and 1940.[28] To be a successful tire builder in the latter year, for example, "you should be active, agile, deft with your hands, not too heavy in weight in order to be able to stand all day and . . . have very good coordination of your hands and feet."[29] A similar statement, except for the reference to weight, would have been apropos two decades before. One result of this continuity was that many men hired during the boom years of the 1910s worked at similar jobs forty or fifty years later.

Fourth, mass production greatly reduced the physical demands of tire production. Early twentieth-century observers agreed that the most striking feature of tire manufacture was its back-breaking character. The work was most arduous in the assembly room and curing room or "pit," the two largest departments. Before 1920 the difficulty of performing the work restricted many jobs to a physical elite of large, muscular men. Employers imposed informal height and weight requirements; 200 pounds was the minimum for a job in the pit. In hiring, they preferred farm laborers. "Let's see your hands," was the employment manager's first question to prospective workers.[30] Because of these restrictions, the wages of machine operators were much higher than those of other rubber workers and factory operatives. During World War I, when the draft severely curtailed the supply of job seekers, employers tried to use women in the tire rooms. Despite special tools and efforts at work simplification, they enjoyed only limited success.[31] The Akron labor force actually declined during the peak of the war-time boom.

The new technology gradually but profoundly lessened the strain of tire manufacture. State's 1909 tire-building machine, the first major breakthrough, enabled the individual builder to quintuple his output, improve quality, and avoid much of the fatigue of handbuilding.[32] The succeeding innovations had similar effects. In the mid-1930s, Clifton

[28] See Drucker, *Rubber Industry* for descriptions of each job.

[29] Ibid., p. 34.

[30] This statement is based on interviews with veteran workers.

[31] See "Some Labor Problems in the Rubber Industry," *India Rubber Review*, 62 (Apr. 1, 1920), p. 413; "War News of the Rubber Industry," *India Rubber World*, 59 (Oct. 1, 1918), p. 6; "War News of the Rubber Industry," *India Rubber World*, 59 (Aug. 1, 1918), p. 645; Memorandum, Labor Department File, Firestone Tire & Rubber Co. Archives.

[32] Snowden B. Redfield, "The Making of Automobile Tires," *American Machinist*, 32 (July 29, 1909), pp. 191–97.

Slusser, vice president for manufacturing at Goodyear and an industry veteran, claimed that all jobs were "white collar" by comparison with the 1910s.[33] Another measure of change was the growth of the female labor force in the 1920s. In the Akron plants, which employed practically no women in the tire departments before 1920, the total reached 20 percent by 1930.[34] By 1940 one-third of the occupations involving machine operations were open to women.[35]

Fifth, as the work became less laborious, it became more demanding in another sense. The new machines were expensive and could operate faster than any individual could work. As a result, managers attempted to spread their investment over a larger product, as Chandler would predict, but with techniques that do not fit the stereotype of mass production. They systematically eliminated machine-paced jobs, which limited the output of the fastest workers, placed as many employees as possible on piece rates, and used time and motion study and frequent rate cuts to raise the norm and eliminate the slowest operatives. These methods were controversial and often counterproductive. Though they helped reduce unit labor costs and stabilize real wages after 1924, they gave the industry a reputation for ruthlessness and insensitivity.[36] "They just run the life out of a man," concluded one worker, summarizing the consensus.[37] The fast pace encouraged informal resistance, worker-imposed earnings limitations and internecine warfare between time-study men and production workers. Rate cuts accounted for virtually all of the strikes that beset the industry during the 1920s and 1930s. In the late 1930s and 1940s union militants made production limitations an effective rallying cry against both the managers and the union establishment and brought productivity growth to an abrupt standstill.[38]

Sixth, mass production altered the balance between worker and foreman. In 1910 tire workers, like most factory employees, occupied an ill-defined substratum of the company hierarchy whose principal ties to the organization were through the foreman. But as machinery costs increased, conveyors created new possibilities for the integration and synchronization of production and worker protests clarified the costs of haphazard shop management. Manufacturers responded with two sets of conceptually distinct but simultaneous innovations. They introduced contemporary industrial engineering techniques to ensure optimum

[33] Text of Local 18282 Conference with Goodyear, March 20, 1935, NLRB Papers, RG 25, Box 15, File 333.

[34] Amy Maher, *Ohio Wage Earners in the Manufacture of Rubber Products, 1914–1928* (Toledo, 1930), pp. 34, 38, 47.

[35] See Drucker, *Rubber Industry*.

[36] Gaffey, *Productivity of Labor*, p. 140.

[37] Whiting Williams to T. G. Graham, Oct. 14, 1934, Whiting Williams Papers, Western Reserve Historical Society, Box 4, Folder 1.

[38] Gaffey, *Productivity of Labor*, p. 116.

utilization of the new machinery and personnel management to combat absenteeism, turnover, and unrest. Besides achieving most of their immediate objectives, they created additional loci of authority and conduits of information to and from the workers. In every company the practical effect was to decrease the supervisor's authority and to empower workers to a degree unknown except among tiny groups of highly skilled industrial employees. Thereafter, workers' entitlements grew slowly but steadily.[39] In the 1930s, when the old physical elite became the core of the AFL and CIO unions that organized the industry, union demands featured frequent calls for the extension and refinement of the personnel programs of the preceding two decades.[40]

The experiences of tire workers thus appear to provide little support for the deskilling and control hypotheses. Mass production raised the quality of the labor force, increased the demand for skilled labor—as the term was understood in the industry—sustained the tradition of high wages and increased the workers' power vis-à-vis the foreman, the only executive who played a meaningful role in their day-to-day activities. It also created opportunities for able workers at the expense of laborers and, to a lesser degree, the large, powerful men who had dominated the industry before 1920. There were problems and tensions but not those commonly emphasized by recent critics of mass production. Most labor conflicts during the 1920s and 1930s resulted from the failure of executives to administer the wage system in the prescribed fashion. Since the proper use of industrial engineering methods was the preoccupation of the management movement during that period and the likely results of "speed ups" and rate cutting were well known, their derelictions are difficult to explain.[41] In that area, if in no other, they failed to grasp the implications of the new technology.

If the history of mass production is to be more than the history of Ford or the auto industry, it will be necessary to expand upon Chandler's pioneering analysis. The pattern of technological change in "mass production" industry after the 1890s, and the relationship between technical innovation and the evolution of the labor force over an extended period are among the most inviting possibilities. One salient example, the history of tire manufacture, suggests elements of continuity and change in the pattern that Chandler and historians of the auto industry have described. If the experiences of tire manufacturers

[39] This process can be followed in the company histories—Allen's *The House of Goodyear; Alfred Lief's The Firestone Story* (New York, 1951); and Babcock's *U.S. Rubber*—and in Daniel Nelson, "The Company Union Movement, 1900–1940: A Re-examination" *Business History Review*, 56 (Autumn 1982), pp. 335–57.

[40] See Sanford M. Jacoby, *Employing Bureaucracy, Managers, Unions and the Transformation of Work in American History, 1900–1945* (New York, 1985), p. 242.

[41] See E. C. Cowdrick's critique of Goodyear wage policy. Memo, July 16, 1936, Special Conference Committee Records in U.S. Senate, Subcommittee of the Committee on Education and Labor, *Hearings*, 76th Cong., 1st sess. (Washington, D.C., 1939), p. 17,071.

differed only in degree from those of steel, tobacco, and match producers a quarter-century earlier, the experiences of tire workers suggest a more complex and less apocalyptic pattern than Braverman and others have advanced. Though they worked differently, tire employees of the 1940s were no more degraded or controlled, no less skilled, and certainly no less feisty than their predecessors of the 1910s.

# [15]

## *Ford Eagle Boats and Mass Production during World War I*

David A. Hounshell

*There are many instances in which inventions made in the private sector find their way into military use. A prominent example, discussed in the previous essay, is radio, although one can also point to various types of aircraft, internal combustion engines, tracked vehicles, and telephones, to mention only a few. In this chapter David Hounshell examines the Ford Motor Company's ill-fated attempt to mass-produce Eagle boats for the U.S. Navy during World War I. The effort foundered for a number of reasons, but among the most prominent were the company's unbridled confidence in the wide applicability of its assembly-line methods as well as its failure to recognize that marine engineering involved design problems and construction techniques different from auto making. Hounshell provides valuable insight into the limits of technological systems by placing these issues in historical perspective.*

Writing in 1934 after "the war to end all wars," Lewis Mumford identified the military as the source of—and indeed the salvation of—mass production. Mumford saw that mass production had grown out of the standardization and quantity production of military weapons in the late eighteenth and early nineteenth centuries; mass production was at once the means of support for and an extension of the regimentation and quantity consumption of soldiers outfitted in standardized uniforms consuming standardized goods. The army was, Mumford argued, the perfect "pattern not only of [mass] production but of ideal consumption under the machine system. . . . Quantity production must rely for its success upon quantity consumption; and nothing ensures replacement like organized destruction." As he concluded, "War. . . is the health of the machine."[1]

Mumford arrived at this view not only through his own study of the history of modern technology but also by considering the arguments of the noted American anthropologist Otis T. Mason and especially the overarching and penetrating analysis of Werner Sombart in his *Krieg und Kapitalismus.* Mason concluded as early as 1895 that "war. . .stands forth pre-eminently as an incitement to the genius of invention and discovery," while less than a generation later Sombart argued that the rise of modern industrial capitalism had resulted in large measure from Western military enterprise.[2]

Since the appearance of Mumford's classic work *Technics and Civilization,* a number of scholars have published analyses that have supported, in broad outline, Mumford's conclusions about the general effect of warfare and military enterprise on technological or industrial development.[3] Most recently, William H. McNeill, in his sweeping survey *The Pursuit of Power: Technology, Armed Force, and Society since A.D. 1000,* has focused on the relationship between military enterprise and technologi-

1. Lewis Mumford, *Technics and Civilization* (New York: Harcourt, Brace & World, 1934), pp. 93–94.

2. Otis T. Mason, *The Origins of Invention* (1895; reprint ed., Cambridge, MA: The MIT Press, 1966), p. 409; Werner Sombart, *Krieg und Kapitalismus* (München: Dunker and Humblot, 1913).

3. See, for example, J. A. A. van Doorn, *The Soldier and Social Change: Comparative Studies in the History and Sociology of the Military* (Beverly Hills, CA: Sage Publications, 1975) and Albert G. Mumma, "Technology's Motivating Force," *Mechanical Engineering* 89 (April 1967): 18–21. In many respects, the present volume is but a continuation of this tradition. It should be noted that there has been one major study arguing against the Sombart/Mumford thesis: John U. Nef, *War and Human Progress* (Cambridge, MA: Harvard University Press, 1950).

cal development. Like Mumford, McNeill specifically addresses the subject of mass production and identifies warfare as the principal force in the creation and diffusion of mass production in the economies of Western Europe. World War I was especially critical, McNeill argues, for stimulating the adoption of mass production technology in Europe for turning out war materiel. Subsequent to the war, the technology of mass production was applied to domestic purposes. Concludes McNeill:

> As so often before, military demand thus blazed the way for new techniques, and on a very broad front, from shell fuses and telephones to trench mortars and wrist watches. The subsequent industrial and social history of the world turned very largely on the continuing applications of mass production whose scope widened so remarkably during the emergency of World War I. Anyone looking at the equipment installed in a modern house will readily recognize how much we in the late twentieth century are indebted to industrial changes pioneered in near-panic circumstances when more and more shells, gunpowder, and machine guns suddenly became the price of survival as a sovereign state.[4]

McNeill's analysis, like Mumford's, deserves careful attention, and a case study of military enterprise and mass production in the World War I era will provide an additional perspective on their overarching views. Henry Ford is often and rightfully associated with the rise of mass production in America. But in the context of World War I, he is perhaps better remembered for his support of the much-ridiculed Peace Ship, dreamed up in late 1915 by Rosika Schwimmer to bring about a negotiated peace among the European belligerents, than for any significant involvement with the production of war materiel.[5] Yet two days after President Woodrow Wilson ended American diplomacy with Germany, the once pacifist Ford declared that he would place the productive resources of the Ford Motor Company "at the disposal of the United States government and will operate it without one cent of profit."[6]

Ford's entry into the preparedness movement was almost as newsworthy as Wilson's diplomacy. For many Americans it was far more concrete and, therefore, more understandable. The

4. William H. McNeill, *The Pursuit of Power: Technology, Armed Force, and Society since A.D. 1000* (Chicago: The University of Chicago Press, 1982), p. 331.

5. On Ford's Peace Ship venture, see Allan Nevins and Frank Ernest Hill, *Ford: Expansion and Challenge, 1915–1933* (New York: Charles Scribner's Sons, 1957), pp. 26–54.

6. *New York Herald,* February 6, 1917; *New York Times,* February 6, 1917.

productive resources of the Ford Motor Company were enormous and growing still greater. Ford had sought to manufacture as many automobiles as he could, while lowering the price of the Model T and paying unprecedented wages to his employees. His industrial empire had turned out almost 600,000 Model T automobiles in 1916 alone, using methods that had been widely publicized in both the technical literature and the popular press. These production techniques and the philosophy that lay behind them would soon become widely known as "Fordism." Not until a decade later, when an article on mass production attributed to Henry Ford appeared in the Sunday *New York Times* and the *Encyclopaedia Britannica* did "mass production" supplant "Fordism" and become the generic description of the methods and philosophy the Ford Motor Company had pioneered.[7] "Mass production is not merely quantity production," wrote Ford, "for this may be had with none of the requisites of mass production. Nor is it merely machine production, which also may exist without resemblance to mass production. Mass production is the focussing upon a manufacturing project of the principles of power, accuracy, economy, system, continuity, and speed."[8] Although vague, this definition pointed to the assembly line, first developed by Ford in 1913, as the critical ingredient of mass production. And the mere existence of a moving assembly line presupposed that behind these assembly techniques lay a manufacturing system where abundant, accurately machined parts flowed to the right place at precisely the correct time. Although the techniques of mass production would be refined by the Ford Motor Company in the years after World War I, they certainly had been mastered by the company by the time the United States entered the war. The company had achieved levels of output and productivity considered highly impressive, if not incredible.

Given Ford Motor Company's achievement and technological virtuosity, it is interesting to examine the company's efforts in World War I. Although Ford's war work ranged from the manufacture of twelve-cylinder aircraft engines, soldiers' helmets, and tanks to the production of gun caissons and armor

7. On Ford and the rise of mass production in America, see David A. Hounshell, *From the American System to Mass Production, 1800–1932* (Baltimore: Johns Hopkins University Press, 1984), pp. 1, 217–61.

8. Henry Ford, "Mass Production," *Encyclopaedia Britannica*, 13th ed., Suppl. Vol. 2 (1926), p. 821 and *New York Times*, September 19, 1926, Sec. 10, p. 1.

plate, one project in particular is noteworthy: the construction of submarine patrol boats, called Eagles. The history of this project provides an excellent opportunity to study how the company that pioneered mass production responded to the war emergency, how the military viewed the promising new technology of mass production, how the public imagination was captured by Ford's work, and what the limitations of the technology actually were.

Henry Ford's account of the Eagle boat project, which appeared in his widely read *My Life and Work* (1922), is rather simple. He pointed out that the Eagle had been designed by the Navy and noted that "on December 22, 1917, I offered to build the boats for the Navy." A contract was let on January 15, 1918, and Ford casually recorded the ensuing events:

> On July 11th, the first completed boat was launched. We made both the hulls and the engines, and not a forging or a rolled beam entered into the construction of other than the engine. We stamped the hulls entirely out of sheet steel. They were built indoors. . . . These boats were not built by marine engineers. They were built simply by applying our production principles to a new product.[9]

Ford's tone suggests that in his view the new product could just as easily have been a vacuum sweeper as a 200-foot long patrol boat; his firm's mass production technology was universally applicable.

The postwar account of Secretary of the Navy Josephus Daniels is equally matter-of-fact. Discussing the German submarine menace during World War I, Daniels noted that the Navy had worked very hard and effectively on developing detection devices but that they fell short of protecting American shipping lanes. What was needed, Daniels argued, was the design and construction of submarine chasers: "I sent an SOS to Henry Ford and at his plant were built a number of the ideal type of submarine-hunting vessels. They were called Eagle boats and were of high speed and noiseless operation."[10]

Both accounts suggest in a most perfunctory way that the Ford Motor Company agreed to manufacture submarine chasers using the same techniques of mass production as employed

9. Henry Ford and Samuel Crowther, *My Life and Work* (Garden City, NY: Doubleday, Page, & Co., 1922), pp. 246–47.

10. Josephus Daniels, *The Wilson Era: Years of War and After, 1917–1923* (Chapel Hill: The University of North Carolina Press, 1946), p. 136.

for the Model T and then readily carried out this task, but such was not the case. And although contemporary technical and nontechnical journalists hailed Ford's submarine chaser venture as "one of the most extraordinary manufacturing activities of the war,"[11] the Eagle boat venture cannot be considered successful either by the standards of the war era or of the present. Rather, Ford's attempt to mass-produce ships using the latest and best production techniques points up some of the inherent limitations of mass production and, more fundamentally, the importance of deep-seated, traditional "craft" knowledge within the institution of the Navy. This latter point is particularly noteworthy because the Navy of World War I was the "new" Navy that had been carefully built over a thirty-year period by a group of scientifically and technologically trained naval officers. In pushing the Navy into the twentieth century, these men had brought about a virtual technological revolution in ship design, construction, and armament. The "craft" knowledge they possessed in 1917 was that obtained through years of experimenting with designs, testing, and creating performance and production standards. Ford and his engineers believed that, with their mass production technology, they could dispense entirely with such experience. The Eagle boat venture is important if for no other reason than it identifies both a certain "vanity" in Henry Ford and his company's leading production engineers in 1917 and 1918 and the failure of the Navy—or more particularly, the Secretary of the Navy—to appreciate the role tradition played in equipping its fleet.

The Eagle boat project had roots in a meeting in February 1917 between Ford and Secretary Daniels in which Ford offered to turn over the full productive resources of the Ford Motor Company to the Navy. Following this meeting, Ford informed an eager press, "I can build 1000 small submarines and 3000 motors a day."[12] Although Ford's claims may not

11. Fred E. Rogers, "Ford Methods in Ship Manufacture—VII," *Industrial Management* 58 (1919): 11.

12. *New York Times,* February 6, 1917. See also Ford's ideas on a small "U-Flivver" in *New York Times,* February 10, 1917. Why Ford devoted his company's resources principally to the Navy rather than the Army is an interesting question. I suspect that Thomas A. Edison's relationship with Josephus Daniels was the determining factor. Edison, whom Ford literally worshipped, served as head of Daniels's Naval Consulting Board, a body of well-known, independent inventors and well-established professional engineers and scientists intended to screen military inventions coming to the Navy from Yankee-type inventors throughout the United States. Scholarship on Daniels and the Naval Consulting Board suggests that Daniels had a nineteenth-century mentality

have inspired or excited U.S. naval officers, who declined to comment on them,[13] both the public and Daniels were captivated by Ford's proposal. President Wilson was, too. During the summer of 1917, Wilson called Ford to Washington to recruit the Flivver King for membership on the United States Shipping Board, the body established to direct America's wartime merchant shipping efforts. Ford's chief administrative assistant, E. G. Liebold, advised him against accepting the appointment because, he contended, the Shipping Board had drawn negative reaction from the public, particularly about the performance of the board's Emergency Fleet Corporation. But Ford and Liebold met with the chairperson of the Shipping Board, Edward N. Hurley, and discussed how Ford could contribute to the war effort. According to Liebold, Hurley showed Ford data on the tonnage of vessels sunk by German submarines and said that unless something changed, the Germans would shortly be sinking ships faster than they could be built. Liebold suggested that rather than merely building more ships to be sunk, the United States should devise some type of submarine chaser to root out and eliminate the menacing submarines. The suggestion stuck, recalled Liebold, and soon the Navy and the Ford Motor Company were on their way to the Eagle venture.[14]

In fact, the Navy had already taken bold steps toward eliminating preying U-boats. Without fanfare, the Navy's Bureau of Construction and Repair, in cooperation with a number of yacht builders, designed a 110-foot, 75-ton submarine chaser. To minimize conflict with other (both merchant and naval) shipbuilding efforts, the Navy settled upon a wooden boat powered by a gasoline engine, which could be built by a large number of maritime firms on the Atlantic,

---

about how technological research and development were done. Specifically, he believed that all American technology had been the product of independent inventors—of the Thomas Edisons of America—and failed utterly to see that modern technology had become the province of formal, bureaucratic research and development organizations. Daniels's views led to the complete failure of the Naval Consulting Board, and they probably were the reason why Daniels pushed hard—perhaps over the objections of seasoned naval officers—for Ford's manufacture of the Eagle boats. On Daniels and the Naval Consulting Board, see Thomas P. Hughes, *Elmer Sperry: Inventor and Engineer* (Baltimore: Johns Hopkins University Press, 1971), pp. 243–74 and Daniel J. Kevles, *The Physicists* (New York: Knopf, 1978), pp. 102–38.

13. *New York Times,* February 10, 1917.

14. E. G. Liebold, Reminiscences, Ford Archives, Edison Institute, Dearborn, Michigan.

Pacific, and Gulf coasts as well as on the Great Lakes. Some 440 such submarine chasers were rapidly built. Able to cruise at almost seventeen knots and equipped with a depth charge launcher, two machine guns, and a 3-inch, 23-caliber boat gun, these submarine chasers nevertheless appeared to be something more than a military vessel. Their traditional, clean lines and their mahogany construction, complete with brass fittings, spoke as much to the yachtsman as to the naval commander. Although these boats achieved several of the Navy's objectives, the submarine menace continued into the late summer and fall of 1917, and the Navy determined that it needed to design a steel ship midway between the wooden chasers and the modern destroyers that distinguished the "new" Navy. These vessels would be designed to have the same speed as the wooden chasers but would have a far greater range and would be more heavily armed.[15]

Henry Ford's decision in November 1917 to accept Wilson's plea that he join the United States Shipping Board dramatically influenced the Navy's planning of these new submarine chasers. Officially, Ford served as a special assistant to the vice president of the Shipping Board's Emergency Fleet Corporation. The *New York Times* reported that Ford's "particular task will be to introduce into shipbuilding the multiple production methods he has used with such marked success in producing automobiles." The same article made clear that the Emergency Fleet Corporation was undergoing a "complete reorganization," with businessmen displacing naval officers in the major departments of the corporation.[16] In a prepared statement delivered the day after accepting his new duties, Ford said that more thought had to be given to machinery in the war effort. He argued that the war would be won by the nation "that knows the secret of quantity production through standardization on one model," and he expressed his disappointment that the U.S. government had not yet standardized its instruments of war.[17]

By December of 1917 Ford had offered the services of his company to the Shipping Board for the purpose of ship construction. Ford had two of his top engineers, William B. Mayo

15. Benedict Crowell and Robert Forrest Wilson, *The Armies of Industry* (New Haven: Yale University Press, 1921), 2: 467–69. See also J. A. Furer, "The 110-Foot Submarine Chasers and Eagle Boats," *U.S. Naval Institute Proceedings* 45 (1919): 743–52.

16. *New York Times*, November 17, 1917.

17. *New York Times*, November 18, 1917.

and William S. Knudsen, study the shipbuilding situation and visit leading shipyards in Newport News, Virginia; Camden, New Jersey; and Philadelphia. In their report to Ford, Mayo and Knudsen made several important recommendations that would soon be hallmarks of the Eagle boat venture. Above all, they recommended standardization of ship design, including their power plants. Mayo and Knudsen also argued for much of the "rough angle bending and smithing work" to be done in factories away from shipyards, for the invention of a new type of rivet heating furnace, for the adoption of a "continuous travelling system" for conveying steel parts to their point of use, for the adoption of electric welding, and for the construction of naval destroyers on "new temporary ways leaving the present ways to be utilized for merchant shipping."[18]

Ford's promotion of these recommendations in Washington intersected with the work of the Navy's Bureau of Construction and Repair on their projected steel submarine chaser. On December 24, 1917, the automobile manufacturer formally offered to build for the Navy a new ship called a "submarine detector-destroyer" with the same quantity production techniques that had been so successful for the Model T. Ford claimed that his company could build six such boats with mass production methods for every one built by conventional methods and at savings of more than 80 percent.[19] At once, Secretary Daniels requested Ford to dispatch a team of his best production engineers to Washington to confer with the Navy's Bureau of Construction and Repair.[20] Through a series of conferences, the Ford experts and the Navy's chief designers arrived at basic design principles for the new submarine chaser. By mid-January the chief of the Bureau of Construction and Repair, Admiral David W. Taylor, and his assistant, Captain Robert Stocker, had carried the design and specifications sufficiently far to allow Ford Motor Company to make a formal offer to manufacture the boats. By this time they had become known as Eagle boats thanks to a *Washington Post* editorial

18. W. B. Mayo and W. S. Knudsen, "Report of Engineers to Mr. Henry Ford Relative to Shipbuilding," December 17, 1917, Accession 572, Box 26, Ford Archives.

19. As quoted above, Henry Ford claimed that his offer was made on December 22, 1917. But Crowell and Wilson, *Armies of Industry*, 2, p. 469, noted the date as December 24. I have not found evidence for either date in the Ford Archives, but I tend to trust Crowell and Wilson rather than Ford and his ghost writer Crowther. Ford's quantity and price claims appear in Crowell and Wilson, *Armies of Industry*, 2, p. 471.

20. Ibid., pp. 469–70.

Ford's assistant Liebold had written for Ira Bennett to convince Congress of the necessity of building these vessels.[21]

Ford's offer reiterated the bureau's specifications and estimated a cost of $275,000 per boat. More important, it laid out a production schedule for the boats to be built at a new factory on Ford's property along the River Rouge near Detroit: one Eagle by mid-July, ten by mid-August, twenty by mid-September, and twenty-five each month thereafter (or roughly one Eagle per working day).[22] This schedule became part of the formal contract between Ford and the Navy executed on March 1, 1918.[23] A few days before submitting the formal offer, Henry Ford had actually told his engineer W. B. Mayo that his company could build one Eagle within "four months then one per day for two months, then two per day for three months."[24] But the schedule in the final proposal was more modest. Three days after Ford's formal proposal went out, Daniels telegraphed Ford, "PROCEED WITH ONE HUNDRED SUBMARINE PATROL VESSELS . . . DETAILS OF CONTRACT TO BE ARRANGED AS SOON AS POSSIBLE."[25] The Eagle project had taken flight.

When the Bureau of Construction and Repair had finished its rough design of the Eagle, it projected a ship that rivaled destroyers built a decade earlier. In fact, at 200 feet in length and 500 tons displacement, the Eagle boat exceeded the size of any U.S. destroyer built prior to 1903. Its 2000-horsepower, steam-driven turbine also testified to its size.[26] But the Eagle represented a radical departure for the Navy. Assisted by

21. E. G. Liebold, Reminiscences, Ford Archives. See also Crowell and Wilson, *Armies of Industry*, 2, p. 470. The names of Taylor and Stocker appear in "Submarine Chasers Made by Quantity-Production Methods," *American Machinist* 49 (1918): 842 and Fred E. Rogers, "Ford Methods in Ship Manufacture—I," *Industrial Management* 57 (1919): 3. However, the *New York Times*'s account of the Eagle's design downplays the active role of Taylor and Stocker (February 1, 1918). The *Washington Post* editorial, "A Challenge to America," appeared December 23, 1917.

22. Ford Motor Company to Josephus Daniels, January 14, 1918, Accession 572, Box 26, Ford Archives.

23. A copy of the contract is in Accession 38, Box 42, Ford Archives. Details appear in Nevins's and Hill's notes in Accession 572, Box 26, Ford Archives.

24. Telegram, W. B. Mayo to E. G. Liebold, in Accession 62, Box 58, Ford Archives, quoted in Nevins's and Hill's notes, Accession 572, Box 26, Ford Archives.

25. Telegram, Josephus Daniels to Ford Motor Company, January 17, 1918, Accession 572, Box 26, Ford Archives.

26. The comparative size of the Eagle boat was pointed out in J. A. Furer, "The 110-Foot Submarine Chasers and Eagle Boats," p. 766 and Crowell and Wilson, *Armies of Industry*, 2, p. 470. The Eagle was designed to cruise at 18.3 knots and have a radius of 3500 miles.

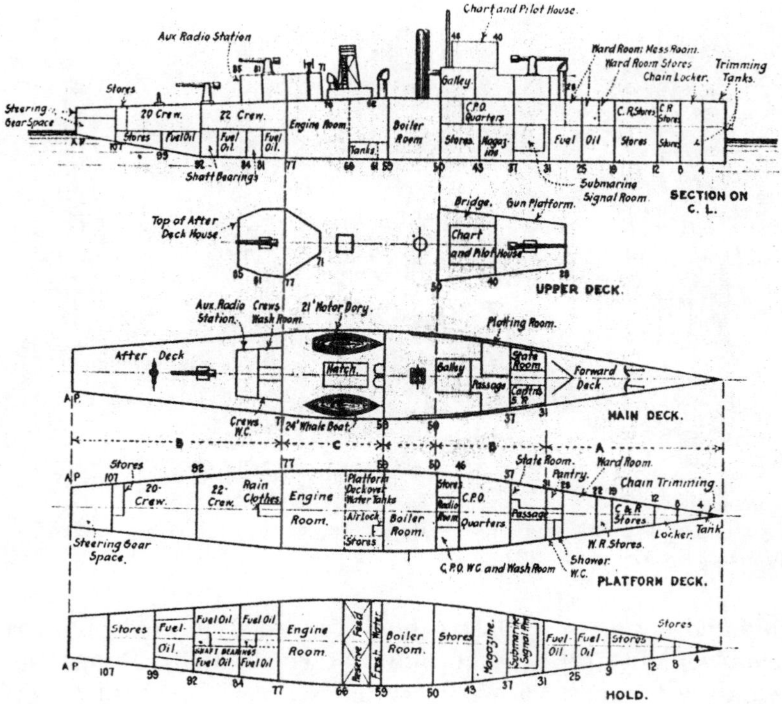


**Figure 1**
Section and plan views of the Eagle Submarine Chaser. The top view is the section along the center line. The view below shows the plans of the hold, the platform deck, the main deck, and the upper deck. The numbers that appear are frame numbers rather than dimensions. *Source: Industrial Management* 57 (1919): 293.

Ford engineers, the Navy designed the Eagle for quantity production by an automobile manufacturer—for mass production—rather than for construction by a shipbuilding firm. Consequently, in the place of smooth curves and rolled members were straight lines, flanged plates, and structural angles. A *New York Times* reporter was not wrong when he compared the Eagle boat with the already famous Ford Flivver (figures 1, 2).[27]

Because the Eagle boat was to be mass-produced, the Ford Motor Company immediately moved in two important ways. It began construction of a full-scale model or pattern of the Eagle in the main craneway at Ford's Highland Park factory (figure 3), where the assembly line had been born five years earlier, and it

27. *New York Times*, July 7, 1918. See also *New York Times*, February 1, 1918.

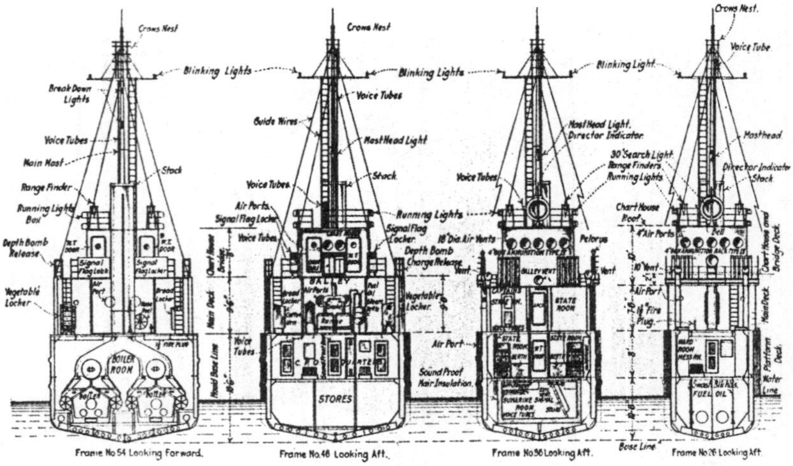


**Figure 2**
Cross sections of the Eagle at frames 54, 46, 36, and 26. *Source: Industrial Management* 57 (1919): 291.

initiated design and construction of the B-Building or boat assembly plant and other facilities at Ford's River Rouge site. Building the pattern accomplished several important things. First, construction of a model gave Ford engineers and naval officers assigned to the task a chance to revise their initial, rough design. Second, as with musket manufacture at Springfield Armory in the antebellum period, the model served as the basis for the design and construction of special jigs, fixtures, templates, and other patterns for parts of the Eagle boat.[28] The model also enabled the designers and production men to decide the position of all rivet holes. Finally, erecting the model provided Ford's production experts with the opportunity to write operations sheets—minutely detailed specifications for the order or sequence of steps in the manufacturing process.[29] Operations sheets had been brought to a high degree of perfection in the production of Model T's and were essential to the

28. On the use of a model-based jig, fixture, and gauging system in the antebellum federal armories, see Hounshell, *From the American System to Mass Production*, pp. 6, 28–29, 34, 41–42 and Merritt Roe Smith, *Harpers Ferry Armory and the New Technology: The Challenge of Change* (Ithaca, NY: Cornell University Press, 1977), pp. 92–93, 109–10, 225–27.

29. Ford's use of the model is discussed in Furer, "The 110-Foot Submarine Chasers and Eagle Boats," p. 766; Rogers, "Ford Methods in Ship Manufacture—I," p. 3; "Submarine Chasers Made by Quantity-Production Methods," p. 843; and William C. Klann, Reminiscences, Ford Archives.

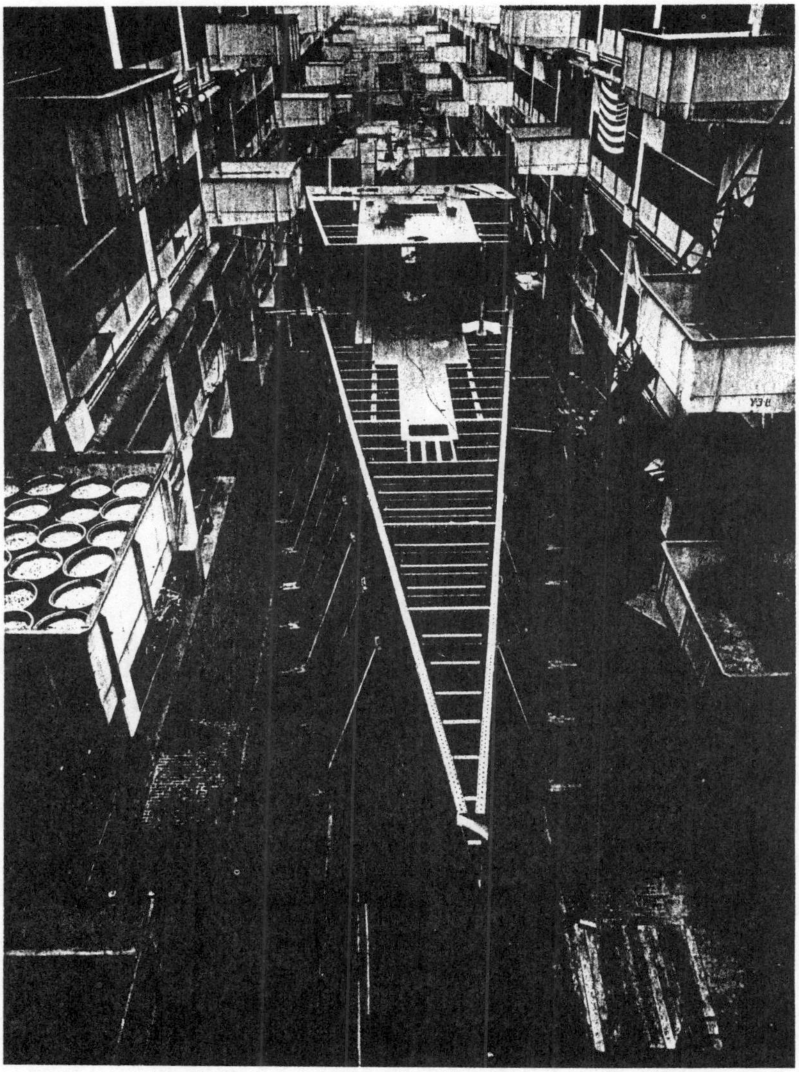

**Figure 3**
Full-scale Eagle model being constructed in main craneway of Ford's Highland Park Factory. *Source:* Ford Archives, Edison Institute, Dearborn, Michigan.

production methods of the Ford organization.[30] By the time the model had been finished and fully utilized, the plant at the River Rouge site had been partially erected, and hulls were beginning to be fabricated there.

The construction of the River Rouge facility in about four months was a remarkable achievement. Henry Ford had purchased the 200-acre site in 1915 with an ill-formed and (in retrospect) ill-conceived idea of building a fully integrated manufacturing establishment there, of bringing in raw materials and shipping out finished automobiles and tractors.[31] Though the war interrupted Ford's plans, it gave him an opportunity to build and to build big. For Eagle boat manufacture, the company hastily designed and constructed (with public funds) the fabricating shop and tool room, or A-Building (150 by 600 feet) where the steel sheets were formed and where many other Eagle parts were fabricated; the main assembly, or B-Building (350 by 1700 feet), which accommodated three parallel assembly lines, two large outfitting buildings, and an elaborate transfer and launching table that moved the Eagle boats out of the assembly building and into the water (figure 4).[32] (After the war, Ford would quickly convert the fabricating shop and the assembly building into a tractor and automobile factory.) While the buildings awed visitors with their size, the transfer and launching tables captivated observers for what they did and how they operated. Designed by Ford engineer W. B. Mayo, the transfer table, 202 feet long and supported on eleven railroad tracks, carried Eagle boats out of the assembly building and then moved them perpendicularly 300 to 600 feet away from this building into position with the launching table. Once moved onto the launching table, Eagle boats were then lowered gently into the water by means of hydraulic jacks (figures 4–7).[33]

Although the size of the Rouge structures and the speed with

30. On the importance of operations sheets at Ford, see Hounshell, *From the American System to Mass Production*, pp. 224, 270–71. For an example of operations sheets for the Eagle boat, see Fred E. Rogers, "Ford Methods in Ship Manufacture—III," *Industrial Management* 67 (1919): 190–97.

31. On Ford's purchase and early plans for the Rouge site, see Nevins and Hill, *Ford: Expansion and Challenge*, pp. 200–216.

32. Rogers, "Ford Methods in Ship Manufacture—I," pp. 2–6. See also "Building the 'Eagle's' Nest," *The Ford Man* 2 (May 17, 1918): 1.

33. Ibid. See also "Submarine Chasers Made by Quantity-Production Methods," pp. 841–43 and Crowell and Wilson, *Armies of Industry*, 2, pp. 471–72.

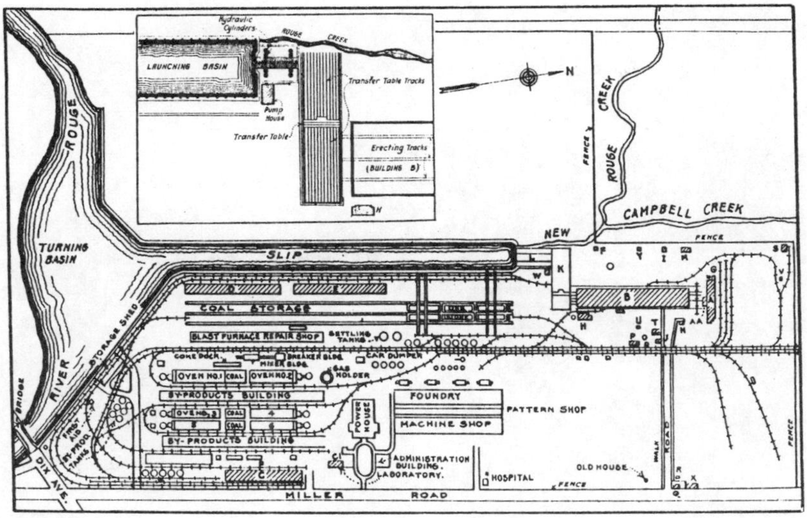


**Figure 4**
Plan of River Rouge factory. Note the structures related directly to Eagle boat production: A. Fabricating building, B. Assembly building, D. Outfitting shop, E. Outfitting shed, K. Transfer table and tracks (see inset for greater detail), L. Launching table. The length of the B-building (1700 feet) provides an idea of the scale of the Rouge's operations. *Source: Industrial Management* 57 (1919): 6, 456.

**Figure 5**
Eagle 1 moving out of assembly building onto transfer table. *Source:* Ford Archives, Edison Institute, Dearborn, Michigan.

**Figure 6**
Eagle moving off transfer table onto launching platform. *Source:* Ford Archives, Edison Institute, Dearborn, Michigan.

**Figure 7**
Eagle on launching platform. Note the large hydraulic jacks to lower the Eagle gently into the launching basin. *Source:* Ford Archives, Edison Institute, Dearborn, Michigan.

which they were erected were impressive, the haste in their construction as well as excessive confidence on the part of Ford's production engineers immediately led to problems. No doubt when Henry Ford first broached the idea of mass-producing ships for the Navy, he, his engineers, and everyone else envisioned the Eagle boats being swiftly assembled from parts as the keel (not unlike the frame of the Model T) moved smoothly and deliberately down an assembly line. That image was soon shattered. Although assembly lines would be built, the boats would not be moved along them continuously. Rather, they would be stopped and worked on at seven distinct assembly stations (figure 8). After plant construction had begun, Ford's engineers found that seven stations would be inadequate, and they were forced to add another 200 feet onto the B-Building to house an initial preassembly operation.[34] Moreover, the engineers rapidly concluded that a ship was such a complex and heavy piece of machinery that it would be impossible to complete it while on the assembly line. Unlike the Model T when it came off the assembly line, the Eagle boat would be fitted out after being launched and with techniques that bore little relationship to Ford's much-hailed "work in motion" or "flow production" methods (figure 9).[35] Moreover, major unanticipated problems arose once production got underway.

In a show of technological virtuosity, the Ford Motor Company moved quickly to lay the keel of the first Eagle. Sixteen weeks after Josephus Daniels gave Henry Ford the go-ahead to build Eagle boats, the keel-laying event took place. As Liebold wired Daniels, "Keel for the first boat has been laid on blocks and side frames are now ready to go up."[36] Two months later, on July 11, 1918, with carefully orchestrated press coverage, the complete hull was launched via the transfer table. To the awestruck American public, Ford's announcement of this event implied that Eagle 1 was ready to chase menacing submarines. Moreover, Ford promised that twelve more boats would be launched within two or three days and that soon there would be

34. "Submarine Chasers Made by Quantity-Production Methods," pp. 841–44.

35. Fred E. Rogers. "Ford Methods in Ship Manufacture—VII," *Industrial Management* 58 (1919): 8–9.

36. Telegram, E. G. Liebold to Josephus Daniels, May 7, 1918, Accession 572, Box 26, Ford Archives.

ORDER OF OPERATIONS

Progressive erection of the fabricated parts and unit assemblies is carried through in the following order:

OPERATION 1

1. Assemble ship truck
2. Keel, keel butt straps *C V K* and bar angle ties
3. Bulkhead keel collars
4. *P* strake
5. *A* strake and lines; *B* strake and liners
6. All athwartship and longitudinal bulkheads below platform deck
7. Frames entire length of boat
8. *D* strake
9. Upper section bulkheads 50, 59, 77 and 92
10. Bow assembly
11. Platform deck beams
12. Transom deck
13. Riveting *C V K*.

OPERATION 2

1. Platform deck plating forward
2. Keelson and stringers on floors
3. Sub-signal room foundations
4. *F* strake, and liners from 6 to 15
5. Engine and boiler foundations
6. Stanchions below platform deck
7. Bulkheads from platform to main deck
8. Platform deck and engine room tank plating
9. Engine and boiler room stringers
10. Reaming and riveting.

OPERATION 3

1. Forward deck plating outside plates
2. After deck plating outside deck
3. Gunwale bar, 4 x 4 inches, and collars
4. Frames to main deck and gussets
5. Main deck beams and all longitudinal ties
6. *E* strake and liners; balance of *F* strake and liners
7. *G* strake and liners
8. Main deck, 4 x 4 inches, gunwale bars
9. *C* strake
10. Bilge keel
11. Rudder brackets
12. Reaming and riveting.

OPERATION 4

1. Oil trunks
2. All stanchions to main deck
3. Windlass and steering engine foundations
4. Fender bars
5. Air lock
6. Reaming and riveting.

OPERATION 5

1. All main deck plating
2. Combing deck bars
3. Deck fittings
4. Scuttle and manhole; ventilator trunks
5. Watertight doors and non-watertight doors
6. Main deck doubling plates
7. Reaming and riveting
8. Caulking.

OPERATION 6

1. Forward deckhouse
2. Breakwater
3. After deckhouse
4. Strut and all sternbearings to stuffing-boxes
5. Boring all stern bearings
6. Bilge and drainage piping
7. Oil compartment tunnel
8. Submarine signal tube and fittings
9. Reaming and riveting.

OPERATION 7

1. Complete reaming and riveting
2. Caulking
3. Sea cock zinc and strainer
4. Steering bearings and stuffing boxes
5. Bitumastic compound fresh water tank
6. Air testing oil tanks
7. Steering engine anchor crane sockets
8. Ladder stairways
9. Rudder and bearings
10. Installation screw gears, oil pumps, coolers, auxiliary pumps and air compressor
11. Painting ship and anti-corrosive on bottom
12. Installation of the shafting and propellor
13. Concreting stern of boat
14. General inspection.

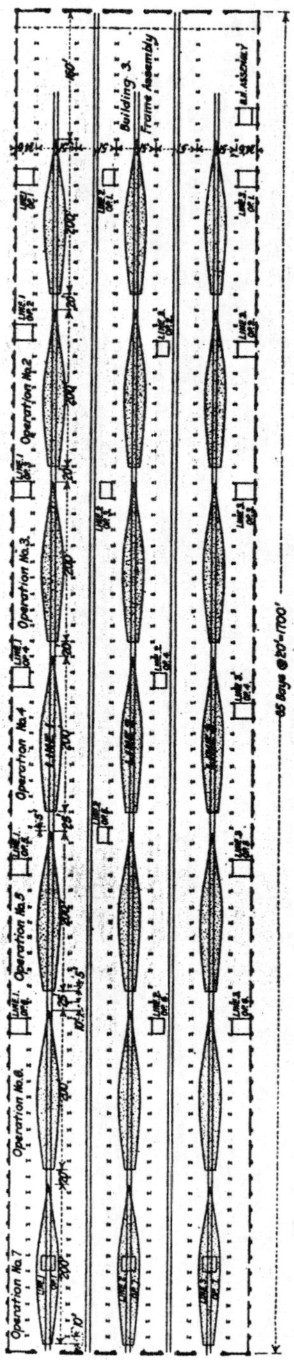


**Figure 8**

Operations sheet for Eagle boat manufacture. *Source: Industrial Management* 57 (1919): 192.

**Figure 9**
Eagle boats being outfitted in launching basin. *Source:* Ford Archives, Edison Institute, Dearborn, Michigan.

one Eagle per day.[37] But those notions quickly vanished. Not until October 30, 1918, was Eagle 1 commissioned. The last of the next dozen promised was not finished until mid-April 1919 (figure 10).[38]

Problems plagued these first Eagles, many of them growing out of inferior production techniques. By far the worst problem centered around riveting. Although Ford proudly claimed that the Eagle boats had been built without the help of seasoned marine engineers, such was not entirely the case. The Navy hired Charles C. West of the Manitowoc Shipbuilding Company as its superintending constructor. West tried to help the Ford engineers work out hull manufacturing methods but met with pronounced resistance.[39] His powerful position allowed him to force Ford to hire additional marine engineers. But he noted in September 1918 that these men were being ignored.[40]

37. *New York Times*, July 12, 1918. Eagle 1 was launched on July 11, as indicated in a telegram from E. G. Liebold to Josephus Daniels, July 11, 1918, Accession 572, Box 26, Ford Archives, and "Launching of Eagle No. 1," *The Ford Man* 2 (July 17, 1918): 1.

38. "Eagle Boat Record," Accession 572, Box 26, Ford Archives.

39. The article "Submarine Chasers Made by Quantity-Production Methods" (see note 21), which was first printed in *Engineering News-Record*, called West "one of the most energetic shipbuilders on the Great Lakes" (p. 842), while Furer regarded him as "a skilful and experienced shipbuilder, [who] made it possible for a company with no previous shipbuilding experience to build successfully seven ships of 500 tons displacement in nine months." "The 110-Foot Submarine Chasers and Eagle Boats," p. 766.

40. Charles C. West to Ford Motor Company, September 9, 1918, Accession 572, Box 26, Ford Archives.

**Figure 10**
Keel for Eagle boat. *Source:* Ford Archives, Edison Institute, Dearborn, Michigan.

Ford engineers and workers found riveting to be more difficult than they had anticipated. In particular, West and his inspectors discovered that because of poor riveting, the Eagle boats were not water-tight or oil-tight (the Eagle was designed with ten separate fuel oil compartments). West reported that the compartments of the first four Eagle boats were leaky and that workmen had "experienc[ed] considerable difficulty" in getting them sealed. In an official complaint, he argued that these problems had arisen because the advice of skilled shipbuilders had not been followed. For example, West had observed workers trying to bolt steel plates together prior to riveting. Using short-handled wrenches while standing on ladders, they were unable to bring the plates together tightly. Metal shavings lodged between the plates and made it virtually impossible for the rivets to pull the plates together for a tight seal.[41] One of the riveters complained independently to Henry Ford about the ladders and urged him to use scaffolding in-

41. Ibid.

stead. This was done eventually, but ladders were not entirely abandoned (figures 11, 12).[42] West reported other problems as well, most notably the poor quality of electric arc welding, which was a technique not used in Model T production. The superintending constructor specifically requested Ford to "do as little electric welding on oil and water tight bulkheads as possible as your welders are so inexperienced that the welds are both defective and porous."[43]

West continued to issue formal complaints about the poor quality of the work on Eagle boats. Moreover, he and other naval officials also began to worry about *quantity*. As late as May 18, 1918, Josephus Daniels continued to believe that Ford would produce all 100 Eagles by January 1, 1919.[44] Ford had, after all, promised to do so. Soon after the initial agreement between Ford and Daniels, Ford Motor Company's engineers had concluded that a plant almost identical to the River Rouge factory would be necessary to achieve the 100 Eagle boats as projected. Quickly approved by the Navy, this plant was to be built near Newark, New Jersey, and would cost another $2.5 million.[45] After giving the go-ahead for this plant, Daniels must have been confident of Ford's meeting the production schedule. But by late August 1918, when only two Eagle boats had been launched and none completely outfitted, the Navy officers in the Bureau of Construction and Repair asked Ford for an accurate revision of its Eagle production schedule. The company projected that only 26 boats would be finished by the end of December rather than the promised 112. (The Italian government had ordered 12 Eagles in addition to the 100 ordered by the Navy.) The remainder would be finished by the end of

42. Riveter to Dear Sir [Henry Ford], September 23, 1918, Accession 572, box 26, Ford Archives. Lt. Julius A. Furer, who observed the Eagle venture, wrote that Ford's policy of not consulting sufficiently with shipbuilding experts "is considered by some to have been a mistake, as the work could have been very materially accelerated had [Ford] employed a sufficient number of men skilled in the various shipbuilding trades to act as instructors for his men." "The 110-Foot Submarine Chasers and Eagle Boats," p. 765.

43. West to Ford, September 9, 1918.

44. Josephus Daniels to Ford Motor Company, May 18, 1918, Accession 62, Box 8, as noted in Nevins's and Hill's notes, Accession 572, Box 26, Ford Archives.

45. The first mention of the New Jersey plant is in a letter from W. B. Mayo to Josephus Daniels, June 27, 1918, which states that Ford Motor Company will proceed with the construction of a plant at Newark Meadows which "will be very nearly a duplicate of that at our Detroit plant." Accession 572, Box 26, Ford Archives. See also Ford Motor Company to Graham Egerton, Solicitor, Navy Department, October 14, 1918, Accession 572, Box 26, Ford Archives.

**Figure 11**
Working on the Eagle from ladders. *Source: Industrial Management* 57 (1919): 290.

**Figure 12**
Working on the Eagle from scaffolding. *Source:* Ford Archives, Edison Institute, Dearborn, Michigan.

April 1919.[46] As with Ford's initial schedule, this proved to be unrealistic.

At the end of 1918—more than a month after the armistice—Ford had delivered only seven boats,[47] and all of them had problems. West reported that all seven Eagles had "oil and machinery troubles." Apparently, the oil problem stemmed from the leakage of oil out of the holding tanks, a problem all too familiar to West. In a letter to Ford, he quoted a communication from Navy headquarters saying that the Bureau of Construction and Repair had gotten what it had expected: "Considering all the circumstances surrounding the initiation of the Eagle program, the Bureau was prepared to accept a quality of workmanship below standard in some particulars and on the earlier vessels delivered."[48] (The riveter who had written to Henry Ford complaining about having to use ladders put the matter more bluntly: the boats had been only "half-built" and then only at a rate of one a month rather than one a day.)[49] The Bureau of Construction and Repair informed Ford that work on the remaining Eagles would have to be "brought up to standard." Especially important was rectifying the problem with leaky fuel oil compartments.[50]

Soon *Detroit Saturday Night* came out with an article that pronounced Ford's Eagle boat venture a failure—one that the Navy would not openly admit. The article noted that the "boasted one-a-day program dwindled and shrank to about one a month" and labeled the reasons: "Ford workmen found shipbuilding a more difficult art than 'flivver' building. Most serious of all was their inability to rivet." Although the Navy and Ford had agreed to reduce the final output of Eagles to sixty, the

46. David W. Taylor to Ford Motor Company, n.d. [received August 26, 1918], Accession 572, Box 26, Ford Archives. Ford's reply came in a letter to Admiral D. W. Taylor, September 10, 1918. The first draft of this letter is dated September 4, 1918. Both are in Accession 572, Box 26, Ford Archives. As late as September 27, 1918, Ford Motor Company was claiming that it would turn out one Eagle per day and it argued that "this is an accomplishment unequalled in the industrial annals of the world, and is a fair illustration of Ford efficiency when carried out by Ford initiative, unobstructed or impeded by outside influences, suggestions and orders." "Patrol Boats," in "Summary of War Activities of the Ford Motor Company," September 27, 1918, Accession 572, Box 26, Ford Archives.

47. "Eagle Boat Record."

48. Charles C. West to Ford Motor Company, December 30, 1918, Accession 572, Box 26, Ford Archives.

49. Riveter to Dear Sir, September 23, 1918.

50. West to Ford Motor Company, December 30, 1918.

Navy, according to the article, showed "no sign of abandoning the reduced program."[51] The Navy's refusal to terminate the Eagle venture after the armistice in spite of increasing Congressional criticism stemmed from Josephus Daniels's enthusiasm for Ford's work—a factor that probably had been decisive in Ford being given the contract in the first place. Daniels argued that the Ford plant was ideal for manufacturing the Eagle and "had the war continued, the Navy would have obtained 500 more of them."[52]

It is difficult to imagine the Rouge factory and its twin plant in New Jersey turning out that many. On May 1, 1919, when Ford had delivered only seventeen Eagles, naval officials wrote the company that it was highly unlikely that Ford would complete the sixty Eagles by November 15, 1919, the deadline Daniels had established when he reduced the Navy's order from one hundred to sixty Eagles. The letter called for "energetic action" on Ford's part lest the company fall short by some twenty boats.[53] Less than a month later, Whitford Drake, who had succeeded West as superintending constructor, wrote the Bureau of Construction and Repair that "unless a very radical improvement is made immediately the program will not be completed this year, and it is very doubtful if it can be completed in any event."[54] Ford Motor Company did indeed marshal the necessary energy and manpower to meet the deadline. The sixtieth boat was delivered on October 15, 1919, slightly less than a year after Henry Ford had initially promised it. As a publicity ploy, the company put all of its riveters to work on the hull of the last keel laid (which turned out to be Eagle 59) in April 1919. Only ten days were required from the laying of the

51. "Why Congress is Investigating the Building of Ford Eagles," *Detroit Saturday Night* 13 (January 11, 1919): 3, as noted in Nevins's and Hill's notes, Accession 572, Box 26, Ford Archives. It is unclear exactly when Daniels cut back the Eagle program to sixty boats. A report signed by Whitford Drake, May 29, 1919, alludes to a telegram of November 1918 from the Navy Department to Ford Motor Company saying that "the sixty boats" were to be completed by November 15, 1919. Accession 38, Box 42, Ford Archives.

52. *New York Herald,* January 5, 1919. The congressional controversy about the Eagle boat program in light of the declared peace began immediately after the armistice. See *New York Times,* November 20, 1918, for the first of the salvos.

53. Carlos Bean [Inspector of Machinery] and Whitford Drake [Superintending Constructor] to Edsel Ford, May 1, 1919, Accession 572, Box 26, Ford Archives. Drake had succeeded West sometime earlier in 1919.

54. Inspector of Machinery and Superintending Constructor to Bureau of Steam Engineering and Bureau of Construction and Repair, May 29, 1919, Accession 38, Box 42, Ford Archives. Drake signed this document.

keel to the launching of the hull. But it took until the end of August to finish the boat. By contrast, the keel of Eagle 60 was laid earlier, on March 30, 1919; the boat was launched August 13 and, as noted, commissioned October 15, 1919.[55]

Without question, the outfitting operations for Eagle boats—installing the turbine, boilers, piping, wiring, armament, and all other equipment—required more time on average than the construction of the hulls. Though the Eagle had been standardized, its very nature and its cramped interior spaces prevented a massed assault on the finishing operations. The detailed operations sheets used for hull fabrication and assembly were not applicable here. Rather, specifications became general, such as "set boilers, turbine, piping flanges, [and] voice tube flanges."[56] These operations, involving a good deal of fitting, mediated against the rapid manufacture of Eagle boats. Certainly the boats were not "mass produced," if one adopts Henry Ford's own criterion for mass production: "In mass production," he later wrote, "there are no fitters."[57] In terms of quantity manufactured, time required, and means of production, Ford's Eagle boat venture did not live up to the promise that Henry Ford had made during the height of the submarine menace.

How did the Eagle boats perform after they were commissioned and what became of them? As indicated, the Navy reported that the initial seven Eagles had had oil and water leakage problems. But at the same time it noted that "Eagles 1, 2, and 3 encountered considerable heavy weather [on their maiden voyages] and their behavior in a seaway has been pronounced excellent."[58] After the six-month guarantee on Eagle 1 expired, the boat's commanding officer reported that "there were no defects in the hull, hull fitting, and equipment of this vessel, which were due to material or workmanship before the commissioning of this vessel." He pointed out that the vessel had been used heavily, particularly on maneuvers in the icy waters off the Russian coast: "It speaks very well for the construction of this vessel that in bucking heavy ice, no damage was

55. "Eagle Boat Record." For an example of the publicity surrounding the big push on Eagle 59's hull, see Fred E. Rogers, "Ford Methods in Ship Manufacture—VI: Launching and the Record of Eagle No. 59 Erected in Ten Days," *Industrial Management* 57 (1919): 456–64.

56. Rogers, "Ford Methods in Ship Manufacture—VII," p. 8.

57. Henry Ford, "Mass Production," p. 822.

58. West to Ford Motor Company, December 30, 1918.

experienced to hull, frames or bulkheads."[59] By contrast, in May 1919 West's successor Drake complained about the quality of Eagles being produced. He noted "the very extensive poor work done in the early stages of the program, and the considerable amount still being done." Reliance upon unskilled supervisors and unskilled mechanics had led to work that continually had to be "patched up or done again."[60] Later, in 1924, Theodore Roosevelt, Jr., reported for the Navy Department that Eagles 1, 2 and 3 had been used until 1920 and had operated "without serious incident . . . although they were very uncomfortable at sea."[61]

Exactly what being "uncomfortable at sea" meant is a matter of definite import. Some charged that the Eagle boats were unseaworthy.[62] Frank A. Cianflone reported that the Eagles handled "like the four-stack destroyers in high seas," and he noted that Eagle 25 capsized in a squall on the Delaware River below New Castle, Delaware, on June 11, 1920. Since nine men died in the accident, the Eagle's reputation of unseaworthiness may have begun here. By 1924 two other Eagles had been lost at sea, enhancing the negative reputation. Thirty of the Eagles had been decommissioned by this date, twenty-two were being used in training naval reservists, and five had been turned over to the Coast Guard. By World War II only eight Eagles remained, and these were used only for duty within United States continental waters. Seven were decommissioned after the war; the eighth, Eagle 56, was the victim of a German torpedo.[63]

The Ford Motor Company cannot be held solely responsible for the Eagle's lack of seaworthiness, if indeed the Eagle had such a problem. The Navy's engineers, David W. Taylor and Robert Stocker, took credit for the Eagle's design, though

59. Statement of Commanding Officer, U.S.S. Eagle One, Brest, France, August 17, 1919, Accession 6, Box 258, as quoted in Nevins's and Hill's notes, Accession 572, Box 26, Ford Archives.

60. Inspector of Machinery and Superintending Constructor to Bureau of Steam Engineering and Bureau of Construction and Repair, May 29, 1919.

61. T. Roosevelt to Gentlemen [Ford Motor Company], April 3, 1924, Accession 572, Box 26, Ford Archives.

62. The first published claim about the instability of the Eagle appeared in "Why Congress is Investigating the Building of Ford Eagles." Josephus Daniels chalked up this and other criticism of the Eagle program to politics related to the senatorial election of Truman H. Newberry. See E. David Cronon, ed., *The Cabinet Diaries of Josephus Daniels, 1913–1921* (Lincoln: University of Nebraska Press, 1963), entries of January 5 and 21, 1919, pp. 364, 369.

63. Frank A. Cianflone, "The Eagle Boats of World War I," *U.S. Naval Institute Proceedings* 99 (June 1973): 76–80.

Ford's production engineers most certainly contributed their ideas to make production easier. Tests of the model conducted by Taylor and Stocker in the tank at the Washington Navy Yard obviously seemed satisfactory to them.[64] But the full-size ship may have had problems. If so, these clearly grew out of Eagle's being designed for mass production by an automobile maker. The tacit knowledge of what constituted a good hull design was consciously disregarded in favor of production priorities. Given the poor performance by the Ford Motor Company, which had been exacerbated by the company's failure to heed the advice of experienced shipbuilders, this tradeoff was probably not worth it in the end.

The Eagle boat venture was, therefore, an ill-fated attempt to bring the supposed power of mass production technology to bear on the problem of procuring ships for warfare. Ford's venture failed to revolutionize shipbuilding techniques. Rather, it pointed up the difficulty of transferring the methods used to manufacture a high-volume consumer durable like an automobile into an area like shipbuilding, which had its own tradition of knowledge and skills and which produced a comparatively small number of units. Moreover, because in many respects the Eagle project was foisted upon the Navy by Josephus Daniels and the wartime shipping crisis, its failure and the "success" of other, non-Ford programs suggests that a close coupling of and careful interplay between military project managers and civilian contractors is essential to military enterprise. The relationship between Ford and the Navy was shaky at best, held together only by the bond between Daniels and Ford.

When in World War II Henry J. Kaiser undertook to "mass-produce" the infamous Liberty ships, he faced many of the same problems as the Ford Motor Company had in the First World War. Like Ford's Eagle venture, Kaiser's project captured the imagination of Americans, but his techniques did not radically change shipbuilding practice either. The Liberty ship, like the Eagle, was an expedient, a stopgap measure born of a wartime emergency. Although allowing for rapid production, Kaiser's techniques had their costs: many, many of the Liberty ships cracked up during normal sea operation, and at least a

64. On February 19, 1918, Daniels recorded in his diary that he had been to the Navy Yard to see the model tests for the Eagle. Cronon, ed., *Cabinet Diaries*, p. 281. See also *New York Times*, July 7, 1918.

third of those that did not "suffered major fractures."[65] Like the history of Kaiser and the Liberty ship venture, the case of Eagle boat manufacture does not represent a failure of mass production but rather the failure of Henry Ford and his engineers and Josephus Daniels and the Navy's engineers to realize the inherent limitations of mass production technology. The technology was not, indeed, universally applicable. Shipbuilding remains an industry where craft knowledge and traditional design and construction continue to play important roles that have long since disappeared in the mass production of consumer durables such as the automobile.

65. Unlike the Eagle boat, the Liberty ship was based on a design that had been well-tested by the British. But the construction techniques worked out by Kaiser, who could speak only of the "front end" and "back end" of a ship, were the key to the Liberty program. Most important among them was the all-welded hull. See L. A. Sawyer and W. H. Mitchell, *The Liberty Ships: The History of the "Emergency" Type Cargo Ships Constructed in the United States during World War II* (Newton Abbot: David & Charles, 1970). Quotation appears on p. 19.

# [16]

# *Flexibility and Mass Production at War: Aircraft Manufacture in Britain, the United States, and Germany, 1939–1945*

JONATHAN ZEITLIN

The relationship between military enterprise and mass production is a central theme in the history of technology.[1] Military demand, military finance, and military ideology, it is widely agreed, provided a crucial stimulus for the technological breakthroughs necessary for mass production. The origins of the "American system of manufactures" during the 1830s and 1840s, as Merritt Roe Smith has shown, lay in the U.S. Army's determined quest for a rifle with interchangeable parts, itself inspired by the earlier visions and experiments of military engineers in 18th-century France.[2] During the late 19th and early 20th centuries, as David Hounshell has demonstrated, American industrialists successfully adapted the techniques of "armory practice"—notably, sequential manufacture of interchangeable parts using special-purpose machinery, jigs, fixtures, and gauges—to high-volume production of light, standard-

DR. ZEITLIN, associate professor of history and industrial relations at the University of Wisconsin—Madison, is completing a book on flexibility and mass production in the British metalworking industries, 1830–1990, with support from the Guggenheim Foundation and the German Marshall Fund of the United States. He is grateful to the following friends and colleagues for generous provision of advice, comments on previous drafts, research materials, and unpublished manuscripts: Herrick Chapman, David Edgerton, John Guilmartin, Howell Harris, Gary Herrigel, Paul Hirst, I. B. Holley, Jr., David Hounshell, Nelson Lichtenstein, Ned Lorenz, Eric Schatzberg, Jim Tomlinson, and Jacob Vander Meulen. The usual disclaimers apply.

[1]For a general survey, see Merritt Roe Smith, "Introduction," in Merritt Roe Smith, ed., *Military Enterprise and Technological Change: Perspectives on the American Experience* (Cambridge, Mass., 1985), pp. 1–38.

[2]Merritt Roe Smith, *Harpers Ferry Armory and the New Technology* (Ithaca, N.Y., 1977), and "Army Ordnance and the 'American System' of Manufacturing, 1815–1861," in Smith, ed., pp. 39–86; Ken Alder, "Terror and Technocracy: The French Origins of Interchangeable Parts Manufacturing" (paper presented at the Society for the History of Technology meeting, Uppsala, Sweden, August 16–20, 1992). For a dissenting view that argues for the independent origins of the American system in the private sector, see Donald R. Hoke, *Ingenious Yankees: The Rise of the American System of Manufactures in the Private Sector* (New York, 1990).


0040-165X/95/3601-0002$01.00

ized equipment such as sewing machines, typewriters, agricultural implements, bicycles, and ultimately automobiles.[3] Both military and economic historians also concur that the resulting capacity of civilian firms to manufacture large numbers of standardized weapons became increasingly central to the conduct of industrialized warfare, while the great munitions drives of World War I in particular played a key part in diffusing mass-production methods from the United States to other combatant powers such as Britain, France, and Germany.[4] World War II, by all accounts, marks the apogee of this symbiosis between mass production and military prowess. For the Allied victory surely depended, as contemporaries themselves believed, on the ability of American (and Soviet) mass-production industry to turn out military aircraft, tanks, and other weapons systems in unprecedented quantities.[5]

Historians of technology are well aware that international military rivalries have often centered on complex weapons systems such as naval battleships produced to customer specifications in small batches or single units using skill-intensive methods. And they have likewise documented the importance of military objectives and material support for the development of flexible, general-purpose technologies such as computers and numerically controlled machine tools.[6]

Yet the prevailing view has been that the military's overriding concern with what Roe Smith calls "uniformity and order" has shaped and reinforced two central premises of modern industrial management: control and standardization. Insofar as military procurement objectives are recognized to diverge from the mass-production model—for example, by emphasizing performance over costs—this has often been seen as an institutional distortion of normal economic processes that inhibits defense contractors from competing successfully in civilian product markets. For contemporary critics of the "baroque" technolo-

[3]David A. Hounshell, *From the American System to Mass Production, 1800–1932: The Development of Manufacturing Technology in the United States* (Baltimore, 1984).

[4]On the importance of interchangeable-parts manufacture for national military power after 1850, see William H. McNeill, *The Pursuit of Power: Technology, Armed Force, and Society since* A.D. *1000* (Chicago, 1982), chaps. 7–9, esp. pp. 232–36, 330–31, 355, 358–59.

[5]For overviews, see Alan S. Milward, *War, Economy and Society, 1939–1945* (London, 1977), chap. 6; R. J. Overy, *The Air War, 1939–1945* (London, 1980), chap. 7.

[6]On the naval arms race before World War I, see McNeill, chap. 8; Jon Tetsuro Sumida, *In Defence of Naval Supremacy: Finance, Technology, and British Naval Policy, 1889–1914* (London, 1990); Clive Trebilcock, *The Vickers Brothers: Armaments and Enterprise, 1854–1914* (London, 1977). On the military role in the development of computers and numerically controlled machine tools, see Kenneth Flamm, *Creating the Computer: Government, Industry, and High Technology* (Washington, D.C., 1988), chap. 3; David F. Noble, "Social Choice in Machine Design: The Case of Automatically Controlled Machine Tools," *Politics and Society* 8 (1978): 313–47, and *Forces of Production: A Social History of Industrial Automation* (New York, 1984).

gies nurtured by postwar patterns of military demand, World War II likewise represents the high-water mark of convergence with civilian production methods from which subsequent procurement policies have systematically deviated.[7]

This view of the relationship between military and civilian technology clearly rests on normative assumptions about the superiority of mass production as a model or paradigm of modern industrial efficiency. But as a growing body of literature has argued, mass-production firms have often been successfully out-competed by practitioners of an alternative strategy—sometimes termed "flexible specialization"—based on the manufacture of a changing array of customized goods using flexible, general-purpose equipment and skilled, adaptable workers. Where demand patterns are uncertain and technologies rapidly changing—as in much present-day industry—such flexible production methods give manufacturers a vital edge in adjusting swiftly to shifting competitive conditions and bringing new products to market. As in the past, too, many firms now seek to avoid the potential disadvantages associated with exclusive reliance on either mass or flexible production by combining elements from each model in varying proportions.[8]

This article reexamines aircraft production in Britain, the United States, and Germany during World War II from a perspective informed by current upheavals in markets, technologies, and industrial organization but attentive to contemporary debates among the historical actors themselves. In each of these countries, I argue, aircraft production, like that of other weapons systems such as tanks or naval warships, required not only the capacity to manufacture complex products in vast numbers, but also considerable flexibility. As fighters, bombers, and other types of aircraft were tested against one another under real battlefield conditions, military authorities demanded continual modifications to overcome design flaws and incorporate new tactical and technical ideas.

[7]For the influence of military concerns with uniformity and order on modern management, and the primacy of performance over price in weapons performance, see Smith, "Introduction" (n. 1 above), pp. 7, 17, 20–22. For contemporary criticisms of military production as "baroque technology" by contrast to civilian mass production, see Mary Kaldor, *The Baroque Arsenal* (London, 1982); Ann Markusen and Joel Yudken, *Dismantling the Cold War Economy* (New York, 1992).

[8]On flexible specialization and hybrid forms of productive organization, see Michael J. Piore and Charles F. Sabel, *The Second Industrial Divide* (New York, 1984); Charles F. Sabel and Jonathan Zeitlin, "Historical Alternatives to Mass Production: Politics, Markets, and Technology in Nineteenth-Century Industrialization," *Past and Present*, no. 108 (1985), pp. 133–76; Paul Hirst and Jonathan Zeitlin, "Flexible Specialization versus Post-Fordism: Theory, Evidence and Policy Implications," *Economy and Society* 18 (1991): 1–55; Charles F. Sabel and Jonathan Zeitlin, eds., *Worlds of Possibility: Flexibility and Mass Production in Western Industrialization* (Cambridge, 1996, in press).

Pervasive but unpredictable shortages of raw materials, components, and labor likewise involved constant improvisation to avoid costly bottlenecks. Established mass-production methods, such as those pioneered by the automobile industry, typically proved too rigid for the high level of uncertainty and rapid pace of innovation imposed by the war economy. Successful aircraft manufacturers therefore needed to find new ways of reconciling the high throughput of mass production with the adaptability of the craft workshop. The result was a proliferation of hybrid forms of productive organization, which anticipated in many respects more recent innovations in flexible manufacturing such as those developed by the Japanese. At the same time, however, the methods used to balance the quality and quantity of aircraft output varied widely across the three countries, as did their relative effectiveness, reflecting deeper national differences in geopolitical position, resource endowments, productive capabilities, and civil-military relations, among other factors.

### *Britain: Quantity Production and Continuous Improvement*

In Britain, as in each of the major combatant powers of the 1930s and 1940s, rearmament and total war demanded vast quantities of munitions, from shells, guns, and small arms to ships, tanks, and transport vehicles, along with the machine tools to produce them. Such a massive increase in output inevitably required both the expansion of established arms manufacturers and the conversion of civilian engineering firms to military production. Nowhere was the transformation more dramatic and far-reaching than in the production of aircraft, the weapons system accorded the highest priority in British strategic planning: on the eve of rearmament in 1933, the industry employed some 20,000 people; at its peak in 1944, the workforce involved had mushroomed to 1.8 million, or more than one-third of the entire manufacturing sector.[9]

The interwar aircraft industry, as these figures suggest, was very much a small-scale affair. From its wartime high of 268,000 employees in 1918, aircraft manufacturing in Britain contracted sharply during the 1920s. With commercial demand limited by the relatively slow development of civil aviation within the British Empire, the industry was largely sustained by the Air Ministry, which sought to maintain its military potential during a period of relative disarmament by sharing out small orders among a "ring" of approved contractors. As a result, the Royal Air Force (RAF) in 1931 was using forty-four different types of airframe and

[9]For these figures, see Peter Fearon, "Aircraft Manufacturing," in *British Industry between the Wars,* ed. Neil K. Buxton and Derek H. Aldcroft (London, 1979), p. 216; M. M. Postan, *British War Production* (London, 1952), p. 310; David Edgerton, *England and the Aeroplane* (London, 1991), p. 72.

thirty-five different types of engine, though a smaller number of more successful designs predominated. Airframe manufacture was thus fragmented into fifteen to twenty small and medium-sized firms, operating primarily as defense contractors, while aeronautical research was conducted by the Royal Aeronautical Establishment and the National Physical Laboratory, which functioned as collective resources for the industry and the RAF.

Before 1935, the airframe manufacturers remained principally a collection of design teams and jobbing workshops, with limited experience of working with jigs and fixtures or other volume-production methods, although orders for individual models might run as high as fifty to 100 in any given year. Aero engine production was concentrated in the hands of a smaller number of larger firms, several of which also made luxury cars. But even here, where interchangeability was more important for technical performance, production runs were short and opportunities for standardization limited. On both airframes and engines, therefore, "bench methods" based on general-purpose tools and extensive hand fitting remained the rule, and skill requirements ranked among the highest in British engineering.[10]

All this began to change significantly with Prime Minister Stanley Baldwin's commitment to achieve air parity with Germany in 1934 and the decision to reequip the RAF with a new generation of modern aircraft in 1936. A key step in the expansion of aircraft production was to be the construction of a series of purpose-built "shadow factories," financed by the state but managed by outside firms, notably, automobile and electrical manufacturers, working to designs supplied by established airframe and engine companies. This scheme, devised by Lord Weir, the air minister's industrial advisor and a longstanding advocate of mass production, was based explicitly on the assumption that the efficiency of aircraft manufacture could be substantially increased by the infusion of management skills and high-volume techniques from the shadow firms. The shadow scheme was forced through by the Air Ministry over the vehement opposition of the Society of British Aircraft Constructors (SBAC), which argued that manufacturers of standardized goods such

[10]On the development of the British aircraft industry to 1935, see Edgerton, *England and the Aeroplane,* chaps. 1–3, who substantially revises the negative view of its performance during this period advanced by Fearon, "Aircraft Manufacturing," pp. 216–40, "The Formative Years of the British Aircraft Industry, 1913–24," *Business History Review* 43 (1969): 476–95, "The British Aircraft Industry and the State, 1918–35," *Economic History Review* 27 (1974): 236–51, and exchange with A. J. Robertson in *Economic History Review* 28 (1975): 648–62, and "The Vissicitudes of a British Aircraft Company: Handley Page Ltd. between the Wars," *Business History* 20 (1978): 63–85. I am grateful to David Edgerton for new information about the size of RAF orders for particular models during this period.

as automobiles lacked the necessary flexibility to accommodate the frequent modifications in design inherent in aircraft production. Despite their claims to superior adaptability, however, the SBAC's own members soon showed themselves only too anxious to slow down the introduction of new types of planes in hopes of obtaining lucrative continuation orders on well-established models.[11]

Whereas some shadow firms like Standard Motors and (later) English Electric adapted easily to the close tolerances, continual design changes, and complex coordination tasks involved in aircraft production, other prominent volume manufacturers proved less successful. Thus Austin was demoted from its preeminent position as the sole assembler of complete aero engines in the second shadow scheme of 1938, while the management of the giant Spitfire shadow factory at Castle Bromwich was transferred from the Nuffield (Morris) Organisation to Vickers-Armstrong in 1940. As a result of such experiences, the Air Ministry and its advisors reluctantly acknowledged that despite the automobile industry's vital importance for the expansion program, its methods and equipment were "far less appropriate" for aero engine manufacture than had been originally anticipated. Too much reliance had been placed on the organizational skills of the motor vehicle firms, argued Sir Charles Craven of Vickers—not without a strong measure of self-interest—since aircraft manufacture was "more of a quantity production type than a mass production problem." Automobile industry practice, it appeared, could not be directly transferred to aircraft manufacture, though shadow firms like Austin, Rover, and Standard did eventually become more adept at adapting techniques such as "straight flow" machining, assembly conveyors, and time study to the production of aircraft components and subassemblies.[12]

Subsequent procurement policy accordingly favored extensions to established aircraft firms over management by outside contractors, with few major exceptions such as Ford's Merlin engine plant. By the end of the war, government expenditure on such extensions exceeded that on shadow factories by a margin of £225 million to £146 million, respectively. Even more striking, only 22 percent of wartime aircraft by weight

[11]On the planning of air rearmament, the development of the shadow scheme, and conflicts between the Air Ministry and SBAC, see Postan, chaps. 2–3; William Hornby, *Factories and Plant* (London, 1958), chaps. 6–8; Edgerton, *England and the Aeroplane*, pp. 68–77, and "State Intervention in British Manufacturing Industry, 1931–51: A Comparative Study of Policy for the Military Aircraft and Cotton Textile Industries" (Ph.D. diss., University of London, 1986), pp. 106–42, 193–202; David Thoms, *War, Industry, and Society: The Midlands, 1939–45* (London, 1989), chap. 1; Robert P. Shay, *British Rearmament in the 1930s: Politics and Profits* (Princeton, N.J., 1977).

[12]Thoms, pp. 9, 17–20, 29–30, 52–53, 56–58, quotations from p. 57.

TABLE 1

BRITISH, AMERICAN, AND GERMAN AIRCRAFT PRODUCTION, 1939–45

| | 1939 | 1940 | 1941 | 1942 | 1943 | 1944 | 1945* |
|---|---|---|---|---|---|---|---|
| A. Number of aircraft: | | | | | | | |
| United Kingdom.......... | 7,940 | 15,049 | 20,094 | 23,672 | 26,263 | 26,461 | 12,070 |
| United States............. | 5,856 | 12,864 | 26,277 | 47,836 | 85,898 | 96,318 | 49,761 |
| Germany ................. | 8,295 | 10,247 | 11,776 | 15,049 | 24,807 | 39,807 | 7,540 |
| B. Structure weight (millions of pounds): | | | | | | | |
| United Kingdom.......... | 29 | 59 | 87 | 134 | 185 | 208 | 95 |
| United States............. | ... | ... | 82 | 275 | 651 | 952 | 430 |
| Germany ................. | ... | ... | 88 | 114 | 163 | 199 | ... |

Source:—R. J. Overy, *The Air War, 1939–1945* (London, 1980), p. 150.
*United Kingdom = January–September; United States = January–June.

was produced outside the air firms themselves, though the proportion was substantially higher for certain important types such as heavy bombers. Two large groups, Vickers-Armstrong (Supermarine and Vickers Aviation) and Hawker-Siddeley (Armstrong-Siddeley, Armstrong-Whitworth, Gloster, Hawker, A. V. Roe), alone accounted for nearly half the total output of British aircraft during World War II.[13]

While the initial phases of rearmament proceeded slowly, not least because of the reluctance of Baldwin's successor Neville Chamberlain and his advisors to place the economy on a war footing and to negotiate over dilution (the substitution of unskilled men and women for skilled craftsmen) with the engineering unions, the expansion programs launched in 1936–38 did ultimately generate large increases of output. Thus by 1940, as table 1 shows, Britain was producing 15,000 military aircraft, fifteen times as many as five years earlier (thirty times more in terms of weight) and 50 percent more than Germany.[14]

After the fall of France, aircraft procurement programs were drastically stepped up, and a much wider range of engineering firms were drawn into the industry as subcontractors and component suppliers. This vast manufacturing archipelago was overseen by a separate Ministry of Aircraft Production (MAP), which developed a complex and sophisticated statistical planning apparatus for coordinating the flow of raw materials, components, and labor among some 12,000 firms while

[13]Edgerton, *England and the Aeroplane*, pp. 72–73, 75; Overy, *Air War* (n. 5 above), p. 165; Hornby, *Factories and Plant*, pp. 214, 222–23; Postan, p. 388.

[14]On the slow pace of prewar rearmament and the problem of dilution, see also Peggy Inman, *Labour in the Munitions Industries* (London, 1957), chap. 2; R. A. C. Parker, "British Rearmament 1936–9: Treasury, Trade Unions and Skilled Labour," *English Historical Review* 96 (1981): 306–43. For the quantitative expansion of production and the comparison with Germany, see Edgerton, *England and the Aeroplane*, p. 71; Overy, *Air War*, pp. 21, 150.

constantly adjusting production schedules for different types of aircraft to take account of supply bottlenecks and changing operational requirements. Sir Stafford Cripps, who became minister of aircraft production in 1942, was an enthusiastic exponent of mass production and scientific management who had run a government munitions factory during World War I. Under his direction, MAP deployed a variety of instruments to encourage the adoption of high-volume methods in aircraft manufacture, notably, the Production Efficiency Board and the Technical Costs Branch, which advised firms on factory layout, tooling, rate fixing, quality control, and labor utilization, placing particular emphasis on the diffusion of motion-study techniques (overseen by a representative of the Metropolitan-Vickers Electrical Co., its leading domestic practitioner). Together with the Labour Supply Inspectorate of the Ministry of Labour, MAP pressed its suppliers to increase dilution, and the proportion of female labor in the industry as a whole reached 36.5 percent by mid-1943. Where aircraft manufacturers did not measure up to these standards of efficiency, finally, MAP used its powers to force through changes directly, as in cases such as Napier's, whose management was transferred to English Electric; Fairey Aviation, where a government controller was installed as managing director and deputy chairman; or, most radically, Short Bros., which was nationalized after repeated failures to meet agreed production schedules.[15]

The central dilemma of wartime aircraft manufacture was the need to balance the qualitative gains obtainable through design modifications against the quantitative losses resulting from interruptions to continuous production runs. Since the development of new aircraft types took some four to seven years, any design might prove strategically obsolete by the time it reached the stage of large-scale manufacture, and periodic revisions were essential to take account of battlefield experience. Insofar as the development of new types could be accelerated through abridged procedures such as ordering "off the drawing board" and what would now be called "simultaneous engineering" of design and tooling, modifications during the course of production became even more crucial in removing bugs from untested models. But the more radical the design change, the longer the time required for its introduction, and the greater the loss in current output arising from alterations to

[15]On MAP and the organization of wartime aircraft production, see Postan (n. 9 above), pp. 303–22; Inman, *Labour in the Munitions Industries*, pp. 42–81, 428–33; Alec Cairncross, *Planning in Wartime: Aircraft Production in Britain, Germany and the USA* (London, 1991), pt. 1; Edgerton, "State Intervention" (n. 11 above), pp. 202–11, and "Technical Innovation, Industrial Capacity and Efficiency: Public Ownership and the British Military Aircraft Industry, 1935–48," *Business History* 26 (1984): 247–79; Nick Tiratsoo and Jim Tomlinson, *Industrial Efficiency and State Intervention: Labour, 1939–1951* (London, 1993), chap. 2.

tooling and falling labor productivity on unfamiliar operations. The solution arrived at by the British was a policy of incremental improvements and continuous modifications that could be "spliced in" more quickly and with less disruption to production schedules through careful coordination among airframe, engine, and accessory manufacturers than could a completely new design. Only at moments of extreme military urgency, as at the onset of rearmament in 1934, after the Munich crisis in 1938, and during the Battle of Britain in 1940, was the RAF's so-called doctrine of quality relaxed in order to secure larger quantities of existing types, while current designs were temporarily frozen.[16]

This policy of incremental improvement and continuous modification allowed the British to maintain a very high average quality of aircraft and leapfrog over German designs by constantly incorporating new tactical and technical ideas. Between 1938 and 1944, for example, the Supermarine Spitfire, one of the most successful and versatile fighters of the war, went through more than twenty revisions significant enough to be denoted by a separate mark number, none of which required more than a fraction—typically a small one—of the person-hours spent in developing and tooling up the original design. These revisions, in turn, were responsible for dramatic advances in the Spitfire's performance, which increased its maximum speed from 356 to 460 miles per hour, its altitude ceiling from 34,000 to 42,000 feet, and its rate of climb from 2,500 to 5,000 feet per minute, while also enhancing its firepower, maneuverability, flying range, strength, and suitability for specialized tasks such as photographic reconnaissance, naval work, and even bombing. Similar gains in performance and adaptability were achieved through piecemeal modification of other major fighter and bomber designs such as the Hawker Hurricane, the de Havilland Mosquito, the Avro Lancaster, and the Vickers Wellington. At the same time, moreover, output continued to rise steadily if not at the rate envisaged by the overambitious programs of the early war years: between 1940 and 1944, as table 1 shows, the weight of aircraft produced more than tripled, from 59 to 208 million pounds per year, notwithstanding the radical shift in its composition from fighters to bombers during 1942–43.[17]

Despite the beneficial impact of design improvements on the quality of British aircraft, such changes were greatly disliked by aircraft manufacturers and those within MAP responsible for meeting output targets,

[16]Postan, pp. 322–45; M. M. Postan, D. Hay, and J. D. Scott, *Design and Development of Weapons* (London, 1964), pp. 1–9, 139–53, 159–74. For a parallel analysis of "the dilemma of mass production: more airplanes or better?" see Irving Brinton Holley, Jr., *Buying Aircraft: Matériel Procurement for the Army Air Forces* (Washington, D.C., 1964), pp. 312–18.

[17]Postan, pp. 326–27, 339–41; Postan, Hay, and Scott, pp. 159–68, 535–36; Edgerton, *England and the Aeroplane* (n. 9 above), pp. 71–72; Overy, *Air War*, p. 150.

since they undercut the effectiveness of mass-production techniques. Thus a 1942 study conducted by Sir Ernest Lemon, MAP's director-general of production and a pioneer of high-volume rolling-stock manufacture on the prewar London, Midland and Scottish Railway, showed that while full-scale jigging and tooling-up could be justified on an uninterrupted series of 1,500 components, the continuous spate of modifications meant that aircraft like the Spitfire were largely produced in batches of 500 or less, on which bench methods would have been more economical. Similarly, prewar manufacturers of standardized consumer durables like the American-owned Hoover Ltd. (vacuum cleaners) complained bitterly that fluctuating order quantities and frequent changes in the design and specification of aircraft components—which MAP itself regarded as unavoidable—largely undermined their ability to realize the benefits of flow production. Even where mass-production firms were able to concentrate uninterruptedly on long runs of standardized equipment for the aircraft program, the inflexibility of their approach could nonetheless prove a serious handicap. Thus Ford's shadow factory turned out single-stage Merlin engines faster and more cheaply than Rolls-Royce's home plant at Derby but unlike the latter could not change over to the two-stage version (which was responsible for major improvements in speed and altitude on the Spitfire and other aircraft) without complete reequipment of both machining and assembly shops.[18]

*The United States: Mass Production and Retrospective Modification*

These trade-offs between quality and quantity of wartime aircraft production were resolved rather differently across the Atlantic. Despite the growth of civil aviation in the United States during the 1920s and 1930s (stimulated by government-subsidized airmail contracts), military demand nonetheless remained predominant as in Britain, and the prewar American industry, like its transatlantic counterpart, "was still making airplanes largely by hand," in the words of I. B. Holley, the historian of U.S. air force procurement.[19] As in Britain, too, automobile

[18]Postan, pp. 341–42, 410–11; Postan, Hay, and Scott, pp. 168–69; Edgerton, "State Intervention" (n. 11 above), pp. 128–30, 133–35; Cairncross, pp. 30–31, 64, 68; Sir Ernest Lemon, "Reduction in the Time of Change-Over to a New Type of Aircraft," August 1942, London, Public Record Office (PRO), BT 28/423/IC/114/2; C. B. Colston (chairman and managing director of Hoover Ltd.), "Simplified or Rationalised Production in the Light Engineering Industry," report to MAP, February 1942, and comments of MAP officials in PRO, BT 28/410/IC/105; Ian S. Lloyd, *Rolls Royce: The Merlin at War* (London, 1978), pp. 120–23.

[19]On the prewar development of the U.S. aircraft industry and the nature of its production methods, see Jacob Vander Meulen, *The Politics of Aircraft: Building an American Military Industry* (Lawrence, Kans., 1991); Holley, *Buying Aircraft*, chap. 2, and "A Detroit

manufacturers played a key part in the expansion of U.S. aircraft production after President Roosevelt's 1940 call for 50,000 planes within two years. But the application of automotive methods to aircraft manufacture was carried much further in the United States, with its greater experience of mass production, its larger reserves of labor and industrial capacity, and its geographical insulation from enemy bombing attacks. Although the Army Air Corps had decided in 1938 to employ only established aircraft firms as prime contractors for final assembly of airframes, influenced by Britain's disappointing experience with the intial phases of the shadow factory scheme, automobile manufacturers and other outside subcontractors ultimately accounted for some 40 percent of wartime aircraft production by weight, including 48 percent of all aero engines and some two-thirds of all combat engines. Automobile company executives, reluctant to forgo current car sales and preparations for new models in the booming market of 1940–41, claimed initially that only 10–15 percent of their machine tools could be converted to aircraft production, but by June 1942 the effective proportion redeployed had reached 66 percent.[20]

The most extreme experiment with the application of Detroit-style mass production to military aircraft was Ford's vast greenfield plant at Willow Run, Michigan, constructed for the manufacture of the Consolidated B-24 bomber. In June 1940, Henry Ford and his associates had proclaimed their ability to produce one thousand fighter planes a day within eight months, provided that the design was frozen at the outset and changes during manufacture were "tabooed." Despite the urgent need for vast quantities of combat aircraft—more than a year later the U.S. Army's aerial striking force was still judged to be at "zero strength"—Ford's remarkable offer was rejected by the military authorities because none of the designs earmarked for production had yet passed beyond the experimental stage. Undeterred by this setback, the company accepted a contract to supply sets of knocked-down parts for the heavy, four-engined, B-24 bomber; risked its own financial and engineering resources to tool up for complete assembly; and retrospectively obtained official sanction—including reimbursement—for the resulting fait accompli. From the

Dream of Mass-Produced Fighter Aircraft: The XP-75 Fiasco," *Technology and Culture* 28 (1987): 578–79, from which the quotation is drawn; Tom Lilley et al., *Problems of Accelerating Aircraft Production during World War II* (Boston, 1947), pp. 5–7, 39–40, 53–55.

[20]On the mobilization of the U.S. automobile industry and its contribution to wartime aircraft production, see Holley, *Buying Aircraft*, pp. 290–92, 304–16, 560–63; W. F. Craven and J. L. Cate, eds., *The Army Air Forces in World War II*, vol. 6, *Men and Planes* (Chicago, 1955; reprint, Washington, D.C.: Office of Air Force History, 1983), pp. 319–30; Lilley et al., pp. 32–36.

outset, Ford engineers planned for mechanization and subdivision of labor on the vastest possible scale. The B-24 was broken down into 20,000 separate operations, while seventy major component sections were to be prefabricated in special areas and moved on conveyors to join the final assembly line. Some 29,000 dies and 21,000 jigs and fixtures (including special fixtures for all assembly operations) were developed at a total cost of $75–$100 million in order to reduce costs, increase accuracy, and ensure interchangeability.[21]

Despite its bold conception and innovative execution—both of which greatly impressed contemporaries—Willow Run was at best a Pyrrhic triumph, as Holley has elegantly demonstrated. As the British had discovered earlier, the difficulties of transferring automotive industry practice to aircraft manufacture proved much greater than Ford had envisaged. In 1941, the B-24 was still a "shop-engineered" aircraft, and all 30,000 of Consolidated's original drawings had to be redrawn to compensate for the more limited technical skills of Ford workers and supervisors. Aluminum airframes turned out to require very different pressing techniques than steel auto bodies, forcing the company to scrap some 14,000 dies and extensively rework 2,400 others before achieving satisfactory results. Nor could the original design long remain stable once exposed to combat conditions: according to Holley, "the last B-24 turned off the Willow Run production line was an entirely different aircraft from the original item" as a result of 130 major and thousands of minor changes to engines, fuel tanks, propellers, armament, and paint which increased its gross weight from 41,000 to 60,000 pounds. The unavoidability of such modifications, coupled with frequent material shortages, blocked the full exploitation of planned economies of scale, and little more than half of all jigs and fixtures constructed by the company at vast expense were ultimately used in production.[22]

By the end of 1944, person-hours per aircraft and manufacturing costs at Willow Run were well below the industry average, while its dedicated assembly fixtures and multioperation machines yielded significant advances in accuracy as well as dramatic reductions in direct labor requirements. But Willow Run was also the slowest of all plants making the Consolidated B-24 to achieve full quantity output—so slow, in fact, that it became widely known as "Will It Run?" Not least among the sources of delay was the problem of attracting a vast new labor force to rural Michigan in the absence of adequate housing, and Ford never succeeded in recruiting more than 40,000 of the 72,000 employees originally planned for the site. By the time Willow Run reached peak

[21]Holley, *Buying Aircraft,* pp. 518–21, 523–24, 526.

[22]Ibid., pp. 521–27; Lilley et al., pp. 49–50; Cairncross (n. 15 above), p. 177.

production, the B-24 itself was partially obsolescent and the factory's potential cost advantage could not be fully realized because of the air force's unwillingness to absorb more than two-thirds of its potential capacity. When military officers came to evaluate the project immediately after the war, they understandably concluded that the Ford approach was "not flexible enough" to use on a product that had not been "completely engineered and ready for mass production," though the B-24 was no different in this respect than other similar aircraft such as the B-25.[23]

To be sure, Ford's experience was by no means representative of American aircraft production during the Second World War. More than 90 percent of all airframes, nearly half of all engines, and a substantial proportion of components and accessories were manufactured by the established aircraft constructors in government-financed extensions and branch facilities, as well as their original home plants.[24] Despite the relatively short runs of individual planes predominant before the war, American aircraft firms, Jacob Vander Meulen argues, were "poised for mass production" because of congressional procurement policies based on price competition and the denial of intellectual property rights in new designs, together with the ideological influence of the automobile industry as a paradigm of modern manufacturing practice.[25] Once massive orders for military aircraft became available on financially attractive terms, therefore, these firms moved rapidly to apply line production methods such as progressive assembly, special-purpose machinery, systematic production control and scheduling, and careful balancing of individual operations, especially at their new purpose-built branch plants. At the same time, however, the aircraft manufacturers remained much more sensitive than the automobile firms and other outside subcontractors to the inevitability of frequent design changes, and, according to a postwar study conducted by the Harvard Business School, they were "ultimately able to introduce a high degree of flexibility" into line-production techniques. Among the most important such innovations was the development of "multi-line assembly" at the Boeing plant in Wichita, Kansas, whereby heavy bombers were broken down into an ensemble of major sections, each of which could be

[23]Holley, *Buying Aircraft*, pp. 326–27, 527–29; Lilley et al., p. 82–83; Overy, *Air War* (n. 5 above), pp. 164, 178. Ford's archrival General Motors failed completely in its attempt to mass produce a new fighter plane from preexisting standardized components: see Holley, "A Detroit Dream of Mass-Produced Fighter Aircraft."

[24]Holley, *Buying Aircraft*, pp. 560–62; Craven and Cate, eds. (n. 20 above), pp. 328–30, 354–56; Lilley et al., pp. 33–36.

[25]Vander Meulen, *Politics of Aircraft* (n. 19 above), esp. chaps. 7–8.

"pre-completed" (and therefore more easily modified) in separate areas of the factory rather than taking shape along a single moving "chassis."[26]

Although American aircraft manufacturers, like their British counterparts, accepted the necessity of periodic design modifications during the course of production, U.S. procedures for accommodating such changes differed significantly from those current on the other side of the Atlantic until relatively late in the war. In most American aircraft factories, designs were temporarily frozen to allow the production of uninterrupted batches of up to 1,500 planes at a time, with indispensable changes introduced retrospectively in twenty special "modification centers" scattered around the country. An entire center at Birmingham, Alabama, for example, was occupied nearly full-time with modifications to B-24s from Willow Run. Such retrospective modification was normally "quick and dirty," extremely costly, and highly labor-intensive: the number of person-hours involved could approach those spent in constructing the original plane, and Holley estimates that "anywhere from 25 to 50 percent of the total labor spent in turning out military aircraft was actually performed at the centers." By 1944, therefore, the U.S. military authorities had become increasingly concerned to rationalize the procedures governing design changes and to reduce the volume of retrospective modifications by making aircraft manufacturers responsible for the costs and administration of the modification centers on the one hand and by helping them to develop more effective techniques of production control on the other. Such reforms, which brought American aircraft manufacturers closer to contemporary British practice, were apparently so successful that during the Korean War a firm like Republic Aviation proved able to incorporate an average of 315 modifications per week in manufacturing the P-84.[27]

[26]Holley, *Buying Aircraft* (n. 16 above), pp. 290–304, 320–24; Craven and Cate, eds., pp. 306–18, 332–34; Lilley et al. (n. 19 above), pp. 18–20, 39–48, 53–56, quotation from p. 41. For "multi-line assembly" at Boeing Wichita and its diffusion to other B-29 plants, see Jacob Vander Meulen, "Flexible Mass Production at War: Heavy Bombers and Large Aircraft Engines in the United States, 1940–45" (paper presented at the Society for the History of Technology meeting, Washington, D.C., October 14–17, 1993), pp. 10, 14.

[27]Holley, *Buying Aircraft*, pp. 528–38; Craven and Cate, eds., pp. 334–37; Air Historical Office, *The Modification of Army Aircraft in the United States, 1939–1945*, Army Air Forces Historical Studies no. 62 (Washington, D.C., August 1947). Better control of production modifications was also necessary to avoid "flooding the tactical units with a heterogenous collection of equipment," which could unduly complicate supply, maintenance, and training procedures. During the autumn of 1943, for example, local commanders frantically demanded a host of design modifications to the P-40 fighter to improve its tactical effectiveness but later complained that "the logistical difficulties imposed by these changes nullified the gains anticipated." See Holley, *Buying Aircraft*, pp. 517–18.

In sheer quantitative terms, the accomplishments of the American mass-production approach to wartime aircraft were nothing short of extraordinary. In numerical terms, as table 1 shows, U.S. output of military aircraft overtook the British in 1941; by 1944 it was nearly four times that of Britain and more than twice that of any other belligerent. Measured in terms of structure weight, the British lead lasted until 1942, but within two years U.S. output was four-and-a-half times greater than that of Britain and nearly five times greater than that of Germany. Willow Run alone turned out one bomber every hour, an output equivalent in weight to 50 percent of the entire German aircraft industry at its peak.[28]

In qualitative terms, however, the results were less impressive. On the eve of Pearl Harbor in December 1941, as an internal air forces account put it, "not one of the new plants authorized after June 1940 . . . had yet produced a single plane; and none of them was destined to get into full production until 1943," by which time the military tide had already turned in Europe. Heavy, four-engined bombers took even longer to reach quantity output, and deliveries of the long-range B-29 needed for the war with Japan had barely begun by the spring of 1944. Willow Run, as we saw, was so slow in reaching full stride on the B-24 that the military value of its vast output was significantly reduced, while the fastest transitions to quantity output of heavy bombers were achieved by the home plants of Boeing and Consolidated, where, as in the case of aero engine manufacturers like Pratt and Whitney or Curtiss-Wright, skill levels remained higher and production methods closer to prewar job shop practice than in the branch and licensee factories.[29]

The American system of temporarily frozen designs and retrospective modifications rather than continuous improvement during the course of production likewise exerted a negative impact on both the quality and quantity of combat-ready aircraft, as Sir Michael Postan argued in his official history of British war production. Where tactical experience was accumulating rapidly, as in the case of bombers in the European theater in 1942–43, Postan observed, the "United States Air Force demanded urgent improvements all the time, and the 'modification centres' were soon choked up with aircraft." The flow of planes to squadrons was thus reduced well below the remarkable numbers

[28]Overy, *Air War* (n. 5 above), pp. 150, 164; Holley, *Buying Aircraft*, pp. 552–56; Craven and Cate, eds., pp. 329–31, 350–51.

[29]Air Historical Office, *Expansion of Industrial Facilities under AAF Auspices*, Army Air Forces Historical Study no. 40 (Washington, D.C., n.d.), p. 107, quoted in Craven and Cate, eds., p. 314; Holley, *Buying Aircraft*, pp. 324–25, 552–53; Lilley et al., pp. 1, 15–20, 25–30, 54, 81–83.

leaving the factory gates, as too was the average quality of aircraft in operation.[30]

British military planners and industrialists were well aware of the U.S. achievement in increasing aircraft output, and a series of technical missions toured American plants during the war to assess the applicability of their methods for securing productivity improvements under domestic conditions. Three central contrasts stood out among the findings of these missions: the larger average size and greater vertical integration of American aircraft factories; their bigger design and planning staffs predominately composed of college-educated engineers; and their more extensive use of moving conveyors, time study, and other line-production methods. Although the British missions were greatly impressed with the efficiency of American practice, recommending both the expansion of domestic technical education and the introduction of moving assembly lines "where output warrants it," they saw little opportunity for the direct transfer of U.S. methods to domestic aircraft production during the war itself. A 1942 mission concluded:

> The fundamental difference in the average size of the plants—large units in U.S.A. and dispersal into small units forced upon us by conditions in U.K.—vitally affects the methods of production, and the extent to which American ideas can be copied in U.K. Some methods of construction and layout of shops in America which are superior to those used in the U.K., are only possible in the relatively unlimited space available and on the scale on which they can then be employed; they could not be used in U.K. except perhaps in a few of the larger plants.[31]

As subsequent studies showed, larger plants and longer production runs meant that labor productivity in American airframe manufacture—measured in terms of structure weight per person-day—was 75 percent higher than that of Britain in 1943–44. But in those relatively few cases where British production was maintained for long enough to take

[30]Postan (n. 9 above), pp. 342–43. The proportion of planes subject to retrospective modification varied significantly by type: thus nearly 100 percent of heavy bombers and transports passed through the modification centers before delivery to the armed forces but only 30–50 percent of fighters. See Holley, *Buying Aircraft,* p. 537.

[31]"Report of British Mission to United States of America to Study Production Methods," September–October 1942, PRO, AVIA 10/104, esp. pp. 3, 9–10, from which quotations are drawn; "Report of the Fedden Mission to the United States," June 1943, PRO, AVIA 10/99, and précis of its findings, April 22, 1943, PRO, AVIA 10/106. For discussions of the findings of these missions, see Postan, pp. 335–36; Cairncross (n. 15 above), chap. 8; and, more tendentiously, Corelli Barnett, *The Audit of War: The Illusion and Reality of Britain as a Great Nation* (London, 1986), pp. 148–54.

advantage of the "learning curve"—fifty to sixty fighters or twenty to thirty bombers per week—output per worker was roughly equivalent and overall costs lower than in the United States.[32] Corelli Barnett claims in his influential polemic *The Audit of War* that the British produced 1.19 pounds of aircraft structure weight per man-day in 1944, compared to 2.76 pounds in the United States, implying a substantially larger productivity gap between the two countries. But Barnett's British calculations are based on employment figures for November 1943, rather than the actual number of man-days worked in 1944, although the labor force in the aircraft industry declined by more than 14 percent during this period.[33] Even more misleading, the figure given by Barnett for American labor productivity is actually a synthetic index of relative efficiency devised by the U.S. Strategic Bombing Survey (USSBS) in which raw data on pounds of weight per employee per day were corrected for variations between countries in product mix (types of aircraft) and scale of production (units per day). Were it possible to calculate a similar index of relative efficiency using British data, such corrections would undoubtedly reduce the initial disparity in person-days per pound of structure with the U.S., as the USSBS found in the cases of both Germany and Japan.[34]

In other respects, too, such aggregate statistical comparisons arguably exaggerate the American productivity advantage. Rapidly rising output per worker in U.S. aircraft factories—direct person-hours per pound of airframe accepted in July 1944 had fallen to 43 percent of those required in January 1943—were partially offset by subsequent reworking in modification centers, which as we have seen could add a further 25–50 percent to the total labor inputs involved.[35] In Britain, unlike the

[32]Eric Mensforth, "Airframe Production," *Proceedings of the Institution of Mechanical Engineers* 156 (1947): esp. 35–37. Subsequent analyses of the "learning curve" have tended to reduce the contribution of longer production runs per se to productivity gains as opposed to related improvements in product design, equipment, tooling, materials, and management: see Louis E. Yelle, "The Learning Curve: Historical Review and Comprehensive Survey," *Decision Sciences* 10 (1979): 302–28; R. M. Bell and D. Scott-Kemmis, "The Mythology of Learning-by-Doing in World War II Airframe and Ship Production," *Industrial and Corporate Change* 4 (1994), in press. I am grateful to David Hounshell for these references.

[33]See Barnett, pp. 145–46; Edgerton, *England and the Aeroplane* (n. 9 above), pp. 79–80; Overy, *Air War* (n. 5 above), p. 171.

[34]For the method used in constructing the original efficiency index, see USSBS, Aircraft Division, *The Japanese Aircraft Industry* (Washington, D.C., May 1947), reprinted in *The United States Strategic Bombing Survey*, ed. David MacIsaac (New York, 1976), 7:27–29, and *Aircraft Division Industry Report: Strategic Bombing of the German Aircraft Industry*, 2d ed. (Washington, D.C., November 1947) (hereafter cited as *Strategic Bombing of the German Aircraft Industry*), reprinted in MacIsaac, ed., vol. 2, figs. 6-11 and 6-12; Jerome B. Cohen, *Japan's Economy in War and Reconstruction* (Minneapolis, 1949), pp. 218–19. The origins of Barnett's error can be traced to Overy, *Air War*, table 15, p. 168, quoted in Barnett, p. 146.

[35]Holley, *Buying Aircraft*, pp. 564, 532.

United States, repair of damaged aircraft was performed on the same facilities as new construction, reducing the apparent productivity of the latter but increasing the overall effectiveness of war production: 48 percent of all planes delivered to the Metropolitan Air Force (that portion of the RAF based in Britain itself) were repaired during the war, using one-third to one-fifth of the raw materials and labor needed for equivalent new construction.[36] Heavy bombers and large transport planes constituted a much larger proportion of American than British output during the final years of the war—40 percent of all planes delivered to the army air forces in 1944 compared to 24 percent of those received by the RAF, and an even larger disparity in terms of weight—but required fewer person-hours per pound of structure than lighter aircraft such as fighters.[37] Structure weight per person-hour, as American aircraft manufacturers and British military planners both insisted, was in any case a highly imperfect indicator of relative efficiency since it failed to capture the impact of variations between firms—and national economies—in factors like labor quality, degree of vertical integration, frequency of modifications, and manufacturability of individual designs, not to mention the tactical value of the planes themselves.[38]

### *Germany: Quality and Quantity Out of Balance*

Germany, the Allies' main aerial antagonist, appears by all accounts to have managed this trade-off between quality and quantity of aircraft less successfully than either Britain or the United States. All the major combatant powers oscillated between defensive phases, in which design changes were severely restricted in order to overcome a numerical disadvantage by maximizing output of current types, and offensive phases, in which greater emphasis was placed on the incorporation of modifications aimed at achieving tactical superiority over enemy air-

[36]"Report of the British Mission" (n. 31 above), p. 16; Postan (n. 9 above), pp. 316–22. Forty-five percent of American combat aircraft were nonetheless undergoing repair on any given day: see Holley, *Buying Aircraft,* p. 246.

[37]Holley, *Buying Aircraft,* pp. 553–55; Postan, app. 4, p. 485. According to 1940 calculations by the British Air Ministry quoted in Barnett (n. 31 above), p. 146, "a ton of heavy aircraft represented less added value and a smaller industrial effort than a ton of lighter aircraft." The USSBS reckoned that the "unit cost or production hours per airplane varies inversely as the weight to the one-third power." See USSBS, *Japanese Aircraft Industry,* p. 27.

[38]For the views of American aircraft manufacturers, see East Coast and Central Aircraft War Production Councils, letter to T. P. Wright re Indices of Airplane Production Efficiency, May 8, 1944, National Archives, Washington, D.C., Truman Committee Papers, Box 685. I am grateful to Jacob Vander Meulen for supplying a copy of this document. For the views of a leading British aircraft planner, see Ely Devons, "Statistics and Planning," in his *Planning in Practice: Essays in Aircraft Planning in War-Time* (Cambridge, 1950), esp. p. 150: "nobody was really interested in the weight of aircraft produced, but in its fighting power," quoted in Edgerton, *England and the Aeroplane* (n. 9 above), p. 80.

craft. In Germany, however, these swings were particularly violent, largely for political and administrative rather than industrial reasons. After freezing existing models for serial production at the onset of aerial rearmament in 1933–34 and again in the wake of the international crises of 1938, the German military authorities were given free rein to demand frequent and radical design changes with little regard for their impact on aircraft output during the crucial period between 1939 and 1942 when the Allies were building up their vast expansion programs. From mid-1941 on, Field Marshall Erhard Milch as technical director at the Air Ministry sought to reorient the industry toward mass production of existing types at the expense of new models, but his efforts were largely frustrated by Hermann Goering's insistence on the development of heavy bombers to support a renewed offensive against Britain, Russia, and ultimately the United States. In 1944, as Allied forces closed in on the Nazis' European empire, Milch and Albert Speer as minister of munitions finally achieved a quantum leap in output of fighter planes for the defense of Germany, but only by concentrating production on a small number of standard types and imposing a ruthless freeze on design modifications.[39]

These violent oscillations and conflicts within German production policy adversely affected both the quantity and the quality of aircraft output. Despite the remarkable accomplishments of the German aircraft industry and the Luftwaffe during the late 1930s, by the time of the Battle of Britain in 1940, U.K. output was 50 percent higher, as we have seen, while the quality of fighters on the two sides was roughly equal.[40] Between 1939 and 1941, the human and material resources pumped into the German aircraft industry increased by 100 percent, but, as table 1 shows, output rose by only 30 percent, compared to planned growth of 100–400 percent. As Richard Overy has argued, the inefficiency of German aircraft production during the early years of the war can be gauged by the fact that with no more labor and fewer raw materials than in 1941, the industry turned out 31 percent more planes in 1942 and 111 percent more in 1943 (30 percent and 85 percent more in terms of

[39]This sketch of the evolution of German aircraft policy and production during the 1930s and 1940s draws on the following major secondary accounts: Edward L. Homze, *Arming the Luftwaffe: The Reich Air Ministry and the German Aircraft Industry, 1919–39* (Lincoln, Neb., 1976); Overy, *Air War* (n. 5 above), esp. chaps. 1 and 7, *Goering: The "Iron Man"* (London, 1984), esp. chaps. 6–7, and "German Aircraft Production, 1939–1942" (Ph.D. diss., University of Cambridge, 1977); Edward R. Zilbert, *Albert Speer and the Nazi Ministry of Arms: Economic Institutions and Industrial Production in the German War Economy* (London, 1981); Cairncross (n. 15 above), chap. 7; Hans-Joachim Braun, "Fertigungsprozesse im deutschen Flugzeugbau, 1926–1945," *Technikgeschichte* 57 (1990): 111–35, and "Aero-Engine Production in the Third Reich," *History of Technology* 14 (1992): 1–15.

[40]For the qualitative parity of British and German fighters in 1940, see Edgerton, *England and the Aeroplane*, p. 37.

weight); in 1944, with virtually the same resources as in 1940, the Germans produced nearly four times the number of planes (238 percent more than 1941 in terms of numbers and 126 percent more in terms of weight). During the last full year of the war, the Germans overtook the British in numbers of planes manufactured and had nearly caught up with them in terms of structure weight as well. By that time, however, matching the British was no longer a relevant target, as the German "production miracle" of 1944 could not prevent the Allies from securing overwhelming air superiority given their vast combined output.[41] In terms of productivity, too, the German aircraft industry fell further behind the United States even as it improved dramatically on its own past performance. Even though output per person-hour in the German aircraft industry increased by some 200–300 percent between 1941 and 1944, the USSBS's index of relative efficiency (which made allowances for variations in the composition and scale of output) put the American advantage at 110 percent in 1944, compared to 20–40 percent in 1941–43.[42]

Equally important was the relative decline in the quality of German aircraft during the war. Of the new generation of aircraft designed to come on stream in 1942—the Heinkel 177 and Junkers 288 heavy bombers and the Messerschmitt 210 heavy fighter—none proved successful enough for large-scale manufacture, and the Air Ministry was therefore obliged to maintain older types in production much longer than originally planned. Even where the Germans managed to develop a radical new design ahead of the Allies, as in the case of the Messerschmitt 262 jet-engined fighter, they failed to bring it into production fast enough to exploit the plane's military potential, in part because of personal interference from Hitler himself, who wanted it converted into a fighter-bomber. Conversely, where the Germans sought to accelerate development of new types through shortcuts such as making several prototypes simultaneously or using a high proportion of preexisting parts, problems constantly cropped up, as in the case of the Heinkel 162 single-engined jet fighter of which some 100 were built before the end of the war: the firm's chief test pilot was killed on its second flight and the design was plagued by persistent technical weaknesses. By 1943–44, this inability to introduce new models, together with growing restrictions on modifications to existing types, meant that the Luftwaffe had lost any claim to qualitative superiority over the Allies even on conventional aircraft.[43]

[41]Overy, *Goering*, pp. 102, 148–50, 176–77, 185, 190–93, 201–4.

[42]Ibid., pp. 150, 190–91; USSBS, *Strategic Bombing of the German Aircraft Industry* (n. 34 above), p. 85.

[43]Overy, *Goering*, pp. 166–68, 191–93, 197; Cairncross, pp. 130–31, 137–45, 149–53.

The relative failure of the German aircraft industry during World War II has often been attributed to the limitations of craft as opposed to mass production. Thus historians like Richard Overy and Hans-Joachim Braun argue that German industry's failure to match—or even approach—the volume of aircraft turned out by the Allies was partially due to the persistence of "artisanal" (*handwerkerliche*) methods and production ideals. Particularly during the early years of the war, these historians contend, skilled workers, craft-trained foremen or *Meister*, and company managers—like the military authorities themselves—resisted attempts to bypass apprenticeship and dilute the workforce through the introduction of semiskilled labor and large-scale manufacturing techniques. A skilled, adaptable workforce and flexible production methods were also necessary to cope with the frequent design changes demanded by the Luftwaffe. In many firms, therefore, aircraft continued to be built by craftsmen using general-purpose equipment, rather than by less skilled workers using specialized machines, which were themselves in short supply because of the predominance of similar "conservative" attitudes in the German machine tool industry itself. Under these conditions, Overy maintains, "Batch production gave way only slowly to line production, while time-and-motion studies on which American practice rested were either not introduced or where they were, proved unworkable because of the traditional methods of work payment and use of skills. Conveyor belt production and rational factory organization developed only slowly and many factories still had little evidence of modern methods even by 1945."[44]

This portrait of German aircraft production and its shortcomings draws heavily on contemporary criticisms by domestic rationalization advocates and by Allied intelligence officers who surveyed the industry immediately after the war. Thus, for example, a British Intelligence Objectives Sub-Committee (BIOS) team charged with investigating methods of production control in German factories concluded that "management was interested much more with [*sic*] the technical development of their products rather than with techniques of control. . . . With the higher proportion of skilled workers in German industry and the greater effort which they were prepared to give there was not so much demand for good internal production organization."

[44] Overy, *Air War* (n. 5 above), pp. 170 (from which the quotation is drawn), 172–74, *Goering*, pp. 159–62, 184–87, 190–91, and "German Aircraft Production" (n. 39 above), pp. 151–70, 205–19; Braun, "Fertigungsprozesse im deutschen Flugzeugbau," esp. pp. 116–19, 127, which speaks of the continuing predominance of "handwerklichen Produktionsideal orientierte Fertigungsmethoden" even at the end of the war, "Aero-Engine Production in the Third Reich" (n. 39 above), pp. 5–6, 11–13, and *The German Economy in the 20th Century* (London, 1990), pp. 134–38.

Of all the aircraft firms visited, only Heinkel's had created a central office for production control and progress chasing as was increasingly the practice in large British concerns. Even there, control of parts movement within a shop remained the sole responsibility of the superintendant or foreman, while irregular supplies of materials and "constant modifications in program and design" meant that "controlled production was hardly to be thought of, let alone line production," detailed planning, or the use of special-purpose machinery: "Indeed factories had to be in production so quickly to meet Air Ministry demands, that output could only be achieved through flexible organisation and the provision of general-purpose machine tools." Another BIOS report on the German aircraft instrument industry similarly found much less dilution and use of simplified methods (including time-and-motion study) than were common in the United Kingdom; German instrument firms, in contrast to their British and particularly American counterparts, "avoided elaborate mechanisation by relying on the great skill of the labour employed." Although the USSBS placed less emphasis than the British BIOS teams on the distinctiveness of German production methods, it too ascribed much of the productivity gap between the two countries to "inefficient labor utilization" in Germany, notably, lesser use of women workers and "high production, special purpose tools."[45]

Some elements of this contrast between productive organization in German and Allied aircraft factories are clearly well-founded. There is general agreement, for example, among historians and contemporary observers alike that until the later years of the war German aircraft firms, like most of the metalworking sector, relied more heavily than their Allied counterparts on the skills of their workforce. Such policies reflected the fact that, as the USSBS observed, "the supply of competent mechanics in the German labor market was much greater than that available to American aircraft manufacturers," as a result of the broad coverage achieved by a revitalized system of craft apprenticeship combining in-plant manual training with technical instruction in public vocational schools.[46] It is also true that both before and during the war

[45]BIOS, *Investigation of Production Control and Organisation in German Factories,* Final Report 537 (London, 1946), pp. 1, 7–8, and "Appendix: Production Control in the Heinkel Aircraft Organisation," pp. 2, 12–13, 19–20, 23–25; BIOS, *Conditions in the German Aircraft Instrument Industry,* Final Report 881 (London, 1947), pp. 8, 10–11, 13; USSBS, *Strategic Bombing of the German Aircraft Industry,* pp. 85–86. For criticisms by German rationalization advocates, see Overy, *Goering,* pp. 162, 184, and "German Aircraft Production," pp. 154–55, 255–56.

[46]USSBS, *Strategic Bombing of the German Aircraft Industry,* p. 86; cf. also Combined Intelligence Objectives Sub-Committee (CIOS), *A Survey of Production Techniques Used in the*

many German industrialists and military planners were critical of American-style mass-production methods for a combination of economic, technological, and ideological reasons, favoring instead production strategies based on customization, flexibility, and "quality work."[47]

In other respects, however, this opposition between German craft traditionalism and Allied mass-production modernism is seriously overdrawn and misleading. Design modifications and program changes, as Overy himself recognizes, were not an optional luxury irrationally insisted on by technologically fastidious air force officers, but rather an indispensable means of adapting equipment with long development times to rapidly shifting tactical conditions and strategic priorities.[48] Even in the United States, therefore, program and engineering changes in wartime aircraft production were sufficiently common that the USSBS claimed—probably mistakenly—that there was no substantial difference between the two countries in this regard.[49] In Britain, the incorporation of design modifications seems, if anything, to have been more continuous than in Germany, at least for certain sections of the industry. Thus British fighters retained a power edge throughout the war, as Sir Alec Cairncross of MAP found when he interviewed German designers in

*German Aircraft Industry,* report XXV-42 (Washington, D.C., July 1945), pp. 4–6. For the development of the German apprenticeship system between the wars, see Albin Gladen, "Die berufliche Aus- und Weiterbildung in der deutschen Wirtschaft 1918–45," in *Berufliche Aus- und Weiterbildung in der deutschen Wirtschaft seit dem 19. Jahrhundert,* ed. Hans Pohl (Wiesbaden, 1979), pp. 53–73. From 1941–42 on, however, foreign workers, prisoners of war, and concentration camp inmates came to form a growing proportion of the workforce, particularly in the engine plants: see Overy, "German Aircraft Production," pp. 207, 212–13, 270–72; Braun, "Aero-Engine Production in the Third Reich, p. 9; CIOS, *A Survey of Production Techniques in the German Aircraft Industry,* pp. 6–7, 11–12.

[47]For German criticisms of American mass-production methods before World War II, see Heidrun Homburg, "Anfänge des Taylorsystems in Deutschland vor dem Ersten Weltkrieg," *Geschichte und Gesellschaft* 4 (1978): 170–94, esp. 174; Hans-Liudger Dienel, "German Opposition against the American System of Production" (paper presented to the Society for the History of Technology meeting, Uppsala, Sweden, August 16–20, 1992); Gary B. Herrigel, "Industrial Organization and the Politics of Industry: Centralized and Decentralized Production in Germany" (Ph.D. diss., Massachusetts Institute of Technology, 1990), esp. pp. 277–79, 439–40, and "Industry as a Form of Order: A Comparison of the Historical Development of the Machine Tool Industries in the United States and Germany," in *Governing Capitalist Economies: Performance and Control of Economic Sectors,* ed. J. Rogers Hollingsworth, Philippe C. Schmitter, and Wolfgang Streeck (Oxford, 1994), pp. 97–128. For the German military's commitment to the doctrine of "qualitative superiority" in armaments production, see also Alan S. Milward, *The German War Economy* (London, 1965), chap. 5.

[48]Overy, *Air War,* pp. 177–78, and *Goering* (n. 39 above), pp. 180–81; but cf. also Milward, *German War Economy,* pp. 159–60.

[49]USSBS, *Strategic Bombing of the German Aircraft Industry* (n. 34 above), pp. 87–88.

1945, mainly because Rolls-Royce "had found ways of modifying the Merlin [engine] on the production line, with or without a change of Mark number, to yield more and more engine thrust," while firms like BMW "had relied on discontinuous changes in specification as one mark of engine replaced another," so that "any lead they established . . . did not last long."[50]

In both Anglo-Saxon powers, similarly, frequent design changes demanded a substantial measure of adaptability in aircraft manufacturing methods. In the United States, as we have seen, aircraft firms modified standard line-production techniques to obtain greater flexibility, while in Britain a large proportion of aircraft components were turned out in batches small enough to make bench methods more economical than full mass-production tooling.[51] Nor was opposition to rationalization and deskilling a German peculiarity. Many British craftsmen opposed dilution throughout the war, while even in the United States, as a Harvard Business School team observed, old-line aircraft manufacturers experienced considerable difficulty in "persuading foremen and other supervisors, who had grown up under job shop conditions, to accept the engineering discipline that large-scale production . . . required."[52] More fundamentally still, as I have sought to show elsewhere, many British engineers and industrialists, like their German counterparts, remained skeptical before the war about the suitability of American mass-production methods in catering to the diverse and fluctuating markets characteristic of domestic metalworking.[53]

The received contrast between craft traditionalism and mass-production modernism in wartime aircraft manufacture is misleading from the German as well as the Allied perspective. During the 1930s, as the USSBS noted, "German manufacturers had made an intensive study of aircraft tooling . . . and had evolved techniques of 'series' or line production which were advanced for that time." According to Edward

[50]Cairncross (n. 15 above), pp. 16, 148–49; on Rolls-Royce's commitment to flexibility and continuous improvement, see also Lloyd, *Rolls Royce* (n. 18 above), pp. 23, 72–74, 118–25.

[51]The CIOS survey rated German manufacturing methods superior to those of the U.S. insofar as "new products could be put into production or designs changed more quickly and easily than they could in America," while also noting that "it is possible that British methods were even more flexible than the German": see *A Survey of Production Techniques Used in the German Aircraft Industry* (n. 46 above), p. 5.

[52]For British opposition to dilution, see Parker, "British Rearmament" (n. 14 above); Inman (n. 14 above), chaps. 2–3. For difficulties in reorienting job-shop supervisors in American aircraft plants, see Lilley et al. (n. 19 above), p. 54.

[53]See Jonathan Zeitlin, *Between Flexibility and Mass Production: Strategic Debate and Industrial Reorganization in British Engineering, 1830–1990* (Oxford, 1996, in press). An article covering the period before 1914 will appear in Sabel and Zeitlin, eds., *Worlds of Possibility* (n. 8 above).

Homze, a leading historian of German aerial rearmament, one such technique, the *Baukastenflugzeuge*, a standardized basic airframe which could be combined with different engines and equipment to serve a variety of missions, was carried so far as to inhibit the search for new design solutions within the industry. Another innovation of the 1930s, the so-called hole system (*Lochbauweise*), which avoided the need for elaborate assembly jigs to secure interchangeability by mass prepunching of airframe sections, anticipating the methods introduced by Ford at Willow Run, proved so inflexible during the war that large quantities of special tooling had to be scrapped every time there was a significant change in production programs.[54]

At the end of the war, too, the USSBS claimed that German aircraft manufacturing techniques "corresponded rather closely with American practice," including the use of "moving belts, with the work along them carefully planned at definite stations"; the main difference was that "operations were broken up among buildings in the same location or dispersed over a wide area" as a defensive measure against bombing attacks. From 1941 on, as Braun has documented, leading industrialists like William Werner (who had worked for Chrysler in Detroit before becoming technical director of Auto Union) and the noted designer Willy Messerschmitt formulated ambitious plans for expanding output of aero engines and airframes, respectively, by specializing suppliers on individual components and by extending the use of modular construction on the *Baukasten* or "Meccano-set" principle. Each of these proposals was eventually rejected, however, principally because they would have increased the vulnerability of aircraft production to disruption by Allied bombing.[55]

Finally, like their British and American counterparts, German aircraft manufacturers devised innovative techniques for realizing some of the benefits of large-scale production without excessive loss of flexibility. Prominent among these were the widespread use of "universal" assem-

[54]For the general point about serial production in the 1930s, see USSBS, *Strategic Bombing of the German Aircraft Industry* (n. 34 above), p. 21. On the *Baukastenflugzeuge,* see Homze (n. 39 above), p. 119; on the "hole system" and its disadvantages, see Braun, "Fertigungsprozesse im deutschen Flugzeugbau" (n. 39 above), pp. 115–16, and *German Economy in the 20th Century* (n. 44 above), pp. 137–38; Zilbert, *Albert Speer* (n. 39 above), pp. 226–27.

[55]USSBS, *Strategic Bombing of the German Aircraft Industry,* pp. 19–20, 97; Braun, "Fertigungsprozesse im deutschen Flugzeugbau," pp. 121–31, and "Aero-Engine Production in the Third Reich" (n. 39 above), pp. 7–8. British investigators, on the other hand, claimed that "strict line production was certainly not possible in Germany" during the war, with the partial exception of Opel, General Motors' prewar German subsidiary: see BIOS, *Investigation of Production Control and Organisation in German Factories* (n. 45 above), p. 7, and "Appendix: Production Control in the Heinkel Aircraft Organisation" (n. 45 above), pp. 13–14.

bly jigs, which "could be adjusted within certain limits to permit fairly wide changes in design," and Heinkel's combination of a modified form of line production (known as "repetition" or "rhythm" work) in the earlier and later stages of airframe manufacture with batch production in other shops.[56]

The central weakness of German aircraft production, on this analysis, lay not in the fact of program changes and design modifications, or even their frequency, but rather in the way they were planned and administered. Unlike in Britain or the United States, alterations in programs and designs were not screened for their potential impact on aircraft output by officials with detailed knowledge of current industrial conditions. Nor did the Germans devote comparable effort to minimizing the impact of such changes on aircraft output by careful phasing of their introduction (whether continuously as in Britain or discontinuously as in the United States) and by systematic coordination among manufacturers of airframes, engines, components, and accessories. Product selection decisions, too, were taken without realistic appraisal of the lead times necessary for the development of new weapons systems, while output targets were often based on mistaken estimates or even deliberate misrepresentations of German and Allied productive capacity. As a special commission established in 1943 to review the problems of the German aircraft industry concluded, the most important cause of lost production during the course of the war was the lack of "clear planning" and "precise allocation of tasks" within the Air Ministry.[57]

This failure of German aircraft planning, as Overy above all has demonstrated, can be traced directly to the decision-making structure of the Nazi regime. Responsibility for aircraft development and production was dispersed among a shifting array of overlapping and competing centers of power—from individual firms, military officers, and government ministers on up to Hitler himself. The punitive response to bad news or pessimistic forecasts within the Nazi political system likewise discouraged upward transmission of accurate but unwelcome informa-

[56]USSBS, *Strategic Bombing of the German Aircraft Industry,* pp. 20, 88; BIOS, "Appendix: Production Control in the Heinkel Aircraft Organisation," pp. 12–13. For experiments at Daimler-Benz during the 1930s with hybrid forms of modified flow production based on mixed-model assembly and conveyor-belt "islands," see Michael Stahlman, "Management, Modernisierungs- und Arbeitpolitik bei der Daimler-Benz AG und ihren Vorläuferunternehmen von der Jahrhundertwende bis zum Zweiten Weltkrieg," *Zeitschrift für Unternehmensgeschichte* 37 (1992): 161–63.

[57]See Overy, *Air War* (n. 5 above), pp. 154–59, 177–80; *Goering* (n. 39 above), pp. 105–6, 159–62, 169–72, 179, 187–88 (from which the quotation is taken), 192–93, 197; Cairncross (n. 15 above), pp. 117–22, 129–36, 139, 147–49; USSBS, *Strategic Bombing of the German Aircraft Industry,* pp. 105, 108; CIOS, *A Survey of Production Techniques Used in the German Aircraft Industry* (n. 46 above), p. 11.

tion by subordinates. In contrast to Britain and the United States, no independent civilian ministry or department was established for the administration of German aircraft production to temper the military's demands for new and improved designs with a parallel concern for increased output; even inside the Air Ministry itself air strategy and aircraft development were largely planned in isolation from one another until midway through the war. As in the other Axis powers, the military retained a dominant influence over the procurement and specification of aircraft, while little effort was made to involve industry in the planning and modification of production programs. All this began to change from 1941–42 with the reforms introduced by Milch and Speer, which included the representation of firms on a series of councils, committees, and "rings" aimed at improving the coordination of aircraft production. But internal conflicts and contradictory objectives within the planning apparatus continued to inhibit the Germans from achieving a satisfactory balance between the quality and the quantity of aircraft output until the scales tipped decisively toward the latter in the desperate conditions of 1944–45. If the air war of 1939–45 was among other things a contest between competing models of productive organization, this contest was overshadowed and overdetermined, as contemporaries were all too acutely aware, by a larger contest between competing political regimes with sharply different patterns of civil-military relations.[58]

## *Conclusions*

Aircraft manufacture did not exhaust the full range of military production during World War II. Analysis of other major weapons systems would show different balances between quality and quantity across the three countries. In the case of tanks, for example, the Germans appear to have been rather more successful in maintaining qualitative superiority over the Western Allies through frequent innovations in technology and design, while also sustaining higher levels of output than the British (but not the Americans) throughout the conflict.[59] The central element of German qualitative superiority in

[58]This is Overy's central argument in *Goering,* chaps. 6–7, *Air War,* chap. 7, and "German Aircraft Production" (n. 39 above), passim. For similar judgments about the primacy of political and administrative factors in the failure of the German aircraft industry during the war, see also Braun, "Fertigungsprozesse im deutschen Flugzeugbau," pp. 119, 127, and "Aero-Engine Production in the Third Reich," pp. 6–7; USSBS, *Strategic Bombing of the German Aircraft Industry,* p. 97; "Final Report of the Farren Mission to Germany: A Survey of a Cross Section of German Aircraft, Aircraft Engine and Armament Industries," 1945, PRO, AVIA 9/85.

[59]For brief comparative overviews of tank production during World War II, see Milward, *War, Economy and Society* (n. 5 above), pp. 181–82; R. A. C. Parker, *Struggle for Survival: The History of the Second World War* (Oxford, 1989), pp. 133–34; György Ránki, *The Economics of the Second World War* (Vienna, 1993), chap. 3.

tanks was the rapid introduction of new heavy types such as the Panther and the Tiger after encountering the Soviet T-34s in 1941–42. Rationalization and coordination of production were also more effective in tank manufacture than in that of aircraft, in part because of its higher priority in German military planning during the late 1930s and early 1940s.[60] Although the British had invented the tank during World War I and pioneered in its development through the early 1930s, their production during the subsequent decade was impeded by a combination of unresolved ambiguities about the army's military role, the low priority accorded to its rearmament, initial selection of weak firms as shadow manufacturers, and a misconceived attempt to combine drawing-board orders with severe restrictions on design modifications—a near-total contrast with aircraft procurement policies.[61] The Americans were more successful than the British in converting locomotive and motor vehicle firms to tank production after 1940, in part because they could profit from the latter's experience, but they too encountered major problems in both design and production, leaving U.S. armor clearly inferior to that of Germany at the time of the D-day invasion of 1944.[62]

But aircraft production was the largest single sector of the war economy in all three countries, involving vast numbers of civilian manufacturers as subcontractors and licensees, and Allied air superiority was widely regarded as a decisive factor in defeating the Axis powers. Hence wartime experience of aircraft manufacture—or rather a particular reading of its lessons—exerted a powerful influence on postwar reconstruction planning beyond the industry itself, especially in Britain.

British industrialists and government officials, as we saw earlier, toured both American and German aircraft factories during and immediately after the war, studying the organization of production and design with an eye to the reform of domestic practice. Despite concerns about the inflexibility of American aircraft manufacture and the adequacy of its system of retrospective design modifications, British wartime missions to the United States were more impressed by the enormous productivity gains obtainable through mass production—if only the necessary eco-

[60]Harmut H. Knittel, *Panzerfertigung im Zweiten Weltkrieg: Industrieproduktion für die deutsche Wehrmacht* (Herford, 1988); Milward, *The German War Economy* (n. 47 above), pp. 100–102, 148.

[61]Postan (n. 9 above), pp. 27–34, 160, 183–95; Postan, Hay, and Scott (n. 16 above), chaps. 13–14; Brian Bond, *British Military Policy between the Wars* (Oxford, 1980), chaps. 5–6; David Fletcher, *The Great Tank Scandal: British Armour in the Second World War* (London, 1989), pt. 1.

[62]See Constance McLaughlin Green, Harry C. Thompson, and Peter Roots, *The Ordnance Department: Planning Munitions for War* (Washington, D.C., 1953), pp. 189–204, 236–39, 250–59, 267–72, 275–304; Harry C. Thompson and Lida Mayo, *The Ordnance Department: Procurement and Supply* (Washington, D.C., 1960), chaps. 10–11.

nomic scale of operation could be achieved at home. Postwar missions to Germany, conversely, were impressed by the skills and training of the labor force at all levels, from engineers and technicians to foremen and manual workers, but not by production methods and organization, which they regarded as inferior not only to that of American firms but also of well-run British companies. "Thus as regards the internal organisation of German industrial concerns we have very little to learn," concluded the BIOS investigation of production control; and this finding was echoed in the reports of other BIOS teams dealing with individual metalworking sectors such as aircraft instruments, machine tools, and power presses.[63]

Hence British government planners took U.S. mass production and the "three charmed 'S's" of simplification, standardization, and specialization as their model for the reconstruction of the domestic metalworking industries, a message reinforced by the postwar Anglo-American productivity missions organized in the context of the Marshall Plan. Some domestic manufacturers eagerly embraced this agenda, as in the case of Standard Motors, which reconverted its wartime shadow factory for volume production of standardized cars and tractors with interchangeable engines. But others, including the leadership of the Federation of British Industries, were less enthusiastic about the Labour government's productivity drive for a combination of economic and political reasons.[64]

Yet as the mergers and reorganizations which swept the metalworking industry during the 1950s and 1960s would demonstrate, British businessmen and public officials alike increasingly accepted mass production and American management methods as a transnational standard of manufacturing efficiency. Far from reviving its competitive fortunes, however, this putative Americanization of British industry was associated instead with a rapid loss of market share both at home and abroad, resulting in a steep decline of domestic production and employment. By

[63]BIOS, *Investigation of Production Control and Organisation in German Factories* (n. 45 above), p. 8; cf. also BIOS, *Conditions in the German Aircraft Instrument Industry* (n. 45 above), p. 9, *The German Machine Tool Industry,* Final Report 641 (London, 1947), pp. 18–20; *German Gauge and Tool Industry,* Final Report 632 (London, 1947), p. 2, and *German Power Press Industry,* Final Report 1826 (London, 1949), pp. 3–6.

[64]On reconstruction planning, the postwar productivity drive, and their limitations, see Jim Tomlinson, "Productivity Policy," in Helen Mercer, Neil Rollings, and Jim Tomlinson, eds., *Labour Governments and Private Industry: The Experience of 1945–1951* (Edinburgh, 1992), pp. 37–54, and "Mr Attlee's Supply-Side Socialism," *Economic History Review,* 2d ser., 46 (1993): 1–22; Tiratsoo and Tomlinson, *Industrial Efficiency and State Intervention* (n. 15 above), chaps. 2–7; Zeitlin, *Between Flexibility and Mass Production* (n. 53 above), chap. 10. For the specific case of Standard Motors, see Nick Tiratsoo, "The Motor Car Industry," in Mercer, Rollings, and Tomlinson, eds., pp. 162–85.

the 1980s, ironically, the competitive difficulties of British metalworking firms, like those of the Americans themselves, were frequently attributed to their inability to match the standards of product innovation and productive flexibility set by the Germans and the Japanese in meeting the demands of increasingly diverse and volatile international markets.[65]

In the United States, by contrast, wartime aircraft production appears to have had little impact on domestic manufacturing practice outside the industry itself. Before the war, as a 1947 report by a Harvard Business School team observed, "standardization of design and steadiness of scheduled rate had been considered fundamental prerequisites to line production. The success of the mass production industries was built in large measure on their ability to call a halt to changes prior to large-scale output so that carefully developed plans could be put into operation without any serious interruption." Hence they concluded: "The fact that the aircraft industry was ultimately able to introduce a high degree of flexibility into production procedures, and thereby to make effective use of line production techniques in spite of change, constituted an outstanding contribution to production management."[66]

Yet there is little evidence that this contribution had much subsequent influence on production management in postwar civilian industry. In automobiles, for example, the "Big Three" manufacturers pursued productivity gains at the expense of product innovation through standardization of major components across models, specialization of production facilities, and the introduction of dedicated automation equipment such as transfer machines. Product competition focused increasingly on annual changes in body styling and superficial variations between models in outfitting and trim, while the resulting proliferation of options was accommodated through greater labor intensity and high buffer inventories within the assembly process. By the 1970s and 1980s, these strategies had left American automobile firms extremely vulnerable to competition from foreign manufacturers—above all the Japanese—able to combine lower costs and higher quality with more flexible response to market trends and more rapid introduction of new models.[67]

[65]For preliminary accounts of the transformation of British industrial structure and managerial strategy since the 1950s and contemporary contrasts with German and Japanese practice, see Steven Tolliday and Jonathan Zeitlin, eds., *The Power to Manage? Employers and Industrial Relations in Comparative-Historical Perspective* (London, 1991), esp. pp. 282–86; Paul Hirst and Jonathan Zeitlin, eds., *Reversing Industrial Decline? Industrial Structure and Policy in Britain and Her Competitors* (Oxford, 1989), and "Flexible Specialization and the Competitive Failure of UK Manufacturing," *Political Quarterly* 60 (1989): 164–78.

[66]Lilley et al. (n. 19 above), p. 41.

[67]For synthetic overviews, see William J. Abernathy, *The Productivity Dilemma: Roadblock to Innovation in the Automobile Industry* (Baltimore, 1978); William J. Abernathy, Kim B. Clark, and Alan B. Kantrow, *Industrial Renaissance: Producing a Competitive Future for America* (New

The case of Germany is the most ambiguous of the three countries. Some metalworking firms, inspired by wartime rationalization programs as well as by the American model, moved aggressively into high-volume manufacture during the 1950s and 1960s, restricting and standardizing their product ranges, reorganizing production on flow lines, introducing special-purpose machinery and dedicated automation equipment, and reducing their dependence on skilled labor. The clearest example is that of Volkswagen, whose low-cost Beetle had been planned before the war but was not manufactured on a large scale until after 1945. Less dramatic but equally significant shifts in this direction could be observed among more established motor vehicle manufacturers, as well as among large concerns in other metalworking sectors such as electrical goods, office equipment, agricultural implements, and construction machinery. At the same time, however, sectors like machine tools and textile machinery continued to be dominated by networks of small and medium-sized firms using skilled craftsmen and universal equipment to turn out specialized niche products; and these rapidly expanded their position in world markets during the 1950s at the expense of foreign competitors such as the British. Vehicle manufacturers such as Daimler-Benz who moved some distance toward mass production also maintained some elements of their prewar flexibility, manufacturing a wide range of trucks and buses as well as luxury cars with a combination of standardized and custom parts. At other volume manufacturers, too, the proportion of skilled workers remained higher than in comparable factories abroad, partly because of the abundant supply produced by the German system of craft apprenticeship. Hence German metalworking firms in many sectors adjusted more easily than their British or American counterparts to the turbulent markets of the 1970s and 1980s, while also responding more successfully to the competitive challenge posed by the Japanese.[68]

---

York, 1983); Steven Tolliday and Jonathan Zeitlin, eds., *Between Fordism and Flexibility: The Automobile Industry and Its Workers,* 2d ed. (Oxford, 1992), esp. introduction. For British accounts of production control methods and restriction of design changes in American metalworking factories during the early 1950s, see British Productivity Council, *Production Control: Report of a Visit to the USA in 1951 of a Specialist Team on Production Planning and Control* (London, 1953); B. E. Stokes, "The Organisation of Production Administration for Higher Productivity," *Journal of the Institution of Production Engineers* 31 (1952): 194–218.

[68]This sketch of postwar German developments draws heavily on Herrigel, "Industrial Organization and the Politics of Industry" (n. 47 above), chaps. 4–7, "Industry as a Form of Order" (n. 47 above), and "The Case of the West German Machine-Tool Industry," in *Industry and Politics in West Germany: Toward the Third Republic,* ed. Peter Katzenstein (Ithaca, N.Y., 1989), pp. 185–220. For the success of German machinery and automobile firms in world markets at British expense during the 1950s, see also Alan Kramer, *The West Germany Economy, 1945–1955* (Oxford, 1991), pp. 182–95; Alan S. Milward, *The European Rescue of the Nation State* (London, 1992), pp. 134–67, 396–424. On the performance of the German

Despite the relative success of wartime aircraft manufacturers in Britain—as well as to a lesser extent the United States and Germany—in modifying mass-production techniques to facilitate frequent changes in models and design, these innovations rarely appear to have been taken up by civilian manufacturers in other industries. Insofar as the wartime experience of aircraft manufacture did influence postwar reconstruction planning, its effect was instead to dazzle government officials and progressive industrialists with the vista of extraordinary productivity gains which could be achieved if continuous output of standard models were not interrupted by "abnormal" disruptions to large-scale production programs.

Ironically, however, the flexible production methods developed by wartime aircraft manufacturers, particularly in Britain, anticipated in many respects those deployed so effectively by Japanese firms during the 1970s and 1980s in civilian sectors such as automobiles and consumer electronics. Like wartime British aircraft manufacturers, for example, Japanese firms in these industries have gained competitive advantages in product development through continuous modification and incremental improvement, using fewer off-the-shelf parts on new models while redesigning them less radically than their American or European counterparts. Like wartime British aircraft manufacturers, too, Japanese firms have enhanced their flexibility and improved the quality of their products by turning out components in small batches "just in time" to meet the demands of rapidly changing production programs. And like wartime aircraft manufacturers in all three countries, finally, Japanese firms have also constructed complex mechanisms for orchestrating collaboration with suppliers and subcontractors in product development, process improvement, and logistical coordination.[69]

Such flexible manufacturing methods reminiscent of wartime aircraft production have proved successful not only in consumer-oriented sectors such as automobiles and electronic equipment but also in the

---

automobile industry through the 1980s, see also Wolfgang Streeck, "Successful Adjustment to Turbulent Markets: The Automobile Industry," in Katzenstein, ed., pp. 113–56.

[69]For this view of Japanese manufacturing and product development methods, see Richard J. Schonberger, *Japanese Manufacturing Techniques* (New York, 1982); Michael A. Cusumano, *The Japanese Automobile Industry* (Cambridge, Mass., 1985); Kim B. Clark and Takahiro Fujimoto, *Product Development Performance: Strategy, Organization, and Management in the World Auto Industry* (Cambridge, Mass., 1991); Tolliday and Zeitlin, eds., *Between Fordism and Flexibility*, esp. introduction and chaps. 8–9. For the role of group organization, coordination committees, complexes, and rings in orchestrating cooperation between firms engaged on common projects in the wartime British, American, and German aircraft industries, respectively, see Postan (n. 9 above), pp. 417–22; Holley, *Buying Aircraft* (n. 16 above), pp. 544–46; Homze (n. 39 above), pp. 77–78; Overy, *Goering* (n. 39 above), pp. 149–50, 190, and "German Aircraft Production" (n. 39 above), pp. 254–64; Cairncross (n. 15 above), pp. 124, 131–32, 135–36.

aerospace industry itself. As David Friedman and Richard Samuels have demonstrated, although Japan's aircraft industry, like that of Germany, was dismantled by the Allies after the war, over the past few decades Japanese manufacturers have built up a rapidly growing business as suppliers of aerospace components. Starting out as licensees and subcontractors for Western firms, Japanese aerospace manufacturers have steadily increased their autonomous design capabilities and expanded their role in each successive international collaborative project. Unlike their American or British counterparts, however, aerospace manufacture is not isolated in specialist firms heavily oriented toward defense contracting, but rather integrated alongside civilian production in large, diversified metalworking groups such as Mitsubishi, Kawasaki, and Fuji Heavy Industries. Both civil and military aircraft components are manufactured alongside other commercial products up to the phase of final assembly using easily redeployed general-purpose equipment such as computer numerically controlled machine tools, while design and engineering knowledge are systematically diffused across the group through mechanisms such as project teams, corporation-wide study groups, and technology focus centers for functional area specialists. The result, Friedman and Samuels convincingly argue, has been a high rate not only of technology "spin-off" from aerospace across the Japanese economy but also of "spin-on" of innovations in manufacturing techniques from other sectors to aerospace itself.[70]

The experience of aircraft manufacture during World War II, like that of Japanese industry today, suggests that the imperatives of military and civilian production are more similar than most historians and policy analysts have assumed, but in different ways than others have argued. The evidence presented here supports neither a sharp distinction between performance-oriented flexible production of military goods on the one hand and cost-oriented mass production of civilian goods on the other, nor the subsumption of both sectors under a universal model of productive efficiency based on a common ideological commitment to order and control. Rather it substantiates the view that under conditions of pervasive environmental uncertainty and rapid technological inno-

[70]David B. Friedman and Richard J. Samuels, "How to Succeed without Really Flying: The Japanese Aircraft Industry and Japan's Technology Ideology," in *Regionalism and Rivalry: Japan and the United States in Pacific Asia,* ed. Jeffrey A. Frankel and Miles Kahler (Chicago, 1993), pp. 251–320; Richard J. Samuels, *"Rich Nation, Strong Army": National Security and the Technological Transformation of Japan* (Ithaca, N.Y., 1994). For the isolation of defense contracting from civilian production in the United States and Britain, see also Jacques S. Gansler, *Affording Defense* (Cambridge, Mass., 1989), pp. 244–45; Markusen and Yudken (n. 7 above), chap. 4; AnnaLee Saxenian, "The Cheshire Cat's Grin: Innovation, Regional Development and the Cambridge Case," *Economy and Society* 18 (1989): 448–77.

vation, such as those prevalent today as well as during World War II, military and civilian manufacturers alike must find ways of balancing countervailing objectives—quality and quantity, design change and production continuity, standardization and customization—in adapting their products to shifting competitive conditions at acceptable costs. The more similar the challenges facing military and civilian producers, finally, the more likely it becomes that the most successful firms—or national economies—will be those in which the institutional boundaries between the two sectors remain most fluid.

# [17]

*Cambridge Journal of Economics* 1989, **13**, 273–305

# The transition to flexible specialisation in the US film industry: external economies, the division of labour, and the crossing of industrial divides

Michael Storper*

## 1. Flexible specialisation and post-Fordist industrialisation

Fordism as a type of production organisation appears to be losing its dominance in the advanced industrial economies. A wealth of industrial case studies has been amassed over the past decade documenting the appearance of new ways of organising production systems, labour markets, and labour-capital relations (Wilkinson, 1981; Solinas, 1982; Scott, 1984; Rubery and Wilkinson, 1981; Murray, 1983; Ikeda, 1979; Bagnasco, 1977; Berger, 1981; Christopherson, 1986; Brusco, 1982; Brusco and Sabel, 1983). Most especially, the recent evidence suggests that vertical integration, mass production, and stable oligopolistic market structures—the hallmarks of Fordist industries—are not the necessary culminations of sectoral development, as was often assumed in the early post-war years (Storper, 1985). Since Fordism refers not only to a type of production organisation but also a compatible form of macroeconomic coordination (the welfare state and the mass consumption society) these changes have implications for the political economy of industrial societies generally (Aglietta, 1976; Boyer and Coriat, 1986).

In this paper I shall be concerned with the process of transition from mass production methods to a 'post-Fordist' form of production organisation. At the outset, it should be noted that we remain far from any consensus as to exactly what constitute post-Fordist modes of production organisation, labour processes, or macroeconomic arrangements (cf. Leborgne and Lipietz, 1987). Nonetheless, two broad interpretations of post-Fordist production organisation (each with inumerable variants) have surfaced in the case-study literature. These include neo-Fordism, whether of the neo-Taylorist variety in many labour-intensive industries, or the revival of Fordism itself but with the use of dedicated automatic technologies of production. Counterposed to neo-Fordism is *flexible production*, which includes everything ranging from flexible mass production to flexible

*University of California, Los Angeles. The research on which this paper is based was carried out with support from the John Randolph Haynes and Dora Haynes Foundation, the UCLA Institute of Industrial Relations, and the Research Committee of the Academic Senate of the University of California. Section 2 is drawn from research carried out in collaboration with Susan Christopherson, and I am grateful for her permission to publish this paper, although the conclusions remain my own. I would also like to thank the following students in the UCLA Urban Planning Program for their able research assistance: Larry Barth, Robin Bloch, Gillian Court, Rebecca Day, and Rehema Grey. The comments of two anonymous referees greatly sharpened the arguments.

0309–166X/89/020273+33 $03.00/0

specialisation (Piore, 1987). These are technologically dynamic production systems in which skilled workers use flexible capital equipment to produce a constantly changing variety of goods. Flexible specialisation typically occurs in smaller-scale industries with a greater variety of output than is possible in flexible mass production (Boyer, 1987; Leborgne, 1987; Piore, 1987). Henceforth, this discussion will concern itself with flexible specialisation.

Flexibly specialised industries have three defining characteristics (Piore and Sabel, 1984). First, they produce a wide range of products for highly differentiated markets and they constantly alter these goods in response to changing tastes and in order to expand their markets. The production system has the flexibility to do this because outputs can be changed by altering the mix of participating input suppliers or because these firms themselves are capable of changing their outputs. In more conventional parlance, the system is vertically disintegrated. Unlike the case of industrial dualism within Fordism, however, small firms are not simply shock absorbers for large firms (Berger and Piore, 1980). They occupy central roles in flexibly specialised sectors, participating in a sophisticated network of interfirm relations whereby firms share knowledge and thereby develop new products and production methods together.

Second, individual firms use flexible and widely applicable technologies: general-purpose machines rather than large, dedicated machine systems. Product innovation is not held back by massive capital investments in rigid technologies. Workers in these firms possess the skills to produce and develop a wide range of products because there is less separation of conception and execution than is required by the deep technical division of labour characteristic of Fordist production.

Finally, flexibly specialised industries balance competition and cooperation among firms. Competition encourages perpetual innovation, unlike traditional craft systems, which may use flexible equipment and skilled labour but are not technologically dynamic. Cooperation is accomplished through learned social practices as well as more formal rules which sustain local business communities and labour markets by socialising knowledge and controlling opportunistic behaviour (Storper and Scott, 1988). These practices and institutions tend to be highly regional in nature because flexibly specialised industries are often highly agglomerated in space owing to the existence of dense transactional relations between firms (Storper and Christopherson, 1987; Scott, 1986).

There are many famous nineteenth century industrial districts: Lyon's silk industry, the cutlery and tool sectors of Solingen and Sheffield, and the textile industries in Philadelphia and Pawtucket (Piore and Sabel, 1984; Scranton, 1983). A large case-study literature has also emerged on contemporary examples of flexible specialisation, most notably in the 'Third Italy' (Bagnasco, 1977; Brusco, 1982; Brusco and Sabel, 1981; Russo, 1986; Becattini, 1987; Fua, 1983), other areas of Europe (Scott, 1988; Sabel *et al.*, 1987; Savy, 1986), Japan (Ikeda, 1979), and even in North America in such sectors as high technology, advanced producer services, publishing, clothing, and steel (Scott and Storper, 1987; Scott, 1984B).

### *1.1. History, destiny, and all that*

Piore and Sabel (1984) advance a theory of 'industrial divides', which are shifts from one hegemonic technological-organisational model of production, such as Fordism, to another, such as a putative post-Fordism. They claim that these divides mark major periods in economic history and social organisation within industrial capitalism. Their story is a useful point of departure, and it can be summarised as follows.

In contrast to theories which hold that technologies and production organisation move along a 'narrow track' of history, economies are better characterised as 'gigantic collective experiments'. Very little about the evolution of technology and production organisation is historically predictable, because the precise contexts in which economic decisions are made are themselves actively constructed by agents wielding complex strategies; these agents in turn construct institutions to do their bidding and solidify their gains. As a result, technology and production organisation have histories like a 'branching tree'. The decisions made at certain critical points steer the future along a branch, and cut off other branches (Piore and Sabel, 1984; David, 1975; Nelson and Winter, 1982; Sabel and Zeitlin, 1985). The question is why, at certain junctures, one specific path is taken rather than another.

The first modern industrial divide was initiated in the mid-nineteenth century when circumstances in the US allowed capitalists to put into place the means to enlarge and stabilise industrial markets. The famous 'American system' of mass production was not the necessary culmination of nineteenth-century industrialisation, but a creature of particular American circumstances that only later, and incorrectly, came to be described as the essential destiny of modern industry in general (Hounshell, 1984; Rosenberg, 1969; Urry, 1986; Braverman, 1974). In reality, Fordism originated as an historical particularity. The American circumstances that generated it included the erasure of European tastes for customised goods, a large number of independent farmers with the means to consume manufactured goods, and the development of the large corporation: all these created an environment favourable to the standardisation of products, the control of markets, and the rationalisation and mechanisation of production. Aided by the stagnation of productivity in a number of British industries, American methods triumphed internationally (Rothbarth, 1946). Once in place, mass production enjoyed powerful advantages over any competing approach. In the twentieth century, mass markets were enlarged by a virtuous circle of rising wages in the mass-production industries themselves, and then given a new lease of life in the post World War II period by the welfare state, American international economic hegemony, and high levels of domestic market concentration in many countries.

Why, then, would we expect a 'second industrial divide' to bring the end of Fordism and the rise of flexible specialisation? Because the institutions that once held the Fordist system together—stabilising demand at high levels and with a high level of intertemporal certainty—have been broken down by the shocks suffered in the capitalist economies starting with the late 1960s. Principal among these is the breakdown of international macroeconomic coordination under American dominance (Boyer, 1986). It is argued by a number of people—from both historical and game-theoretic perspectives—that it is unlikely that internationally-coordinated Keynesian macroeconomic stabilisation can be implemented and thus bring back Fordism (Elster, 1982; Mistral, 1986; Piore and Sabel, 1984). There is no single dominant power to provide the central coordination such a system would require. If one accepts this premise, it follows that flexible specialisation—along with flexible mass production—is one of the few workable possibilities at the level of production organisation to the current state of 'disorganized capitalism' (Piore and Sabel, 1984; Offe, 1985).

It is important to appreciate in what this analysis does, and does not, consist. The argument is that initial technological choices are not necessarily dictated by relative efficiencies, but once adopted they tend to persist until a major shock makes them inviable. It is, of course, an empirical question as to whether initial forms of Fordist production were, in static terms, more efficient than the existing competition, an issue which has generated considerable controversy in American and British economic historiography (Habbakuk, 1962; Rothbarth, 1946; David, 1975; Temin, 1966). The point is that mass production technologies were installed against a changing macroeconomic environment which provided the conditions for Fordism *to become* a superior technique. Thus, the analytical core of Piore and Sabel's claim to an historical *theory* of technology and production organisation is that macroeconomic and institutional conditions have a certain independence in history. Or, to put it more pedantically, if there is no general equilibrium, then the fates of industrial sectors are outcomes of a contingent history.

The argument does not, however, theorise the putative *transition* from Fordism to flexible specialisation. It simply reasons that if the first industrial divide was a contingent outcome of particular historical circumstances then a second divide is now possible due to changed circumstances. The argument is conspicuously silent as to the *process* by which initial organisational and technological experiments in response to changing circumstances might eventually cause a divide to be crossed, as the new technologies and organisational forms become superior to the old ones. Moreover, virtually all the existing literature on flexibly specialised industries is based on cases of transition from traditional craft production to flexible methods or on the rise of flexible production *ab initio* in new sectors such as high technology and producer services. None actually documents a transition from the use of mass-production methods to flexible specialisation. In this paper, therefore, I first document the transition from the use of mass-production principles to flexible specialisation in the postwar US film industry. Contrary to its glamorous and artisanal image, large firms once manufactured motion pictures via a production process organised along mass-production principles in stable, oligopolistic markets. The leading capitalists in the industry explicitly used Henry Ford as their role model. In section 2, the detailed processes of competition, technical choice, and reorganisation which effected the transition are described at some length.

Section 3 then builds on the case study to isolate the processes by which the divide between mass production and flexible specialisation may be crossed. The process of transition can best be understood, in our view, as a shift to a new form of production organisation which embodies increasing external economies of scale as the result of the development of the industry's social division of labour. To build our explanation of the transition, we use three theoretical building blocks, concerning: vertical disintegration and the theory of the firm; external economies of scale and the social division of labour; and, technical-organisational choice with increasing returns.

In addition to offering a plausible theory of the transition between one technological-organisational régime of production and another, section 3 sheds critical light on the literatures from which it draws. First of all, it suggests that the theory of the firm stemming from the work of Coase (1937) and recently extended in the New Institutional Economics (or, 'transaction cost economics'), while providing key insights into the problem of the organisation of production, underestimates the role of external economies of scale and their dynamic properties in solidifying organisational régimes of production. The existence of external economies enables us to propose a much richer and more historical conception of the dynamics of production organisation. Likewise, Young's (1928) seminal contribution on the development of the social division of labour as the source of dynamic external economies underestimates the variety of conditions under which these economies are generated; they do not arise only under conditions of smooth upward growth.

Beyond its usefulness in advancing the theory of flexible specialisation, the film industry offers an unusually rich case study. First, it is quite far along in the process of organisational change, having begun to abandon routinised batch production as early as 1948. Second, the whole industry, and not a specialised or marginal segment thereof, has experienced the transformation. Finally, it is an American industry of the twentieth century. It grew up in an environment lacking the artisanal and regional traditions of Europe, and without the nineteenth century antecedents of other industries. Because it emerged in the world centre and at the historical apex of mass production, it provides insight into the way flexible specialisation develops in the specific and important context of late twentieth century America.

## 2. The rise and fall of mass-production methods in the film industry

### *2.1. From craft to mass-production methods: the Golden Age of Hollywood*

The history of the film industry is similar to that of many other industries as they developed in the United States after the turn of the century. It began as a craft but, with the creation of a large, assured market, the product was standardised and the production process rationalised.[1]

The film industry that developed in New York shortly after the turn of the century closely replicated the craft production techniques of the theatre. Production was carried out by small crews in firms such as Lumiere, where artisans worked together to produce an unstandardised product. Despite artisanal production techniques, however, pressure was applied on the industry to adopt mass-production methods by the then-existing entertainment industry—nickelodeons and vaudeville—who provided the speculative capital to produce films. They intended to develop the mass market of new immigrants who populated American cities. Movies came to be an entertainment product rather than strictly an art form.

By the time film-making was established in California in the 1920s, it had become industrialised. While we now classify it as a 'service' industry, one of the earliest studios was named the Universal Film *Manufacturing* Company. Its artificially lighted stage was dedicated by the mass-production capitalist of the age, Henry Ford. Within a year of its establishment in 1918, this one plant had produced 250 films, a figure equal to the feature film production of the entire American film industry today. The other early studios in California, such as D. W. Griffiths' Fine Arts Studio, were established as full-service film production facilities intended to produce a standardised film product. Films were sold by the foot rather than on the basis of content (Hampton, 1970).

Two other figures involved in building the film industry, Thomas Ince and Adolph Zukor, were also responsible for rationalising its production process. Zukor, founder of Paramont Pictures, integrated production and distribution through the use of contracted exhibitors nationwide. From his office in Times Square in New York City, Zukor contracted with 6000 theatres nationwide to provide 3–4 films per week. The US public supported Zukor's marketing network by going to the movies three to four times weekly. Weekly attendance in the US eventually reached 90 million. The downtown theatres typical of that period had seating capacities of 2000, compared to today's average of 300–500 (and still declining owing to the spread of mini-theatres).

Paralleling Zukor's innovations in corporate structure were the production innovations of Thomas Ince. With an assured and stable theatrical market, the entire range of film-making activities could be integrated within one large factory, the studio. Ince set up the

[1] This section draws upon Storper and Christopherson (1985).

studio to fabricate and assemble batches of a semi-standardised product, the 'formula' picture (Storper and Christopherson, 1985). Ince developed a management-oriented model that strictly separated conception from execution. The vehicle for this production process was the 'continuity script', which fragmented the story of a motion picture and reordered it so that each bloc of scenes in a set or location could be filmed at the same time or, alternatively, so that a set of actors could film all the scenes in which they were to be involved in a continuous work session. The continuity script also impeded costly improvisation by giving the producer virtually complete control over film content. Scripts could be ordered from writers according to a desired formula, for a desired length, and producers could insure that directors would be faithful to them. This production management process was honed in 'Inceville', as it was known, under a schedule of two productions per week.

The production process established in this period consisted of: pre-production (selection and preparation of the script and shooting location); production (construction of sets, filming); and post-production (film processing, editing, sound track). Each of the three labour processes was organised according to mass-production principles. For example, the major studies had permanent staffs of writers and production planners who were assigned to produce formula scripts in volume and push them through the production system. Production crews and stars were assembled in teams charged with making as many as thirty films per year. Studios had large departments to make sets, operate sound stages and film labs, and carry out marketing and distribution. A product would move from department to department in assembly-line fashion. The studios endeavoured to maximise capacity utilisation and stabilise throughput.[1] As a result, the internal organisation—or technical division of labour—in each phase of the labour process became increasingly similar to that of true mass production, where routinisation and task fragmentation were the guiding principles. This factory-like organisation and oligopolistic corporate structure, popularly known as the 'studio system', was well-established throughout Hollywood by the mid 1920s, and it prevailed until the late 1940s (Storper and Christopherson, 1985).

This period is viewed by many as the Golden Age of the industry because the volume of production permitted the establishment of something resembling a regional 'mass collective' workforce in Hollywood (cf. Murray, 1983). Workers had the expectation of stable work over a fairly long period of time. Apprenticeships within the patriarchal world of the studios represented admission to a restricted internal labour market. Entry barriers to someone not already in 'the industry' were enormous (Christopherson and Storper, 1989).

The studio system was a concentrated oligopoly: a small number of producers were responsible for the majority of the industry's output and they simultaneously controlled distribution and exhibition. In 1944, for example, the five major studios earned 73% of

[1] The studio system was organised according to mass-production principles even though the number of films produced per year in a studio (between 30 and 200 or so) was small enought to count as batch production by the standards of most other industries. Analogous cases are airframe construction or tanker construction. These are not, as is commonly thought, 'one off' production methods. The latter are found in, for example, the construction of some aerospace or military hardware systems. In those cases, the large firm is a 'systems house', which engineers the final output and issues orders for the production of necessary components, and then assembles the final product, driving the whole system by final demand. The studios, on the other hand, took slightly customised versions of a technologically-similar product and pushed them through the production process, with its well-defined division of labour. The system ran on the assumption of a stable demand and planned output. This is also what occurs in airframe construction once the product is designed and enters serial production. The systems house and mass-production firm organise their labour processes very differently, the latter being geared toward routinisation, the former toward flexibility.

domestic cinema rentals and owned or had interests in 4424 theatres—24% of the US total. The latter statistic, while impressive, understates the extent of the major studios' vertical interests, since their cinemas included 70% of the first-run cinemas in the 92 cities with populations greater than 100,000 and these same cinemas accounted for more than 50% of all US box office receipts. The extent of market concentration at smaller geographical scales was even greater: for example, RKO owned 100% of first-run capacity in Minneapolis and Cincinnati and Warner Brothers owned 90% in Philadelphia. In 46% of US markets, one distributor owned *all* cinemas (Waterman, 1982). It was the destruction of these assured market outlets that spelled the end of the studio system.

*2.2. The crisis of the studio system*

The studio system reached its zenith during the war years. The number of films released peaked in 1942 (Table 1). Direct studio employment in Los Angeles, by this time the centre of the US film industry, reached its high point in 1944, at 33,000. Attendance in US cinemas peaked in 1946, with over 90 million admissions per week.

In the late 1940s and early 1950s, the absolute size of the market for films began to shrink and the remaining market was much less stable. There were two main shocks to the studio system. The first, anti-trust action by the US Supreme Court, made the industry's market less certain. 'The Paramount Decision' (*U.S. v. Paramount Pictures*, 334 US 131, 1948) forced the studios to divest their cinema chains. Since the assured market once enjoyed by the studios was now gone, average returns per picture declined and returns per film began to fluctuate wildly. The second shock to the studio system, the advent of television, altered the industry's market structure and overall growth prospects. Television's success was closely tied to demographic changes, such as the high rate of post-war family formation and, spatially, to the process of suburbanisation. With the diffusion of television ownership, what had been essentially a unified market for filmed entertainment, dominated by one medium, became a segmented market in which different products competed for the consumer's entertainment expenditures. The feature film audience declined by 50% between 1946 and 1956. The size of the box office as a whole declined by almost 40% in real terms between 1945 and 1955 and in 1960 was only 39% of its real level at the end of the War (Table 2). The result, in combination with the Parmount decision, was that the gross revenues of the ten leading companies in the industry fell by 26% in this same period, from $968 million to $717 million. Profits of the leading eight studios declined by more than 50% in real terms (Table 3).

*2.3. Early flexibility strategies: vertical disintegration and product innovation*

The studios responded to the crisis with two strategies designed to increase their flexibility. They initially reduced the number of films produced; later, as we shall see, they turned to vertical disintegration as a way to cut overheads and increase the quantitative and qualitative flexibility of output. The number of US-produced feature films declined by 28% in the ten years after 1946 (Table 1). The average number of films made per year in the US during the 1950s was only 80% of the 1940s level and, by the end of the 1950s, stood at less than half the number produced in the peak production year of 1941 (Table 1).

The industry also began to increase qualitative flexibility to compensate for output reductions, by undertaking to differentiate their products through constant innovation. Initially, the most standardised categories of film-making, short subjects and newsreels, were completely eliminated from the major studios' product range. In turn, studios drew people back into the cinemas with an innovative type of film known as the 'spectacular',

**Table 1.** *Films produced and re-issued in the US, 1930–77*

| Year | New | Re-issued | Total |
|---|---|---|---|
| 1930 | — | — | 355 |
| 1935 | 388 | 3 | 391 |
| 1940 | 472 | 3 | 475 |
| 1941 | 497 | 7 | 504 |
| 1942 | 484 | 8 | 492 |
| 1943 | 426 | 6 | 432 |
| 1944 | 409 | 6 | 415 |
| 1945 | 367 | 8 | 375 |
| 1946 | 383 | 17 | 400 |
| 1947 | 371 | 55 | 426 |
| 1948 | 398 | 50 | 448 |
| 1949 | 406 | 85 | 491 |
| 1940–49 Average | 421 | 25 | 446 |
| 1950 | 425 | 48 | 473 |
| 1951 | 411 | 28 | 439 |
| 1952 | 353 | 33 | 386 |
| 1953 | 378 | 36 | 414 |
| 1954 | 294 | 75 | 369 |
| 1955 | 281 | 38 | 319 |
| 1956 | 311 | 35 | 346 |
| 1957 | 363 | 19 | 382 |
| 1958 | 327 | 25 | 352 |
| 1959 | 236 | 18 | 254 |
| 1950–59 Average | 338 | 36 | 374 |
| 1960 | 233 | 15 | 248 |
| 1961 | 225 | 15 | 240 |
| 1962 | 213 | 24 | 237 |
| 1963 | 203 | 20 | 223 |
| 1964 | 227 | 15 | 242 |
| 1965 | 257 | 22 | 279 |
| 1966 | 231 | 26 | 257 |
| 1967 | 229 | 35 | 264 |
| 1968 | 241 | 17 | 258 |
| 1969 | 241 | 10 | 251 |
| 1960–69 Average | 230 | 20 | 250 |
| 1970 | 267 | 39 | 306 |
| 1971 | 281 | 32 | 313 |
| 1972 | 279 | 39 | 318 |
| 1973 | 237 | 38 | 275 |
| 1974 | 229 | 45 | 274 |
| 1975 | 190 | 40 | 230 |
| 1976 | 187 | 30 | 217 |
| 1977 | 154 | 32 | 186 |
| 1970–77 Average | 228 | 37 | 265 |

*Source*: Motion Picture Association of America.

which recast the feature as a form of entertainment significantly different from television. Technical innovations were aimed at the 'look' of the film in an effort to make the image in motion pictures superior to that of television. Cinerama, Technicolor, and 3D—all

**Table 2.** *US box office receipts ($ million)*

| Year | Nominal $ | 1967 $ |
|---|---|---|
| 1945 | 1450·0 | 2690·0 |
| 1950 | 1376·0 | 1908·0 |
| 1955 | 1326·0 | 1653·0 |
| 1960 | 951·0 | 1072·0 |
| 1965 | 927·0 | 980·9 |
| 1970 | 1162·0 | 999·1 |
| 1975 | 2115·0 | 1312·0 |
| 1980 | 2899·0 | 1174·6 |
| 1982 | 3450·0 | 1193·0 |

*Source*: *International Motion Picture Almanac*, 1983.

**Table 3.** *Profits in the film industry ($ million)*

| | Eight major studios[a] | | Whole industry/post-tax[b] | |
|---|---|---|---|---|
| Year | Nominal $ | 1967 $ | Nominal $ | 1967 $ |
| 1945 | 57·2 | 106·1 | 99·0 | 183·6 |
| 1950 | 38·0 | 52·7 | 60·0 | 83·2 |
| 1955 | 37·4 | 46·6 | 61·0 | 76·0 |
| 1960 | 32·0 | 36·1 | 1·0 | 1·1 |
| 1965 | 72·5 | 76·7 | 39·0 | 41·2 |
| 1970 | −57·2 | −49·8 | 8·0 | 6·8 |
| 1975 | 262·0 | 162·5 | 131·0 | 81·2 |

[a]*Source: The Economist*, as quoted in Steinberg, C., *Film Facts*, 1980, pp. 85–87.
[b]*Source*: US Department of Commerce, Bureau of Economic Analysis, *Census of Service Industries*, 1977.

innovations of the 1950s—were aimed at constituting the film as an event rather than an everyday experience. The increased attention to the individual film meant increased budget for talent, marketing and advertising. As Table 4 indicates, individual production budgets rose rapidly along with the decline in volume of total production in the immediate post-war period. The average cost per picture made by Metro-Goldwyn-Mayer rose from $1·3 million in 1952–53 to $1·8 million in 1954–55. For Paramount, the comparable figures are $1·7 million and $2·5 million.

This strategy of product differentiation increased the need for specialised inputs. The studios began to turn to independent producers to develop these differentiated film products. In the early 1950s, they established a putting-out system for pre-production work, in an effort to encourage innovative ideas. Warner Brothers made their first major advances to outside producers in 1951 to the tune of $6·4 million, and increased them rapidly to $25 million by 1956. As the decade progressed, additional aspects of the

**Table 4.** *Production expenditures and production costs US*

| | Total industry production expenditures[a] | | Average production cost per feature[b] | |
|---|---|---|---|---|
| | $ | 1967 $ | $ | 1967 $ |
| 1941 | NA | NA | 400,000 | 950,000 |
| 1949 | 2,881,600,000 | 4,035,854,300 | 1,000,000 | 1,400,000 |
| 1955 | 2,738,700,000 | 3,414,837,900 | NA | |
| 1960 | 2,966,000,000 | 3,343,855,000 | NA | |
| 1965 | 2,917,000,000 | 3,086,772,400 | NA | |
| 1972 | NA | NA | 1,890,000 | 1,508,370 |
| 1974 | NA | NA | 2,500,000 | 1,550,868 |
| 1978 | NA | NA | 5,000,000 | 2,558,850 |
| 1980 | NA | NA | 8,500,000 | 3,444,084 |
| 1982 | 2,500,000,000[c] | NA | 11,300,000 | 3,908,682 |

[a]*Source*: Standard and Poor's *Industry Survey*, 1961, p. A62, 1966, p. A72.
[b]*Source: International Motion Picture Almanac*, 1983.
[c]*Daily Variety*.

production process were split off from the studios' operations and moved to the external market. Eventually, this meant the end of the 'term contract', under which writers, actors, and skilled production technicians worked exclusively for one studio full-time for a guaranteed period. This strategy of reducing labour overheads necessitated changes in employment contracts. In the late 1950s, writers, actors, producers and directors were all put under project contracts, usually for just one film. The heavily unionised skilled craft-workers, on the other hand, were placed on seniority rosters, effectively shifting the old internal labour markets of the studios to an external, collectivised system. The unions came to serve as hiring halls (Christopherson and Storper, 1989).

Each round of overheads reduction in the face of a stagnant, more unstable and more segmented market created unanticipated new pressures for further disintegration, which the studios could not control. For example, the studio system had functioned as a star creation machine, placing individuals under long-term contracts at favourable prices and then making some of them into very valuable commodities, stars. Under the 'star system', as it was known, if an individual achieved star status, the studio reaped the benefits of having him/her under a long-term contract whose terms had been established prior to stardom. The studios thus obtained a monopoly over the actor's specific human capital by controlling access to training at the port of entry to the labour market for stars. Since the maturation of stardom necessitated long-term investments in specific human capital, the star system encouraged vertical integration (cf. Williamson, 1985).

With the end of long production runs came the end of long-term contracts. It not only became more difficult to control product markets long enough to produce stars, but in the short run stars became another form of the overheads the studios were attempting to shed. Thus, the need for asset specificity was overshadowed by scale, overheads, and other flexibility considerations in the studios' decision-making. Successful stars, however, were no longer bound to long-term contracts and could now demand much higher salaries or very lucrative profit-sharing arrangements in return for appearing in a film. In exercising their newly-found market power, stars shifted distribution of the rents to specific assets in

**Fig. 1** *Domestic and foreign rentals of nine US companies.* (Allied Artists, Columbia, Disney, MGM, Paramount, Twentieth Century Fox, United Artists, Universal and Warner Bros.) ——: US and Canadian rentals, · · · · · : foreign rentals.

*Source*: Wertheimer Associates.

their favour.[1] According to industry sources, a significant proportion of the dramatic increase in production budgets in recent years is for steep increases in stars' salaries (Table 4; Storper and Christopherson, 1985). These increases have placed even greater pressure for cost reduction on other parts of the film budget which, in turn, encourages further vertical disintegration. Thus, the beginnings of vertical disintegration, intended to cut costs and achieve product differentiation and innovation, had in turn unforeseen consequences which promoted further vertical disintegration. The studios could not control the process they had started.

### *2.4. The 1960s: failed attempts at building stable markets*

The 1960s opened with the studios' attempt to restabilise markets at a tolerable minimum size by increasing distribution of American pictures abroad through international integration of exhibition. The market for privatised entertainment products such as television had developed much more slowly in Europe than in the US during the 1950s, and so the market for cinema films continued to grow there. The major studios bought cinema chains in Europe and even in Africa, attempting to replace the loss of their domestic market.

[1] Profit participation arrangements, by definition, are designed to shift distribution in favour of stars, because they are negotiated *in addition to* salaries, not as a replacement thereof. Rising salaries, even without profit participation, probably effected dramatic redistribution of the rents on stardom toward the stars themselves, thus reducing the profit shares of the studios because this was a period of declining market size and very limited demand elasticity. While some spectaculars could have very high earnings, average earnings per film were in decline, yet studios were forced to establish salaries by contract in advance of profit data. It follows that studio's shares were cut in favour of stars' shares.

**Table 5.** *Film releases in the US, 1943–1978*

| Year | Imported | Total | Year | Imported | Total |
|---|---|---|---|---|---|
| 1943 | 30 | 427 | 1960 | 233 | 387 |
| 1944 | 41 | 442 | 1961 | 331 | 462 |
| 1945 | 27 | 377 | 1962 | 280 | 427 |
| 1946 | 89 | 467 | 1963 | 299 | 420 |
| 1947 | 118 | 486 | 1964 | 361 | 502 |
| 1948 | 93 | 459 | 1965 | 299 | 452 |
| 1949 | 123 | 470 | 1966 | 295 | 451 |
| | | | 1967 | 284 | 462 |
| 1950 | 239 | 622 | 1968 | 274 | 454 |
| 1951 | 263 | 654 | 1969 | 180 | 412 |
| 1952 | 139 | 463 | | | |
| 1953 | 190 | 534 | 1970 | 181 | 367 |
| 1954 | 174 | 427 | 1971 | 199 | 432 |
| 1955 | 138 | 392 | 1972 | 147 | 376 |
| 1956 | 207 | 479 | 1973 | 168 | 463 |
| 1957 | 233 | 533 | 1974 | 270 | 550 |
| 1958 | 266 | 507 | 1975 | 242 | 604 |
| 1959 | 252 | 439 | 1976 | 222 | 575 |
| | | | 1977 | 249 | 560 |
| | | | 1978 | 137 | 354 |

*Sources: Film Facts,* 1980: Motion Picture Association of America.

Foreign markets accounted for 50% of total revenues at this time, primarily through pre-sales (Fig. 1).

The most important strategy adopted by the major studios during this transition period was to attempt to dominate televison markets. In 1965, television networks spent $115 million to commission made-for-television films (MFT). The average budget per MFT increased from $169,000 in 1961 to $290,000 in 1966 because of the depletion of film libraries. The studios hoped that capturing the market for MFTs, combined with their growing European cinema market, would enlarge and restabilise their markets. The television networks, however, (as oligopsonistic consumers of MFTs) began to contract with smaller studios (so-called 'mini-majors') in order to force the major studios into competitive bidding over the price of MFTs. Because the studios owned large physical plants ('back-lots', large studios) and because they were almost completely unionised, they were essentially forced out of these more cost-sensitive segments of the market, low-budget cinema films and MFTs, which were then taken over by the same independent production companies whose existence in the first place had been set into motion by subcontracting from the major studios. This blow was aggravated by the fact that European producers' presence in American markets grew dramatically in the early 1960s (Table 5). Thus, even though new sources of market growth arose in the 1960s, they could not be captured or controlled by the major studios.

The failure to reassert mass markets was reflected in a return of the industry's profitability crisis in 1970, when the eight major studios all suffered large losses, and almost no profits were made in the industry as a whole (Table 3). The studios still held large quantities of property (in sound stages, 'back-lots', and film libraries). With low profits but large assets, their stocks were undervalued and they become vulnerable to takeovers.

Some studios, such as Paramount, responded by allowing friendly takeovers to occur, in their case by becoming a division of the conglomerate Gulf & Western. Most of the studios, however, began to trim their facilities and reduce overheads again. They redefined their production facilities as 'profit centers', requiring them to support themselves through rentals to independent producers making films with studio financing. In a number of cases, the studios simply sold their back-lots. Twentieth Century-Fox, for example, sold the land which is now Century City on Los Angeles' Westside—a large office, commerical and residential development.

Increased filming on location also promoted the reduction in physical scale of the major studios. Initially, vertical disintegration encouraged location shooting as a cost-cutting move on the part of independent production companies, and as a product differentiation strategy in the case of some spectaculars. Location shooting, which is a type of change in production technique, began as a direct *consequence of* vertical disintegration; like many such practices; it seems to have reinforced itself in circular and cumulative fashion with the result that the studios can no longer control its use. The inital increase encouraged technological innovations designed to make location work easier. The Cinemobile (a mobile studio) and the panaflex camera (hand-held, but very quiet), make it possible to achieve technical quality on location equal to the studio or back-lot. Location shooting is now a genuine alternative to the studio in most situations, offering a more realistic 'look' and lower overhead costs. Localities, too, have begun to promote themselves by offering better services and, sometimes, considerable subsides to production companies (Storper and Christopherson, 1985).

Location shooting also enables producers to avoid work rules in the industry centres, Los Angeles and New York. West Coast union work rules prevail within a radius of 300 miles from the corner of Beverly and La Cienega Boulevards in Hollywood. Since these rules limit the length of the working day, they can force a film to stay in production for more shooting days than would otherwise be the case, entailing the fixed costs associated with each day of production activity. Even skilled workers who go on location from Hollywood within the radius often waive work rules informally when they are on location. And non-unionised screen extras and construction workers can be secured on location. For all these reasons, the studios' stages and back-lots are now in many ways unable to compete—whether in quality or price terms—with locations.

By the 1970s most of the studios had, in effect, ended their roles as physical movie factories. Even though disintegration had begun with the limited objectives of cost-cutting and product differentiation, in the end specialised firms and non-studio locations proved superior to the studios, not only for low-budget films and MFTs, but for high-budget and high-revenue productions as well. The studios could no longer compete against the independent production companies and specialised contractors they had helped to create, in the very market segments they had hoped to retain. Virtually none of the major studios today can offer the range of specialised services which are obtainable from independent firms; few can even carry out relatively routine tasks at lower costs than the independent supplier firms.[1]

[1] Rigorous production cost analysis is virtually impossible in the film industry, precisely because production has become so much less standardised than it used to be. It is equally difficult to supply data on minimum efficient dimensions of the different phases of the production process. Note, too, that it is not being claimed here that today's independent firms can carry out their activities at lower unit cost than could be done by very specialised departments under the old studio system, but rather that they can produce more specialised outputs at lower costs than the studios are able to do at their *current* output levels. See section 2.6 for further discussion of scale economies and production costs.

**Table 6.** *Number and percentage of productions by organisation type, selected years, 1960–1980*

| Year | 1960 | 1965 | 1970 | 1975 | 1980 |
|---|---|---|---|---|---|
| Independent | 42 (28%) | 40 (21%) | 93 (44%) | 138 (56%) | 129 (58%) |
| Major | 100 (66%) | 130 (68%) | 96 (46%) | 81 (33%) | 69 (31%) |
| Mini-major | 9 (6%) | 20 (11%) | 18 (9%) | 24 (10%) | 24 (11%) |
| Totals | 151 | 190 | 209 | 244 | 223 |

*Source*: Storper and Christopherson, 1985. See p. 286. n. 1

*2.5. The new structure of the film industry*

After their initial experiment in contracting out writing in the 1950s, the studios developed a system of working jointly with independent production companies, also in order to encourage new ideas. Production costs per picture were already rising rapidly in the 1960s, so the studios also used the new system to share investment risks with the independents. In the event of success, profits are divided; but in the case of failure, the studio's investment exposure is minimised because the independent production company usually raises a substantial proportion of the initial investment using the major studio's capital as 'seed money'. Table 6 reports the changes in the types of organisations carrying out film production between 1960 and 1980.[1] The trend toward production by independent companies and away from the major studios is clear. In 1960, 28% of US films were produced by independents or independents combined with others, while 66% were made with the participation of major studios. By 1980, the 1960 proportions were almost reversed: independents produced 58% of films while majors participated in only 31%.

The increase in number of independent production companies, however, was only a small part of the organisational restructuring of the industry. The production system for various kinds of intermediate inputs steadily grew and diversified. Both the studios and the independent production companies turned to specialised, independent supplier firms and subcontractors to carry out the whole range of pre-production functions. We examined the births and deaths of selected filmmaking services and facilities in Southern California. The facilities and services tabulated all increase in absoluted numbers during the period (Tables 7 and 8).[2] Given that film output was stable for the index years, the magnitude of these increases provides convincing evidence of a trend toward vertical disintegration. In all categories of input-supplying firms, the major leap in the number of firms took place between 1974 and 1982, after the crisis of the early 1970s. We also looked

[1] Information on film-making services and facilities in the Los Angeles region was obtained by using service directories for the industry, including the *Hollywood Blu-Book*, the *International Motion Picture Alamanac*, and *The Production Company*. While the firms listed in the *Blu-Book* and the *Almanac* are not necessarily a complete listing of services and facilities, they come closest to sampling the whole industry.

In establishing the numbers of firms in each category, we assumed that a firm not listed in the previous period had been established in the interim. We were careful, also, to eliminate any duplication in directory listings, so as to minimise the possibility of double-counting.

The production companies for films were obtained by following the 'Production Charts' in *Daily Variety* and *The Hollywood Reporter* each week for the entire year of each year tabulated. We then classified the companies as major studios, mini-majors, independents, and so on.

[2] It is irrelevant whether concentration in final markets changes or does not. Vertical disintegration is measured in terms of whether the number of intermediate input producers changes for given levels of final output.

**Table 7.** *Number of Los Angeles area establishments: production firms*

| | 1966 | 1974 | 1982 |
|---|---|---|---|
| Production companies | 563 | 709 | 1473 |
| Rental stages | 13 | 24 | 67 |
| Properties | 66 | 33 | 184 |
| Editing | 4 | 31 | 113 |
| Lighting | 2 | 16 | 23 |

*Source*: Storper and Christopherson, 1985. See n. 1, p. 286.

**Table 8.** *Number of Los Angeles area establishments, post-production*

| | 1966 | 1974 | 1982 |
|---|---|---|---|
| Recording/sound | 20 | 33 | 187 |
| Film processing | 43 | 76 | 55 |
| Film effects | 10 | 27 | 42 |

*Source*: Storper and Christopherson, 1985. See n. 1, p. 286.

**Table 9.** *Establishments—US*

| | | 1968 | 1974 | 1981 |
|---|---|---|---|---|
| SIC 7813: | Motion pictures excluding TV | 666 | 1279 | 1023 |
| SIC 7814: | Motion pictures and tape production for TV | 490 | 978 | 1420 |
| SIC 7819: | Services allied to motion picture production | NA | 716 | 1077 |
| Total | | 1156 | 2973 | 3520 |

*Source*: US Dept. of Commerce, *County Business Patterns*

in detail at the dates of establishment of a sample of 200 independent production companies in all segments of the industry, including commercials, industrial films, animation, special effects, television, and cinema films. Of these, 140 were established between 1971 and 1982 (*The Production Company*, 1984).

These statistics on organisational proliferation are confirmed in Table 9, which demonstrates the dramatic increases in total number of establishments in the film industry. Significantly, SIC 7819 (services allied to motion picture production), which covers virtually all the new, specialised firms in the industry, was only created as a separate reporting category in the mid 1970s in response to the increasing visibility of these enterprises.

**Table 10.** *Employees per establishment*

| | 69 | 74 | Change[a] 58–74 | 91 | Change[a] 74–81 |
|---|---|---|---|---|---|
| SIC 7813 | | | | | |
| CA | 38·98 | 15·89 | 40·76 | 45·42 | 285·94 |
| NY | 13·12 | 7·12 | 54·26 | 8·78 | 123·31 |
| FL | 16·08 | 4·75 | 29·53 | 7·78 | 163·78 |
| TX | 14·42 | 11·58 | 80·3 | 14·72 | 127·11 |
| US | 23·21 | 11·19 | 48·2 | 25·04 | 223·77 |
| SIC 7814 | | | | | |
| CA | 26·25 | 36·28 | 138·20 | 31·48 | 86·7 |
| NY | 23·15 | 11·18 | 48·29 | 21·72 | 194·2 |
| FL | 13·33 | NA | NA | 12·08 | NA |
| TX | 18·3 | NA | NA | NA | NA |
| US | 21·2 | 20·82 | 98·2 | 24·10 | 115·75 |
| SIC 7819 | | | | | |
| CA | NA | 34·02 | NA | 22·7 | 66·9 |
| NY | NA | 13·34 | NA | 10·8 | 81·3 |
| FL | NA | 14·7 | NA | 6·6 | 45·1 |
| TX | NA | 10·07 | NA | 15·60 | 154·9 |
| US | NA | 21·33 | NA | 16·87 | 79·09 |

*Source: County Business Patterns.*
[a]First year = 100.

The deepening of the social division of the labour in the motion picture industry as a consequence of vertical disintegration should be associated with increasing output specialisation on the part of the supplier firms, who should have a narrower *scope* of activity than the integrated major studios they replace. The evidence for this comes from interviews with firms in the industry, with most reporting exactly what is expected theoretically: the smaller firms are generally specialised in a particular type of intermediate output, but within that field, they have a wider range of potential variations—because they are innovative—than the independent firms which existed in the 1950s or the specialised departments in the major studios which once carried out similar phases of the production process. A film lab, for example, will no longer do all kinds of film processing, but in turn will tend to have a much greater variety of techniques it can apply within its specialised domain, and it constantly interacts with production companies and equipment manufacturers to develop new ways of processing film (Storper and Christopherson, 1985; cf. Russo, 1986).

The changing division of labour should be reflected in declining firm size in an industry such as this, where total output is not increasing; in a growing industry, it would be reflected in slower growth in the size of intermediate output producers than in the output of the industry as a whole. Table 10 shows that the size of establishments in the specialised supplier part of the industry (SIC 7819), measured by number of workers, has indeed undergone a steady decline. This decline differs markedly from the cyclical declines and recoveries observed in other parts of the industry. Table 11 confirms the decrease in

**Table 11.** *Receipts per establishment*

| Year | Nominal $ | 1967 $ |
|---|---|---|
| 1963 | 533,869 | 582,190 |
| 1967 | 642,792 | 642,792 |
| 1972 | 607,313 | 484,687 |
| 1977 | 971,033 | 535,004 |
| 1982 | 1,279,827 | 442,693 |

*Sources:* US Department of Commerce, Census of Service Industries Deflator: US Bureau of Labor Statistics, Consumer Price Index.

establishment size by showing that it is also manifested in a decrease in revenues per establishment between 1963–1982.[1]

*2.6. The recomposition of the entertainment industrial complex*

The process of reorganisation does not end with the vertical disintegration of film production and creation of a flexibly specialised production system. Both the major studies and specialised production firms have spread the risk that comes with unstable markets and highly differentiated products by diversifying into related markets. The studios, for example, form financial alliances with firms in other areas of entertainment provision, playing the role of institutional investor. They also finance multiple production packages and advance capital for distribution in return for some intertemporal control over different marketing areas. The entertainment industries, including television, recording and video as well as motion pictures, are witnessing a wave of horizontal integration as new entertainment products are created, enjoying overlapping production processes and markets. A film, for example, has a profitable 'after-life' in outlets other than cinemas, including television, videocassettes, sound tracks, and related promotional devices. This is reflected very powerfully in the changing revenue mix for theatrical films (Table 12).

Input-supplying firms, on the other hand, increase the scale of their operations by specialising in certain generic functions (props and scenery; film editing; sound mixing) but working on a broad variety of final outputs and constantly innovating. In production, firms branch out to serve other final output sectors in the entertainment area, and by exploiting some of the advantages of specialisation. A typical story is that of Cinnabar, a firm established by two individuals to make sets for independent productions, essentially a specialised carpentry shop. Having successfully built sets, they began to work

[1] In passing, it should be noted that vertical disintegration has been accompanied by a dramatic change in the industry's labour demand. The low point in total employment was reached in 1963, reflecting the decline of workforces at the major studios. Beginning with the last and most dramatic stage of vertical disintegration in the early 1970s, the industry's workforce grew very rapidly, so that in 1982 it was 259% of its 1963 level in a period of stable output and declining total production expenditures. Nonetheless, the aggregate quantity of work available (i.e. the total hours of work available) has not increased. Thus, a much larger pool of individuals is competing for the available work, and most of them are now employed intermittently and on a part-time basis. The Writers' Guild is a typical case: membership has increased from approximately 800 in 1970 to 8000 today, but the quantity of scripts commissioned has not risen appreciably. A complete shift in the nature of employment relations in the industry has occurred, characterised by new forms of employment instability and qualitatively new forms of labour market segmentation (Christopherson and Storper, 1989). In fact, the total amount of *work* available in these firms has decreased dramatically, even as the number of workers has grown. This is consistent with their decreasing participation in production.

**Table 12.** *Revenue mix for cinematic films, selected years 1978–84*

| | 1978(%) | 1980(%) | 1983(%) | 1984(%) |
|---|---|---|---|---|
| Rentals | 80 | 74 | 48 | 54 |
| Pay TV | 2 | 5 | 12 | 12 |
| Cassettes | NM | 1 | 15 | 22 |
| Networks and syndication | 14 | 14 | 19 | 8 |
| Other | 4 | 6 | 6 | 4 |
| Total | 100 | 100 | 100 | 100 |

*Source*: Wertheimer Associates; Metro Goldwyn/Mayer.

collaboratively with production companies in designing sets, and in so doing diversified into making the physical props for all kinds of *trompe l'oeil* effects. Currently a 100-person firm, they are now actually producing commercials which require special effects, and themselves subcontracting much of the more standardised carpentry and construction work of set-building because they have outgrown their own facilities. They now work in movies, television, and commercial markets.

This recomposition of the entertainment industries into an entertainment industrial complex is an attempt to reallocate and collectivise risk in the production process. Since demand from the whole complex is greater than that from any particular market, some of the uncertainty that exists in any particular output sector is offset, and greater scale and specialisation on the part of individual firms is supported. This is affirmed by the dramatic shift of the television production industry from New York to Los Angeles in the period under examination. With the exception of network news, the vast majority of television production shifted to the West Coast in the 1970s (Storper and Christopherson, 1987). Television has simultaneously gone through a process of vertical disintegration and it requires inputs similar to those used in films. The geographical consolidation of filmed entertainment production increases the potential throughput for Hollywoods's specialised firms and enables them to trade better on their specialised skills (Marshall, 1919; Becattini, 1987; Bellandi, 1986).

New interindustrial relations, by increasing the variety of potential interconnections and markets to be served by a specialised firm, encourage innovational activity within the firm's specialised product domain. This switch from product *differentiation* in the motion picture industry to product *variety* in the entertainment industry thus has strong dynamic properties. Increases in product variety are typical of both flexible automation and flexible specialisation, whereas only product differentiation was typical of Fordist industries (Boyer and Coriat, 1986; Holmes, 1987). In the end, it can be said that the interindustrial redivision of social labour stems as much from endogenous innovation dynamics 'from below' as it does from corporate planning 'from above'.

## 3. External economies and industrial divides

The history recounted above is puzzling in certain respects. First, why did the studios not simply shed risk in the face of greater uncertainty or instability, and then arrest the process of vertical disintegration at that point by limiting the role of production subcontractors and independent producers? In other words, we have no explanation of why the breakdown in Fordist regulatory institutions in the motion picture industry did not lead to industrial dualism, rather than flexible specialisation. Second, in the many well-documented cases of flexible specialisation in the revitalised craft industries of Western Europe, markets have been enlarged through internationalisation, and a well-developed set of marketing intermediaries now exists to smooth demand fluctuations and permit some production planning. Yet vertical disintegration seems to persist nonetheless. Our existing treatments of flexible specialisation are long on comparative statics, but short on theoretical explanations of its dynamic properties.

In what follows, we argue that once producers begin to adjust to increased inter-temporal uncertainty through vertical disintegration, external economies can, at a certain point, be said to have overtaken internal economies of scale, in both production and transaction costs (sub-sections 3.1–3.3). The existence of strong external economies means that these outcomes need not be intended or planned by large firms; if new production techniques are superior, at a given moment, to what they replace, the path taken can be the outcome of short-term strategies or even accidents (section 3.3). Once taken, however, the path is not necessarily reversible.

*3.1. The division of labour and the theory of the firm*

We may begin to understand flexible specialisation through Coase's general theory of the firm, which holds that the firm is a system of governance which serves as an alternative to the market and that the border between the firm, as a system of internal (non-market) transactions, and the market, as a system of transactions between firms, will very likely be fluid over time (Coase, 1937). At the core of Coase's reasoning is the concept of organisational *scope*: internal economies of scope exist when the various tasks in a production process, even when extensively fragmented into an elaborate technical division of labour *à la* Adam Smith, lend themselves to vertical integration under one managerial hierarchy. When internal economies of scope are minor or negative, however, then the different labour processes may be separated out into independent firms linked by transactions in markets rather than within the firm. Coase emphasises the role of transaction costs in defining economies of scope. This leads to a neoclassical version of scope economies, where firms supposedly internalise transactions until the cost of organising an extra unit of transaction internally is equal to the cost of carrying out that linkage in the market (cf. Scott, 1986). Williamson (1975, 1985) explores the nature of different types of transactions, arguing that some transactions must be internalised because they are subject to pervasive market failure. First, if one firm requires specific assets—e.g. capital equipment dedicated to one particular output—then external firms cannot be relied on to provide this equipment, nor should the consuming firm expose itself to the risk that another firm might fail to produce for it. Second, if one party to a transaction possesses information the other transacting party lacks, then the potential for opportunism exists and the uninformed party would systematically be exposed to loss. In both these circumstances, not only do market transactions fail to protect the parties, but contracts are generally infeasible because of the proliferation of contingent claims; vertical integration is the likely outcome. The insights of Coase and Williamson thus enable us to imagine why real production systems are likely to be governed by a complex amalgam of organisational forms consisting of the 'pure' mechanisms of firms and markets which are, however, merely the extremes of a continuum consisting of many intermediate types of linkage, such as contracts, strategic alliances, and other forms of quasi-integration.

Nonetheless, their definition of the sources of economies of scope is too narrow, for logically we should expect the choice of organisational form to minimise total production costs, and not just one component thereof, i.e. costs associated with transactional activity. Stigler (1951) made an early effort to do just this, by incorporating technology in the guise of scale effects. An increase in the extent of the market affords opportunties to develop an increasingly elaborate technical division of labour, with specialised subprocesses now operated at higher scales. It is commonly the case that, as the market grows, the production of different intermediate inputs reach their mimimum efficient dimensions at different points. Diseconomies of scope may now be said to exist when integration would result in either excess unsold output of one input or its being produced at less than its minimum efficient dimension, given existing technology. These scale-based diseconomies of scope might be greater than the increased transaction costs which would arise if one of the inputs were purchased from another firm. In this case, vertical disintegration would occur even if it resulted in an increase in transaction costs. Likewise, vertical integration might occur where clear technical separabilities exist but transactions are subject to market failure or simply are very expensive to carry out. The advantages of integration and disintegration shift according to changes in the technical division of labour in production, which in turn is not amenable to the marginal adjustments assumed by neoclassical theory.

In this light, there are several principal sources of internal economies of scope. First, there are the actual costs of carrying out transactions; vertical integration tends to reduce many transaction costs. Second, integration concentrates profits from the entire input-output chain and, where oligopolistic conditions prevail, may generate a substantial profit increment in the form of quasi-rents. Third, there are managerial economies associated with integration. Where firm-specific knowledge (as in a patented or company-identified product) or tight technical indivisibilities of different production stages are involved, integration tends to occur. Fourth, where technological interrelations are tight, such that the assets of one activity must be specific and dedicated to another task, even where strict technological indivisibilities do not exist, then transactions are subject to market failure and integration usually takes place.[1] In all, integration is strongly encouraged by inter-temporal certainty and when there are physical or intellectual reasons for combining different labour processes.

There are, by the same token, many reasons for vertical and horizontal disintegration, which now correspond to an expanded definition of internal diseconomies of scope. The necessary condition of disintegration is a technical division of labour, or where labour processes resist synthesis into indivisible machine systems. A number of contingencies then encourage disintegration. First, if output markets are unpredictable or unstable, producers may disintegrate in order to avoid the transmission of uncertainty through the vertical structure of the firm (Berger and Piore, 1980). Second, final output producers may require inputs which can be manufactured more efficiently by firms which devote specialised attention or firm-specific knowledge to the task (Vennin and de Banville, 1975;

[1] This refers to the group of phenomena known as 'principal-agent' problems in transaction cost and information economics (Williamson, 1985). Principals, such as the firm needing to purchase inputs which are highly specific to their own final outputs, would be subject to a form of extortion by agents if they relied on outside firms to provide these inputs or services with no other assurances than the spot market can offer. It is also extremely difficult to write contracts that, *ex ante*, resolve all possible contingent claims that might arise. In the face of these 'moral hazards', vertical integration is therefore likely. The risks, however, go the other way as well: for an independent firm which dedicates its assets to production which can only be used by another particular firm, there must be non-market means by which to ensure that the consuming firm will not renege on its commitment, leaving the producing firm with no alternative markets.

Holmes, 1986). Third, these same inputs may be produced in plants having minimal optimal scales of operation which can only be attained if they serve a multiplicity of downstream firms (Holmes, 1986; de Banville, 1982). Fourth, disintegration may be used as a tool to manage capital-labour relations: large firms with internal labour markets, for example, can avoid union labour contracts that apply to the whole firm by subcontracting certain operations or purchasing inputs from other firms in segmented labour markets (Rubery and Wilkinson, 1981). One form of this is vertical relations between firms in the formal and informal economies (Portes and Benton, 1984).

In the case of the film industry, there was a clear decrease in asset specificity on the part of many input-providing firms. The origin of this phenomenon was in the demand of the studios themselves for product differentiation. Later, it became an endogenous product of the input-providing firms themselves, who began to develop their own internal flexibility so as to increase internal scale economies and diversify to spread risks. With disintegration, however, firm-specific assets were replaced by industry-specific assets in the form of new kinds of flexibly specialised human and physical capital and flexible, but firm-specific knowledge; stated another way, firms now develop their asset specialisation in a flexible way.

*3.2. The social division of labour and increasing returns*

It now remains to set the theory of the firm outlined above into motion. We must comprehend some of the dynamic properties of the division of labour in production if we want to explain how industrial divides may be crossed. We may begin with Adam Smith's notion that increases in the extent of the market afford opportunities to deepen the technical division of labour: the different tasks within the production process can be fragmented out into specialised activities. This conception of specialisation as exclusively tied up with the development of the technical division of labour is too narrow, however, for a growing market is just as likely to trigger the deepening of the *social* division of labour, as should be obvious from the discussion in the previous section. The development of both the technical and the social divisions of labour is associated with what Young (1928) so aptly referred to as the increasing 'roundaboutness' of production (see also Bohm-Bawerk, 1891).

Roundaboutness generally expresses the existence of increasing returns to scale. With increases in the total volume of production, the specialisation of activities and firms and thus their internal efficiency can increase, while simultaneously reducing the costs of transactions between them. The existence and nature of increasing returns is highly contested, both theoretically and empirically (Rowthorn, 1975A, 1975B; Thirlwall, 1980B; McCombie, 1981, 1983; McCombie and Ridder, 1984; Douglas, 1976; Cripps and Tarling, 1973). We cannot do justice to all the related issues here and so will restrict our treatment to two issues, *viz.* whether increasing returns exist; and whether some take the form of external economies.

Much of the scepticism concerning increasing returns in general is an artifact of the theoretical distinction frequently made between changes in production techniques or technical progress, and increasing returns 'without' changes in technique. Increasing returns are then said to attach only to an existing production function, while any other changes in returns are due to changes in technique. In other words, the analysis by definition becomes a static one, and increasing returns disappear by virtue of the decision to consider them exogenous. On the other hand, if an analysis of returns over time is undertaken and increasing returns do happen to appear empirically, they can be formally identified with the returns of particular firms by assuming that the technique changes which generate them are events purely internal to firms; increasing roundaboutness is thus not a change in technique but an epiphenomenon. As a result, little importance is accorded to increasing returns generally, and no role is given to external economies. This has been the case since Sraffa's and Knight's comments on the subject in the 1920s (Thirlwall, 1983).[1]

Yet increasing returns are not reducible to technical progess in the narrow sense because they can be rooted in the organisational dynamics of modern industrial production as a whole. An historical analysis of economic development leads to very different conclusions from static analysis. A view toward dynamics suggests that: (1) increasing returns really do exist and are widespread because (2) they are not due to 'exogenous' changes in technique but to the endogenous and intertwined development of the division of labour and technology; and (3) in this light, the distinction between internal and external economies is often artificial.

Technical progress in the narrow sense of development of better machines and hardware generally depends upon the development of the technical and social divisions of labour. Increasing returns cannot be said to be contained within firms or machines, since both the latter develop historically in a way that is inseparable from shifting inter-task (whether within firms, between firms, or between industries) connections. As the issue was framed by Allyn Young:

> The mechanism of increasing returns is not to be discerned adequately by observing the effects of variation in the size of the individual firm or of a particular industry ... what is required is that industrial operations be seen as an integrated whole (1928, p. 539).

This is obvious in the 'pure' Smithian growth case. Increasing output may give rise to greater firm specialisation (which may exert downward pressure on firm size) while the same specialisation permits an increase in the returns to scale of the particular specialised operation. Take first the case of the technical division of labour, which is normally thought to give rise to increasing *internal* economies of scale. As different operations are divided up, as in Smith's pin factory, the costs of production per unit of output are reduced, yet the returns are likely to be reaped by one firm. Yet even single operations on single machines within a firm would often not be able to realise any of these efficiencies unless they were inserted within a technical division of labour: their own increasing efficiencies are precisely due to, and a product of, their interconnections with other phases of the labour process. Though contained within a firm, the returns are, in effect, 'external' to the particular operations. This is what Marx labelled 'cooperation'. The Coasian view of the firm as a set of transactions among activities vividly captures this notion. The social division of labour is actually just an extension of this initial division, but institutional as well as technical (or task) borders are now drawn between operations. The splitting off of new specialised firms in the social division of labour is inextricably related to the

[1] We need only remember the way that Sraffa illustrated the impossibility of combining increasing and diminishing returns in the same model, when he criticised Marshall's attempts to do so in the 1920s (Sraffa, 1926). Young, too, was adamant in his declarations on the subject, that the kind of external economies to which he was referring were not of the type that Sraffa thought they were, i.e. within one narrowly-defined 'industry' and thus amenable to modelling *via* a conventional production function. There is, of course, an enormous literature on all these points (Kaldor, 1972).

We should also mention the aggregation problem in measuring increasing returns. The problem is to choose the unit of measurement, generally the geographical areas in which income, consumption, and production constitute enough of a closed system for the effects of dense consumption be traced to a particular production system. Since most empirical work on the subject uses aggregate regional or national inputs and outputs, serious estimation problems plague measurement efforts.

development of inter-firm trade within and between industries. The ability for any particular specialised operation to realise internal economies is thus an historical outcome of the totality of its interconnections, intra- and inter-sectoral alike. In both cases, scale economies are external in nature.

Indeed, we might imagine returns to be strictly internal only in the very restricted number of cases where they are literally embodied in one indivisible machine or in one indivisible person. Even here, doubts may be cast, for those units of activity in turn can often exist only as a consequence of their input-output relationships. If increasing returns were strictly internal to the firm or task, we might suppose that diseconomies of scope would be entirely absent and all production would then occur in one gigantic vertically-integrated firm or machine. History rejects this counterfactual, on which the sceptics' claims are based.

The case for specifically external economies is more obvious when we consider an industry whose markets are not growing, as in the film industry in the post-war period. When the second or fourth types of diseconomies of scope referred to in the previous section are present, then inter-firm intertemporal flexibility becomes the precondition for realisation of internal economies of scale in the industry's firms. For example, in the film industry in the 1960s and 1970s, the studio system would have suffered lower returns than the flexibly specialised system, owing to what would have been a chronic condition of excess capacity or dramatically reduced production levels, while its opposite, an entirely artisanal production technique, would have had to forgo the returns to specialisation. Whether these scale economies of flexible specialisation are *greater* than those of the studio system in its heydey is another issue, since we are assuming a change in context.[1] The point is that the industry gains very significant increasing returns owing to the way that flexible inter-firm relations reduce excess capacity and offer greater levels of specialisation than would be possible with artisansal methods. This results in increasing returns for the sector as a whole.

The formation of an electronic entertainment industrial complex, however, suggests even more clearly the existence of external economies of scale. In the face of stagnant and unstable markets, the process of competition in the film industry actually resulted in an inter-industry recomposition of the division of labour, so as to increase the flexibility and through-put of the complex as a whole. It is very likely that the roundaboutness of production in those related sectors increased dramatically through vertical disintegration in each. The horizontal recombination of the tasks of different final output sectors then resulted in inter-industrial scale economies of exactly the sort whose existence Knight and Sraffa denied.

### *3.3. The endogenous dynamics of the division of labour*

In an industry with high levels of vertical disintegration, an industrial complex comes into being involving an elaborate structure of inter-plant transactional relationships such as face-to-face contacts, detailed exchanges of strategic information, long-term and short-term contracting, recontracting, strategic alliances, and various forms of quasi-integration and intermediate forms of governance, material input-output flows, circulation of labour

[1] We do not have direct evidence of scale or the nature of returns in the film industry in this paper. At the moment, it is almost impossible to measure these returns at the industry level or even for the electronic entertainment complex. The organisational dispersal of inputs and outputs and the temporal variability of inputs (e.g. labour time, wage bills, etc), not to mention the problem of choosing appropriate measures (e.g. each film 'unit' can have a very different value), are beyond our current measuring capabilities.

supplies, and so on. Yet, as previously noted, there are many circumstances where market failure in these transactional relations can occur: asset specificity, information impactedness, or monopolies in input or output markets (which destroy bargaining alternatives), and these circumstances are therefore associated with vertical integration. In functioning industrial complexes, however it seems frequently to be the case that communities of trust are formed in which acceptable norms of doing business are socialised and in which producers face formal or informal mutual surveillance (Storper and Scott, 1988). This seems especially to be the case in traditional craft production regions of the Third Italy and France (Ganne, 1983; Camagni and Capello, 1987), and in the 'communal firm' in Japan (Dore, 1987), as well as in the form of strong traditional associations in today's Hollywood (Storper and Christopherson, 1985). As a result, the risks of opportunistic behaviour are severely reduced. Marshall noted this in citing the existence of 'mutual knowledge and trust' as underlying the 'normal arrangements of many transactions', or what has more recently been called the 'social embeddedness' of market relations (Bellandi, 1986; Fua, 1983; Marshall, 1890; Granovetter, 1985; Arrow, 1974). Williamson and all the contract theorists rooted in neoclassical thinking clearly underestimate the value of these social structures and consequently improperly universalise historically-specific types of behaviour (cf. Dore, 1987). Vertical and horizontal disintegration are now *both cause and effect* of existing external economies.

The same is true in the case of geographical concentration of the industrial system. Frequently, the dense transactional relations between firms in a vertically-disintegrated industry cause producers to agglomerate, forming concentrated industrial districts or regions (Scott, 1986). The concentration of many firms in one area can lower the costs of negotiating and carrying out transactional activity and thus encourage further vertical disintegration. The way is also cleared for higher levels of vertical disintegration than would be possible in a Williamsonian world when bargaining alternatives are created by the existence of large numbers of producers from whom buyers can then choose and switch. Again, the social division of labour becomes both cause and effect of external economies. Each of these features of an industry with a complex social division of labour—a collective order, large numbers, and geographical agglomeration—is a source of increasing returns which lie outside the firm and is controllable by no single group of firms or actors. Because of this, they tend to be self-perpetuating through their promotion of increasing specialisation and interconnection.

External economies in the social and organisational structure of the industry also encourage technological innovations which promote further rises in productivity. The social division of labour in an industry contains a rich stock of organisational complexities harbouring many new technological and commerical possibilities. Individuals already caught up in the workaday world of vertically-disintegrated industries are optimally positioned to perceive and take advantage of these opportunities, and thus they tend to continue developing the division of labour through processes of innovation and entrepreneurial activity (Scott and Storper, 1987). Skills are socialised and tastes are honed so that, as Marshall phrased it, the 'mysteries of production become no mysteries'. We might say that there is an intrinsic 'opportunity grid' in this division of labour. With increasing returns of any type, and especially in the case of external economies, the very distinction between 'exogenous' changes in circumstances and 'endogenous' search and choice processes comes into doubt.[1] Or, as Young put it

[1] This includes not only changes in processes, but also the design of the good or service itself, as we have seen in the case of films (see Williamson, 1985, p. 22).

Every important advance in the organization of production alters the conditions of industrial activity and initiates responses ... which in turn have a further unsettling effect. This change becomes progressive and propagates itself in a cumulative way' (Young, 1928, p. 533).

The division of labour, as a central element in the dynamics of technological and product innovations in modern industry, is part of the 'dynamism' of increasing returns in general since it is a central element in the realisation of efficiency gained through experience, specialisation, and diversification. In reality, both 'firms' and 'industries' are being redefined, such that the notion of returns as strictly internal loses its meaning in any dynamic, historical sense. In sum, in functioning industrial systems, both the division of labour and technological innovations tend to be endogenously and dynamically reproduced and are, in turn, mutually reinforcing. The external economies of the system thus have strong dynamic properties.

The temporal course of the division of labour is subject to dramatic and often unanticipated changes as a result of the dynamics described above. To take just one example of simultaneous, multiple, and inter-firm/inter-industry shifts in production techniques, the cheapening of telecommunications and recent advances in information transmission have facilitated interconnection between different firms and units in the producer services and banking industries and other corporations and firms in the economy. Some phases of production have obtained larger scale and become more routinised and more vertically integrated than ever before (e.g. processing of bills for credit cards has been centralised within large corporations, largely at the expense of specialised billing firms). But, in many other cases, the centralisation and specialisation of activities that were formerly scattered throughout the industrial economy has been encouraged by the ease of interconnection. These activities have, in turn, become more specialised, and as a result are now separated out into independent departments or firms, involved in new specialised complexes of producer services and banking activities. It can readily be seen that there is no end in principle to the interaction of scale and specialisation in the industrial economy. This reasoning runs manifestly counter to dominant models of production organisation in the post-war period where it was frequently held that mass production was the culmination of development for all 'mature' sectors (Scott and Storper, 1987; Storper, 1985; cf. Vernon, 1966; Burns, 1934; Kuznets, 1930).

In a sense, the film industry represents an exception which proves the rule of external economies. Unlike the restrictive Smithian case, in which steady growth is the foundation of the division of labour, in this case decline and instability conditions generated innovational activity which, in effect, produced a new industry on the ashes of the old, and it is this new industry which undoubtedly enjoys the increasing returns to which we are referring. Verdoorn effects are likely to result not simply from a linearly advancing social division of labour, but from a *continually differentiating and uneven roundaboutness* in the economy as a whole, whose contours cannot be foretold. Perhaps even Young understated the case for the existence of external economies. Could it be that a wide variety of growth and even sectoral decline conditions lead to reorganisations of the social division of labour and produce increasing returns?

### *3.4. The choice of technique: branching points and increasing returns to scale*

Thus far, we have identified some of the properties of one alternative to Fordist production methods, but we have not established sufficient theoretical reasons for the transition. That is, given a change in technological or institutional circumstances, a theory of the process of selection from possible alternative techniques is needed. Two main points need to be made here.

First, we may ask whether, under the changed institutional and technological circumstances of the 1950s, flexible specialisation became more efficient than the hypothetical technical-organisational régime, industrial dualism. We may imagine a point, probably somewhere in the early 1960s, when the studios could have found a level of production that supported the stable portion of the demand curve and retained that production in-house, while contracting out the rest. The problem with this and all such similar organisational schemes which rest on product and process standardisation is that markets had already become vigorously contested by that point, and one of the major means of contestation was product design (Spence, 1983). This is, of course, precisely what has happened in many formerly stable Fordist oligopolies, especially in consumer durables industries, in the 1980s. The strategy of serving the quantitatively stable portion of the demand curve with a qualitatively stable product would not have worked. Dualism lacks dynamic efficiency properties in contestable markets.

In the second place, it does not therefore follow that the precise system of flexible specialisation which emerges in each case is the first-best (or Pareto-efficient) technique, Rather, it can be shown that, given an initial technological-organisational 'branching point', a variety of alternatives, each with increasing returns and each more efficient than the system being replaced, may exist as possibilities for the sector, but the actual path taken is not necessarily the most efficient, nor is the outcome inevitable from the starting point. Why, then, would we expect it to persist? Because with increasing returns, paths taken are not easily reversible. By describing the historical properties of techniques with increasing returns we can begin to theorise the industrial divide described in our case study. To do so, we can draw on the model of Arthur (1983).

For the moment, consider the restricted case of 'technical choice' from a set of known blueprints. Usually there are several ways of carrying out any given economic task. These technologies may exist in the form of method or information, or they may already be incorporated in physical plant or machinery; they may also be marketed as products. Take the case where two or more superior technologies are competing with each other to replace an outmoded technology (thus, not the standard diffusion case where a new and superior technology competes with an old and inferior one). For example, in the 1890s, the steam engine, the electric motor, and the gasoline engine competed as power sources for the new automobile. In the 1800s and on into this century, spinning mules competed with ring-frames in cotton manufacturing. Recently, different configurations of personal computers, videocassette players, and video recorders, have competed for their respective markets.[1]

For the purpose of this example, think of the pure increasing returns case, where returns rise simply with the numbers who choose the technology. In the standard (neoclassical) textbook case of technical choice, all the technologies competing for adoption show diminishing returns, and market sharing is the result. As demand increases, adoption follows a composite supply curve obtained from the lateral addition of the separate returns curves for each technology. The outcome is completely predictable—we can determine *in advance* the market shares after $n$ choices in this situtation—and it is easy to show that the

[1] Note that the analogy to objects rather than techniques (in the conventional sense) is appropriate here, for in a social division of labour, adoption of a technique by any particular firm is important to the next in the sense literally of what that firm consumes in terms of objects, i.e. intermediate inputs. New interindustrial production complexes, moreover, grow out of simultaneous redefinitions of products and processes.

configuration which emerges is *efficient*. The outcome is also *flexible*: adjustment of either returns curve can always shift the composite supply curve and hence market shares.

Where objects or techniques with increasing returns compete, the market outcome is unforeseeable. With increasing returns, non-convexities appear in the supply curve and multiple equilibria are called into being. Information on preferences, endowments, and transformation possibilities enables us to locate these long-run equilibria *hypothetically* in the form of comparative statics, but this cannot tell us *which* one will actually capture the market. Reconsider the old fable of an island where cars are suddenly introduced. Drivers are free to choose between the left-hand and the right-hand sides of the road and have no in-built bias toward either. Each side possesses increasing returns; as a higher proportion of drivers chooses one side, the very real returns to choosing that side rise rapidly. Common sense tells us that we should observe a good deal of randomness to the proportions initially driving on each side (David, 1975). But, if one side by chance got sufficiently ahead, other drivers would 'fall in' on that side, so that eventually all cars would drive on the same side of the road. The side that 'wins'—comes to dominate the market—cannot be deduced in advance. The outcome is not determined at the outset (Arthur, 1983).

Arthur points out four aspects of these outcomes that are relevant to the conjecture about historical branching points. First, in contrast to the usual diminishing returns situation, the outcome need possess *no efficiency* properties; the side that takes the market need not, from any long-run viewpoint, be the better of the two. Second, driving is now locked-in to the chosen side. The outcome is structurally *rigid*, in that marginal inducements to individual drivers to change sides would likely prove ineffective and policy would have to find other means. Third, even though we know drivers' preferences and possibilities, *ex ante* the outcome is *unpredictable*. Small events outside the model, equivalent to what Piore and Sabel call 'strategies'—perhaps some drivers' reactions, perhaps a dog running into the road, perhaps the timing of certain traffic lights—may be crucial in deciding the outcome. And, fourth, *ex post*, exact causality is hard to assign—it would certainly be a mistake to ascribe it to the innate superiority of the outcome. It is *indeterminate*. When returns are not simply increasing functions of scale but are also dynamic—owing to the specialisation and diversification which is part of the social division of labour—then even this example of the choices between multiple equilibria is too restrictive. There are no real equilibria. The number of possible paths is indeed much greater and even less foreseeable than in this example, and short-run events may be more important than implied in the formal logic.

The transition from Fordism to vertically-disintegrated flexible specialisation might be accounted for as follows. Some initial shock—political, macroeconomic, institutional in nature—dramatically raises the level of uncertainty or instability in an industry's markets. In their initial strategies of self-preservation, firms turn to subcontracting or outwork, thus creating a dualistic production system. They may, in the meantime, begin to search for production techniques which can function under increased uncertainty without a loss of control (as in today's search for flexible automation), and they may also explore avenues for reasserting control over markets. Still, their markets are subject to periodic shocks (overproduction, decline) and these large firms suffer rising costs or periodic losses. Subcontractors may begin by occupying specialised market niches. They proceed to develop new products or services and to diversify their marketing strategies. Horizontal inter-firm networks begin to grow: product variety begins to supplement product differentiation and horizontal inter-industry relations offset some of the risk faced by supplier firms and encourage specialisation. As the efficiency of these networks increases—manifested in higher quality, lower prices, or product innovation—the larger firms increasingly turn to them in crises. In the absence of opportunities to resynthesise labour processes, the internal economies of large firms continue to falter while external economies slowly increase in the production complex as a whole. With each round of crisis, large producers use vertical disintegration as a survival strategy. Each round of disintegration simultaneously strengthens external economies. A process of replacement of internal economies is set into motion, and beyond a certain point the large firms can no longer reverse it, because no single firm can assert enough control for a long enough period of time. Disintegration, in this manner, may begin with subcontracting, but it may end with the appearance of a network of independent supplier firms and a flexibly specialised system (Chaillou, 1977). Short-term events along the way are critical to the unfolding of the whole process. It holds, moreover, that the precise form that flexible specialisation takes is not necessarily the optimal one (in comparative static terms) from the starting point. The particular path of vertical disintegration that is taken, at a certain point, becomes self-confirming. History is produced.

## 4. Conclusions

In this paper, a broad argument has been laid out concerning the process of transition from large batch or mass production to flexible specialisation. The argument has focused on the steady weakening of internal scale economies in favour of external economies, attributable to a shift toward a vertically-disintegrated form of production organisation. Our intention is in no way to detract from the use of a rich method of historical political economy in explaining industrial divides, but rather to develop the analytics of transition processes. Responses to changed circumstances generate counter responses, and these processes have their own internal dynamics. In turn, these same processes may have effects upon the macroeconomic and institutional environments which set them into motion in the first place.

One implication of the reasoning advanced in this paper is that the role of external economies in industrial development has—with only a few exceptions—been severely underestimated. This is the true of the contemporary flexible production debate as well, where it is frequently argued that flexible specialisation is a system with few economies of scale and therefore offers little hope for contributing to a viable post-Fordist macroeconomic regime of accumulation (Boyer, 1897; Leborgne, 1987). As we have argued, however, scale economies are not created only by progressive lengthening of the production run, but can result from different combinations of flexibility and specialisation and interconnection between activities. It is indeed theoretically possible that reductions in both aggregate output and the size of particular production runs could still be accompanied by rapidly increasing returns to scale.[1]

[1] The attention given to the size of the production run is, in our view, something of an analytical hangover from the practices of economists trained in observing Fordist industries. The argument that *only* flexible automation—with its more limited product variety and its longer discrete production runs—has even the theoretical possibility of achieving the productivity gains which would be required to form the basis of a new regime of accumulation in the advanced capitalist economies seems highly questionable in light of the arguments presented in this paper. In the medium and long runs, productivity and innovation are very likely going to depend on maximising the ability to shift between different product ranges—what we used to call 'retooling' and now call 'dynamic flexibility'. This is precisely why the new modal form of organisation of the large, capital-intensive industries is the *systems house* rather than the vertically-integrated manufacturing firm. Even in flexibly automated mass-production industries, the sources of productivity will be not only in internal economies of scale but in the organisation of the production system as a whole so as to maximise convertibility—an external economy.

If we are correct in these claims, then there are general lessons for the explanation of economic history. Most importantly, the organisation of production is not determined strictly by predictable responses to given external circumstances, whether those of neo-classical choice and substitution theories, classical models of heterogeneous capital, or neo-Marxist models of labour control. Rather, in the presence of these circumstances, the future is—in some significant measure—steered by what is already cemented into a complex and highly externalised system of economic relations and organisational forms and practices, and choices made along the way are important to long-run outcomes; they are not just part of the standard deviation around predictable and externally-determined trends (Nelson and Winter, 1982). By no means does this then imply that evolution is everything, i.e. that the system is closed to the possibility of industrial divides. It suggests, however, that divides are not precipitated by marginal alterations in circumstances, but require major changes in institutions, markets, or technological possibilities, such that the basic possibilities for the division of labour are altered. Once set into motion, these divides may be self-perpetuating.

Because we know so little about the role of external economies in industrial development, this framework needs substantial extension in its substantive and analytical terms of reference. The dynamic interactions of technological innovation and choice, the structure of markets, and external economies should be at the centre of further efforts to extend our existing understanding of industrial divides. For example, we need to know whether vertical disintegration—and the dynamic processes of learning-by-doing and development of externalised, embedded, social practices with which it is associated— itself encourages continual diversification of product markets and thus, after a certain point, whether it might therefore impede the kind of technological or organisational innovations that would be capable of resynthesising labour processes and restoring mass or batch production or vertical integration.[1] Especially relevant to the current conjuncture is the question of whether flexible automation offers the potential to respond to current market configurations but without the levels of vertical disintegration implied in the flexible specialisation model (cf. Boyer and Coriat, 1986). Mass production industries such as automobiles and steel have larger blocs of fixed capital equipment than does the motion picture industry. Do greater indivisibilities at each phase of the technical division of labour—even if not inherently associated with economies of scope—limit the possibility that specialised firms could build up their capital stocks under conditions of market uncertainty? Do they therefore allow large firms to control the futures of those industries?

[1] At present the major studios (and the mini-majors) give orders in the film industry. They are capable of advancing funding for a project, whether internally or to an independent production company under contract. They also typically finance the distribution of a film, which for a major cinematic production averages more than $5 million today. As such, final market concentration remains very high and studios retain an enormous influence over the industry's output. In 1987, output of the US industry rose 17% over the previous year, attaining its mid 1970s levels and resulting in a shortage of exhibition screens in major US markets such as New York and Los Angeles. Major distributors such as Paramount or Warner Brothers have formed strategic alliances in the 1980s (i.e. forms of quasi-integration which remained legal after 1948 according to the interpretation of the Reagan anti-trust lawyers), permitting them exclusive access to the most coveted cinemas, while independent producers have a much more difficult time securing screens, especially at certain times of year. The combination of greater output and shortage of screens will most likely lead to the weakening of some of the smaller studios and a new wave of mergers between major studios and mini-majors and independents. Oligopoly is thus securely installed, even in the flexibly specialised film industry. The structure is something like Benetton in the Italian clothing industry, or the US machine tool industry after its most recent wave of mergers and acquisitions. Nonetheless, this continuing and vigorously reasserted financial centralisation in the industry has not led to reintegration of production activity. The latter suggests that the current organisation of production continues to be reproduced by very strong external economies.

Alternatively, will flexible *automation* become practical for a wide variety of production runs, reducing minimal optimal scales and reducing the differences between flexible specialisation and flexibly automated mass production? Does such a reduction lead back to vertical integration through increased economies of scope, or to specialisation and vertical disintegration? All these questions are concerned with whether the inter-firm relationships in flexibly specialised industrial sectors find stable reproductive schemes over the medium term, or whether they are simply temporary aberrations from higher levels of vertical integration.

These questions represent a minimum agenda for moving forward into a scientifically-viable and politically-relevant theory of industrial divides. They go to the very roots of contemporary industrialisation processes in the advanced capitalist countries. New experiments in the organisation of production systems, the development of productive labour, and the structuring of industrial communities are under way. How these experiments evolve and are given form in durable configurations of industrial systems has major implications for the future of industrial societies.

## Bibliography

Aglietta, M. 1976. *A Theory of Capitalist Regulation*, London, New Left Books

Arrow, K. J. 1974. *The Limits of Organization*, New York, W. W. Norton

Arthur, W. B. 1983. 'On Competing Technologies and Historical Small Events: The Dynamics of Choice Under Increasing Returns', Stanford, Calif., Stanford University Department of Economics, mimeo

Bagnasco, A. 1977. *Tre Italie: La Problematica dello Svillupo*, Bologna, Il Mulino

de Banville, E. 1982. La Sous Traitance pour l'Automobile et le Poids Lourd: Constatations et Hypothèses, *Revue d'Economie Industrielle*, vol. 19, 34–40

Becattini, G. (ed.) 1987. *Mercato e Forze Locali: il Distretto Industriale*, Bologna, Il Mulino

Bellandi, M. 1986. 'The Marshallian Industrial District', Florence: Università degli Studi di Firenze, Dipartamento di Scienze Economiche, Studi e Discussioni no. 42

Berger, S. and Piore, M. J. 1980. *Dualism and Discontinuity in Industrial Societies*, Cambridge, CUP

Berger, S. 1981. 'The Uses of the Traditional Sector in Italy', in S. Bechofer and B. Elliot (eds), *The Petite Bourgeoisie*, London, Macmillan

Black, J. 1962. The technical progress function and the production function, *Economica*, vol. 29, 166–170

von Bohm-Bawerk, E. 1891. *The Positive Theory of Capital*, New York, G. E. Stechert

Boyer, R. 1986. 'Les Crises ne sont plus ce qu'elles etaient . . .', in R. Boyer (ed.), *Capitalismes Fin de Siècle*, Paris, Presses Universitaires de France

Boyer, R. 1987. 'What is the Next Socio-Technical System Made of?' Berkely, CA: Paper presented at BRIE Conference on Skills and Automation, September

Boyer, R. and Coriat, B. 1986. 'Technical Flexibility and Macro Stabilisation', Paper presented to the Conference on Innovation Diffusion, Venice, March

Braverman, H. 1974. *Labour and Monopoly Capital*, New York, Monthly Review Press

Brusco, S. 1982. The Emilian model: productive decentralisation and social integration, *Cambridge Journal of Economics*, vol. 6, no. 2

Brusco, S. and Sabel, C. 1981. Artisanal production and economic growth, in: F. Wilkinson (ed.), *The Dynamics of Labor Market Segmentation*, London, Academic Press

Burns, A. 1934. *Production Trends in the United States Since 1870*, New York, National Bureau of Economic Research

Camagni, R. and Capello, R. 1987. 'Italian Success Stories of Local Development: Theoretical Conditions and Practical Experience', Milan, Università Bocconi, Istituto di Economia Politica, mimeo

Chaillou, B. 1977. Definition et Typologie de la Sous Traitance, *Revue Economique*, vol. 28, no. 2

Christopherson, S. 1986. 'Peak Time, Slack Time: Trends Toward Labor Flexibility in the Reported and Unreported Economy', Washington, DC: Report to the U.S. Congress, Office of Technology Assessment

Christopherson, S. and Storper, M. 1989. New forms of labor segmentation and production politics in flexibly specialized industries, *Industrial and Labor Relations Review*, vol. 42, no. 3, 331–347
Coase, R. H. 1937. The Nature of the Firm, *Economica*, vol. 4, 386–405
Cripps, T. F. and Tarling, R. J. 1973. *Growth in Advanced Capitalist Countries, 1950–1970*, Cambridge, CUP
David, P. A. 1975. *Technical Choice, Innovation, and Economic Growth*, New York, Cambridge University Press
Dore, R. 1987. *Taking Japan Seriously*, Stanford, Calif., Stanford University Press
Douglas, P. H. 1976. The Cobb-Douglas production function once again: its testing, and some new empirical values, *Journal of Political Economy*, vol. 84, 903–915
Elster, J. 1982. *Ulysses and the Sirens*, Cambridge, CUP
Fua, G. 1983. Rural industrialization in later developed countries: the case of Northeast and Central Italy, *Banco Nazionale del Lavoro Quarterly Review*, December, 351–377
Ganne, B. 1983. *Gens du Cuir, Gens du Papier*, Paris; Editions du Centre National de la Recherche Scientifique
Granovetter, M. 1985. Economic action and social structure: the problem of embeddedness, *American Journal of Sociology*, vol. 91, no. 3
Habakkuk, H. J. 1962. *American and British Technology in the Nineteenth Century*, Cambridge, CUP
Hampton, B. 1970. *History of the American Film Industry: From Its Beginnings to 1931*, New York, Dover
Hollywood Reporter annual. *Studio Blu-Book Directory*. Los Angeles, Hollywood Reporter.
Holmes, J. 1986. The organizational and locational structure of production subcontracting, in A. J. Scott and M. Storper (eds), *Production, Work, Territory*, London, Allen and Unwin
Holmes, J. 1987. The impact of technological change on the organization and locational structure of automobile production in Canada, in K. Chapman and G. Humphrys (eds), *Technological Change and Industrial Policy*, Oxford, Basil Blackwell
Hounshell, D 1984. *From the American System to Mass Production, 1800–1932: The Development of Manufacturing Technology in the United States*, Baltimore, Johns Hopkins University Press
Houssiaux, J. 1957. Le Concept de Quasi-Integration et le Role des Sous-Traitants dans l'Industrie, *Revue Economique*, vol. 8, 221–247
Ikeda, M. 1979. The subcontracting system in the Japanese electronic industry, *Engineering Industries of Japan*, vol. 19, 43–71
*International Motion Picture Almanac*, 1933–1985, annual. New York, Quigley Publications
Kaldor, N. 1972. The irrelevance of equilibrium economics, *Economic Journal*, vol. 8, 2
Kaldor, N. 1975. Economic growth and Verdoorn's Law: a comment on Mr Rowthorn's article, *Economic Journal*, vol. 85, 891–896
Kelly, J. 1985. Management's redesign of work: labour process, labour markets, and product markets, in D. Knights, H. Wilmott, D. Collinson (eds), *Job Redesign: Critical Perpsectives on the Labour Process*, Aldershot, Gower
Kuznets, S. 1930. *Secular Movements in Production and Prices*. Boston, Houghton-Mifflin
Leborgne, D. 1987. 'Equipements Flexibles et Organisation Productive: les Relations Industrielle au Coeur da la Modernisation—Elements de Comparaison Internationale', Paris: CEPREMAP, Couverture Orange
Leborgne, D. and Lipietz, A. 1987. 'New Technologies, New Modes of Regulation: Some Spatial Implications', paper presented to the Conference on Changing Labour Processes and New Forms of Urbanization, Samos, Greece, Paris,CEPREMAP, Couverture Orange No. 8726
Littler, C. 1985. Taylorism, Fordism and job design, in Knights, Wilmott, and Collinson (eds), *Job Redesign: Critical Perspectives on the Labour Process*, Aldershot, Gower
Marshall, A. 1890. *Principles of Economics*, London, Macmillan
Marshall, A. 1919. *Industry and Trade*, London, Macmillan
McCombie, J. S. L. 1981. What still remains of Kaldor's Laws, *Economic Journal*, vol. 91, 206–216
McCombie, J. S. L. 1983. Kaldor's Laws in retrospect, *Journal of Post-Keynesian Economics*, vol. 5, 414–429
McCombie, J. S. L. and de Ridder, J. R. 1984. The Verdoorn Law controversy: some new empirical evidence using US state data, *Oxford Economic Papers*, vol. 36, no. 2
Mistral, J. 1986. Régime Internationale et Trajectoires Nationales, in R. Boyer (ed.), *Capitalismes Fin de Siècle*, Paris, Presses Universitaires de France

Murray, F. 1983. The decentralisation of production: the decline of the mass collective worker, *Capital and Class*, vol. 19, 74–99

Nelson, R. and Winter, S. 1982. *An Evolutionary Theory of Economic Change*, Cambridge, Mass., Harvard University Press

Offe, C. 1985. *Disorganized Capitalism*, Cambridge, CUP

Ohno, T. 1982. How the Toyota production system was created, *Japan Economic Studies*, vol. X, no. 4

Piore, M. 1987. 'Corporate Reform in America and the Challenge to Economic Theory', paper presented to the Industrial Relations Seminar, MIT, February

Piore, M. and Sabel, C. 1984. *The Second Industrial Divide*, New York, Basic Books

Portes, A. and Benton, L. 1984. Industrial development and labor absorption: a reinterpretation, *Population and Development Review*, vol. 10, 4

*The Production Company: The Complete Guide to Commercial and Industrial Production Companies in Los Angeles* (1984 edn.) Edited by Louis Nicolaides, Los Angeles, Parmorand Publications

Rosenberg, N. 1969. *The American System of Manufactures*, Edinburgh, Edinburgh University Press

Rothbarth, E. 1946. Causes of superior efficiency of USA industry as compared with British industry, *The Economic Journal*, vol. 56, 383–390

Rowthorn, R. E. 1975A. What remains of Kaldor's Laws?, *Economic Journal*, vol. 85, 10–19

Rowthorn, R. E. 1975B. A reply to Lord Kaldor's Comment, *Economic Journal*, vol. 85, 897–901

Rubery, J. and Wilkinson, F. 1981. Outwork and segmented labour markets, in F. Wilkinson (ed.), *The Dynamics of Labour Market Segmentation*, London, Academic Press

Russo, M. 1986. Technical change and the industrial district: the role of interfirm relations in the growth and transformation of ceramic tile production in Italy, *Research Policy*, vol. 14, 329–343

Sabel, C. and Zeitlin, J. 1985. Historical alternatives to mass production: politics, markets, and technology in nineteenth century industrialization, *Past and Present*, vol. 108, 133–176

Sabel, C. F., Herrigel, G., Kazis, R., Deeg, R. 1987. How to keep mature industries innovative, *Technology Review*, April

Savy, M. 1986. Les Territoires de l'Innovation: Technopoles et Aménagement, l'Expérience Française, *Revue d'Economie Régionale et Urbaine*, vol. 1, 41–60

Sayer, A. 1985. 'New Developments in Manufacturing and Their Spatial Implications: From Flexible Manufacturing to Just-in-Time', Lesvos, Greece: Paper presented to the Conference on Spatial Structures and Social Process

Scitovsky, T. 1963. Two concepts of external economies, in Agarwal, A. and Singh, S. (eds), *The Economics of Underdevelopment*, New York, Oxford University Press

Scott, A. J. 1984A. Territorial reproduction and transformation in a local labor market: The animated film workers of Los Angeles, *Environment and Planning D: Society and Space*, vol. 2: 277–307

Scott, A. J. 1984B. Industrial organization and the logic of intrametropolitan location III: A case study of the women's dress industry in the Greater Los Angeles region, *Economic Geography* vol. 60, 3–27

Scott, A. J. 1986. Industrial organization and location: division of labor, the firm, and spatial process, *Economic Geography*, vol. 62, 215–231

Scott, A. J. 1988, *New Industrial Spaces: Flexible Production Organization and Regional Development in North America and Western Europe*, London, Pion

Scott, A. J. and Storper, M. 1987. High Technology Industry and Regional Development: A Theoretical Critique and Reconstruction, *International Social Science Review* vol. 1, no. 12, 215–232

Scranton, P. 1983. *Proprietary Capitalism*, Philadelphia, Temple University Press

Solinas, G. 1982. Labour market segmentation and workers' careers: the case of the Italian knitwear industry, *Cambridge Journal of Economics* vol. 6, no. 4

Spence, M. 1983. Contestable markets and the theory of industry structure: A review article, *Journal of Economic Literature*, vol. 21, 981–990

Sraffa, P. (1926). The laws of returns under competitive conditions, *The Economic Journal*, December, vol. 36, 535–551

Stigler, G. J. 1951. The division of labor is limited by the extent of the market, *Journal of Political Economy*, vol. 59, 185–193

Storper, M. 1985. Oligopoly and the product cycle, *Economic Geography*, vol. 61, no. 3
Storper, M. and Christopherson, S. 1985. *The Changing Location and Organization of the Motion Picture Industry*, Los Angeles, UCLA Graduate School of Architecture and Urban Planning, Research Monograph No. 127
Storper, M. and Christopherson, S. 1987. Flexible specialization and regional industrial agglomerations. *Annals of the Association of American Geographers*, vol. 77, no. 1
Storper, M. and Scott, A. J. 1988. The geographical foundations and social regulation of flexible production complexes, in J. Wolch and M. Dear (eds), *Territory and Social Reproduction*, London, George Allen & Unwin
Temin, P. 1966. Labor scarcity and the problem of American industrial efficiency in the 1850s, *Journal of Economic History*, vol. 26, 272–298
Thirlwall, A. P. 1980. Rowthorn's interpretation of Verdoorn's Laws, *Economic Journal*, vol. 90, 386–388
Thirlwall, A. P. 1983. A Plain Man's Guide to Kaldor's Laws, *Journal of Post-Keynesian Economics*, vol. 5, 345–358
Urry, J. 1986. Capitalist production, scientific management, and the service class, Scott, A. J. and Storper M. (eds), *Production, Work, Territory*, London, George Allen & Unwin
Vennin, B. and de Banville, E. 1975. Pratique et Signification de la Sous-Traitance dans l'Industrie Automobile en France, *Revue Economique*, vol. 26, no. 2
Verdoorn, P. J. 1980. Verdoorn's Law in retrospect: a comment, *Economic Journal*, vol. 90, 382–385
Waterman, D. 1982. The structural development of the motion picture industry, *American Economist*, vol. XXVI, no. 1
Williamson, O. 1975. *Markets and Hierarchies*, New York, The Free Press
Wiliamson, O. 1985. *The Economic Institutions of Capitalism*, New York, The Free Press
Young, A. (1928). Increasing returns and economic progress, *Economic Journal*, December

# Name Index